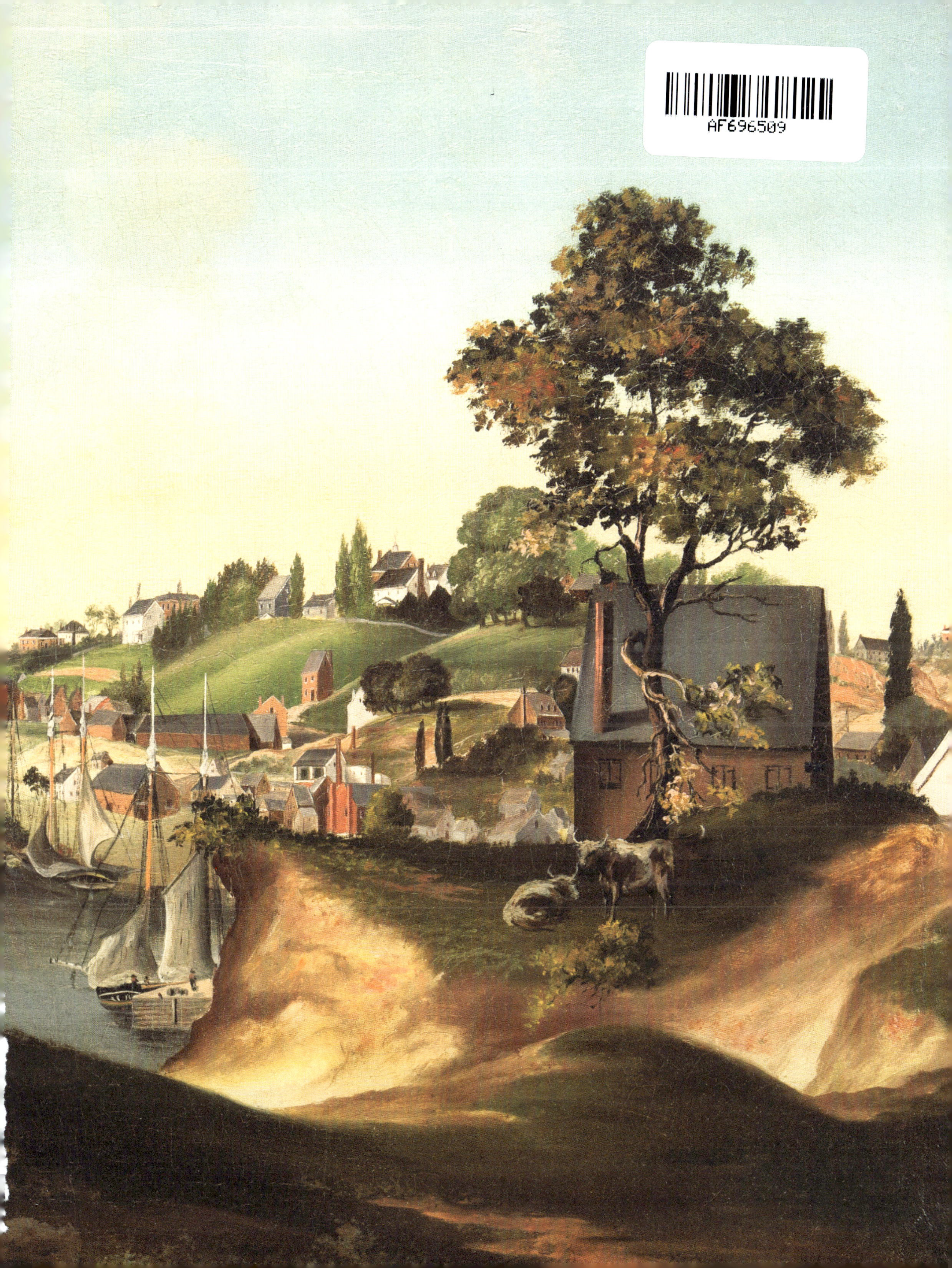

Ceramics in America

CERAMICS IN AMERICA 2013

Edited by Robert Hunter

Published by the CHIPSTONE FOUNDATION

Distributed by University Press of New England

Hanover and London

Cover Illustration: Detail of storage jar, attributed to Stephen B. Sweeney, Henrico County, Virginia, 1838–1863. Salt-glazed stoneware. H. 15". (Private collection; photo, Gavin Ashworth.)

Endpapers: Albert Pleasants, *View of Richmond*, Richmond, Virginia, 1841–1844. Oil on canvas, 24 1/8" x 36 3/16". (Courtesy, Museum of Early Southern Decorative Arts, Gift of Grayco [3543].)

Design and production: Wynne Patterson, VT
Copyediting: Mary Gladue, CT
Proofreading: Mary Cason, NJ
Editorial Assistant: Angelika Kuettner, VA
Type: Aardvark Type, CT
Printing: Meridian Printing, RI

Published by the Chipstone Foundation
Distributed by University Press of New England
1 Court Street
Lebanon, New Hampshire 03766

Printed in United States of America
ISSN 1533-7154
ISBN 978-0-9827722-2-5

Contents

Editorial Statement

Ceramics in America is an interdisciplinary journal intended for collectors, historical archaeologists, curators, decorative arts students, social historians, and studio potters. We seek articles on the broad role of historical ceramics in the American context, including essays on ceramic history, archaeological research, technology, social history, studio pottery, and collecting. Further information about this process can be found at the Chipstone Foundation's website, www.chipstone.org. Authors are encouraged to contact the editor before submitting materials.

The Chipstone Foundation offers significant honoraria for full-length manuscripts accepted for publication, and reimburses authors for photography that is approved in writing by the editor.

If an article is accepted for publication, the author should submit photocopies of the title page and copyright page for all cited published sources. Authors are encouraged to provide digital files for all images, along with permissions to reprint. Low-resolution digital images or photographs of poor quality will not be accepted for publication.

Questions and comments from readers and authors are welcome.

Robert Hunter
Editor, Ceramics in America
P.O. Box 114
Yorktown, Va. 23690

Introduction

Robert Hunter

▼ THEY SAY old pots never die, they just get broken and thrown away—only to be dug up by some ceramics fanatic who writes endlessly about their significance. Okay, perhaps no one has actually said that, but it is apt considering the content of the many articles that have appeared in *Ceramics in America* over the last thirteen years. And I am happy to say that the 2013 volume continues to epitomize the enthusiastic search for meaning in ceramics through dogged research, creative interpretation, and, in some cases, brilliant observation.

Ceramics scholars have been writing about American pottery for well over a century. The legendary Alice Morse Earle, one of the early pioneers writing on ceramics, produced numerous books and articles between 1890 and 1904 on early American material culture, among them *China Collecting in America* (1890). A self-proclaimed "china hunter," Earle published much of the early knowledge about the use of imported Chinese and English wares in domestic settings in New England. Another hero of American ceramics was Edwin Atlee Barber, whose *Pottery and Porcelain of the United States* (1909) is still considered the bible for the history of domestic ceramics production. These authors, and of course others, provided a broad outline of ceramic history in the American context that scholars and collectors continue to fill in.

What is astonishing, though, is the magnitude of the new discoveries and reinterpretations of even well-known ceramics traditions. Most American ceramics research and collecting is regional if not local. Collectors hoard information as well as pots, and the competition for either can be ferocious. Clay-based feuds among collectors frequently result in a resistance to publishing new research, in an arguably counterproductive belief it will protect one's interests. With the advent of the internet, it is auction houses that disseminate regional information, some of which is decades out of date. The more iconic a particular pottery type has been in the seminal literature, the less likely new research will be incorporated into catalog descriptions or exhibit labels.

The nineteenth-century alkaline-glazed face vessels from the Edgefield District of South Carolina are a prime example of an American ceramic genre that has a long history with both collectors and scholars. Barber discussed them at length in his 1909 book and connected their making and use with the slave population of Edgefield whose cultural roots extended back to Africa. While various attempts have been made in the ensuing years to better understand the historical origins and use of these vessels, most of the basic steps in connoisseurship have been not undertaken. A surprising amount of speculation has been put forth in the absence of archaeological findings, documentary evidence, and rigorous typological analysis. In a major effort to remedy this, researchers Claudia Mooney, April Hynes, and Mark M. Newell submit their provocative article "African-American Face Vessels: History and Ritual in 19th Century Edgefield." They further the argument that face vessels were not just functional jugs and cups, but also served as powerful ritualistic objects used specifically in the African-American practice of conjuring.

This essay is but one scholarly product that has resulted from the landmark exhibit "Face Jugs: Art and Ritual in 19th-Century South Carolina," curated by Mooney and sponsored by the Chipstone Foundation. Another important result of the Chipstone initiative was the symposium "Unmasking the Mysteries of Face Jugs," held at the University of South Carolina's McKissick Museum of Art in Columbia, December 7–9, 2012. Beyond the diverse array of speakers, the symposium invited the ceramics community at large to contribute their thoughts regarding the history, uses, and meanings of these culturally charged objects. Mooney, Hynes, and Newell's work adds a significant contribution to the literature on Edgefield face vessels and, perhaps most important, has catalyzed the field to continue the archaeological, typological, and material research needed to fully understand the history and use of the face vessel.

Another long-standing topic related to Edgefield ceramics is examined in Philip Wingard's article "From Baltimore to the South Carolina Backcountry: Thomas Chandler's Influence on 19th-Century Stoneware." The potter Thomas Chandler is well known to collectors of alkaline-glazed stoneware from the Edgefield district of South Carolina. His creations are beautifully thrown and finished with distinctive and aesthetically inventive slip-trailed decorations. One could make a case that Chandler's pots are the most beautiful American stoneware ever produced. Despite the survival of so many of his utilitarian masterpieces, a full history of Chandler's life has never been recorded in the published works on Southern pottery.

This reevaluation of Chandler's early history was triggered in part by the discovery of a Baltimore churn made in the typical blue-and-gray salt-glazed ware of that city's prolific potting history. This object, however, is anything but typical, having been incised with one of the most elaborate scenes of a house and farmyard ever found. The churn was signed and dated by Thomas Chandler in 1829, thus providing a link to his early training in

Baltimore. Wingard, a long-time collector and dealer of Edgefield stoneware, deciphers this "Rosetta stone" of Chandler's work and has written the most complete rendition of this master potter's biography. Along the way, Wingard draws on his years of experience collecting and conducting research on Chandler's marks, glazes, and decorative styles to build a chronology and understanding of the potter's life. Beyond basic connoisseurship, Wingard strengthens his arguments by making use of archaeological and documentary evidence. Perhaps the most intriguing aspect of Wingard's article is the suggestion that Chandler, having been exposed to face vessels in Baltimore in the 1820s, should be fully credited with introducing this form into the Edgefield potteries. The publication of Wingard's research surely will induce archaeological exploration of the many sites in which Chandler worked during his career and a comprehensive exhibition on his life and work.

In sharp contrast to the incomplete knowledge of stoneware production in the Edgefield district of South Carolina are the stoneware products of Alexandria, among the most easily recognized in the country. Highly collectible, the pots are boldly decorated in cobalt blue and virtually always identified with a potter's stamp. In the 2012 volume of *Ceramics in America*, Barbara Magid offered "'Stone-ware of excellent quality, Alexandria Manufacture' Part I: The Pottery of John Swann," an encyclopedic overview of the early potter. This year she continues with "Part II: The Pottery of B. C. Milburn." In this heavily illustrated essay, she provides the most comprehensive typological study of Benedict Cuthbert Milburn's products through the analysis of more than 150 surviving vessels and tens of thousands of archaeological fragments recovered in excavations. Of particular interest is Milburn's use of cobalt slip trailing, which Magid suggests began between 1847 and 1854. The style seems suspiciously similar to what Thomas Chandler was doing contemporaneously in Edgefield, South Carolina, and warrants scrutiny for evidence of a possible connection. Magid's study firmly establishes Milburn's place in the history of American stoneware, documenting both his aesthetically competent decorations on what must be considered utilitarian objects and his ability to provide Alexandria consumers with a signature product for more than forty years.

Artistic vision expressed through ceramics has an ancient history, and the use of the medium to create sculpture is as old as the craft itself. Clay has continued to be relevant well into the space age and, when paired with new technologies, continues to be a vital part of our life. Thomas Folk provides a perfect case study for this assertion with his fascinating article "Waylande Gregory: Science and Ceramics." Waylande Gregory, considered to be one of the leading figures in mid-twentieth-century ceramic art, had a long career designing porcelain tableware for major American brands. However, it is his monumental sculpture "The Fountain of the Atom," made for the 1939 New York World's Fair, that helped seal his critical acclaim. Folk focuses on this particular work to give us a sense of Gregory's interest in the atomic age and its prescient ramifications for mankind as expressed in his metaphoric treatment of the figures that make up this sculptural tableau. Folk organized a retrospective exhibition of Gregory's work, sponsored by the University of Richmond Museums, Virginia, which has traveled to the Cranbrook Art Museum in Bloomfield Hills, Michigan, and will be on view there until March 23, 2014. A biographical catalog accompanies the show and is highly recommended.

Although the science of ceramic making has come a long way in explicating ancient clay bodies and glazes, sometimes the simplest thing can befuddle the most astute expert. Ivor Noël Hume has made a career out of finding overlooked bits of information, and here he questions a tiny pinhole found in a group of English earthenware money boxes—common, everyday objects made by potters working in the Surrey/Sussex area of England just west of London. These ceramic objects have been well known to archaeologists excavating sites in the metropolitan area. Some examples have survived fairly intact, though most, of course, were smashed for the monetary rewards within. These money boxes, like American piggy banks, have slots for the coins, but why the pinhole? In the quest for an answer, Noël Hume enlisted the help of London ceramics specialist Jacqui Pearce and master potter Michelle Erickson, the results of which are presented in a two-part article: "A Hole in One? or, In Search of Piggy Banks and Christmas Boxes" and "Money Boxes: The London Evidence." Whether the function and meaning of these mysterious pinholes are fully resolved I leave to the reader to decide. It is, however, an instance in which the question is as brilliant as the answer.

Given the enormous amount of research and publication in recent years on the topic of eighteenth-century English pottery, one could wonder what is left to present and discuss. Diana Edwards, noted scholar and author on this broad topic, shows us that important collections are available to be mined for ceramic treasures. In her essay, "English and Continental Delft at Temple Newsam House, Leeds," Edwards opens the display cabinets of this English manor house. While the English creamware collections of Temple Newsam are widely known and have been beautifully published, the tin-glazed wares have not received their due attention. The objects included in the article span a hundred-year history and reflect changes in style and social rituals. Readers are encouraged to visit this wonderful house museum and explore its rich furnishings.

There has been no greater icon in the American ceramics world than the American China Manufactory of Gousse Bonnin and George Anthony Morris, who opened their factory in Philadelphia, Pennsylvania, in 1769. Within three years it had closed, leaving behind scattered historical accounts and, as of this writing, a mere twenty intact objects, several archaeological specimens, and a handful of tiny fragments excavated from the site. In contrast to many early endeavors, there is a rich documentary trail that records the rise and fall of this important American factory. In fact, the entire 2007 volume of *Ceramics in America* was devoted to this topic, with the expectation that it would be the final word on the subject. Through careful reading of untapped archival resources, however, ceramics scholar Diana Stradling has gleaned even more evidence of the story. Titled "Bonnin & Morris: Reading between the Lines," the article presents fascinating new information on the identities of those involved in the factory, among them Thomas Byerley, Josiah Wedgwood's nephew and the head painter, and "about a dozen prime hands" from London's Bow porcelain factory who had traveled to be part of the Philadelphia enterprise. Mysteries and questions remain about the factory and its products, but Stradling gives us hope that answers are still available to be found.

In 1978 my colleague Dan Mouer, an archaeologist with Virginia Commonwealth University, took me to a stoneware pottery site deep in the woods in eastern Henrico Country. I was impressed with the mounds of waster fragments piled high, but I didn't fully appreciate the ramifications of the untouched condition of the site. Since then, researching one of the most significant nineteenth-century stoneware complexes in America has been a major preoccupation. Unlike Alexandria stoneware, which is nearly always signed, the salt-glazed potters who worked in the Richmond and Petersburg region of Virginia in the first half of the nineteenth century rarely signed their pots. From about 1805 until the devastation of the Civil War, dozens of stoneware potters plied their trade, creating visually distinctive wares that have long been lumped together as the James River Valley school.

Richmond was at the epicenter of stoneware production for many reasons, but the two most important were clay and transportation. As illustrated by the endpapers chosen for this volume, the deepwater port of Rocketts was clearly a hub for shipping. The scene, by artist Albert Charles Pleasants, was done sometime between 1841 and 1844 and shows the port at it busiest. The building in the foreground may be the location of one or more of the Richmond potteries. If you look closely at the men loading cargo on the dock, you might see the ovoid shapes of Richmond stoneware awaiting their final destination. I would like to thank Robert Leath and Daniel Ackermann of the Museum of Early Southern Decorative Arts for making this image available, and special thanks to Alex Franklin for his research related to this painting.

In the first comprehensive exposé of this incredibly rich and varied regional stoneware tradition, authors Kurt Russ, Oliver Mueller-Heubach, Marshall Goodman, and myself combine decades of expertise, collecting, and research in "The Remarkable 19th-Century Stoneware of Virginia's Lower James River Valley." The reader is also referred to the exhibition "From Kaolin to Claymount: Demystifying James River Valley Stoneware," sponsored by the Museum of Early Southern Decorative Arts, available online at www.mesda.org/onlineExhibits_sprite/mesda_from-kaolin-to-claymount-demystifying-james-river-valley-stoneware.html. Though a great amount of new information is presented in this article, the study is still but one step in organizing future research, which must include the archaeological examination of these well-preserved sites that have lain fallow for years but will be destroyed by development.

The 2013 volume concludes with five insightful book reviews compiled and edited by Amy Earls. Book reviews have been a staple of the journal since its inception, and continue to document the scholarship of the field.

Planning is underway to present more theme-oriented volumes. The 2014 volume of *Ceramics in America*, for example, will be particularly exciting, as I have asked ten authors to present their "Top Ten" lists on assigned topics, from the most important twentieth-century ceramic vessels to those that best reflect London's ceramic history. This volume should elicit quite a bit of discussion and, it is hoped, some controversy over the choices. Readers are encouraged to submit documents, queries, and information about ceramics-related objects to the journal's Facebook page at https://www.facebook.com/pages/Ceramics-in-America/

Ceramics in America

Figure 1 *An Aesthetic Darkey*, from a series of stereoscopic cards by photographer J. A. Palmer of Aiken, South Carolina, 1882. (Courtesy, J. Garrison Stradling.)

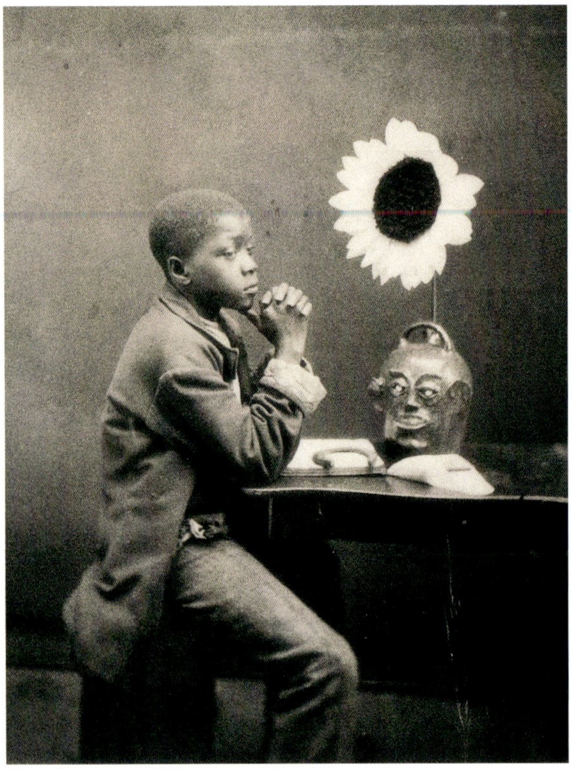

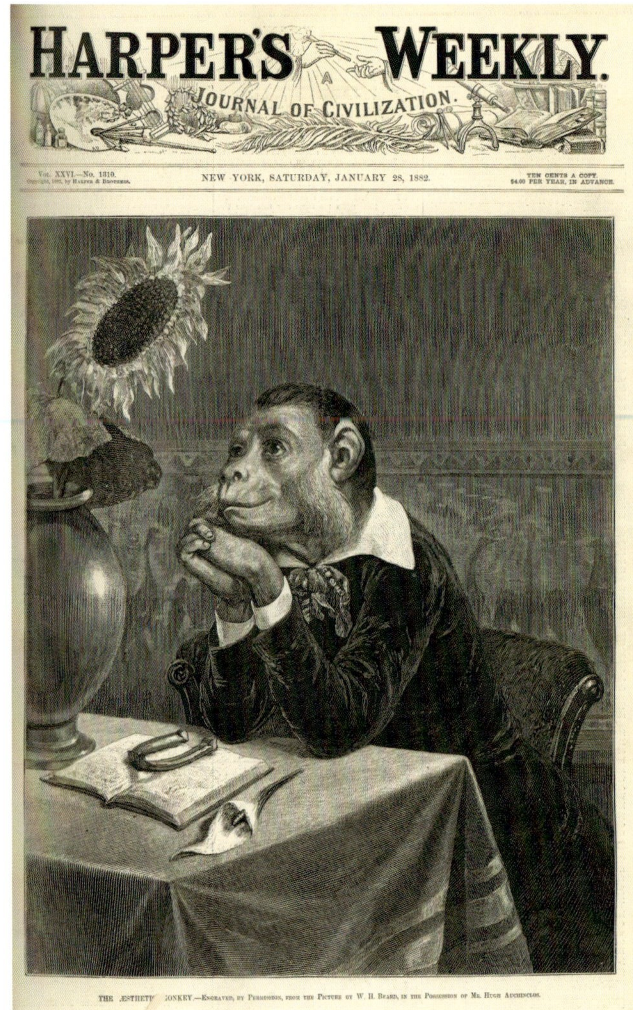

Figure 2 *The Aesthetic Monkey*, W. H. Beard, 1882. (Courtesy, Wisconsin Historical Society, WHS-91892.) This image is a harsh, and possibly homophobic, visual satire of British writer and cultural commentator Oscar Wilde, who embarked on his North American tour in 1882.

Claudia A. Mooney, April L. Hynes, and Mark M. Newell

African-American Face Vessels: History and Ritual in 19th-Century Edgefield

> When my paw, 'Obie' wuz a courtin, a nigger put a spell on him kaise he was a wantin' my maw too. De nigger got a conjure bag and drapped it in de spring what my paw drunk water from. He wuz laid up on a bed o' rheumatiz fer six weeks. . . . His brother . . . went to another old conjure man and axed him to take dat spell off'n paw. De conjure man lowed to paw's brother dat a grapevine growed over de spring, and fer hem to go dar and cut a piece of it six feet long and fetch it to his house at night. When he tuck it to de conjure man's house, de conjure man . . . took de vine in a dark place and done somethin to it—de lawd knows what. Den he tole my paw's brother to take it home and give it to paw.[1]

▼IN 1882 J. A. Palmer, an Irish photographer who settled in Aiken, South Carolina, took a provocative photograph that he titled *An Aesthetic Darkey* (fig. 1).[2] Staged in the manner of historical portraiture, this image features a pensive African-American youth sitting with his elbows resting on a table, a book held open by a horseshoe and a calla lily, and a large sunflower protruding from a face jug.[3] Palmer's composition was inspired by the cover of the January 28, 1882, issue of *Harper's Weekly*, which featured an engraving of William Holbrook Beard's painting *The Aesthetic Monkey* (fig. 2).[4] In an overtly racist gesture, Palmer substituted an African-American for Beard's monkey and replaced his vase with a face jug.

As a late-nineteenth-century resident of Aiken, which was part of the old Edgefield District, Palmer would have known that face vessels were locally made products, but it is doubtful that he had any nuanced understanding of their meaning and use. Indeed, contemporaneous perceptions of these idiosyncratic southern pottery forms varied from community to community and from person to person. Within the white community, face vessels were seen as stylized, and often mocking, representations of African-Americans. Some owners may have even equated the production of these forms with enslaved African-American makers, who worked in Edgefield potteries. But for many of those potters, and members of their community, the distinctive face vessels were ritualistic objects used in the practice of African cultural and religious beliefs. Recent research, including the discovery of face jugs that belonged to or descended in the families of African-Americans from the Edgefield District, has established a link between those pieces and African-American craftsmen (fig. 3), one of whom arrived from the Kongo in 1858. Yet this essay does not purport to establish who introduced the face-vessel tradition to the African-American community in Edgefield. The empowerment of seemingly inanimate objects and the interconnectedness between the human world and the spirit world is the most

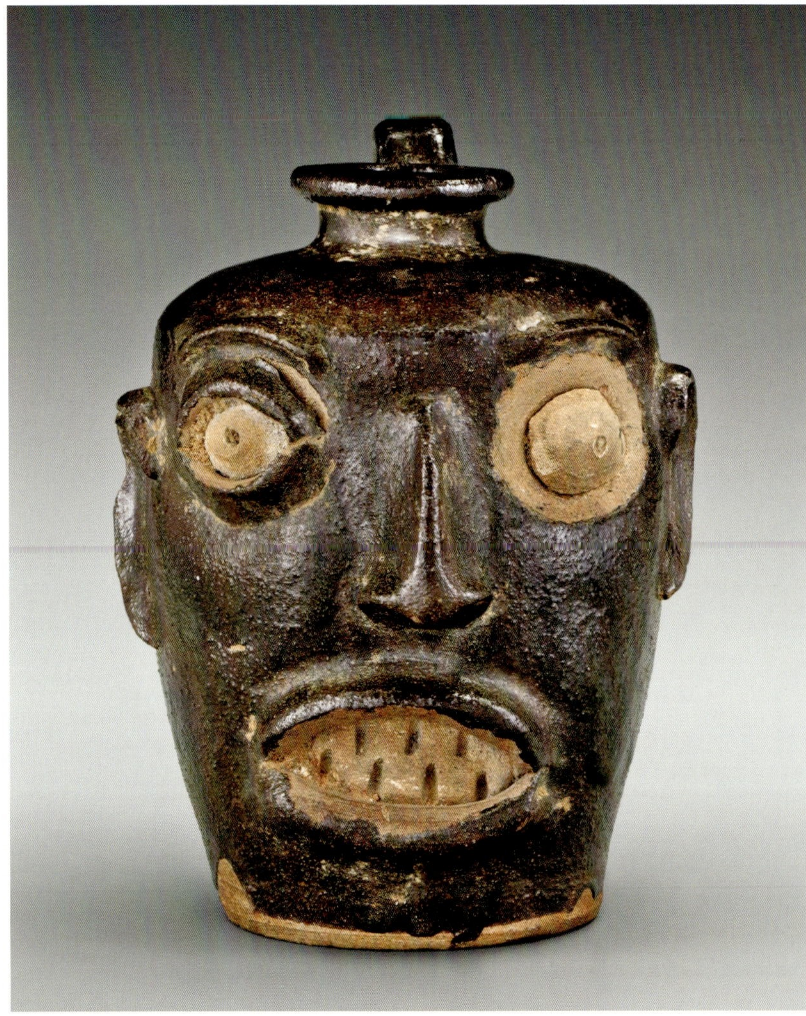

Figure 3 Face jug, Edgefield District, South Carolina, 1860–1870. Alkaline-glazed stoneware. H. 7". (Courtesy, April Hynes; photo, Gavin Ashworth.) Plumber Robert Strang discovered this jug in the 1950s in the Germantown section of Philadelphia, while digging a trench for the foundation and plumbing of the soon-to-be-built Leeds Middle School. The jug was buried at the site of a house that belonged to Leroy Stockton Wingate, who had two live-in servants from the Edgefield District. According to the 1930 U.S. Census, Lewis Gardner was Wingate's chauffeur and Lewis's wife, Leatha, was his cook. A World War I draft card for Lewis Gardner states that he was born in Edgefield. Genealogical records indicate that the Gardners' ancestors were slaves at the plantation of Colonel Thomas J. Davies, a pottery owner interviewed by Edwin Atlee Barber in 1909. "Thomas Jones Davies Bible Records Transcription," *Thomas Jones Davies Bible Records*, available online at http://digital.tcl.sc.edu/cdm/compoundobject/collection/davies/id/75/show/63/rec/19 (accessed March 12, 2013).

important concept underlying this essay's revised interpretation of Edgefield face jugs.

Historiography

In the 1909 edition of *The Pottery and Porcelain of the United States*, Edwin Atlee Barber recounted his conversations with South Carolina, planter Colonel Thomas Davies.[5] The latter owned the Palmetto Fire Brick Works, a pottery in the town of Bath, then part of the Old Edgefield District. Davies recalled that in 1862 local slaves made "homely designs in coarse pottery" including "weird-looking water jugs, roughly modeled on the front in the form of a grotesque human face,—evidently intended to portray the African features" (figs. 4–6). He also noted that face vessels were not made in secret and that the white population in nineteenth-century South Carolina derisively regarded them as having African precedents:

> These curious objects, which I have seen in several collections, labelled "Native Pottery made in Africa," possess considerable interest as representing an art of the Southern negroes, uninfluenced by civilization, and we can readily believe that the modelling reveals a trace of aboriginal art as formerly practised by the ancestors of the makers in the Dark Continent.[6]

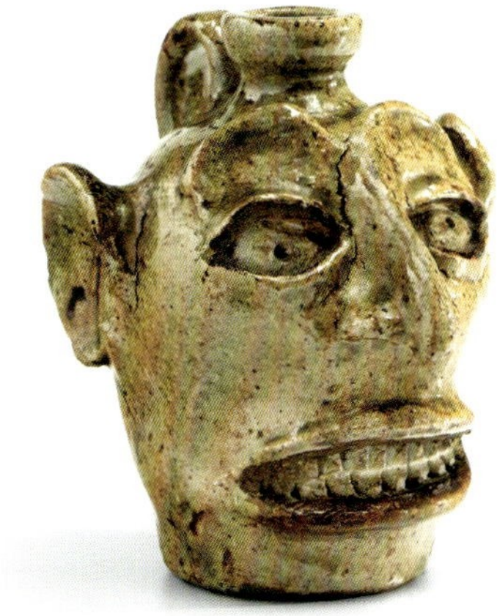

Figure 4 Face jug, Edgefield District, South Carolina, 1860–1880. Alkaline-glazed stoneware. H. 5¾". (Private collection; photo, Gavin Ashworth.) Initially made by slaves and later by free African-Americans, early Edgefield face vessels are wheel-turned from stoneware clay and fired with alkaline glazes that range in color from sandy brown to olive green. The most distinctive feature of these objects is their highly expressive, applied face, which typically includes wide-open eyes, mouths with bared teeth, and idiosyncratic ears, noses, and eyebrows. On the earliest forms, the eyes and teeth are made from white kaolin clay, often left unglazed. Art historian Robert Farris Thompson notes that "the white . . . eyes and teeth, set against the glaze, make the finest Afro-Carolinian face vessels appear to roar where works three times their size merely whisper. . . . [N]othing in Europe [is] remotely like them, for the use of kaolin insets into the body seem[s] peculiar to the Afro-Carolinian and his imitators." Robert Farris Thompson, "African Influence on the Art of the United States," in William R. Ferris, ed., *Afro-American Folk Art and Crafts* (Boston: G. K. Hall, 1983), p. 39.

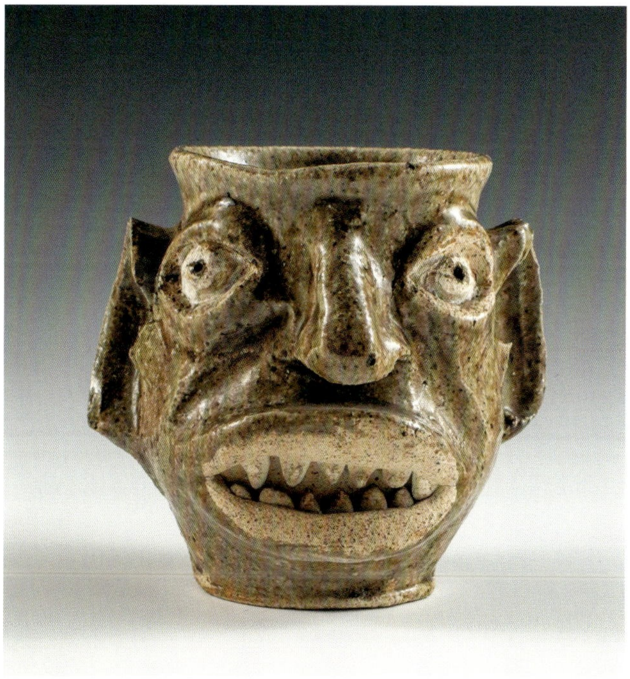

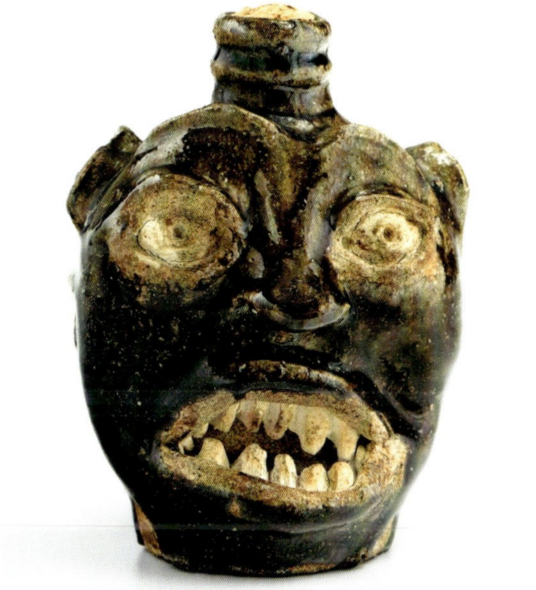

Figure 5 Face cup, Edgefield District, South Carolina, 1860. Alkaline-glazed stoneware. H. 4½". (Private collection; photo, Robert Hunter.)

Figure 6 Face jug, Edgefield District, South Carolina, 1860–1880. Alkaline-glazed stoneware. H. 6". (Private collection; photo, Gavin Ashworth.)

Aside from Barber's interview, nothing of substance was published on Edgefield face jugs until the early 1980s, when they captured the attention of Robert Farris Thompson, a prominent art historian at Yale University and an authority on Yoruba art and the African diaspora. In his book *The Four Moments of the Sun: Kongo Art in Two Worlds* and essay "African Influence on the Art of the United States," Thompson cautioned readers not to ignore the distinctive cultural context in which the makers of face vessels worked or the possibility that African traditions influenced their work—an oversight that continues to constrict a fuller understanding of those ceramic forms. He described the potteries' dependence on slave labor, pointing out that African-Americans were responsible for much of Edgefield's production, and attributed face vessels to three anonymous potters: the Master of the Davies Pottery; the Master of the Diagonal Teeth; and the Master of the Louis Miles Pottery. In sum, early Edgefield face vessels were not the creation of a single potter but rather an artistic concept based on a shared background and belief system among certain makers and users.

Much of Thompson's work centered on slave potters that originally came from the Kingdom of Kongo (Democratic Republic of Congo, Congo-Brazzaville, and Angola) and on visual relationships between Edgefield face jugs and Bakongo wooden figures, which were often made with additional materials including different clays, mirror fragments, and cowrie shells, and,

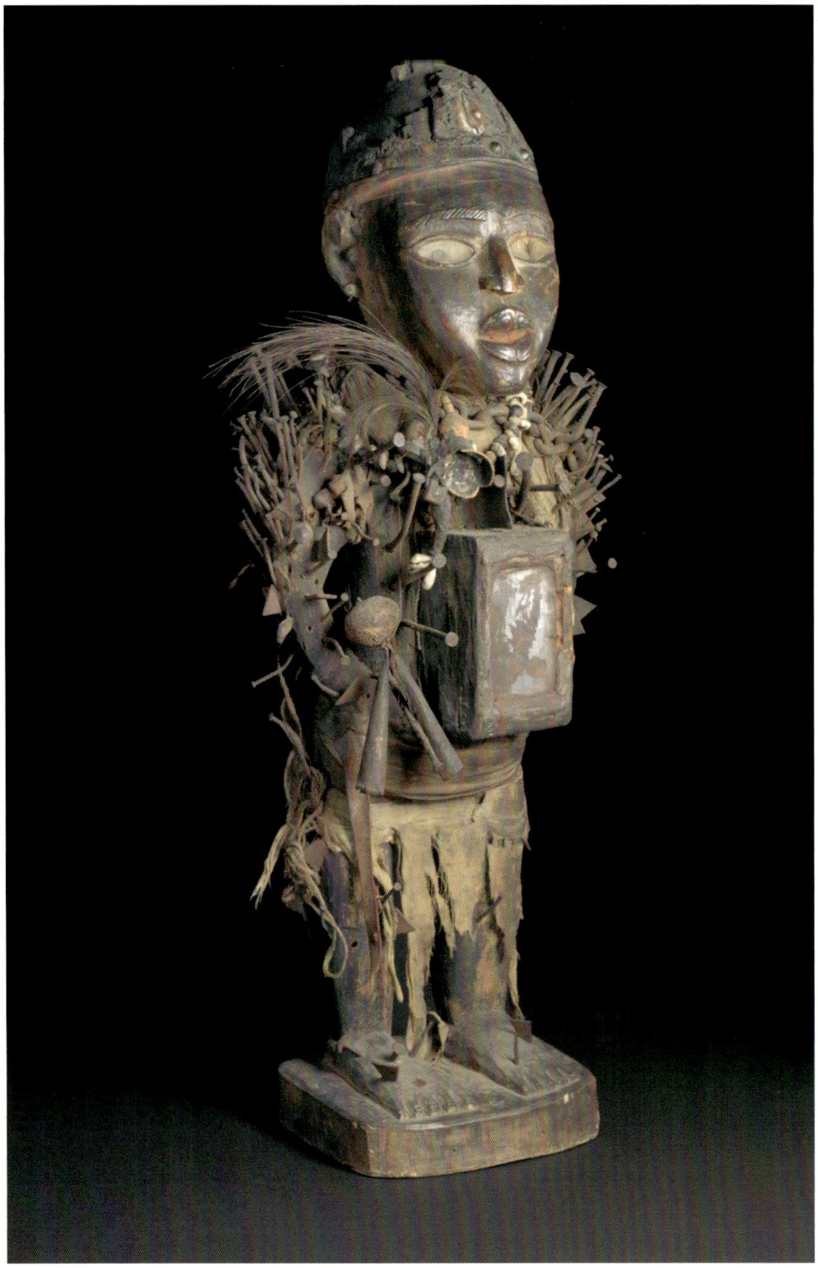

Figure 7 Power figure (*nkisi nkondi*), Vili, Republic of the Congo, early- to mid-nineteenth century. Wood, metal, glass, fabric, fiber, cowrie shell, bone, leather, gourd, and feather. H. 28¼". (Courtesy, Ada Turnbull Hettle Endowment, 1998.502, The Art Institute of Chicago; photo © The Art Institute of Chicago.)

Figure 8 Postcard of a Kongo village with three *minkisi* (sing. *nkisi*) in foreground. Reproduced from Hans-Joachim Koloss, ed., *Africa: Art and Culture* (Munich: Prestel for the Ethnological Museum, Berlin, 2002), p. 126.

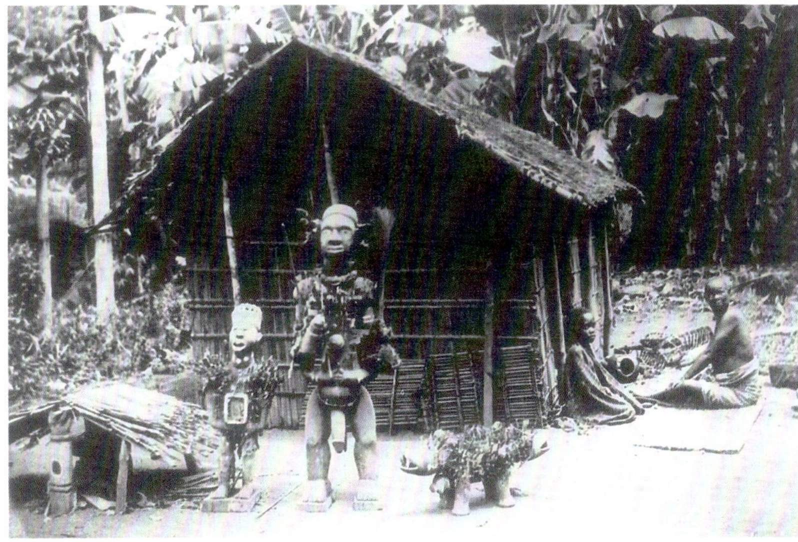

in the case of certain forms, augmented with nails, bones, teeth, bells, and other objects applied during the course of use (figs. 7–8).[7] Thompson noted that the use of kaolin to make the eyes of Edgefield vessels white relates to the ancient northern-Kongo practice of "inserting porcelain fragments within the eye sockets of carved human figures"[8] as well as to West African figural carvings, which occasionally have cowrie shells representing eyes. The use of white or reflective materials in that context facilitated and reflected spiritual possession, what Thompson describes as the "flash of the spirit":

> The vital spark or soul within the spirit-embodying medicine may be, according to the Bakongo, an ancestor come back from the dead to serve the owner of the charm, or a victim of witchcraft, captured in the charm by its owner and forced to do its bidding for the good of the community (if the owner is generous or responsible) or for selfish ends (if he is not).[9]

One class of object Thompson singled out for comparison with Edgefield face jugs are *minkisi* (*zinkisi* south of the Kingdom of the Kongo; sing. *nkisi*),

Figure 9 Photograph of a man sitting on a mat with a variety of *minkisi*. The image is stamped "Congo France" and dated October 20, 1909. Reproduced from Raoul Lehuard, *Art BaKongo: Les centres de style*, 2 vols. (Arounville, France: Arts d'Afrique Noire, 1989), 1: 81.

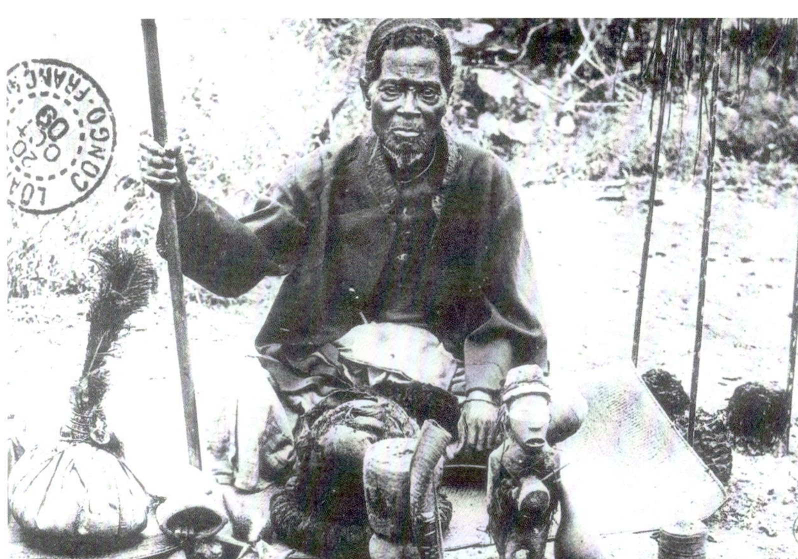

horns, shells, gourds, bundles, ceramic vessels, figures, or other objects infused with the powers of dead spirits (fig. 9). Kaolin and red ochre were important in the composition of some *minkisi* since the white color of kaolin was associated with the dead (fig. 10) and red ochre was associated with "blood and danger because the dead have the power to both afflict and cure the living."[10] About 1900, Mu-Kongo Nsemi Ikaki defined a *nkisi* as:

> [t]he name of the thing we use to help a person when that person is sick and from which we obtain health; the name refers to leaves and medicines combined together. . . . Thus an nkisi is also something that hunts down illness and chases it away from the body. An nkisi is also a chosen companion, in whom all people find confidence. It is a hiding place for people's souls, to keep and compose in order to preserve life.[11]

Figure 10 Power figure, Democratic Republic of the Congo. (Private collection; photo, Marc Felix.) The kaolin on this figure signifies the white skin of the dead.

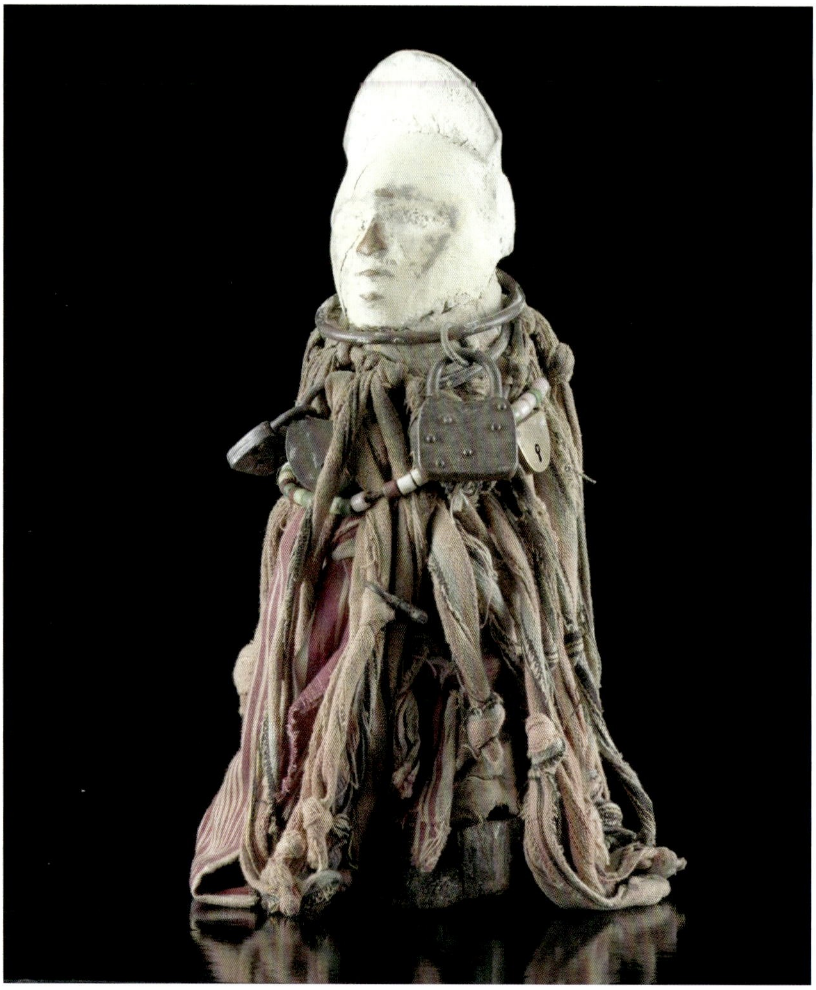

Nkondi minkisi, which are among the known variety in the West, are wood figures, usually male, with a hollow stomach in which the magical items were placed. These figures, as well as other forms of *nkisi*, were owned and operated by *nganga*, who had the power to activate a *nkisi* by summoning ancestral spirits to the container.[12] These activating substances included clays from the land of the dead (graves of prominent individuals or sacred sites), items chosen because their names sounded similar to that of the

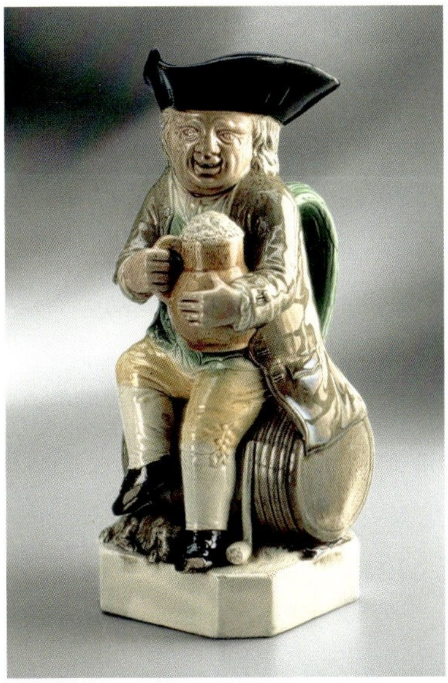

Figure 11 Toby jug, Staffordshire, England, ca. 1785. Pearlware. H. 10". (Chipstone Foundation; photo, Gavin Ashworth.)

action to be performed, and metaphorical materials. In the case of the *nkondi nkisi*, the figure was pierced with nails, blades, screws, or other metal objects to spur the spirit into action and/or to represent different covenants and resolutions.

Although there is no evidence that Edgefield slaves made *nkondi minkisi* or any related wooden forms, their production of "conjure bags," which performed similar ritualistic functions, is well documented. This lends credence to Thompson's assertion that early Edgefield face jugs were related to the more sculptural Kongo *minkisi*. However, it is likely that face jugs served a variety of ritualistic functions, since the African-American community in Edgefield included people from many other areas of the Continent.

As Thompson has observed, the adaptation of Western ceramic art had precedent in the Kongo. During the last quarter of the eighteenth century, British Toby jugs (fig. 11) were sent to the Kingdom of Kongo, where they were used as a surrogate for the human skulls commonly used by Kongo lords to drink palm wine during rituals that reified their terrifying might and powers over life and death in the eyes of other community members. As such, the Bakongo transformed the relatively inexpensive Toby into a prestige object with a ritualized role. By the mid-nineteenth century, the Bakongo apparently started making their own Toby-like figures out of wood and terracotta (fig. 12). Ultimately, Thompson put forth the notion that Edgefield pots are not direct descendants of the British Toby jugs; at best, they are distant cousins that share certain visual features and have links to African ritualistic objects.[13]

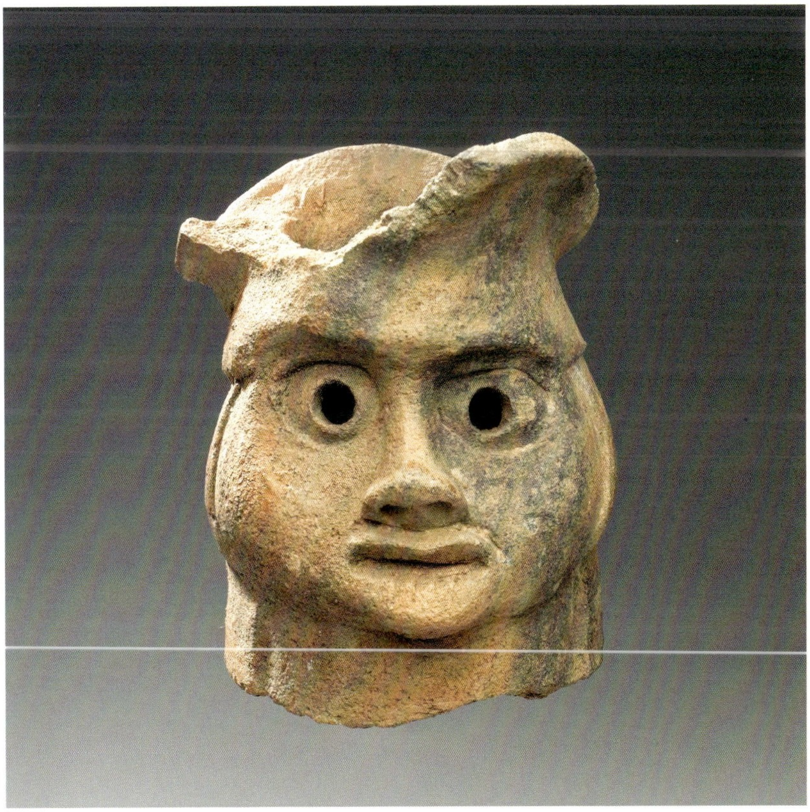

Figure 12 Kongo terracotta head. (Courtesy, Royal Museum of Central Africa, Tervuren, Belgium, EO.1949.1.18; photo, J. Van de Vyver, © RMCA Tervuren.)

Following in Thompson's interpretative footsteps is the scholarship of historian John Michael Vlach, who explored both West and Central African influences on face jugs. He also built on Barber's interview with Colonel Thomas Davies, noting that face jugs were initially made by slaves and then later by free African-Americans. Vlach may have overreached when he posited that face vessels were made at all of the Edgefield District potteries owing to the exchange or movement of slave workmen, but it is logical to assume that this occurred to some degree. He also noted that these enigmatic forms were found as late as 1940 in African-American households in Aiken County, where they still were regarded as objects imbued with power. Ultimately, Vlach agreed with Thompson that early Edgefield face vessels had direct links to Bakongo culture. The former also cited a historical event that would explain how West and Central African cultural traditions came to South Carolina:

> One of the last slave cargoes brought into the United States was [aboard] . . . the *Wanderer* [which landed] in 1858 on Jekyll Island, Georgia. Most of that group were Kikongo-speaking, and they ended up near Edgefield. They were to be the last contribution of African heritage to the area, and their presence most likely provided a stimulus for sustained appreciation of face vessels by local Blacks.[14]

This event is critical to understanding early Edgefield face jugs, their relation to African ritualistic objects like Bakongo *minkisi*, and the significant role played by slaves who arrived on the *Wanderer* in 1858.

As Vlach noted, Edgefield slaves and, later, free African-Americans lived and worked in a community defined by the manufacture of pottery. Therefore their appropriation of alkaline-glazed-stoneware technology and forms for the production of ritualistic objects is logical. Thompson provided evidence of other African cultures whose materials changed, or evolved, over time. The oldest medium for representing the Yoruba god Eshu is thought to be lateritic red clay cone-shaped sculptures, used as early as 1659. The figure of Eshu was later carved out of wood, and in the twentieth century Yoruba descendants in the Americas began creating likenesses of Eshu out of concrete.[15] As Vlach suggested, those objects must be understood as creolized forms incorporating non-African pottery-making technology with African stylistic and ritualistic designs, and, in the use of kaolin, materials. Again, this can be traced to the dynamic interactions between African-American slave potters in Edgefield and the newly arrived Bakongo slaves from the *Wanderer*.

In 1985 art historian Regenia Perry posited that Edgefield face vessels must be seen as symbolically rich objects, laden with vital cultural implications about the makers' backgrounds and daily life. In forwarding this theory, Perry also drew an art historical distinction between African-American–made and Anglo- or Eurocentric face vessels. In *Spirits or Satire: African-American Face Vessels of the 19th Century*, she proposed that early Edgefield face vessels fulfilled roles as potent, ritualistic objects, which differentiates them from the emulative and nonritualistic face vessels made later by white potters: "While the slave-made vessels appear to pulsate with an inner

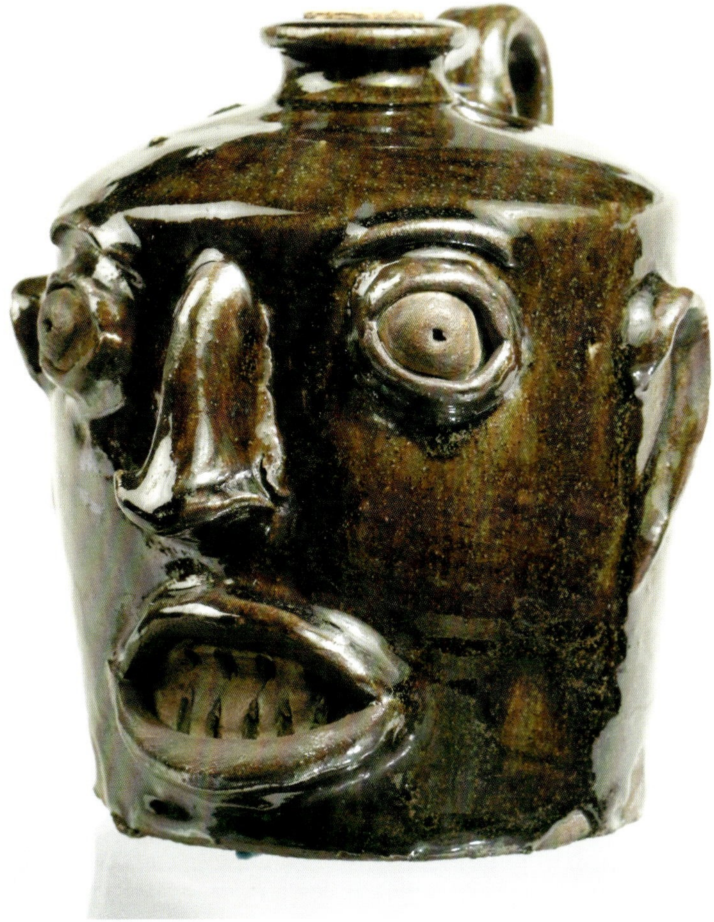

Figure 13 Face jug, Edgefield District, South Carolina, 1860–1880. Alkaline-glazed stoneware. H. 7⅞". (Courtesy, New York Historical Society, 1937.141.)

Figure 14 Face jug, Lanier Meaders (American, 1917–1998), 1987. Glazed low-fired stoneware. H. 10". (Courtesy, Milwaukee Art Museum, Gift of Ruth and Robert Vogele, M2000.84.)

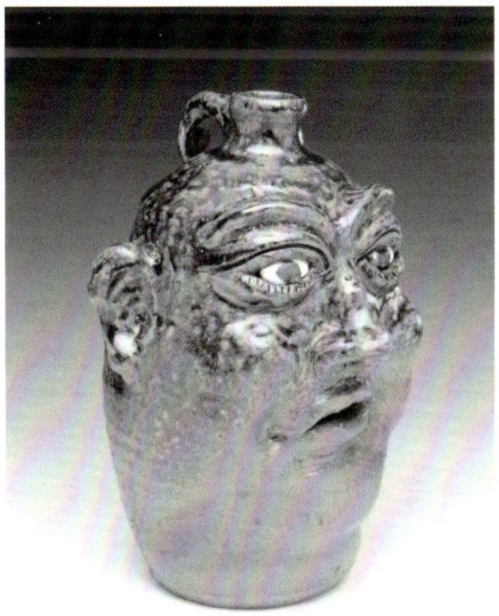

life, those made by Lanier Meaders and other Anglo-American imitators are very caricature-like and superficial in spirit" (figs. 13, 14).[16] "Considering the traditional African belief in spirits and souls," Perry wrote, "it is entirely possible that African-American face vessels were commissioned and owned by individual slaves and served as their only tangible connection with the spirit world."[17] She also noted that a number of Edgefield face vessels have histories of being recovered along the route of the Underground Railroad.[18] Building on Thompson's ideas about the cultural centrality of spiritual practice and ritualized objects, Perry suggested that this history of migration further supports the idea that the Edgefield face vessels were connected to the spirit and ancestral worlds, as slaves would not have taken whimsical or satirical portraits with them when fleeing the plantation or, after freedom, moving north in search of work.[19]

In *Great and Noble Jar: Traditional Stoneware of South Carolina*, Cinda Baldwin concurred with previous scholars regarding visual similarities between the Edgefield face vessels and African sculpture and the importance of slave labor in local potteries. She also agreed with Vlach's assertion that slaves from the *Wanderer* had an impact on the face-vessel tradition and added that one of the Bakongo aboard that ship, a man named Romeo, was

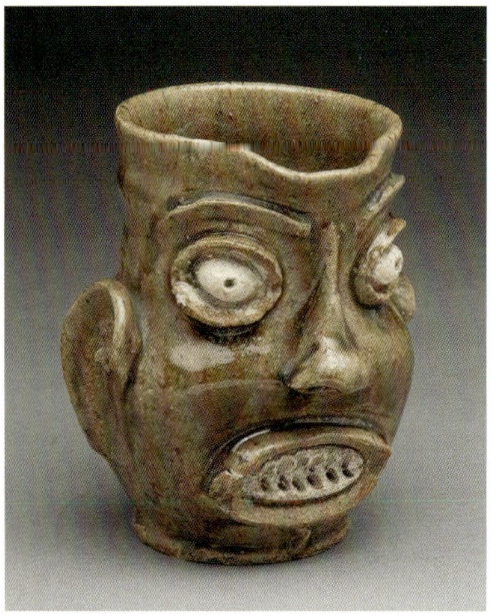

brought to Edgefield and worked at the Palmetto Fire Brick Works.[20] As did Thompson and Vlach, Baldwin discussed the possibility that the Edgefield face vessels were indirectly connected to the Anglo-American Toby-jug tradition. Vlach specifically suggested that potters from Bennington, Vermont, who came to Kaolin, South Carolina, in 1858, might have brought the Toby tradition to that area.[21] Copies of the British Toby form were common in New England, and a pottery built by transplanted Bennington potter Anson Peeler was located near the Palmetto Fire Brick Works. Vlach's theory was based on the aesthetic similarity between the spouts of the Toby jug's tricorn hat and two known Edgefield face pitchers (fig. 15). However, at this time there is no evidence that the Bennington potters brought Toby jugs with them on their journey south or that the Edgefield slaves ever saw any Toby forms.[22]

Figure 15 Face pitcher, Edgefield District, South Carolina, 1860–1880. Alkaline-glazed stoneware. H. 5½". (Courtesy, David B. Ward; photo, Jim Wildeman.)

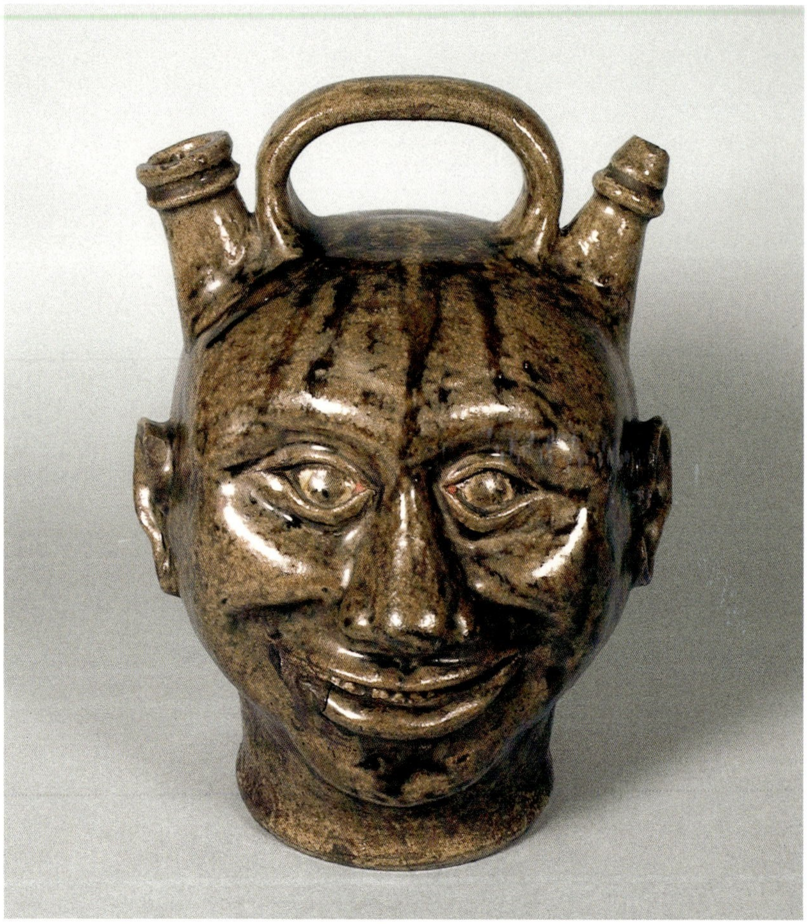

Figure 16 Face jug, Thomas Chandler, Edgefield District, South Carolina, ca. 1850. Ash-glazed stoneware. H. 11". (Private collection; photo, McKissick Museum, University of South Carolina.)

One instance of a northern pottery tradition influencing Edgefield production is a stoneware harvest jug made at Thomas Chandler's manufactory around 1850 and first published in Jill Beute Koverman's exhibition catalog *Making Faces* (fig. 16).[23] Although this large-scale harvest jug is figural and predates the face vessels that are the subject of this article, its design emanates from a northern ceramic tradition. The eyes are not made of kaolin and the face is less stylized, making it difficult to discern whether the

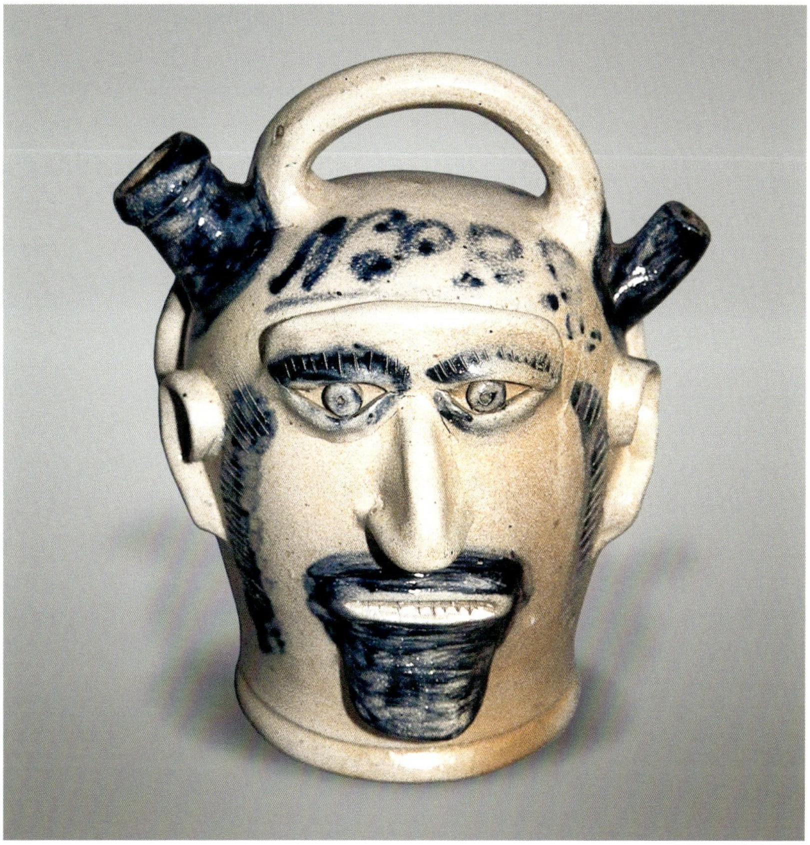

Figure 17 Two-face monkey-form jug, attributed to Henry Remmey Jr. or his son Richard, Philadelphia, Pennsylvania, 1858. Salt-glazed stoneware. H. 11". (Ex-collection of Tony and Marie Shank; photo, Tony Shank.)

subject is representational or comical in the manner of a circus figure. Moreover, the strap-handled form and facial features are reminiscent of face vessels attributed to the Remmey family of potters from Philadelphia, which is not surprising given that Chandler trained in Baltimore in the 1820s and may have worked alongside Henry Remmey Sr. (fig. 17).[24] In overall form and date of production, and likely in function as well, Chandler's 1840s harvest jug differs significantly from the African-American Edgefield face-vessel tradition. It is aligned not with any African customs, but rather with an earlier northeastern pottery design that in turn was directly borrowed from British customs. As such, the influence of the Chandler jug on the subsequent African-American examples should be considered cautiously. If a larger body of Chandler-made versions ever is found then it might well be logical to suggest that the face jug may have been something of a local form and played a role in the general design of the later African-American examples. For now, however, the Chandler jug is both stylistically and ritualistically different.

The most recent scholarship on Edgefield face jugs is Mark M. Newell and Peter Lenzo's "Making Faces: Archeological Evidence of African-American Face Jug Production," which introduced the first archaeological evidence for the manufacture of that genre of work in the Edgefield District. In addition to publishing fragments recovered from the Joseph G. Baynham site in Eureka, South Carolina, the authors surveyed previous theories pertaining to face-jug production and suggested that early Edgefield examples were tied to African practices including root magic. They also illustrated and discussed a small, circa-1900 face vessel made by South Carolina potter Mark Baynham, whom they credited with starting the "tradition of white Southern potters copying the African-style jug."[25]

All of the aforementioned scholars agreed that slaves and, later, free African-Americans made early Edgefield face vessels, and most asserted that those objects have aesthetic and ritualistic connections to African culture. John Burrison, though, presents the problematic contention that Edgefield face jugs are part of an Anglo-American tradition. This is initially refuted by Colonel Thomas Davies's firsthand account of slave potters working at the Palmetto Fire Brick Works as well as more recent research that has identified specific African-American potters. In "Afro-American Pottery in the South," Burrison minimized the contributions of African-American artisans, stating that "the majority of known Black American folk potters were concentrated in Edgefield District, South Carolina, but even so they do not appear to have numbered many more than a handful, and information on them is scanty."[26] From this questionable observation he deduced that white artisans likely made some of the pieces attributed to black potters.

> Assuming, of course, that this ware was in fact made by Blacks; the documentation seems to be limited to interviews conducted by the Charleston Museum in this century, and my own research has taught me to accept oral history data only with caution. I am not seriously doubting this documentation; however, I intend to establish that there was a fairly early tradition of Anglo-American stylized face vessels, and it does not seem impossible that white potters operating in Edgefield District would have been responsible for some of the pieces attributed to Blacks.[27]

Burrison incorrectly claims that only a handful of African-Americans worked in the potteries. As with so many other industries around the time of the Civil War, when white males went off to battle, the Edgefield potteries relied heavily on slave labor, and African-Americans were in all of the potteries performing duties ranging from gathering the clay to turning vessels.[28] As Cinda Baldwin observed:

> Slaves constituted a significant portion of the labor force in nineteenth-century Edgefield industry. From 1800 to 1820 the number of slaves increased four times while the white population decreased. . . . During the same period the local pottery industry, which employed a significant amount of slave labor, was established to satisfy the needs of plantations for food storage and preservation.[29]

Burrison's thesis, which by his own admission lacked a firm evidentiary basis, ignores the face vessels' visual similarities to African sculptural traditions and does not consider possible uses within the African-American community.[30]

The Wanderer *Connection*

By the nineteenth century, the slave population in South Carolina was fairly diverse, although a high proportion hailed or descended from Africans of the Congo-Angola region. Daniel C. Littlefield writes, "The cosmopolitan nature of colonial South Carolina is painted in the advertisements for slave runaways." He goes on to describe advertisements for slaves from Angola,

Bambara, Guinea, Limba, and Timmene.[31] The Africans from these different cultures brought their own traditions and rituals to the United States, many of which survived and evolved. Given the prevalence of figural terracottas in African ritualistic practices (figs. 18, 19), it is likely that more than one culture contributed to the development of Edgefield's face-vessel tradition. Indeed, objects in that tradition are best understood as creolized forms whose use varied depending on the cultural background of the maker and/or owner.

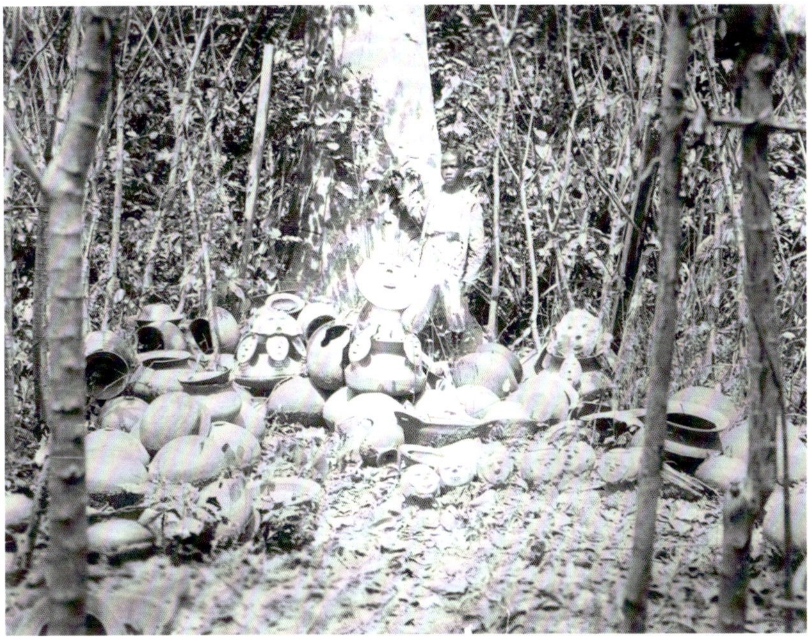

Figure 18 "Place where the pots of the deceased are kept," photographed by Friedrich August Louis Ramseyer, Ghana, 1888–1908. 4¾ x 6½". (Basel Mission Archives D-30.23.022; www.bmarchives.org/items/show/57131; photo, courtesy Basel Mission Archives.)

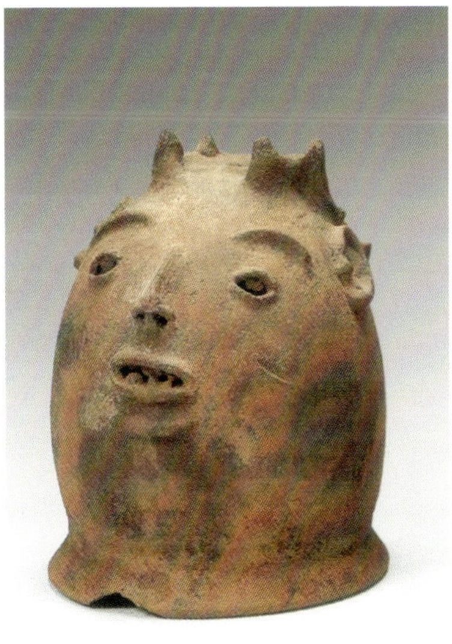

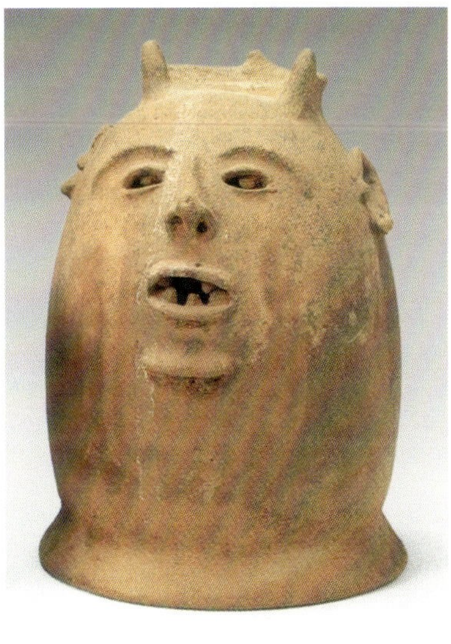

Figure 19 Ewe terracotta pots. Reproduced from Karl-Ferdinand Schaedler, *Earth and Ore: 2500 Years of African Art in Terra-cotta and Metal* (Eurasburg, Germany: Panterra Verlag, 1997), p. 191.

Face vessels may have been made in the Edgefield District before 1850, but the arrival in 1858 of the slave ship *Wanderer* off the coast of Jekyll Island, Georgia, was a significant event in the development of that regional tradition. Although the 407 Africans aboard were illegal cargo, a group of

Southern planters, led by Charles Lamar of Savannah, Georgia, conspired to acquire the slaves. He and his cohorts hoped their actions would lead to a reopening of the slave trade and secession from the Union.[32] Many of the *Wanderer* slaves were put on barges and trains going to Alabama, Florida, Louisiana, and Texas. About half were sent aboard the steamboat *Augusta*, which sailed up the Savannah River to a plantation owned by Charles Lamar's cousin Thomas Lamar. From there, many were sent to work in northern Georgia and South Carolina, including in the Edgefield area.

In 1908 anthropologist Charles Montgomery interviewed seven of the *Wanderer* Africans living in the Edgefield County, Aiken County, and Augusta area: Tucker Henderson, Tom Johnson, Lucy Lanham, Ward Lee, Katie Noble, Romeo Thomas, and Uster Williams.[33] Based on Montgomery's writings, Kikongo language scholars John Thornton and Linda Heywood assert that Tom Johnson, Lucy Lanham, Katie Noble, and Romeo Thomas came from the same area of Madimba in the valley of the Mbidizi River in the Kingdom of Kongo (fig. 20).[34] This establishes a direct connection between the *Wanderer* Africans and Kongo cultural traditions. Romeo Thomas appears to be a central figure in the *Wanderer* story as well as the face-vessel story. Originally named Tahro, he was twenty-seven when he arrived and the eldest of the *Wanderer* group sent to Edgefield. At the time of his death, he claimed to have been a leader in the Kingdom of Kongo, calling himself "Tahro, King of Kuluwaka, Africa."[35]

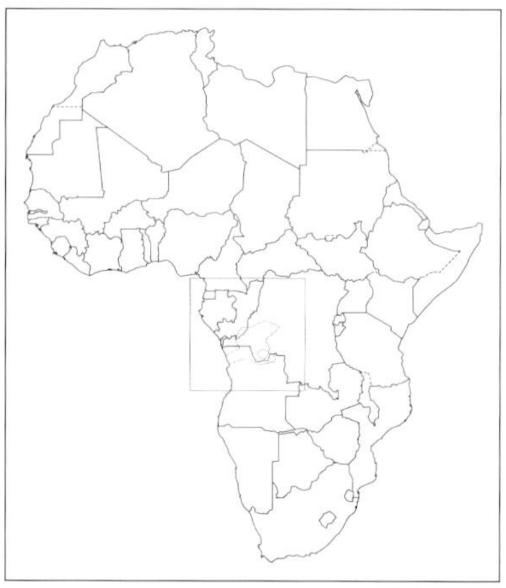

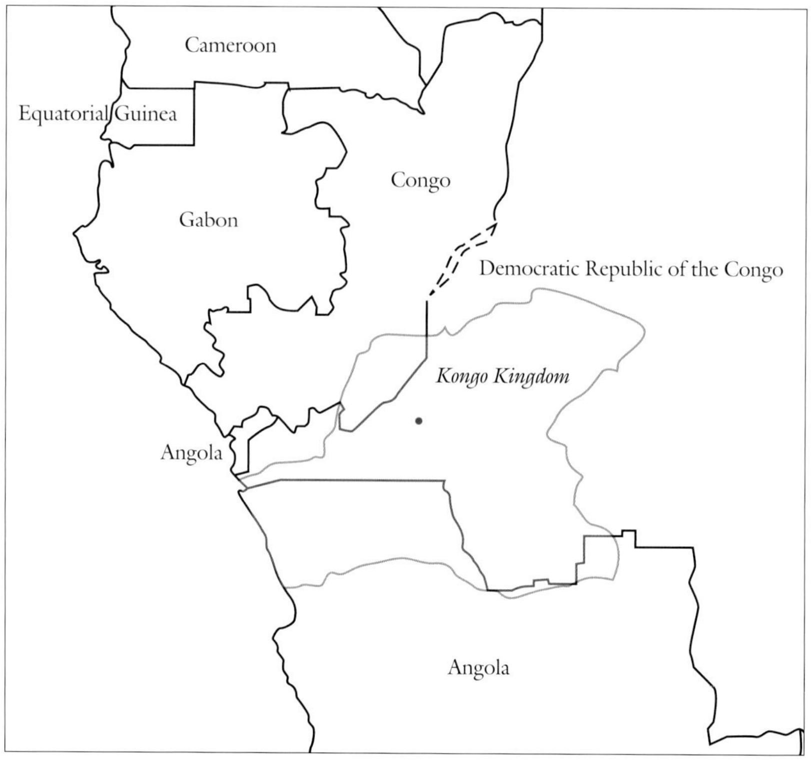

Figure 20 Map of Africa. Area in gray showing the Kingdom of Kongo, ca. 1850. Red dot indicates Madimba in the valley of the Mbidizi River in the Kongo Kingdom. (Map by Wynne Patterson.)

The arrival of the *Wanderer* received national attention. On November 21, 1859, the *New York Herald* reported that a beautiful silver goblet "offered by Col. Hunt, for a specimen of native Africans to be exhibited at the State-Agricultural Fair, was taken yesterday by Dr. Bland of Edgefield. . . ."[36] Although whites appear to have been interested in the *Wanderer* slaves as "wild Africans" and curiosities, it is unclear how members of Edgefield's slave community perceived them. Historian Joseph Holloway suggests that area blacks may have seen the newly arrived Africans as living conduits to their own cultural traditions.[37]

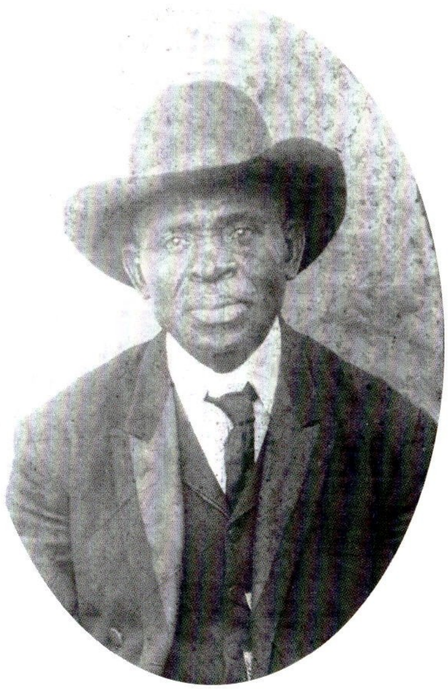

Figure 21 Photograph of Ward Lee, Edgefield County, South Carolina, ca. 1900. (Courtesy, Lee family.)

Romeo Thomas was hired at the Palmetto Fire Brick Works on at least three separate occasions.[38] Another member of the *Wanderer* group was Ward Lee (fig. 21), who it appears was connected with the Edgefield potteries as well. Oral tradition maintains that he was a potter, and a small white face jug that descended in his family relates very closely to sherds recovered at the Baynham site (fig. 22).[39] Reputedly made by Lee, this jug establishes a direct connection between Edgefield face vessels and a slave from the *Wanderer*. The object is covered in a kaolin wash that appears to have been applied after firing. As Thompson and others have noted, white substances and materials like kaolin had profound meaning in Kongo culture, and African culture in general. Evidence of the latter can be observed in the African-American practice of decorating graves with white objects such as china, clocks, and porcelain chickens.[40]

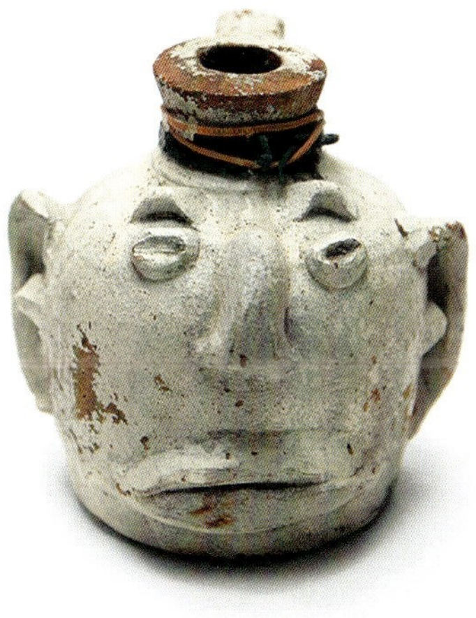

Figure 22 Face jug attributed to Ward Lee, Edgefield District, South Carolina, ca. 1880. Unglazed stoneware. H. 5". (Courtesy, Georgia Archeological Institute.)

In the most general terms, anthropological and historical studies suggest that cultural traditions rarely survive unaltered when carried from one place to another. Those of the incoming group, in this case African slaves who were forcibly moved across the Atlantic, tended to merge with traditions already present in the new environment and to create new, syncretic customs. Although based on West and Central African traditions, the rituals and customs of the *Wanderer* slaves were adapted and modified by countless new influences and forces—the staggering restrictions of slavery, the strongly Eurocentric cultural conventions in South Carolina, the myriad traditions of African-Americans within Edgefield's slave communities, and the simple fact that the *Wanderer* slaves were brought to a region that was largely defined by the production of utilitarian stoneware. This creolization of ideas and traditions is central to understanding Edgefield face vessels, which themselves are based in African practices adapted to a South Carolina context.

A growing body of evidence suggests numerous ways in which the early Edgefield face vessels are directly linked to African-inspired rituals and religious practices—including conjure, hoodoo, or root magic—that survived the trip across the Atlantic and were re-created in America. The jugs should be viewed as the product of a migrated African tradition and also as objects whose distinctive history has yet to be fully told. In South Carolina's highly restrictive slave communities, deeply rooted African or African-American cultural ways, including the ritualistic use of face vessels, often had to be practiced covertly. Certain Africanisms—language, dress, teeth sharpening, scarification, and body piercing—were seen by many whites as evidence of the culturally backward traditions among slaves and were regarded as a threat to plantation authority. Likewise, African-inspired language, dress, mannerisms, music, and dance were viewed not only with disdain but also trepidation. Many forms of African-American social activity fit into a larger pattern of slave resistance that, among other things, provided the slave with a degree of control over his or her own life, and that kept alive distinctive cultural ways that otherwise would be stifled. Resistance, an essential survival strategy adopted by many African-Americans, took on many forms, ranging from literal, sometimes violent, acts of defiance against white rule to more subtle patterns of opposition. Examples of the latter include working at a deliberately slower pace, leaving tools out in the rain, purposefully hurting oneself, letting animals out of the barn, and breaking tools and equipment. When slaves cooperated with each other while performing these quiet acts of defiance, they were at times successful in lessening their burden. As historian John Boles explains, "The institution of slavery was a constant process of accommodation and mutual adjustment, with masters always having the upper hand but with slaves controlling a larger fraction of their lives than the system would logically seem to allow."[41] The relative success of these and other behavioral strategies—and by extension the ritualistic power of the Edgefield face vessels—was due, in no small part, to the belief of many plantation owners and others that slaves were incapable of such well-considered tactics.[42] However, strategies of resistance were not simply about breaking the bonds of white oppression. They also kept alive modes of cultural activity, secret communication, and spiritual belief that otherwise would not have survived.

The arts of conjure, root magic, charms, spells, and other practices that were so central to life in Edgefield and other communities during the period of slavery and beyond reflect the direct continuation of time-honored African customs. Western cultural and religious practices had an impact on certain Kongo customs long before the *Wanderer* slaves arrived in America. With regard to Christianization efforts among the Kongo people in the sixteenth century, John Thornton notes, "Christianity was accepted, not as a new religion, but as a syncretic cult, fully in keeping with other cults in the Kongo and deriving from Kongo and not from European or Christian cosmology."[43]

If Christian precepts were present in the religious and ritualistic practices of the *Wanderer* slaves, they might have been imperceptible to the white population. Indeed, whites likely considered African rituals to be threatening and potentially destabilizing.[44] Scholars of African-American history have documented many distinctive social and artistic customs that had to be practiced covertly because they did not reinforce the status quo or provided too much cultural freedom to slaves. New slaves were prohibited from speaking their native language when being "broken in," and many forms of African expression were suppressed or eliminated:

> There was even a fleeting existence of African architecture, on St. Simons Island, when an African-born man named Okra suddenly built a habitation in the manner of his ancestors. The building is said to have measured approximately twelve by fourteen feet in plan, had a dirt floor, daub-and-wattle walls ("the side like basket weave with clay plaster on it"), and a flat brush and palmetto roof. In perfect evocation of the windowlessness of much sub-Saharan traditional architecture, the only source of light was the single door. But "Massa made him pull it down. He say he ain't want no African hut on he place."[45]

In general, many African customs were not practiced openly until the end of the Civil War. After being emancipated, former Bakongo slave Tahro (Romeo Thomas), built a small, windowless house in the style of a Kongo home or idol house (figs. 23, 24). Vlach argued that African-style dwellings served as the model for later American "shotgun" houses—the narrow, linear houses found throughout the South. Romeo's house may have been similar to those built after the Civil War by *Wanderer* Africans Umwalla and Cato in nearby Hamburg, South Carolina. In an article in the *Augusta*

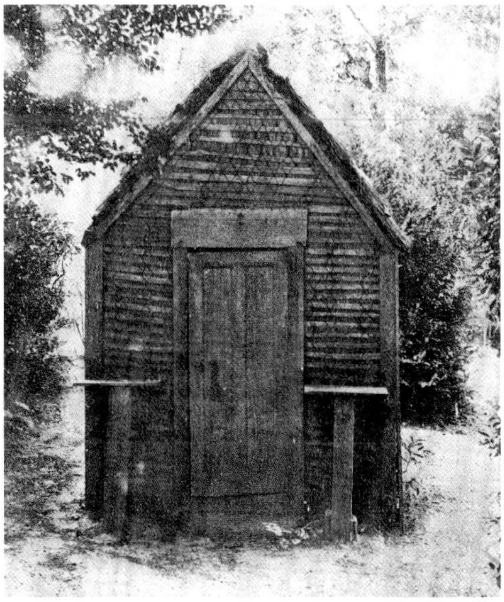

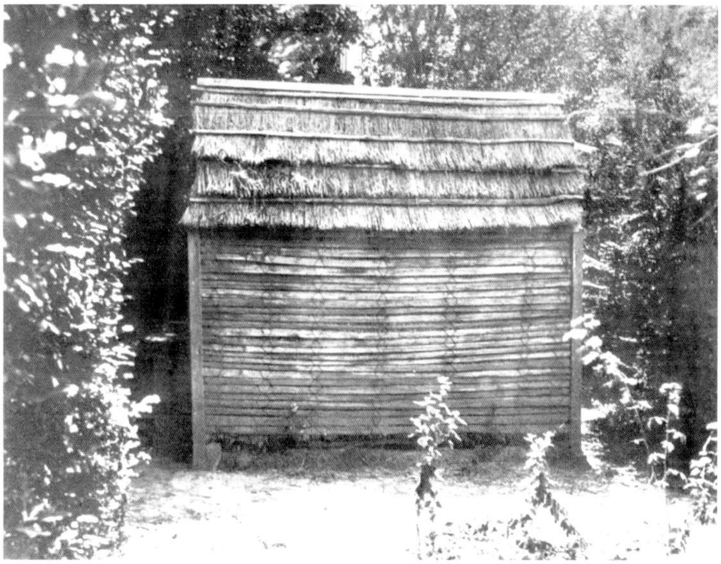

Figure 23 Photographs showing the front and right side of the Romeo Thomas House, Edgefield District, 1908. Reproduced from Charles J. Montgomery, "Survivors from the Cargo of the Negro Slave Yacht Wanderer," *American Anthropologist*, n.s., 10, no. 4 (1908): between pages 616–17 and 618–19.

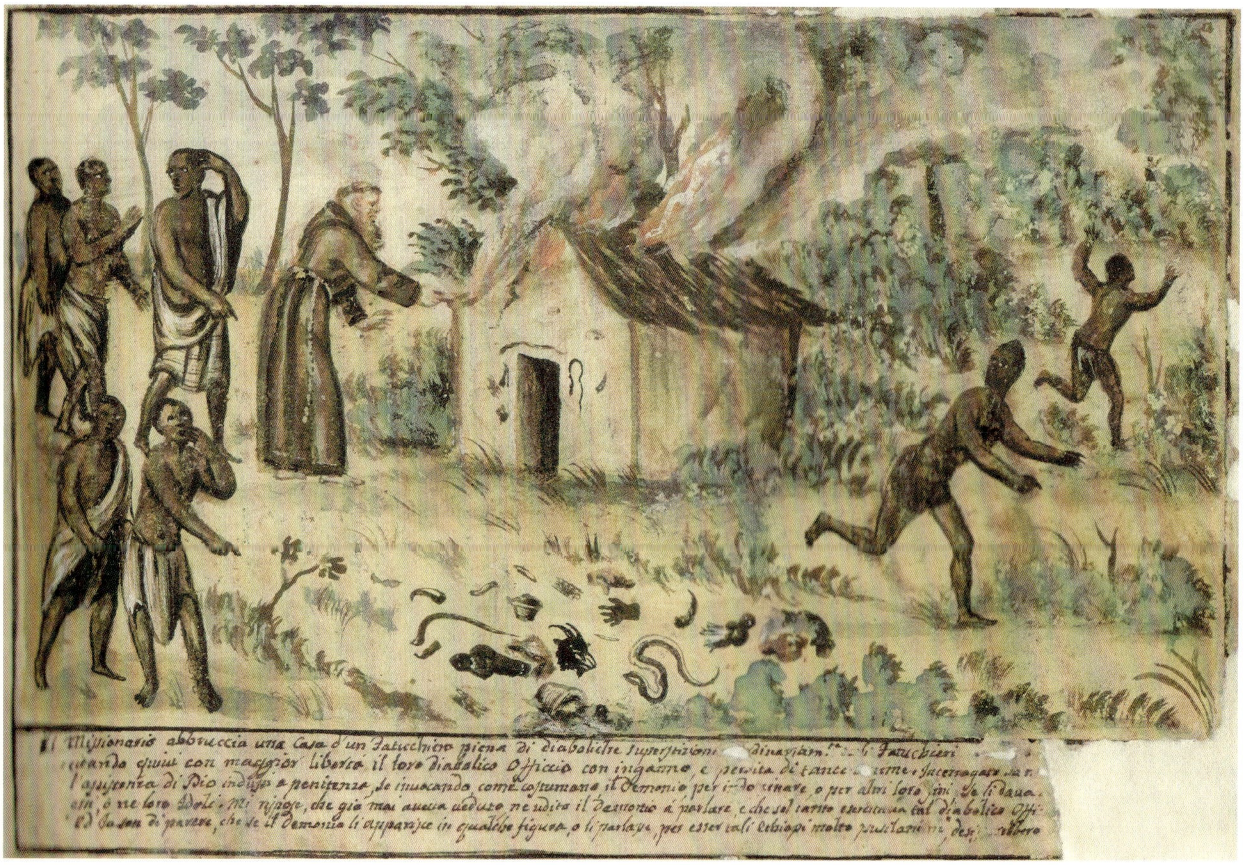

Figure 24 "Catholic Priest Burning Idol House, Sogno, Kingdom of Kongo, 1740s," reproduced from Paola Collo and Silvia Benso, eds., *Sogno: Bamba, Pemba, Ovando . . .* (Milan: Franco Maria Ricci, 1986), p. 163. (Photo, courtesy Biblioteca Civica Centrale di Torino.)

Chronicle in 1888, Umwalla stated that he was born in Guinea, in West Africa, and that he regularly visited other *Wanderer* Africans on nearby plantations, where he enjoyed watching performances of African-American dance because they reminded him of a type of African dance.[46] The songs sung by the field hands were also familiar. When several voices joined in unison he was "carried across the ocean to the primitive scenes of his youth." Thirty years after his forced relocation to America, Umwalla said that he was not homesick because so many of the customs in the slave quarters were very much like those of his people in Africa.[47]

The covert retention of African traditions and the widespread strategy of resistance to preserve essential cultural ways and modes of communication are seen in other ways as well. In a recent interview, ninety-six-year-old Georgia Scott, an African-American teacher born in 1916, well past the abolishment of slavery, recounted the continued use of drums for sending messages from one community to another in Aiken and Edgefield counties: "I remember as a young girl, the sound of the drum travelling across the Savannah River, telling us that someone had died or that there was to be a special meeting of a dance."[48] This strategy was especially vital in antebellum South Carolina, where reading and writing was forbidden to slaves and general communication between slave communities was achieved by coded means such as drumming. The *Wanderer* Africans also used drums to communicate. Tonie Houston, a preacher from Tatemville, commented on their use of this instrument in 1940: "For their meetings, he said, Golla Tom or another would beat the drum—signaling them to gather; then all would sing and dance in a circle to the accompaniment of the drum. The drums of death would also sound, summoning the 'setting-up or wake.'"[49]

The practice of "coding," through the use of deception and double meanings in southern African-American communities, extended well beyond the use of drums. Secret messages or rituals were carefully inserted into seemingly innocuous African-American cultural practices, which kept slave owners from objecting and prohibiting them. In his *Narrative of the Life of Frederick Douglass, an American Slave*, for example, Douglass described slaves singing hymns while walking through the woods on their way to a neighboring plantation: "This they would sing, as a chorus, to words which to many would seem unmeaning jargon, but which, nevertheless, were full of meaning to themselves."[50] Spirituals were also used as vehicles to convey secret messages. For example, since African-Americans were not allowed to hold religious meetings without the presence of a white man, many met in secret. They conveyed the meeting time and place through the spirituals. Former slave Wash Wilson recalled: "When the niggers go round singin' 'Steal Away to Jesus,' dat mean dere gwine be a 'ligious meetin' dat night. De masters . . . didn't like dem 'ligious meetin's, so us natcherly slips off at night, down in de bottoms or somewhere. Sometimes us sing and pray all night."[51] These songs were also used to warn slaves of the presence of patrollers or other dangers encountered in their daily lives, and operated as a critical response to slavery. Former slave and activist Booker T. Washington commented:

> As the day [of emancipation] grew nearer, there was more singing in the slave quarters than usual. It was bolder, had more ring, and lasted later into the night. Most of the verses of the plantation songs had some reference to freedom. True, they had sung those same verses before, but they had been careful to explain that the "freedom" in these songs referred to the next world, and had no connection with the life in this world. Now they gradually threw off the mask; and were not afraid to let it be known that the "freedom" in their songs meant freedom of the body in this world.[52]

Multiple levels of meaning went undetected in part because many slaveholders and overseers believed that the singing connoted contentment, and they assumed that the songs were simple forms of entertainment. A former slaveholder, recalling a slave song about a fox, stated that it was "typical in that it shows the Negro's fondness for dramatic dialogue and his interest in animals."[53] In fact, this slaveholder completely misconstrued the song's significance, unaware that in African-American folklore the fox often represents an oppressor who is outsmarted by the rabbit trickster. By the same token, white southerners did not fully understand the true meanings and uses of the face vessels.

African-American folk tales reveal a similarly strategic utilization of double hidden meanings. In the African-American folk tradition, animals were used as fictional surrogates for real-life plantation figures. The rabbit is an especially popular character in these works, and is distinguished by cleverly triumphing over stronger characters. For example, in Joel Chandler Harris's

Brer Rabbit's Riding Horse, the character of Brer Rabbit pretends he is so sick that Brer Fox must put on a saddle in order to carry him over to Miz Meadow's house. In this way Brer Rabbit flips the master/slave relationship while also humiliating the dominant character.[54] The slaves identified with the Brer Rabbit character, and through these stories were able to imagine a world in which they prevailed over their masters. As Lawrence Levine explains in his book *Black Culture and Black Consciousness: Afro-American Folk Thought from Slavery to Freedom*, "The trickster's exploits, which overturned the neat hierarchy of the world in which he was forced to live, became their exploits; the justice he achieved, their justice; the strategies he employed, their strategies. From his adventures they obtained relief; from his triumphs they learned hope."[55] What slaveholders perceived as whimsical tales were in actuality searing critiques of plantation life; and what were perceived as whimsical objects were highly ritualistic objects.

The African-influenced potters of Edgefield did much the same thing in taking the basic utilitarian jug form and converting it into a carefully coded spiritual object with hidden meanings. "We can be certain that these vessels were not frivolous or whimsical items," wrote Vlach, "they are sculptures willfully created to convey a message, one known at least among a community of slave artisans and probably by the surrounding slave community."[56] In other words, the Edgefield face vessels need to be understood as bicultural or syncretic objects that were influenced by both African and distinctly African-American cultural traditions.[57] An example of the syncretic nature of the face vessel is their possible function as water containers. The Bakongo land of the dead is believed to reside under lake bottoms and riverbeds, connecting the spirits to water. African-American history specialist Jason Young has stressed the importance of water for both the Bakongo and the enslaved African-Americans in South Carolina. As an example, he pointed to the ritual of complete water immersion in African-American baptisms, which was viewed as symbolizing a death of the old self and a birth of the new, more pure self.[58] Young argued that this ritual grew out of West African initiation rituals of complete water immersion and the Bakongo beliefs in *simbi* spirits. Vlach likewise drew the connection between water and the land of the dead in African-American burial practices; slave graveyards contained white seashells, as well as broken pottery and glassware, a phenomenon also found in Bakongo burial sites. He explained that "Afro-American graves may thus be read as a kind of cosmogram: the world of the living above; the dividing line of shells [or broken pottery]; the realm of the spirits, which is not only under ground but also under water."[59]

Mirrors or other reflective objects placed on graves served as a way to speak to those who had passed away, which grew out of a pre-nineteenth-century Bakongo practice in which they communicated with the dead by looking into a pool of water. Based on these beliefs about the power of water, it is significant that the Edgefield face jugs are in fact hollow vessels, and not flat dishes or solid sculpture. They are all containers that could have held water, and, according to oral history, have been found in African-American graveyards.[60]

The vessel form was likely a purposeful choice based on the known conduits to the spirit world, whose true meaning was kept hidden from the white community. Edwin Atlee Barber wrote that in 1862 Colonel Thomas Davies witnessed his slaves making face jugs, which he described as "weird-looking water jugs."[61] The key point is that face jugs fall into the tradition of coding and what was simply a "water jug" to Davies might have been something very different to people within the African-American community. Davies's comment shows that their making was not hidden from the view of the white pottery owners. The African-Americans may very well have encouraged the misinterpretation of the form as simply an odd drinking vessel, leaving the owners unaware of its hidden meanings and uses. Vlach alludes to this deliberate deception:

> In the presence of the ferocity and energy expressed by the best of these Afro-Carolinian vessels, one senses a shift in the attitude, a craft based on the self-generated incentive of a vital culture, standing apart from the nature of most pre-Civil War Southern Afro-American industry. The distinction, the writer would guess, stems from the fact that the Afro-American craftsmen made these vessels for themselves and their people for traditional reasons of their own. Under the noses of their masters they succeeded in carving out a world of aesthetic autonomy.[62]

The complexity of African-American religious and ritualistic practices cannot be overstated, and the covert or concealed nature of many of the traditions makes their interpretation today all the more difficult. In *Come Shouting to Zion: African American Protestantism in the American South and British Caribbean to 1830*, authors Sylvia Frey and Betty Wood argue that "the passage from traditional religions to Christianity was arguably the single most significant event in African American history. . . . It provided Afro-Atlantic peoples with an ideology of resistance and the means to absorb the cultural norms that turned Africans into African Americans."[63] This is not to say that traditional African religious beliefs were abandoned. Rather, African-Americans often customized Christian practices to cope with the harsh realities of enslavement. Some aspects of African religions were adapted to Christianity, whereas others were practiced in addition to Christianity, such as conjure and root work. Slave-religion expert Albert J. Raboteau expounded that "adaptability, based upon respect for spiritual power wherever it originated, accounted for the openness of African religions to syncretism with other religious traditions and for the continuity of a distinctively African religious consciousness."[64] The aforementioned practices—drumming, storytelling, music, and the like—reveal a process of adaptation and assimilation of African beliefs into African-American beliefs and rituals.

Visual and historical evidence strongly suggests that the early Edgefield face vessels were directly inspired by African culture, which was reintroduced to this area with the landing of the *Wanderer*, and also by local African-American cultural traditions as practiced in slave and free black communities. To gain fuller insight into the vessels' formal and ritualistic attributes, it is essential to explore the African beliefs from which they came.

In most African societies, ritualistic pottery vessels are visually complex, just as jugs are more ornate than contemporaneous utilitarian jugs. In *Four Moments of the Sun*, Thompson discusses a type of terracotta vessel called a *diboondo* (plural *maboondo*) (fig. 25). The *diboondo* resembles a cylindrical or quadrangular hollow column divided into horizontal registers and is placed on the tombs of important Bakongo. The *maboondo* are perforated with triangles, squares, and diamond shapes and are incised with different designs as well as sometimes decorated with applied figures. These vessels were meant to be spiritual containers. As with so many examples of material culture, these vessels blur the line between art and culture, form and function.

Thompson's pivotal scholarship emphasized the distinctive and symbolically charged use of kaolin, a pure-white river clay that in a West African context was widely regarded as having magical properties and can be linked to many ritualistic customs. The purposeful addition of kaolin to ceramics and other ceremonial objects in Africa—as in Edgefield, South Carolina—mimics that used to activate a connection to the spirit world. Diviners in Baule culture (Côte d'Ivoire), for example, rub kaolin over their eyes and mouths in order to initiate communication with the spirits while in a trance.[65] Carved Baule divination figures that depict humans mirror this practice by also featuring kaolin around the eyes (fig. 26). In West Africa, Vodun worshipers "who desire to enter the spirit world, similarly cover their bodies with white kaolin powder considered 'food for the gods'. This dramatic form of body painting allows the devotee to attain the highly sought state of possession, whereby a deity may enter his body and empower him."[66] The Fang similarly use kaolin to connect to the spirit of the dead, and the Bembe apply kaolin over their masks' eyes to suggest the ability to see into the spiritual world. The implications of the use of kaolin on the Edgefield face vessels are even stronger in light of other African cultural traditions.

Figure 25 *Diboondo*, Democratic Republic of the Congo, ca. 1800. Terracotta. H. 17¹¹⁄₁₆". (Courtesy National Museum of African Art, Smithsonian Institution, purchased with funds provided by the Smithsonian Collections Acquisition Program, 89-13-7; photo, Franko Khoury.)

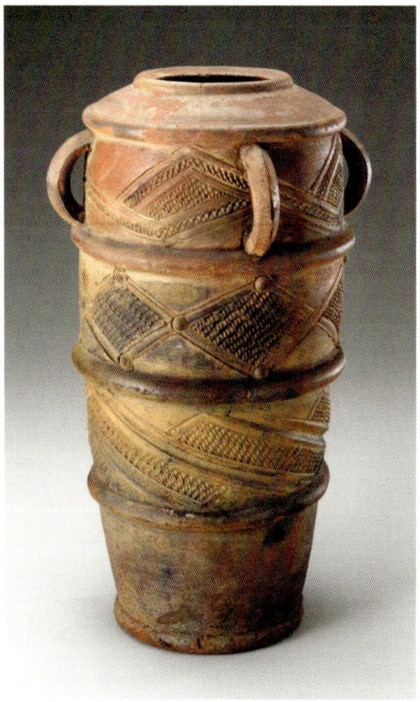

As Gary Zabel writes, for example, the Bakongo believe that the land of the dead is a mountain made of kaolin, and that white is the color of the spirit world:

> In that extraordinary land of kaolin, in that invisible looking-glass realm, those who have led a basically upright life clear through to its conclusion have the opportunity to purge themselves of remaining moral impurities. They then enter the cycle of existence once again by being reincarnated in the form of their own grandchildren. Those who have conducted themselves in an especially righteous way are transformed into immortal simbi spirits who take the shape of pools, waterfalls, stones, and mountains.[67]

Rubbing kaolin on or placing it inside *minkisi* was a necessary way, both literally and symbolically, to establish a connection to this purified place. As with the Baule, Bakongo diviners would also rub the white kaolin around their eyes and mouths to better see and speak to the spirits.[68]

Many West African cultures place particular importance on the head as the location of identity and force of life. The Yoruba created pots that featured human faces and were used in a variety of rituals. A Songye power figure called a *nkishi* (similar to a *nkisi*; plural *mankishi*) is carved in the form

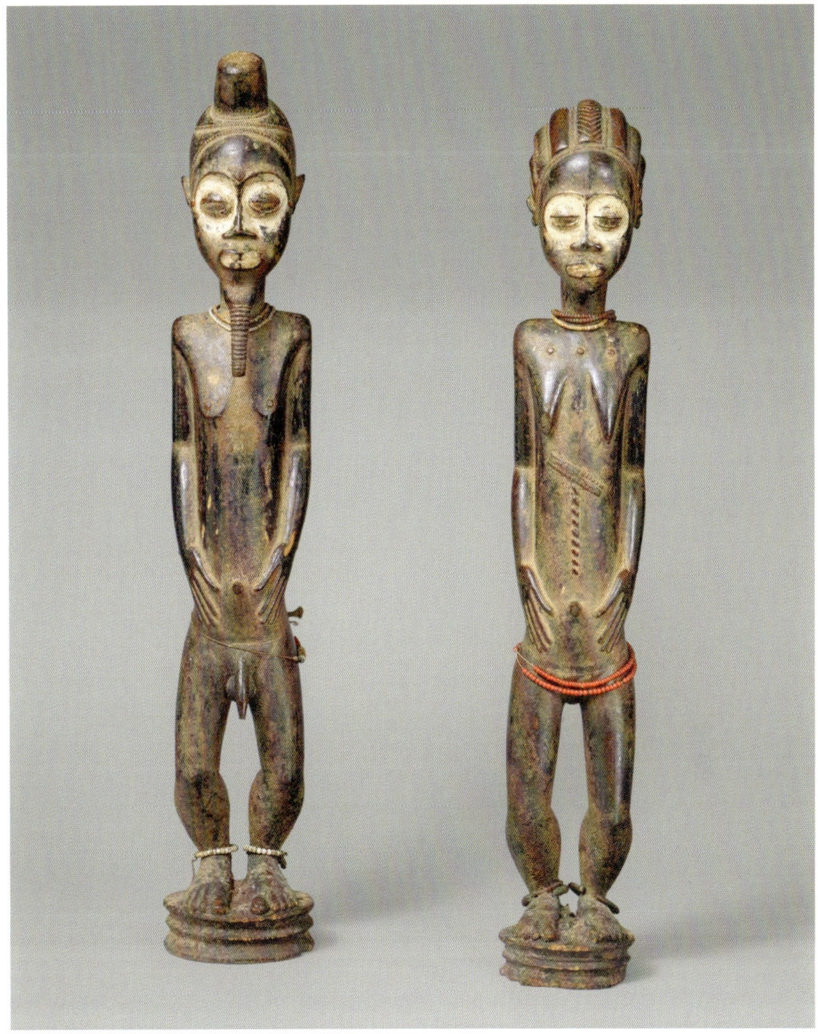

Figure 26 Male and female divination figures, Baule, Côte d'Ivoire, nineteenth or early twentieth century. Wood, pigments, beads, and iron. H. of male 21⅞"; h. of female 2⅝". (Courtesy, The Metropolitan Museum of Art, New York, The Michael C. Rockefeller Memorial Collection, Gift of Nelson A. Rockefeller, 1969, 1978.412.390, .391.)

of a male. Alisa LaGamma explains that "*mankishi* enhance the overall quality of life by assuring that universally shared concerns, such as procreation, protection against illness, witchcraft, and war, are addressed through divine intervention."[69] The physical attributes of each sculptural form are first determined by a conversation between the *nganga* and the ancestral realm or spirit world. As LaGamma notes, "the physiognomy and regalia of a community *nkishi* emphasize the influential character of its subject, conveying a degree of physical strength and social rank commensurate with the spiritual powers it commands."[70] Typically, the heads of these figures are disproportionately large in relation to the body, and also feature wide, almond-shaped eyes, an unusually broad forehead, and a jutting beard. But again, these attributes are emphasized for a reason. To the Songye people, the broad forehead symbolizes omnipresence, the beard symbolizes leadership, and the large eyes symbolize contemplation, all to convey the notion that the spirit occupies the figure.[71]

The kaolin-decorated Edgefield face-vessel forms are heads, rather than full bodies in the manner of carved wooden *nkisi* figures (fig. 27), which is fully consistent with keeping intact an African tradition of honoring the head as the force of life and container of spiritual powers. As with most African sculptural forms related to the human body, the Edgefield face vessels are not meant to be individualized portraits or caricatures. Rather, they function in a totemic way and their features have symbolic meaning. The large kaolin eyes and teeth connote the ability to see and talk to the spirits (fig. 28); the beards, present on some face vessels, denote wisdom.

The Art of Conjure and Edgefield Face Vessels

Among the many African cultural rituals that persisted during and after slavery was the practice of conjure—also root magic or root medicine—a widespread African-influenced system of magic still practiced in the South.[72] Born out of West and Central African divination and spiritual practices, conjure was prevalent in North American slave communities in the nineteenth century. In fact, the 1900 census for the Schultz Township of Aiken, South Carolina, lists Richard Phillip as working in root medicines (fig. 29).[73] Aiken was known as a center of conjure in the nineteenth and early twentieth centuries.[74] Conjure was used for diverse purposes, from protection and healing to causing harm to others. It was perceived as providing an additional level of protection for its practitioners. Raboteau explained the function of conjure in this way:

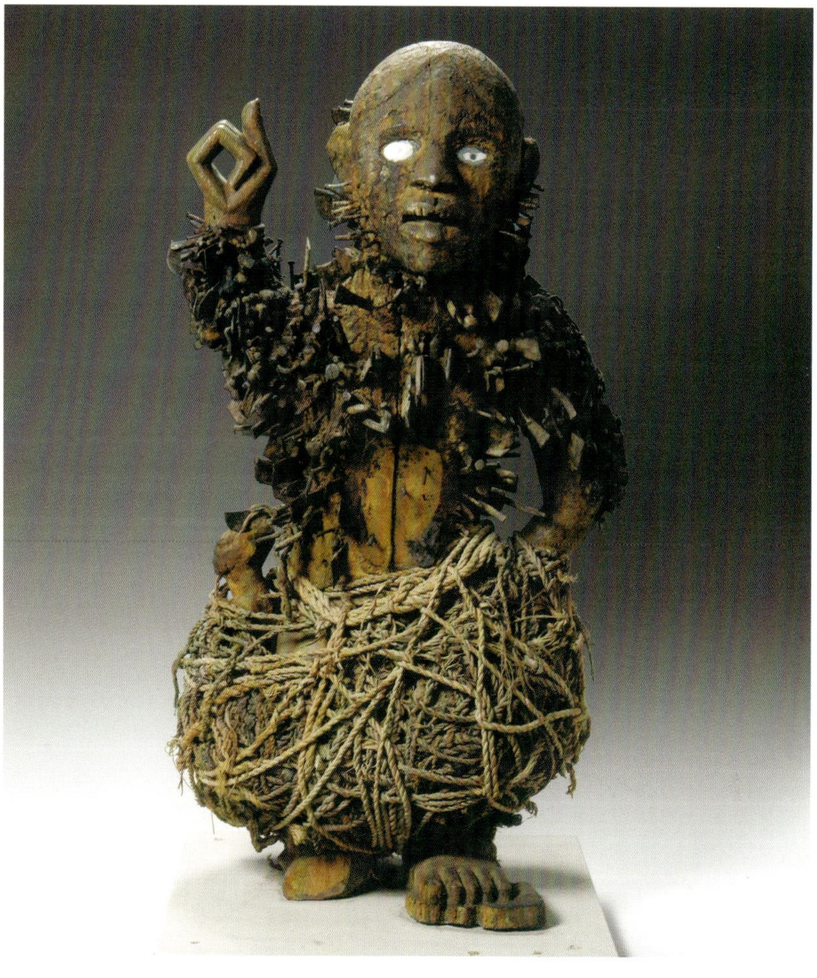

Figure 27 *Nkisi nkonde* with tongue, Lekela Kongo (Yombe), Democratic Republic of Congo, 1913(?). (Courtesy, Coll. G. Rosselli-Lorenzini, Museo Nazionale Preistorico Etnografico "Luigi Pigorini," inv. 84205; photo © S-MNPE "L. Pigorini" Roma EUR, courtesy MiBAC.)

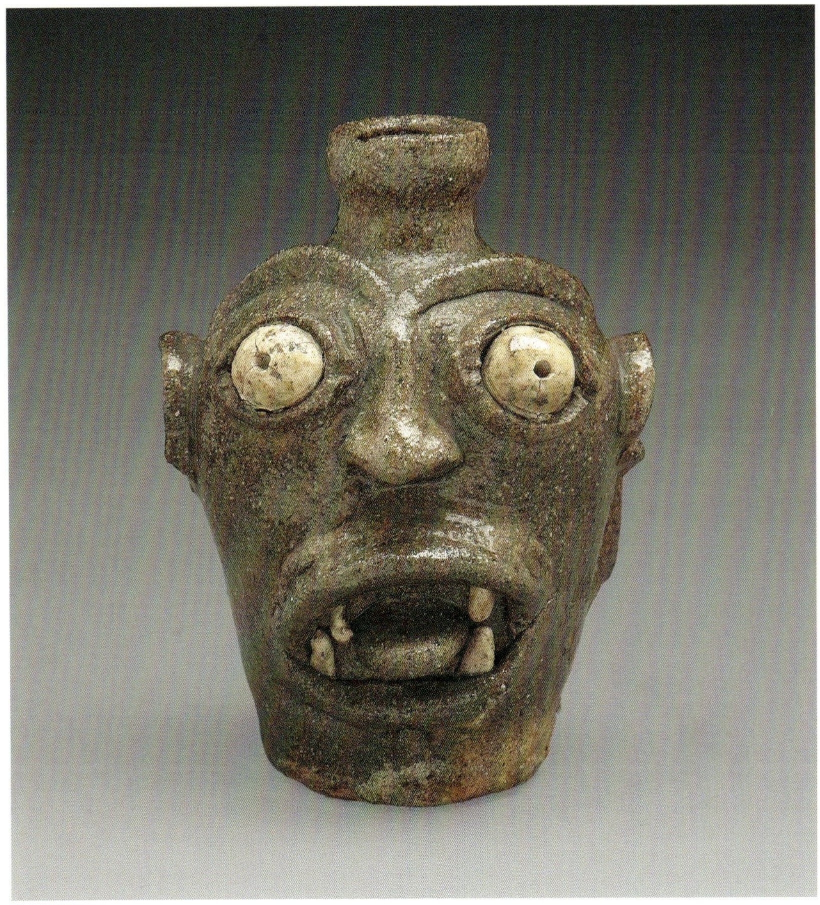

Figure 28 Face jug, Edgefield District, South Carolina, 1860–1880. Alkaline-glazed stoneware. H. 4½". (Chipstone Foundation; photo, Jim Wildeman.)

> Primarily, conjure was a method of control: first, the control which comes from knowledge—being able to explain crucial phenomena, such as illness, misfortune and evil; second, the control which comes from the capacity to act effectively—it was "a force . . . by which man can (or thinks he can) achieve almost every desired end" . . . third, a means of control over the future through reading the "signs"; fourth, an aid to social control because it supplied a system whereby conflict, otherwise stifled, could be aired.[75]

The actual practice of conjure, which in Edgefield might have included the use of face vessels, helped slaves cope with the realities of plantation life by providing them with a degree of power. It not only functioned outside white authority but also could be activated by slaves against their owners. A former slave from South Carolina interviewed for the Federal Writers' Project of the Works Progress Administration stated: "Conjin 'Doc.' say dat he done put a spell on ole Marse so dat he wuz 'blevin ev'y think dat us tole him bout Sa'day night and Sunday morning."[76] Slaves seem to have turned to conjurers for help with everything from love spells to revenge against another slave to protecting themselves against the violence that white Christian preachers were telling them was justified.

Figure 29 Entry from 1900 U.S. Census, Schultz Township, Aiken, South Carolina.

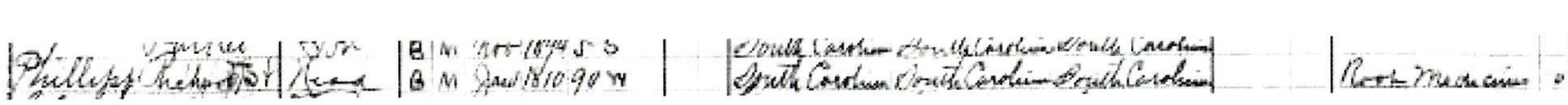

Conjurers, or conjure doctors, tended to be elderly slaves or newly arrived Africans.[77] They were considered to be wiser and more closely tied to the supernatural world, and emphasized the use of charms and countercharms to protect or harm. Conjurers created charms by combining certain magical materials inside a container. The most common materials included in these conjure bags were hair, nail clippings, personal objects belonging to the victim, grave dirt, roots, and powders. There can be little doubt that this practice was well established long before Bakongo slaves arrived on the *Wanderer*. Many of the *Wanderer* slaves, including Romeo Thomas and Cudjo, claimed to be kings, queens, or princesses—an approximation, most likely, for African terms that denoted tribal leaders, chiefs, and *ngangas*.[78] In addition to the claims of prior privilege and leadership, there are many recorded instances of *Wanderer* Africans being treated as persons of privilege by their peers and of them practicing magical rites and root medicine, both associated with persons of special stature in African communities. Bettis C. Rainsford recounted a story wherein Sarah Bull Barnwell, widow of John Gibbs Barnwell of Beaufort, reportedly purchased two *Wanderer* slaves—a man and a woman—paying for them in gold. The woman had been a princess in Africa and the man subsequently became her husband.[79]

Evidence suggests that members of slave communities in the South believed that native Africans had special powers, such as the ability to hoe a field without picking up a tool, or to rise up and fly home to Africa.[80] Prince Sneed of White Bluff, Georgia, recalls his grandmother telling him of how Africans shouted, "'*Kum buba, yali kum buba tambe, dum kunka yali kum kunta tambe*.' Then they rised off the ground and fly away."[81] As noted earlier, powerful Kongo Africans fulfilled the roles of *nganga*, or medicine men, in their villages. They were charged with assessing situations in which magic was believed to have inflicted harm. They would then use charms, spells, and spiritual objects such as *nkisi* to resolve the matter.[82]

The beliefs and practices associated with conjure were largely imported with enslaved Africans into America, the Caribbean, and South America. In addition to the use of traditional charms and spells, African slaves and their descendants incorporated other objects into their belief systems—such as the use of a silver coin around the ankle to ward off evil or give warning of a malefic conjure (fig. 30). Slaves also turned to root magic, as the medicinal counterpart to conjure. It was consistently practiced in South Carolina—sometimes in addition to traditional Western medicine, sometimes instead of it. Africans imported by the *Wanderer* into the southeastern United States, especially in the Edgefield area, were adept in the practice and application of root magic.[83]

The Edgefield and Aiken areas are reported to have been at the center of conjuring activity.[84] A specific example of conjure from 1868 was published in the *Herald*, which related a story of Randall and Lizzie Williams on the Woods Plantation near Aiken. Lizzie had fallen ill and, after receiving no benefit from the attendance of several physicians, was led to believe she had been "conjured." As a last resort, Randall contacted a "Voodoo Doctor," who sent them to another alleged "sorcerer in Edgefield County." The

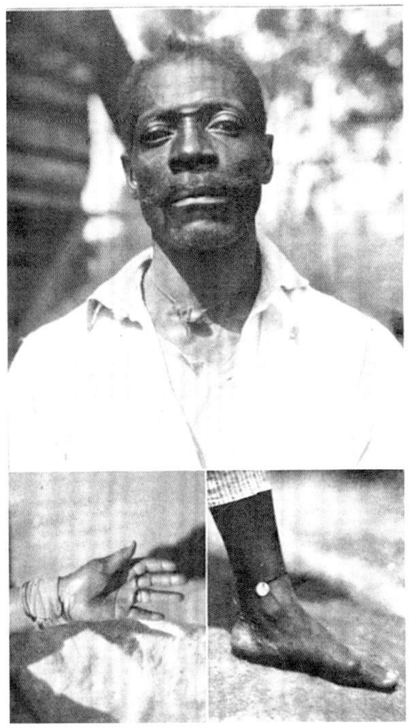

Figure 30 "Nutmeg, red flannel and silver," in Newbell Niles Puckett, *Folk Beliefs of the Southern Negro* (Chapel Hill: University of North Carolina Press, 1926), between pp. 314–15. (Courtesy, Schomburg Center for Research in Black Culture / General Research and Reference Division.)

sorcerer said that a diligent search of the property would reveal the cause of his wife's troubles. The "charm" was subsequently found and said to have been "hand made, and when found the next day was the wonder of all the African Americans in the neighborhood."[85] Next, by direction of the Edgefield sorcerer, Williams took the charm to the river (water being where spirits lived) and, walking backward, approached the water and tossed the charm over his head into the current. It was reported that from the day the charm was found, Lizzie Williams began to improve "and is now as active and sprightly as any young girl on the plantation."[86]

Twenty years later a group of African-Americans gathered in Augusta, Georgia, to observe a conjure. By the doorstep of a house were two objects made of red flannel containing a rabbit foot, some dried herbs, and a substance a woman said was a powdered frog. At the corner of the house was another charm—a tomato can stuffed with roots, red flannel, and two or three pieces of hide with hair attached—which this woman believed had been used by her landlord to terrorize his tenants into paying on time. It was reported that the cure for the charm was that it "must be buried on the first moonlit night at 12 o'clock to destroy its power."[87] These narratives and more prove that conjure was prevalent within the African-American community in South Carolina.[88]

In conjure, charms are often buried under the doorstep, to be activated by the person as he or she steps through the doorway or onto the front step. As Jeffrey Anderson explained in *Conjure in African American Society*, "all that an enemy has to do is get some of his victim's hair, his nails or water in which he has bathed and have a witch doctor make a concoction which buried in front of the victim's door or secretly hung in his room will bring sure death."[89] In addition, the doorway was believed to be the spot one would place a broom to keep a witch from entering the house.[90] This has a West African precedent. Newbell Niles Puckett has explained that the Susu-speaking people believe "witches are born, not initiated, they put inside a man's house 'medicine' in a pot, consisting of rice, groundnut sesame and fundi, which is buried inside the door and causes him to get a bad crop."[91]

Oral tradition around Edgefield and in other areas reveals that face jugs were purposefully placed in the doorways of African-American homes. For example, Ward Lee's great-granddaughter, Mrs. Pat Bishop, related that in addition to the jug illustrated in figure 22, a blue-and-white face vessel was placed at the door of her father's house.[92] In 1956 Mrs. Frances Harris, another Edgefield resident who was related to a *Wanderer* slave, recalled a face vessel that her mother-in-law kept by the door. It was, she said, "kept all shined up and looking pretty by the door."[93] The Reverend Alexander Pope, also of Edgefield County, remembered seeing face jugs used as doorstops.[94] These instances suggest a ritualistic function of the face jug in proximity to the spiritual portals of doorways.[95]

When asked specifically about the historical use of face vessels in the community, Edgefield historian Wayne O'Bryant stated, "We have always known what face vessels were used for, they were conjure jugs. No one has ever asked us before."[96] The face jug illustrated in figure 31 might be a "conjure jug." Coated with a brownish olive-green glaze, the jug features a unibrow and ears with a defined tragus, kaolin eyes and teeth that are glazed

Figure 31 Face jug, Edgefield District, South Carolina, 1860–1880. Alkaline-glazed stoneware. H. 6⅝". (Chipstone Foundation; photo, Robert Hunter.)

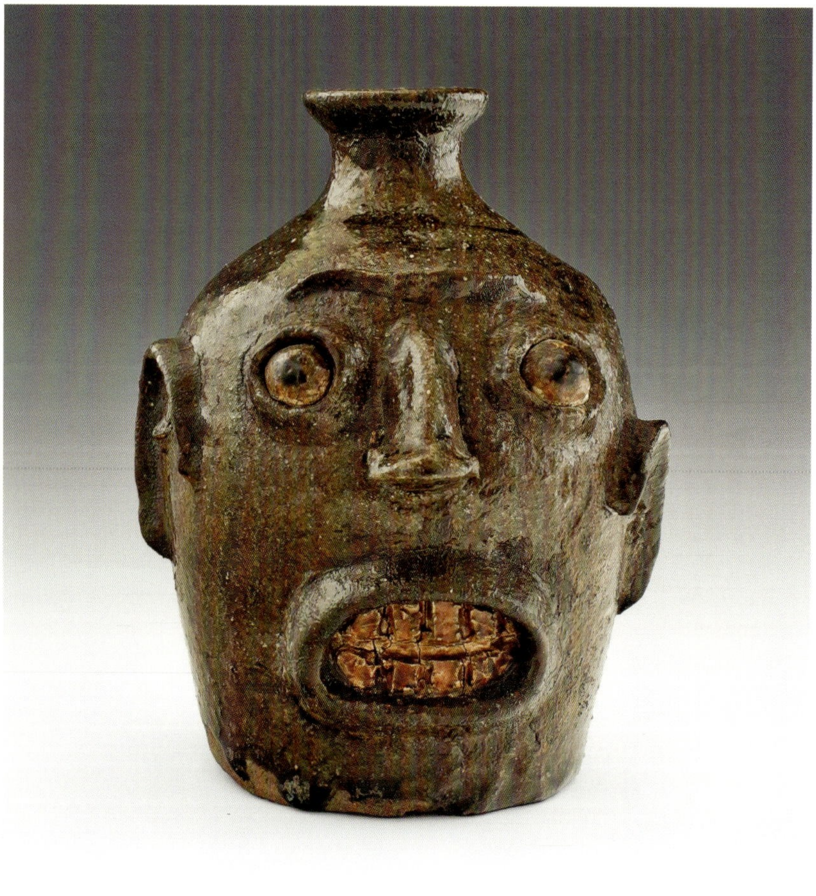

(those areas were waxed on most Edgefield face vessels to prevent the glaze from adhering), black pupils, and a rare two-word inscription on the back (fig. 32). The first word is "Squire," which, with "Esquire," was a common first name in South Carolina during the nineteenth century. The second word is less legible, although some observers have interpreted it to read "Pope" or "Peter." The 1880 U.S. Census for the township records an African-American by the name of Squire Pope living alongside pottery workers in Shaws Township, Edgefield.[97] "Squire" and "Peter" were also found in the Freedman's Savings Bank records of a former slave by the name of Glasco Middleton (b. 1835).[98] Middleton had two children named Squire and Peter.[99] However, when the inscription is examined in raking light, the second word appears to be "Pofu," which could refer to a number of things. Pofu is the name of a town in the Democratic Republic of Congo. Swahili dictionaries define *pofu* as "blind" or "to make blind," and a late-nineteenth-century Kikongo dictionary defines *mpofo* as "blind."[100] In *The Afro-American Tradition in Decorative Arts*, John Michael Vlach points out that Kikongo words were in use in South Carolina in the nineteenth century.[101] Many of the *Wanderer* Africans also spoke Kikongo.[102] If the second word on the jug is "Pofu," the inscription would translate "the blind Squire."

Like the inscription, the glazed eyes on this jug may indicate that it was intended for conjures to either cause or cure blindness. Blindness was one of the most common conjures in South Carolina. Baldwin referred to the

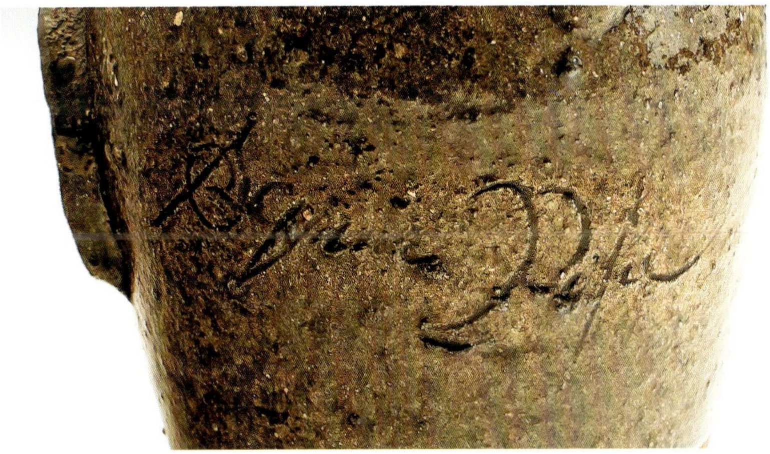

Figure 32 Detail of the inscription on the back of the face jug illustrated in fig. 31.

former *Wanderer* slave Uster Williams, who believed he had been blinded by a spell.[103] In *Folk Beliefs of the Southern Negro*, Newbell Niles Puckett recounted:

> Another conjurer tells me that anvil-dust (he calls it "Allerkerchie") is mixed with a person's food or put in his hatband. That, in his food, will "rot his guts and make snakes inside of him," while that, on his hatband, will make him go entirely blind. The reptile-dust is the common mode, however; and kept in a little vial in my conjure collection always won respect and opened wary mouths.[104]

African traditions rarely survive unaltered in the New World. Therefore, even though Edgefield face vessels were influenced by the Bakongo culture, they were not straightforward *minkisi* nor did they solely perform the same function. Rather, the *minkisi* inspired a new conjure-vessel form; one that may have been considered more powerful since it had kaolin integrated into the form.[105] Jason Young's work supports this notion. He states that, "black American conjure traditions were not mimetic; rather, they established a ritual space where Africa was both remembered and (re)imagined. In the end slaves elaborated on the Kongo minkisi and created new ritual objects, replete with their own construction and with their own manner of invocation."[106]

Although there is much evidence to support the argument that the early Edgefield face vessels are polyritualistic objects derived from a Kongo belief system brought over by the slaves on board the *Wanderer*, research on these iconic vessels continues. "Face Jugs: Art and Ritual in 19th-Century South Carolina" and "Unmasking the Mysteries of Face Jugs," recently hosted by the Columbia Museum of Art and the McKissick Museum of Art, are examples of ongoing interest in the face jug topic. New archaeological research focusing not only on face-jug sites but also on the *Wanderer* slaves is being conducted by a consortium from the University of Illinois at Urbana-Champaign.[107] Material analyses of the face vessels have been initiated by the Chipstone Foundation in order to better understand their chemical composition, and cultural research continues. One thing is already clear, though—the Edgefield face vessels have sparked an incredible response from a wide variety of researchers, collectors, and enthusiasts. While the utilitarian and ritualistic functions of these vessels might never be fully understood, there is no question that face jugs were important objects in the lives of the nineteenth-century Edgefield community of enslaved and freed African-Americans.

1. Sallie Layton Keenan, "Born in Slavery: Slave Narratives from the Federal Writers' Project, 1936–1938, South Carolina Narratives, Volume XIV, Part 3," p. 78, available online at http://memory.loc.gov/cgi-bin/ampage?collId=mesn&fileName=141/mesn141.db&recNum=0 (accessed December 3, 2013).

2. Harvey S. Teal, *Partners with the Sun: South Carolina Photographers, 1840–1940* (Columbia: University of South Carolina Press, 2001), p. 125. Palmer is remembered as an artist who took photographs throughout Florida, Georgia, and South Carolina, specializing "in scenes from the black community, including not only portraits, but also their cabins and 'at work' scenes in cotton fields and other locations."

3. A similar image in the series features an African-American woman.

4. At the time, this iconography was popularly associated with the Aesthetic Movement. Wilde was a well-known proponent of this controversial style, which was derided by many social and style leaders for its effeminate and immoral character.

5. The section on Edgefield face jugs as well as the conversation with Colonel Davies are not in the two previous editions of the book.

6. Edwin Atlee Barber, *The Pottery and Porcelain of the United States and Marks of American Potters*, rev. and enl. ed. (New York: Feingold & Lewis, 1909), p. 466. Some scholars have questioned this recollection on the basis of it being recorded forty years after the fact; it remains the only firsthand written account referencing Edgefield face jugs, however, and there is no evidence it should be distrusted. In *Early American Pottery and China* (New York: Century Company, 1929), author John Spargo referenced Barber's description but added that "we know practically nothing" about the "grotesque stoneware jugs" (p. 347).

7. Thompson suggested that "a connection between Congo-Angola arts and South Carolina pottery made by Afro-Americans is a distinct possibility and would seem likely in view of the particular slaving history of the area. This correspondence, if proved, would make definite the links between the Bakongo and Aiken County pottery." Thompson, *Flash of the Spirit: African*

and Afro-American Art and Philosophy (New York: Random House, 1983), p. 41. It is our contention that the correspondence can be proved. *Bakongo* is a term coined in the 1910s to describe the peoples who speak the Kikongo language. The inhabitants of the Kingdom of Kongo called themselves Essikongo.

8. Robert Farris Thompson and Joseph Cornet, *The Four Moments of the Sun: Kongo Art in Two Worlds*, exh. cat. (Washington, D.C.: National Gallery of Art, 1981), p. 158.

9. Thompson, *Flash of the Spirit*, p. 118. Thompson and Cornet, *Four Moments of the Sun*, pp. 157–62.

10. Monica Blackmun Visonà, Robin Poynor, and Herbert M. Cole, *A History of Art in Africa*, 2nd ed. (Upper Saddle River, N.J.: Pearson/Prentice Hall, 2008), pp. 359–60.

11. Thompson, *Flash of the Spirit*, p. 117.

12. The authority of a *nganga* was precarious and was constantly challenged.

13. Thompson, *Four Moments of the Sun*, pp. 157–62.

14. John Michael Vlach, *The Afro-American Tradition in Decorative Arts*, exh. cat. (1978; repr. with new preface, Athens: University of Georgia Press, 1990), p. 86.

15. Thompson, *Flash of the Spirit*, p. 24.

16. Regenia Alfreda Perry, *Spirits or Satire: African-American Face Vessels of the 19th Century*, exh. cat., Gibbes Art Gallery, Charleston (Charleston, S.C.: Carolina Art Association, 1985), pp. 7–8.

17. Ibid., p. 10.

18. Ibid.

19. An example is an Edgefield face vessel that was given as a gift about 1917 by the son of a former Southern slave to the manager of a "poor farm" in Hingham, Massachusetts. The son was destitute and the jug was his only possession; the informant stressed that the vessel held much importance for the son and that he held it in high regard: "The vessel was in perfect condition and obviously well cared for. This was never in a field or used for utilitarian use. I have worked in a field and this jug would not be enough to quench a worker's thirst nor could it have stayed so well preserved. There is not a scratch on it." Pat Van Der Lind, personal communication with April Hynes, April 10, 2012.

20. Cinda K. Baldwin, *Great and Noble Jar: Traditional Stoneware of South Carolina* (Columbia: McKissick Museum of the University of South Carolina; Athens: University of Georgia Press, 1993), p. 82.

21. Kaolin was part of the Edgefield District.

22. Baldwin acknowledged that the Bakongo Africans newly arrived on the *Wanderer* might have inspired the potters in Edgefield to create Toby-like vessels. Specifically, she references Thompson and his argument that Toby jugs were appropriated by chiefs in the Kongo to serve a magical ritualistic purpose. Baldwin, *Great and Noble Jar*, p. 81.

23. Jill Beute Koverman, *Making Faces: Southern Face Vessels from 1840–1990* (Columbia: McKissick Museum of the University of South Carolina, 2001), p. 7. Koverman cites a Romanesque nose found at the Phoenix Factory site by Stephen Ferrell as evidence that Chandler may have made other face vessels.

24. See, in this volume, Philip Wingard, "From Baltimore to Edgefield: The Influence of Thomas Chandler on 19th-Century Stoneware," pp. 38–76.

25. Mark M. Newell and Peter Lenzo, "Making Faces: Archeological Evidence of African-American Face Jug Production," in *Ceramics in America*, edited by Robert Hunter (Hanover, N.H.: University Press of New England, 2006), p. 126.

26. John A. Burrison, "Afro-American Folk Pottery in the South," in William R. Ferris, ed., *Afro-American Folk Art and Crafts* (Boston: G. K. Hall, 1983), p. 336.

27. Ibid..

28. There is evidence that slaves worked in the potteries before the Civil War as well. Dave Drake, for example, produced signed pots for the Lewis Miles pottery as early as 1840. Arthur F. Goldberg and James P. Witkowski, "Beneath His Magic Touch: The Dated Vessels of the African-American Slave Potter Dave," in *Ceramics in America*, edited by Robert Hunter (Hanover, N.H.: University Press of New England, 2006), p. 71.

29. Baldwin, *Great and Noble Jar*, pp. 71–72. Baldwin makes multiple references to slaves working in the Edgefield potteries.

30. Michael D. Hall also made the argument that the Edgefield face vessels were made by white potters. He goes a step further than Burrison, speculating that face vessels were neither ritualistic nor whimsies, but rather were linked to the nationwide Temperance Movement, a theory that lacks supporting evidence. Hall, "Brother's Keeper: Some Research on American

Face Vessels and Some Conjecture on the Cultural Witness of Folk Potters in the New World," in *Stereoscopic Perspective: Reflections on American Fine and Folk Art*, Contemporary American Art Critics 11 (Ann Arbor, Mich.: UMI Research Press, 1988), pp. 195–225. Burrison does not discount the possibility of the face jugs being inspired by an African tradition, although he suggests they also were inspired by European forms. "Germany and Africa, then, are two possible sources for American face jugs, with a third possibility that they arose independent of any Old World influence." John A. Burrison, "Fluid Vessel: Journey of the Jug," in *Ceramics in America*, edited by Robert Hunter (Hanover, N.H.: University Press for the Chipstone Foundation, 2006), p. 113. Following a discussion of a Nigerian portrait pot and the Thomas Chandler face jug, Burrison stated, "Did Chandler then introduce the concept to Edgefield slave potters, or were they working in a separate, perhaps African-based tradition? We now know that the face jugs of Cheever and Lanier Meaders were part of a continuous Anglo-Southern tradition, but whether ultimately inspired by slave-made examples, as Thompson suggests, might never be learned." Ibid., p. 115.

31. Daniel C. Littlefield, *Rice and Slaves: Ethnicity and the Slave Trade in Colonial South Carolina* (Urbana: University of Illinois Press, 1991), p. 115.

32. In 1808 the federal government made the importation of slaves directly from Africa illegal.

33. Charles J. Montgomery's article states that Tom Johnson came from the east coast of Africa "[w]here you can see the water just drippin' out o' the sun." Based on Professor John Thorntorn's unpublished research ("Enslavement of Wanderer," unpublished manuscript), though, it is more likely that he came from the Kingdom of Kongo. Uster Williams came from the "Bezy River" in Africa. He recalled his African customs in remarkable detail and believed a "Spell" was placed on him through "witchycraft." Ward Lee, also known as Cilucangy, came from the village of Cowany. Tucker Henderson was eighteen years old when he arrived from Africa. Tahro, also known as Romeo, was the oldest of the group and came from Kuluwaka Africa. Montgomery, "Survivors from the Cargo of the Negro Slave Yacht Wanderer," *American Anthropologist*, n.s., 10, no. 4 (1908): 613–14.

34. It is likely they were partisans of Pedro V, who lost a battle at São Salvador at about the same time the *Wanderer* arrived.

35. Luweka means "to arrive" in Kikongo, whereas *ku* is a locative particle. Kuluwaka may thus mean "the place of arrival." These groups, called *kanda* in Kikongo, were often connected by kinship or common purpose. John K. Thornton, "Elite Women in the Kingdom of Kongo: Historical Perspectives on Women's Political Power," *Journal of African History* 47, no. 3 (2006): 439.

36. There are several newspaper articles that describe the wonder and excitement produced by the *Wanderer* Africans. One of the sources is the November 21, 1859, issue of the *New York Herald*, which states: "The premium offered by Col. Hunt, for a specimen of native Africans to be exhibited at the State-Agricultural Fair, was taken yesterday by Dr. Bland of Edgefield who brought two on the grounds. Their arrival created quite a sensation with the large crowd assembled in the amphitheater. The premium was a beautiful silver goblet."

37. Joseph E. Holloway, ed., *Africanisms in American Culture* (Bloomington: Indiana University Press, 1990), p. 39.

38. Davies Personal Papers, South Carolina Department of Archives and History, Columbia.

39. Several descendants of the Ward Lee family still live on land purchased and farmed by Lee after the Civil War. His granddaughter relates that in addition to the face vessel now in the family, another vessel was placed at the door of her father's house.

40. University of Southwestern Louisiana, *Baking in the Sun: Visionary Images from the South*, exh. cat. (Lafayette: University Art Museum, 1987), pp. 159–61.

41. John B. Boles, *Black Southerners: 1619–1869* (Lexington: University Press of Kentucky, 1984), pp. 83–84.

42. Former slaves have admitted that they dared not utter their private thoughts to a white man, and considered deception to be one of the best weapons of self-defense available to them. Ronald Takaki, *A Different Mirror: A History of Multicultural America* (Boston: Back Bay Books/Little, Brown, 1993), pp. 114–15.

43. John Thornton, "The Development of the African Catholic Church in the Kingdom of Kongo, 1491–1750," *Journal of African History* 25, no. 2 (1984): 151.

44. These practices were more likely to be allowed if they were viewed as born of ignorance.

45. Thompson, "African Influence on the Art of the United States," p. 43. Although slaves were prohibited from speaking African languages, it is known that they continued to do so, as referenced above. The quote within Thompson's quote comes from the Georgia Writers'

Project [of the Work Projects Administration], *Drums and Shadows: Survival Studies among the Georgia Coastal Negroes* (Athens: University of Georgia Press, 1940), p. 179.

46. Although many slaves aboard the *Wanderer* were from the Kingdom of Kongo, some, like Uwalla, were from West Africa.

47. *Augusta Chronicle*, October 1, 1888.

48. Georgia Scott, personal communication with the authors, October 2012.

49. Georgia Writers' Project, *Drums and Shadows*, p. 60.

50. Frederick Douglass, *Narrative of the Life of Frederick Douglass, an American Slave* (New York: Modern Library, 2000), p. 27.

51. Albert J. Raboteau, *Slave Religion: The "Invisible Institution" in the Antebellum South* (1978; updated ed., Oxford: Oxford University Press, 2004), p. 213.

52. Ibid., p. 249.

53. Dorothy Scarborough, *On the Trail of Negro Folk-Songs* (Cambridge, Mass.: Harvard University Press, 1925), p. 70.

54. The notion of being "ridden like a horse" is also found throughout slave narratives and writings on slave beliefs. It was common thinking that witches would ride their victims during the night. If a slave woke up with scratch marks (supposedly from the saddle) or felt very tired in the morning, it was believed that s/he had been visited by a witch the night before. Newbell Niles Puckett, *Folk Beliefs of the Southern Negro* (1969; repr., facsimile ed.; Whitefish, Mont.: Kessinger Publishing, 2003), pp. 151–53.

55. Lawrence W. Levine, *Black Culture and Black Consciousness: Afro-American Folk Thought from Slavery to Freedom* (New York: Oxford University Press, 1978), pp. 113–14.

56. John Michael Vlach, "International Encounters at the Crossroads of Clay: European, Asian, and African Influences on Edgefield Pottery," in Catherine Wilson Horne, ed., *Crossroads of Clay*, exh. cat. (Columbia: McKissick Museum, University of South Carolina, 1990), p. 29.

57. The Bakongo had experienced the teachings of Catholic missionaries beginning in the sixteenth century. According to Professor John Thornton, these beliefs were propagated and shaped by the Kongo elite as well as educated Kongo laypeople. Catholicism was adapted and combined with traditional beliefs, resulting in a complex syncretic religious system. Personal communication with the authors.

58. Jason R. Young, *Rituals of Resistance: African Atlantic Religion in Kongo and the Lowcountry South in the Era of Slavery* (Baton Rouge: Louisiana State University Press, 2007), p. 85.

59. Vlach, *Afro-American Tradition in Decorative Arts*, p. 143.

60. During an interview with April Hynes at the Bettis Academy in Edgefield on August 20, 2012, the Reverend George Brightharp recalled that many small face jugs were exhumed when an African-American graveyard in Shultz Hill, South Carolina, was relocated.

61. Barber, *Pottery and Porcelain of the United States*, p. 466.

62. Robert Farris Thompson, "African Influence on the Art of the United States," in William R. Ferris, ed., *Afro-American Folk Art and Crafts* (Boston: G. K. Hall, 1983), p. 41.

63. Sylvia R. Frey and Betty Wood, *Come Shouting to Zion: African American Protestantism in the American South and British Caribbean to 1830* (Chapel Hill: University of North Carolina Press, 1998), p. 1.

64. Rabouteau, *Slave Religion*, pp. 4–5.

65. Alisa LaGamma, *Art and Oracle: African Art and Rituals of Divination*, exh. cat. (New York: Metropolitan Museum of Art, 2000), p. 23.

66. Carol Beckwith and Angela Fisher, *The Painted Bodies of Africa: African Body Painting, Tattoos, and Scarification* (New York: Rizzoli International Publications, 2012), p. 1.

67. Gary Zabel, "Kongo and the Iconoclastic Controversy," *Phantasmagoria, Part One: Primitive Accumulation*, chap. 2 (available online at www.faculty.umb.edu/gary_zabel/Unpublished%20Work/Phantasmagoria/Part%201,%20ch.%202.htm (accessed July 15, 2013). It should be noted, however, that not all Bakongo share this belief.

68. This detail is absent in many of the Anglo imitators' works.

69. LaGamma, *Art and Oracle*, p. 33.

70. Ibid., p. 34.

71. A similar practice is found in the Igbo culture, where wood sculptures of gods and goddesses also are distinguished by their enlarged heads that symbolically reflect the god's intelligence as well as its spirit. On the other hand, the figures might have short legs, or these might be completely neglected, as they are unimportant to the sculpture's ideological impact.

72. As explained by Yvonne P. Chireau, "[b]y the nineteenth century the terms for African American Conjurers would change. 'Hoodoos' and 'root workers' were the most common

vernacular expressions that depicted persons who were believed to be able to manipulate unseen forces or 'work the spirits.'" Chireau, *Black Magic: Religion and the African American Conjuring Tradition* (Berkeley: University of California Press, 2003): 21.

73. 1900 U.S. Census, Aiken, Shultz Township, Aiken County Courthouse, Aiken, S.C.

74. Leonora Herron and Alice M. Bacon, "Conjure & Conjure Doctors," in Alan Dundes, ed., *Mother Wit from the Laughing Barrel: Readings in the Interpretation of Afro-American Folklore*, Critical Studies on Black Life and Culture (1973; repr., Jackson: University Press of Mississippi, 1990), pp. 359–68.

75. Raboteau, *Slave Religion*, p. 286.

76. "Born in Slavery: Slave Narratives from the Federal Writers' Project, 1936–1938, South Carolina Narratives, Volume XIV, Part 1," p. 2, available online at http://memory.loc.gov/cgi-bin/ampage?collId=mesn&fileName=141/mesn141.db&recNum=0 (accessed May 25, 2013).

77. Romeo/Tahro would have fit this description when he proclaimed he was a king.

78. Cudjo, a *Wanderer* African enslaved in Augusta, Georgia, stated that he was the son of a Congo king. *Augusta Chronicle*, November 26, 1899.

79. Bettis C. Rainsford Sr., "The Story of the Pine House," May 13, 2013 (from a wedding announcement published by the *Edgefield Historian*). This story was told to Bettis Rainsford ca. 1970 by Mrs. W. L. (Emily Strother) Dunovant Jr. (1892–1972) of Edgefield, granddaughter of Rev. and Mrs. Walker.

80. Georgia Writers' Project, *Drums and Shadows*, pp. 47, 73, and 93.

81. Ibid., p. 93.

82. Another way of thinking about the "flash of the spirit" is in terms of life force. Richard J. Gehman explains: "All good or ill experienced by men and woman is due to such influences. It is the explanation for success in battle or a harvest failure of the same. Life force is vital for heath, wealth, world power and success. In face, everything that makes life at all worth living. To have it is to be healthy and strong. To lose it is to grow weak and die. It exists in greater concentrations in famous men, strong charms, revered fetishes and powerful gods. Gods, humans (both living and departed), animals and plants have life only as they have this vital force. Families, lineages and tribes have corporate vital force and their happiness and prosperity depend on conserving and strengthening it." Gehman, *African Traditional Religion in Biblical Perspective* (1989; rev. ed., Nairobi: East African Educational Publishers, 2005), p. 86.

83. Congo Ben was a *Wanderer* African living on a plantation near the Bayou Teche, which was near New Orleans. He was famed for his skill as a root worker and snake charmer. A newspaper account relates how he cured a snakebite with a special poultice made of roots, "picked at just the right phase of the moon." When a local physician examined Congo Ben's work, he said "I would offer him $50 if he would tell me just what he used. These people have secrets that would be priceless to the pharmacopeia, if only they would reveal them." *Grand Fork Herald*, October 7, 1909. Similar reports show that a number of *Wanderer* Africans were skilled in the use of root medicine, such as Old Bulgee and Abdoula, both from Montgomery, Alabama.

84. Herron and Bacon, "Conjure & Conjure Doctors," pp. 359–68.

85. *The Herald*, October 18, 1868.

86. Ibid.

87. *Augusta Chronicle*, May 12, 1888.

88. In 1905 it was reported that these beliefs were still so widespread within the African-American community that Dr. Silas Floyd, a well-known Augusta minister at Ebenezer Baptist Church, told his congregation to reject these beliefs. He admonished those who sought conjurers as a way to find a thief or become lucky in love, saying: "I appeal to you today not to pay out your hard-earned money to these frauds, fortune tellers, conjurers, root-workers and the like." *Augusta Chronicle*, August 28, 1905.

Georgia Scott shared with the authors numerous conjuring activities she had observed in the area during her life. Other conjuring activity was reported to the authors by the Reverend George Brightharp, who experienced it within his own church sanctuary and in the community. Video interview with April Hynes and Mark Newell, August 20, 2012.

89. Jeffrey E. Anderson, *Conjure in African American Society* (2005; paperback ed., Baton Rouge: Louisiana State University Press, 2007), p. 53.

90. Ibid.

91. Puckett, *Folk Beliefs of the Southern Negro*, p. 148.

92. Mrs. Pat Bishop, personal communication with Mark Newell and April Hynes, August 21, 2012.

93. Mrs. Frances Harris, personal communication with April Hynes, August 20, 2012. It is interesting to note that Mrs. Harris's mother-in-law told her that the face jug had been given to her by her grandmother Moriah Nobles, whose sister-in-law, Katie Noble, was one of the *Wanderer* slaves. 1870 U.S. Census, Butler Township, Edgefield County.

94. Rev. Alexander Pope, telephone conversation with April Hynes, March 21, 2013.

95. The practice of placing face jugs by the door was not limited to Edgefield. In an interview with Mark Newell in 2012, Lynnae Thandiwe revealed that a face jug in her family's possession since the 1930s was kept at the doorway of her Arkansas family home.

96. Wayne O'Bryant, interview with Mark Newell, Bettis Academy, August 20, 2012.

97. 1880 U.S. Census, Shaws Township, Edgefield, South Carolina.

98. The Freedman's Saving and Trust Company, also called the Freedman's Savings Bank, was a banking system set up by the United States government to foster the economic development of the African-American community. It functioned between 1865 and 1874.

99. 1870 Census, Edgefield County.

100. Rev. W[illiam] Holman Bentley, *Dictionary and Grammar of the Kongo Language, as Spoken at San Salvador, the Ancient Capital of the Old Kongo Empire, West Africa*, bk. 7, s. 721–1052 (London: Baptist Missionary Society, 1895), pp. 815, 932, 1015.

101. Vlach, *Afro-American Tradition in Decorative Arts*, p. 86.

102. In a personal communication with April Hynes on August 30, 2012, Professor John Thornton recognized the language samples in the Charles Montgomery article as Kikongo.

103. Baldwin, *Great and Noble Jar*, pp. 82–83.

104. Puckett, *Folk Beliefs of the Southern Negro*, p. 252.

105. The face cups and pitchers could have been associated with conjure, albeit in a different form. A common affliction attributed to conjure was a sensation that spiders or other small animals were crawling under one's skin. The afflicted would often seek out a conjure doctor who would cure him/her by cutting their skin, pressing a cup to the spot, and proceed to bleed them until the animal fell into the cup, or by having the affected party strain as if throwing up into a tub. Georgia Scott recalls an event in which she witnessed a conjurer curing this affliction using a tub to catch the offending animals: "Many years ago, when my husband was alive, he was a Reverend and had been called to the home of a woman who was convinced she had been conjured. I drove him there so I was present. The conjurer came to her house and told her she had a snake inside her. He had her sit in front of a tub filled with water. He asked her to lean over it, close her eyes and strain as if about to throw up. Sure enough, there in the tub when she finished was a long snake. The conjuror said he would have to come back several times and repeat the process in case any small creatures were left inside her. This he did, finding a frog in her the next time. He charged her money each time. For this reason she was unable to pay her rent and she had called my husband in the hopes that the church would help her out." Georgia Scott, personal communication with Mark Newell and April Hynes, Bettis Academy, August 20, 2012.

106. Young, *Rituals of Resistance*, p. 125.

107. The group has made available images of the digs at www.facebook.com/groups/168330813202140/.

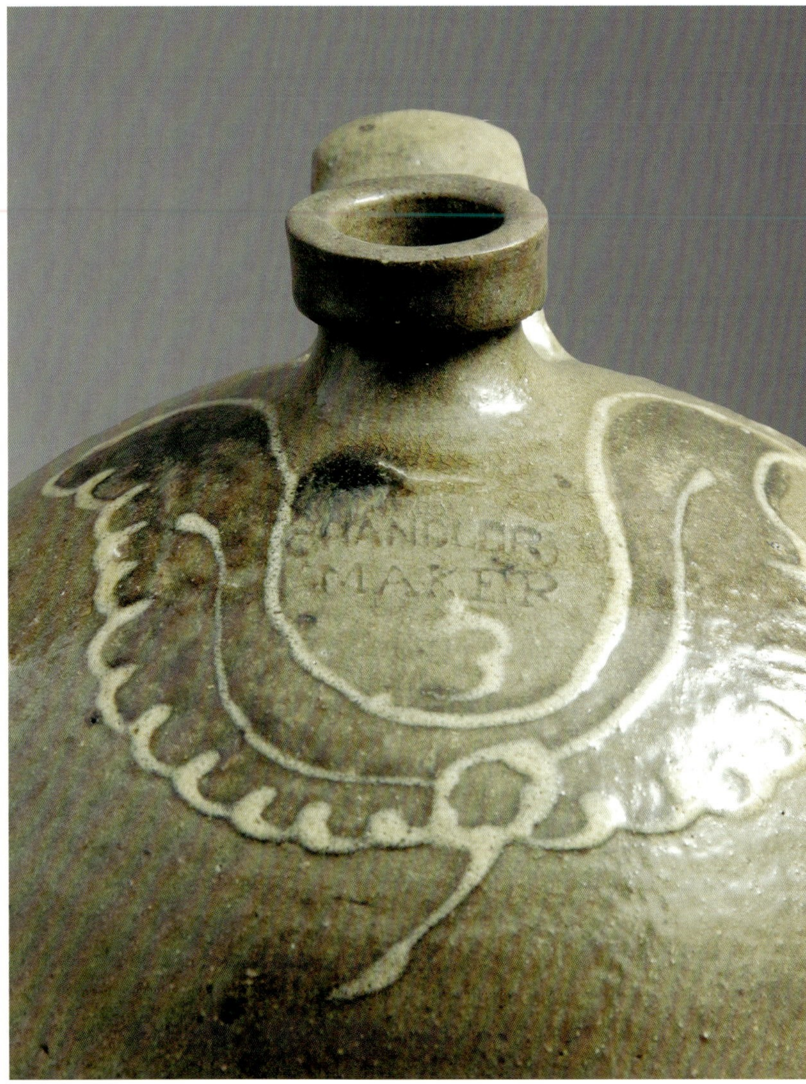

Figure 1 Detail of jug, Thomas Chandler, Edgefield District, South Carolina, 1847–1852. Alkaline-glazed stoneware. Capacity mark: 3. Stamped mark: "CHANDLER MAKER" (Author's collection; photo, Jan Todd.)

Philip Wingard

From Baltimore to the South Carolina Backcountry: Thomas Chandler's Influence on 19th-Century Stoneware

▼ IN THE WINTER of 1980, during a tour of homes in my town, I saw for the first time an alkaline-glazed jug. It was dark olive green with rivulets of dripped glaze running down on all sides. The glaze reminded me of the candles my mother loved that would melt down to nothing during supper and, in the process, cover the old wine bottle we used for a holder with layers of different-colored dripped wax. The alkaline-glazed jug was from a region in North Carolina known as the Catawba Valley, and its elegance struck me to the core. "It is unfortunate," I said to my host, "that such a wonderful example of nineteenth-century stoneware could not have been made in South Carolina." "You're so wrong," my friend informed me. "Extraordinary stoneware *was* made in South Carolina, at a place called Edgefield" (fig. 1).

A few years later, I found myself face-to-face with an example of Edgefield stoneware. It was a honey-colored, two-gallon storage jar decorated with a large number 2 inside a scalloped cartouche, all in a dark slip under a glossy glaze. The decoration appeared to have been applied with a brush. The simple design had a mystery to it that impelled me to look for more examples made by the same hand, that of potter Thomas Chandler. Being a South Carolina native, I had studied my state's history, but had never been taught about South Carolina–made stoneware fashioned during a time when farms and families depended on it, or about its significance to commerce and industry or craftsmanship. How could historians have ignored such important utilitarian artifacts so essential to our ancestors' lives? More important, to me, who was Thomas Chandler, and where did he learn to pot?

In this article, I explore Thomas Chandler's background and the forces that led him from his early beginnings as a potter in Baltimore, Maryland, to become one of the premier potters in Edgefield, South Carolina. I will show that many of the decorative elements in the Mid-Atlantic salt-glazed traditions Chandler was exposed to during his training in Baltimore are later manifested in his considerable body of alkaline-glazed stoneware. It is my contention that Thomas Chandler introduced the face vessel to Edgefield—a form adopted and adapted by many of the African-American potters working in the district during the mid-nineteenth century.

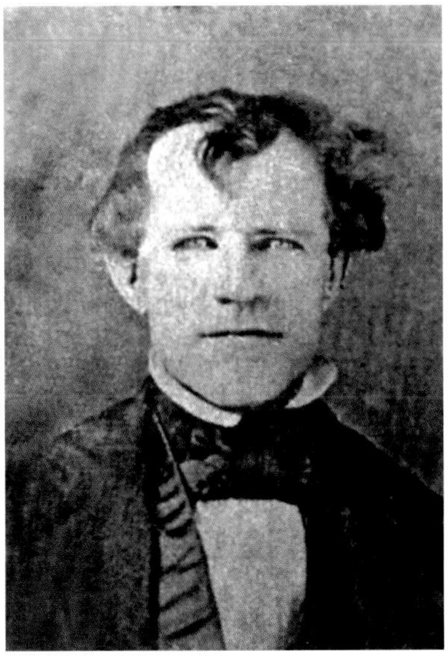

Figure 2 Daguerreotype of Thomas Chandler Jr., Edgefield District, South Carolina, ca. 1850. (Courtesy, William Robert Mathis, a direct descendant.)

Thomas Mitchel Chandler Jr.'s Early History and Training

Thomas Mitchel Chandler Jr. (fig. 2) was born in 1810 to a family of tradesmen and farmers on the Eastern Shore of Virginia in Drummondtown

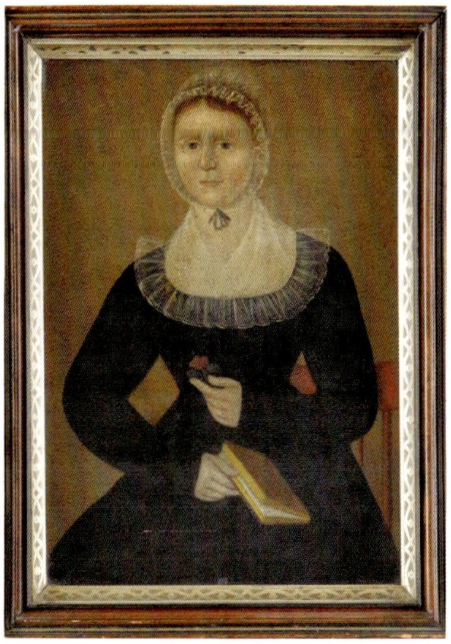

Figure 3 Portrait of Elizabeth Wise Chandler, attributed to Joshua Johnson, ca. 1815, Accomack County, Virginia. Oil on artist board. 19½ x 13½". (Courtesy, Freeman's Auctioneers and Appraisers, Philadelphia; photo, Elizabeth Field.) Elizabeth Wise Chandler was Thomas Chandler Jr.'s mother.

(now Accomack), Virginia.[1] The Chandlers were industrious in Drummondtown in the eighteenth and nineteenth centuries. Thomas Chandler Sr. was born there about 1780 and, when he was orphaned in 1796, an uncle, Littleton Chandler, became his guardian. The following spring, at age seventeen, Thomas Sr. was sent to Baltimore, where on May 18, 1797, he was bound out to Caleb Hannah, an established Windsor chair maker.[2] After completing his apprenticeship at the age of twenty-one, he married Elizabeth Wise in Baltimore on July 16, 1802 (fig. 3), and later that year returned with her to Drummondtown to ply his trade as a chair maker.[3] On August 30, 1803, he took on orphans Jacob Phillips and John Drummond as apprentices.[4]

Thomas Sr. and his brother Elisha (who had also apprenticed as a Windsor chairmaker in Baltimore) may well have taught their older brother Mitchell the trade, because Mitchell Chandler is listed in the 1850 Accomack census as a chairmaker, aged seventy, living with his older sister Rosey Chandler Fitzgerald, a widow.[5] By 1817 Thomas Sr. and his family were back in Baltimore, perhaps influenced by the Tariff Act of 1816 and its positive economic effect on local manufacturing.[6]

Thomas Sr. prospered in Baltimore during this period. He had a portrait painted of his wife (and, perhaps, one of himself) by the freed black artist Joshua Johnson.[7] Thomas Sr.'s Windsor chairmaking shop was on the north side of Pitt Street, near Eden Street. In *C. Keenan's Baltimore Directory for 1822 & 1823*, there are four important pottery manufacturing concerns listed very near this location: master potter Henry Remmey had a stoneware shop on the northwest corner of Bond and Pitt streets; potters Thomas Morgan and William Amoss were on the corner of Pitt and Green streets; potter David Parr was on Eden Street, and his brother Elisha Parr was on Pitt Street; and the important Baltimore Stoneware Manufactory (same shop as the Remmey shop) was on Pitt and Bond. A map of Baltimore by Fielding Lucas Jr. (fig. 4) places Thomas Sr.'s Windsor shop within a block or two of these four potteries, in the midst of one of the most dynamic stoneware-making environments in the country. In addition to stoneware, Morgan and Amoss, the Parr brothers, and Henry Myers were also manufacturing earthenware in 1822.[8]

The Baltimore Stoneware Manufactory on Pitt and Bond streets had been established circa 1812 by William Myers, a china merchant who hired master potter Henry Remmey Sr. from New York to run the operation (which Remmey did until about 1829).[9] By 1822 Henry Myers, nephew to William, had a china retail outlet on Baltimore Avenue, where the citizens of Baltimore, as well as young Thomas Chandler Jr., could see the finished products of all the major potters working in the area.[10] Although too young to become an apprentice when he arrived with his family in Baltimore in 1817, he was free to glean knowledge from any of the stoneware makers, perhaps doing odd jobs and chores for them in exchange for the opportunity to learn the craft.

The recent discovery of a salt-glazed churn signed and dated by Thomas Chandler confirms that he did indeed make stoneware in Baltimore, Mary-

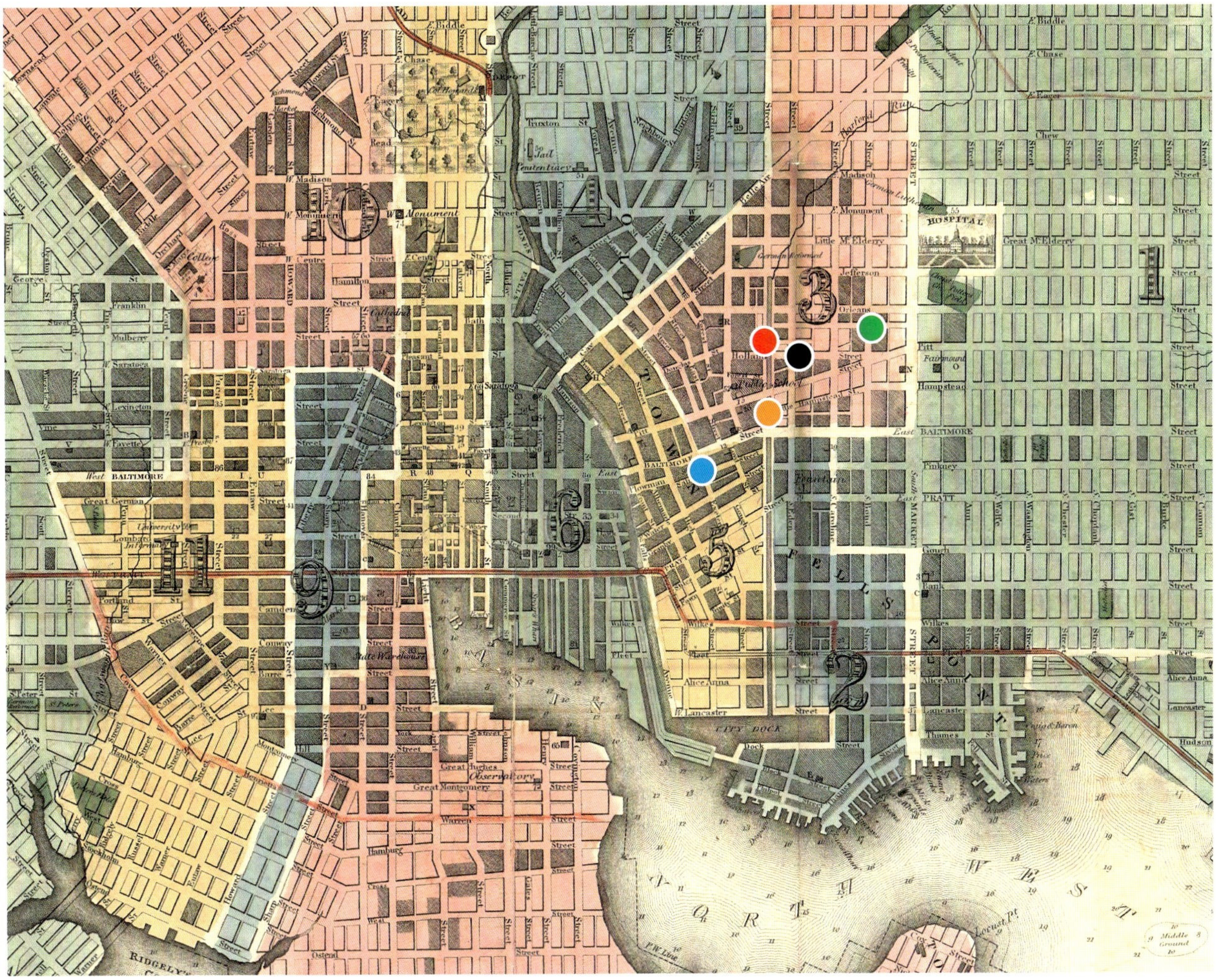

Figure 4 *Plan of the City of Baltimore* compiled from actual survey by Fielding Lucas Jr., 1822, improved 1836. (Courtesy, Library of Congress, Geography and Map Division.)

- *Black* Thomas Chandler's Windsor chair shop, north side of Pitt Street, east of Eden.
- *Green* Henry Remmey stoneware factory, northwest corner of Bond and Pitt streets.
- *Orange* David Parr and James Burland stone and earthenware pottery, west side of Eden Street, north of Dulaney Street.
- *Red* Elisha Parr stone and earthenware pottery, north side of Pitt Street, east of Hartford Run.
- *Blue* William Morgan stoneware pottery, northeast corner of Pitt and Green streets. Henry Myers, China Merchant, on Dulaney Street, one block south of Pitt.

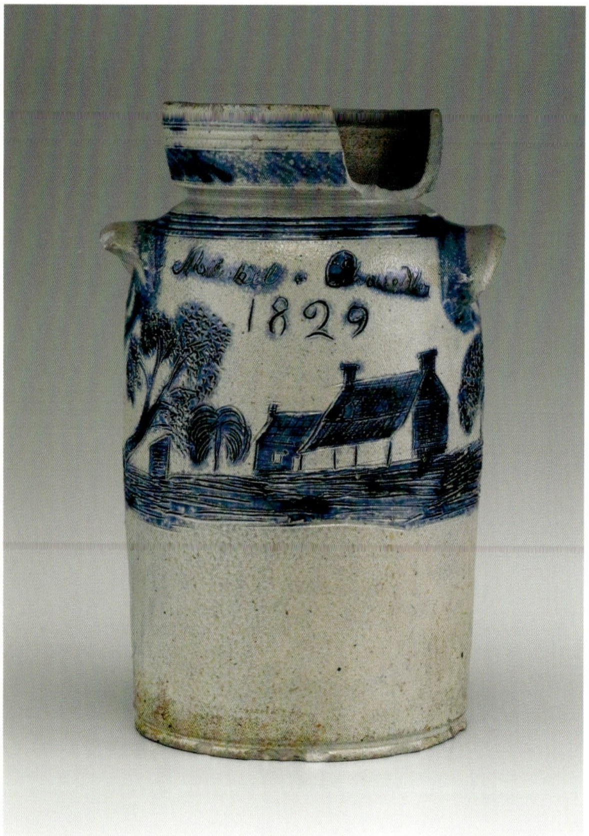

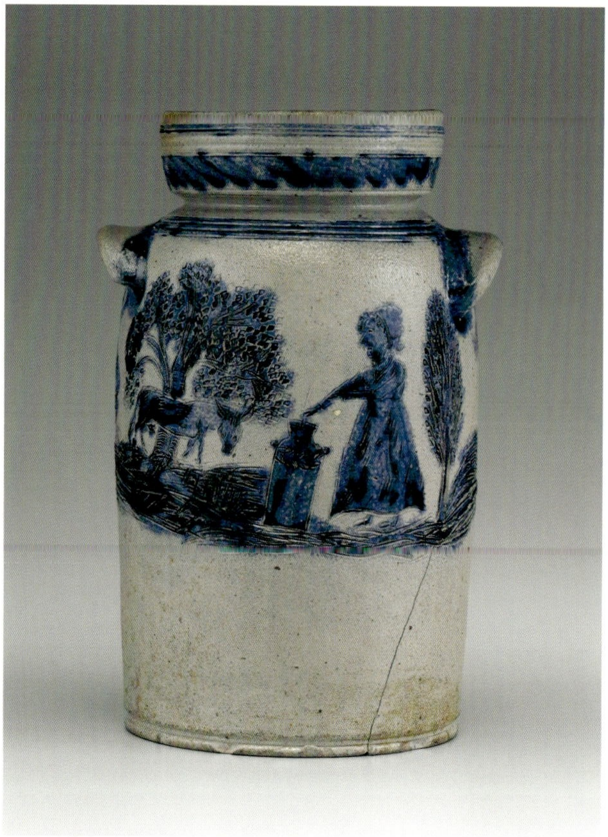

Figure 5 Churn, Thomas Chandler Jr., likely Baltimore Stoneware Manufactory, Baltimore, Maryland, 1829. Salt-glazed stoneware. H. 14½". Incised: "Mitchel · Chandler / 1829" (Private collection; photo, Gavin Ashworth.) The scene is believed to be the Mitchell Chandler homestead in Accomack. The handles, rim, and overall form are much like churns made by Henry Remmey while he was working for Henry Myers in Baltimore ca. 1825–1829.

land. The churn is incised with the scene of a farmhouse and a woman churning butter among a highly detailed landscape of trees, bushes, and outbuildings, all accented with brushed cobalt blue (figs. 5, 6): "Mitchel Chandler" and "1829" are incised in script on the upper body; on the bottom is the signature "Thomas M. Chandler / Maker / Baltimore / August / 12 / 1829" (fig. 7). Thomas Chandler would have been eighteen or nineteen years old when he made this churn, the age at which most stoneware apprenticeships began. Much as a hand-stitched sampler reflects the skills attained by a young girl in finishing school, the churn represents the skills Chandler had attained in his training to become a potter: fine turning, glazing, firing, decoration, form, and function. The churn appears to have been a presentation piece to Chandler's uncle and his sister, Mitchell Chandler and Rosey Chandler Fitzgerald; it descended in the family and had been in Accomack for many years.[11]

It is clear that from about 1817, when he arrived in Baltimore, until about 1829, when he left, Thomas Chandler Jr. learned all the details necessary to become a master potter. It is likely that he made and/or decorated additional stoneware produced in Baltimore from 1825 until 1829, but examples have not been firmly identified. Yet in 1829, perhaps because his father was no longer supporting the family, Thomas Jr. applied for a grocer's license in Baltimore.[12] According to Matchett's *Baltimore City Directory* for 1831, Chandler's mother was operating a grocery store on "Monroe Street near Spring" to support the family.

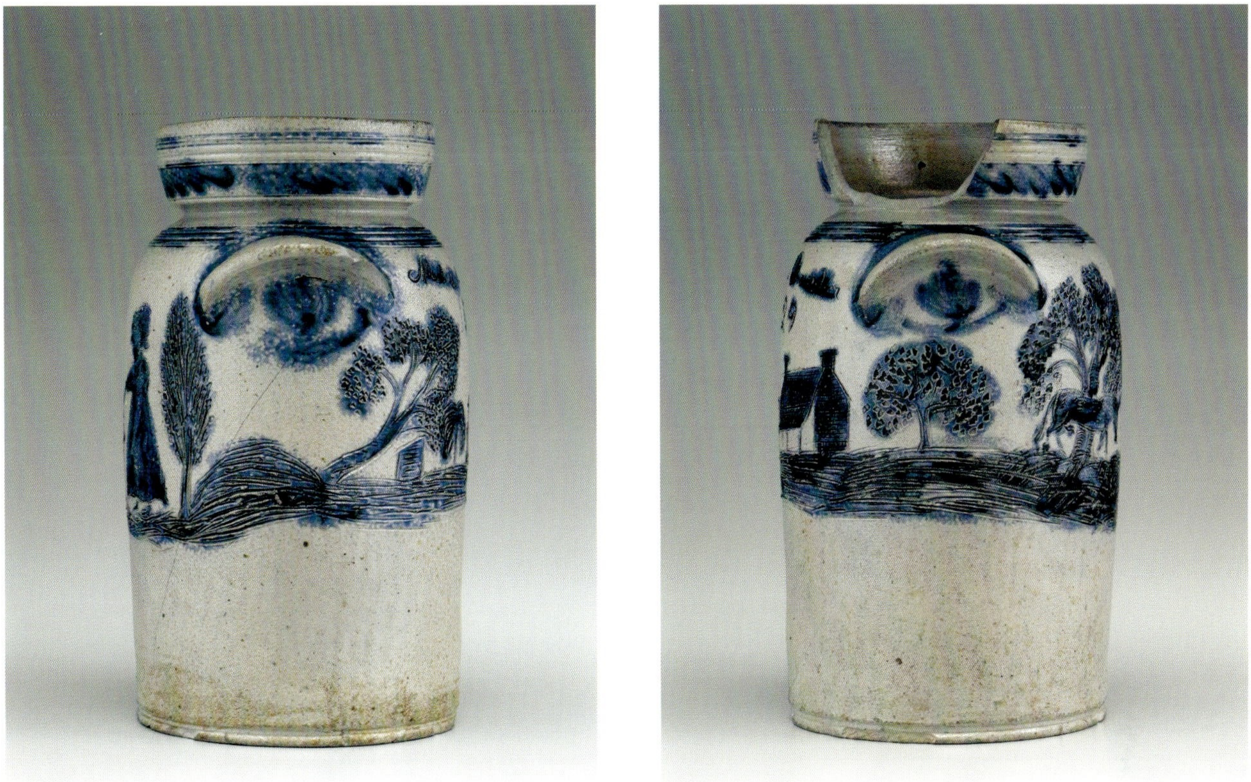

Figure 6 Side views of the churn illustrated in fig. 5.

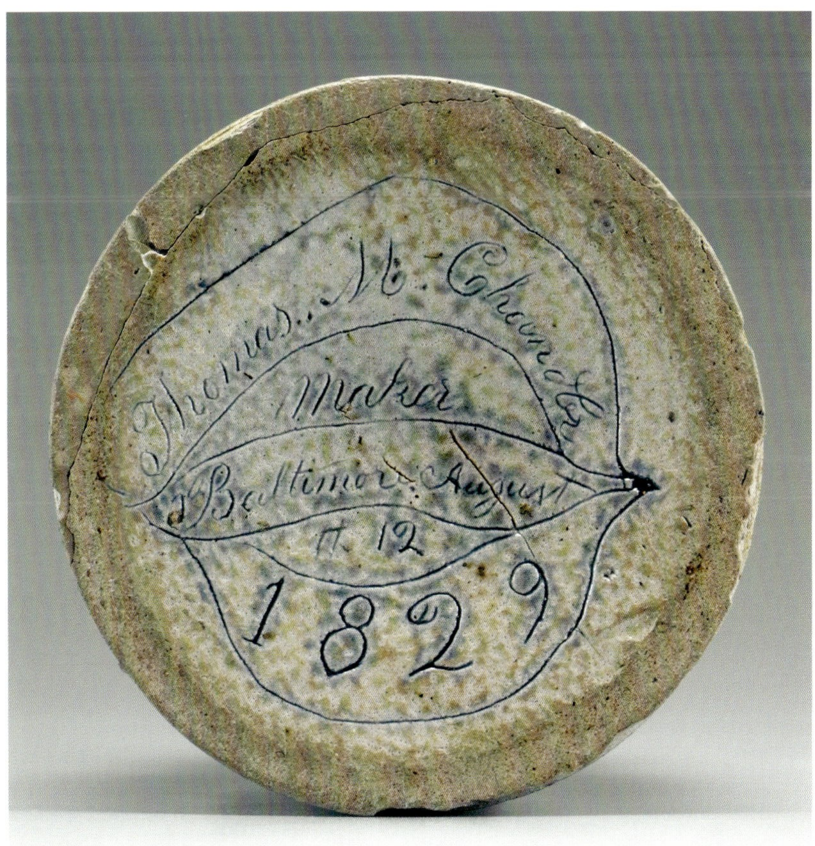

Figure 7 Incised inscription on the bottom of the churn illustrated in fig. 5: "Thomas M. Chandler / Maker / Baltimore August / 12 / 1829"

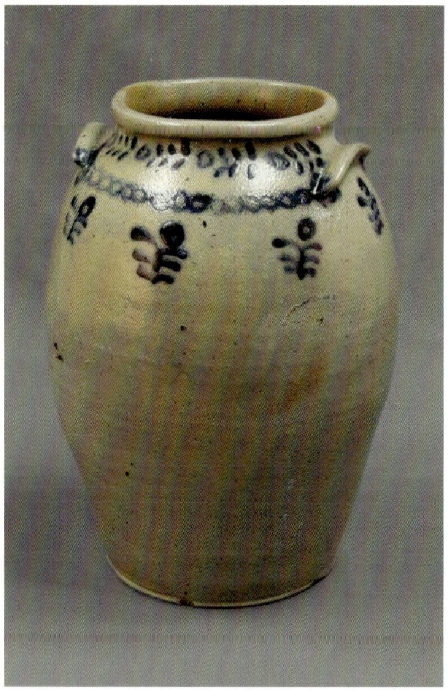

Figure 8 Storage jar, attributed to Thomas Chandler, Baltimore, Maryland, or possibly Richmond, Virginia, 1820s. Salt-glazed stoneware. H. 14". (Author's collection; photo, Jan Todd.)

On October 22, 1831, at age seventeen, Thomas Jr.'s younger brother George was bound out and apprenticed to Henry Myers to become a stoneware potter.[13] George's apprenticeship verifies a strong bond between the Chandler family and Henry Myers, one of the most important figures in Baltimore stoneware at that time and situated near the Chandler household.

The great potter Henry Remmey had left Baltimore about 1829, and potter David Parr Sr. died in 1832 from cholera.[14] Between late 1829 and 1832 it is possible that Thomas Chandler was making stoneware with John P. Schermerhorn and/or other potters in Richmond, Virginia (fig. 8). During this period, several Baltimore potters had matriculated through the Richmond and Henrico potteries, among them Thomas Amoss (see, in this volume, Kurt C. Russ et al., "The Remarkable 19th-Century Stoneware of Virginia's Lower James River Valley," pp. 200–258). While Chandler's name has yet to be firmly associated with the Richmond and Henrico potteries, the 1830 census for Henrico County, Virginia, records a twenty- to thirty-year-old man living with Schermerhorn. This might refer to Thomas Chandler, who would have been approximately twenty-one years old. An intriguing clue to this possibility is the presence of a "Chandler Road" in the vicinity of the Henrico County factory complex where other Baltimore potters are known to have worked.

By 1830 the advantage American manufacturers maintained against English imports ended with the repeal of some of the stricter tariffs on English goods, whereupon England renewed sending cheaper finished goods to America. Port manufacturing centers such as Baltimore and Richmond were affected immediately by the sudden increase in lower-priced imported stoneware. Some Baltimore stoneware makers began to close their shops.[15] Perhaps due to economic conditions, by late 1832 Thomas Chandler had moved to Albany, New York, an area that had close connections to both Richmond and Baltimore.[16]

The frontier was one market not affected by the economic chaos in the east. The expansion of the backcountry into Alabama, Florida, and Georgia increased demand for stoneware and related products, and it is possible that finding a place to produce stoneware near the edge of the frontier was on Chandler's mind. After completion of the Erie Canal in 1825 created a continuous waterway from New York City to the Great Lakes, Albany was situated as a place from which to head west, and perhaps Chandler moved there in 1832 for this reason. It is possible he worked for a short time for Paul Cushman, who was the owner of several major pottery factories in Albany in the 1820s and early 1830s. However, perhaps because he had no other prospect for employment, on December 14, 1832, Chandler enlisted in the U.S. Army at Albany, joining the 4th Infantry Company A under Major Dade for a term of five years. His enlistment records describe him as 5' 5" tall, with blue eyes, light hair, and a fair complexion. His profession is listed as musician, which suggests he had abandoned, at least temporarily, the avocation of stoneware potter, or perhaps being a musician paid a higher enlistment bounty; regardless, joining the army brought Chandler south.

Thomas Mitchel Chandler Goes to Edgefield

In November 1832 a state convention in South Carolina passed the Ordinance of Nullification, which declared that the federal tariff acts of 1828 and 1832 were unconstitutional and unenforceable in South Carolina and would not be obeyed by the state after February 1, 1833.[17] The tariffs were designed to improve manufacturing in America but hurt the centuries-old exchange between England and South Carolina of raw material for finished goods. By February 1833 President Andrew Jackson, a native of South Carolina but a staunch Federalist, had pushed a bill through Congress authorizing the use of force against South Carolina if the state refused to comply with federal law. Concurrently, led by Henry Clay, Congress passed significant reductions in the tariffs, which Jackson also signed into law, thereby essentially ending the conflict.[18]

With this crisis over, Jackson turned his attention to the growing Indian rebellion in Alabama, Florida, and Georgia. In 1832 Augusta, Georgia, was on the edge of the southern frontier, at the head of the Meriwether Trail and the Indian trade routes leading southwest into southern Alabama and Seminole and Creek Indian lands. Several Indian nations in the southeast were rising up against the United States in a dispute over land that the federal government considered U.S territory, ceded by Spain after the Louisiana Purchase. In late December 1832 Chandler's infantry company was deployed to Augusta and arrived by January 20, 1833.[19] However, the resolution of the nullification issue in February changed the status of the 4th Infantry and, by early March, it was ordered to Fort Mitchell, Alabama, in support of a U.S. response to the Indian conflict.

On March 30, 1833, 4th Infantry Company A left the Augusta Arsenal headed for Fort Mitchell, Alabama, and five days later, on April 5, Chandler deserted and was AWOL for six months, until October 5.[20] A pottery established in 1818 by Cyrus Cogburn in southeast-central Washington

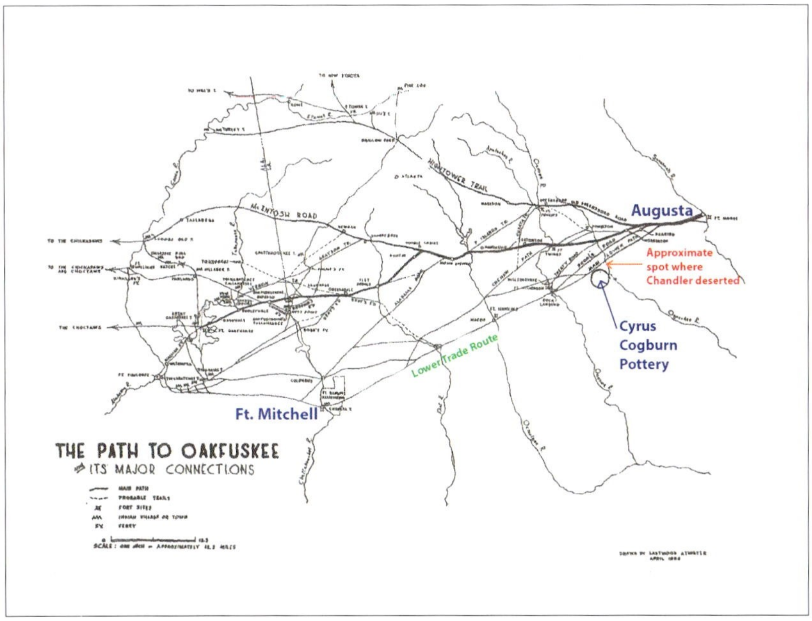

Figure 9 "The Path to Oakfuskee," drawn by John H. Goff, reproduced in *Georgia Historical Quarterly* (March 1955): inserted between pages 4 and 5. The lower trade route between Augusta, Georgia, and Fort Mitchell, Alabama, is shown, with the parts pertinent to Thomas Chandler highlighted. The circle notates the proximity of the Cyrus Cogburn pottery to the lower trade path, which was used by the U.S. Infantry in March–April 1833. The arrow shows the approximate place of Chandler's desertion.

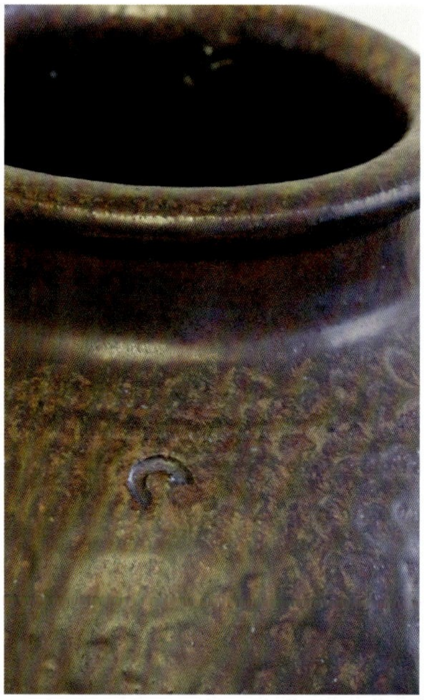

Figure 10 Detail showing production mark "C" on iron-slip alkaline-glazed stoneware jar from Edgefield District, South Carolina, attributed to Thomas Chandler, 1836–1840. (Author's collection; photo, Jan Todd.) This mark is associated with Thomas Chandler at various production sites, but it has not been found in archaeology.

County, Georgia, was within a mile of the Indian trade route the infantry used to march to Fort Mitchell during the first quarter of the nineteenth century.[21] The route, also known as the Meriwether Trail, was the most direct one from Augusta to Fort Mitchell.[22] Based on a study of the number of days it typically took an army regiment to march this distance on the trail, I have concluded that the area near Cogburn's pottery is approximately where Chandler deserted (fig. 9).[23]

Cyrus Cogburn was in Edgefield before 1818, working with Abner Landrum at Pottersville and possibly with John Landrum at Horse Creek. By 1827 he had moved to Upson County, Georgia, but the Washington County pottery he established in 1818 was still in operation in 1833. If Chandler did indeed work there during his desertion, learning about the alkaline glaze may have earned him acceptance as a potter in Edgefield a few years later. Fragments marked with a small incised "C" have been found at the site and might be associated with Chandler's work there.[24] There is a strong indication that Chandler used the production mark "C" on his ware at Pottersville, Trapp and Chandler, and Shaw's Creek in Edgefield District (fig. 10).

After Chandler rejoined his company in October 1833, he was held in confinement for a full year at Fort Oglethorpe, Georgia, three miles south of Savannah. He was tried in a court-martial proceeding and was acquitted, but remained in confinement. A month later Chandler was granted a transfer to the 2nd Artillery Regiment, Company E, stationed at the Augusta Arsenal. This transfer, which took effect upon Chandler's release from confinement in November 1834, placed him back at the Augusta Arsenal.[25]

From November 1834 until the early spring of 1836, Company E was held in reserve at the Augusta Arsenal, giving Chandler his second opportunity to work with the potters of the region, especially in the Edgefield District. In the spring of 1836, however, several 2nd Artillery companies, including Company E, were ordered south to protect the flank of the forts and posts on Florida's west coast. Company E saw action on June 9 at Micanopy, Florida, when Seminole Indians led by Chief Osceola attacked the post. It was likely in this battle that Chandler was wounded in his thigh, as his widow stated years later in her claim for pensioner's status in 1857.[26] In the two years between January 1835 and February 1837, nine officers and 103 men of the 2nd Artillery were killed in action, or died of wounds received or disease contracted in Florida.[27]

Chandler's second and third absences from duty occurred in 1836 and were relatively brief: July 14–August 12 and November 8–December 4. They occurred shortly after the action at Micanopy and were likely for recuperation rather than desertion, as I can find no record of Chandler being confined or disciplined for them.[28] Chandler's artillery captain, C. S. Merchant, officially discharged him on July 7, 1837, the result of "inabilities of a local nature from injuries received while in the service of the United States," and he listed Chandler's occupation as stoneware potter.[29] This would indicate that between his enlistment in 1832 and his discharge in 1837

Chandler had found his way back into the fraternity of stoneware potters. According to his service record, Chandler's last absence occurred on October 10, 1837, after his official discharge. The final notation on Chandler's service record was made on July 10, 1838, just nineteen days before his marriage on July 29 to eighteen-year-old Margaret Durham, daughter of Edgefield potter John Isaac Durham.[30] Apparently Thomas Chandler and the Army had parted ways on good terms.

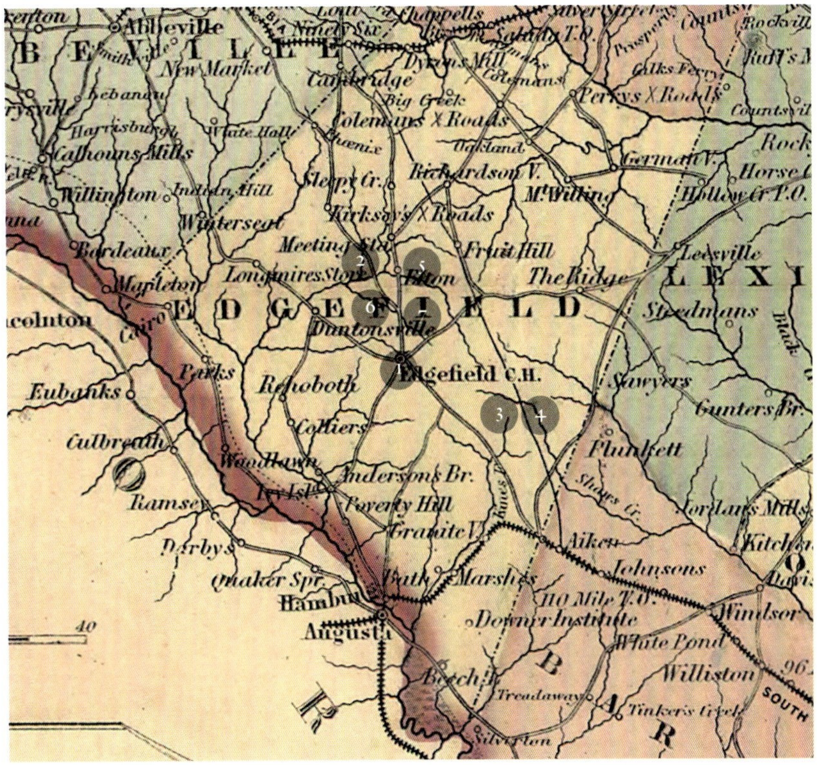

Figure 11 Sites of potteries where Thomas Chandler worked in the Old Edgefield District of South Carolina between 1835 and 1853, superimposed on J. H. Colton & Co.'s 1855 map of South Carolina.

1 38ED11, *1835–1837*: Chandler appears to first work with Rueben Drake and Collin Rhodes at Pottersville.

2 38GN169, *1837–1838*: Chandler is found working with John Isaac Durham north of Edgefield.

3 38AK495, *early 1839*: Chandler has moved to a site on Shaw's Creek and begins experimenting with iron and kaolin slip decoration on alkaline-glazed pots.

4 38AK495, *early 1840–1844*: When Phoenix Factory is created as a partnership at Shaw's Creek in early 1840, Chandler becomes primary turner and decorator of stoneware produced there. When the partnership dissolves in late 1840, Chandler remains at Shaw's Creek until ca. 1844, likely working for Collin Rhodes. By 1844 he no longer turns pottery at site 38AK495, but decorates for Rhodes part time until about 1851.

5 38GN16, *1844–1848*: It appears Chandler has opened and is working at "Chandler Site I." This is near the Durham/Trapp/Chandler site 38GN169.

6 38GN169, *1848–1850*: Chandler enters into partnership with John Trapp at the old Durham/Mathis pottery, now known as the Trapp/Chandler/Presley pottery, where he works for about two years.

7 38GN343, *1850*: Chandler opens his second shop, "Chandler Site II," about a mile south of site 38GN169 on the old Martintown Road. By 1853 it appears Thomas Chandler has left Edgefield and moved to Mecklenburg County, North Carolina.

Thomas Chandler Pottery Sites in Edgefield

I believe Thomas Chandler made his first appearance in Edgefield in the mid-1830s, but it is difficult to pinpoint where he first worked. The map presented in figure 11 provides an overview of the locations where Chandler is known to have produced or decorated stoneware. The absence of decorated stoneware in the Edgefield District before 1836 is good evidence that he

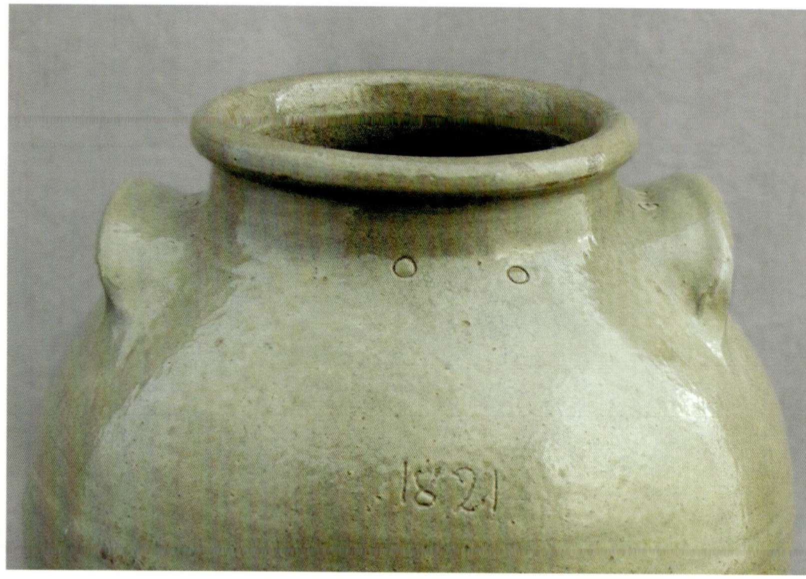

Figure 12 Storage jar, John Landrum's Horse Creek Pottery or Abner Landrum's stoneware factory at Pottersville, Edgefield District, South Carolina, 1821. Alkaline-glazed stoneware. H. 11". (Private collection; photo, Jan Todd.) This represents some of the earliest alkaline-glazed ware from Edgefield.

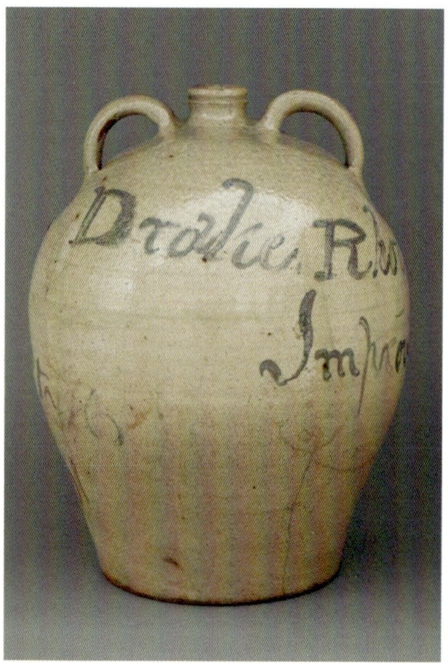

Figure 13 Two-handle jug, Rueben Drake and Collin Rhodes Pottery, Pottersville, South Carolina, 1836. Alkaline-glazed stoneware. H. 17½". Inscribed in cobalt: "Drake Rhodes & Co Improved Stoneware Edgefield Ct H S.C. 1836" (Courtesy, Museum of Early Southern Decorative Arts.)

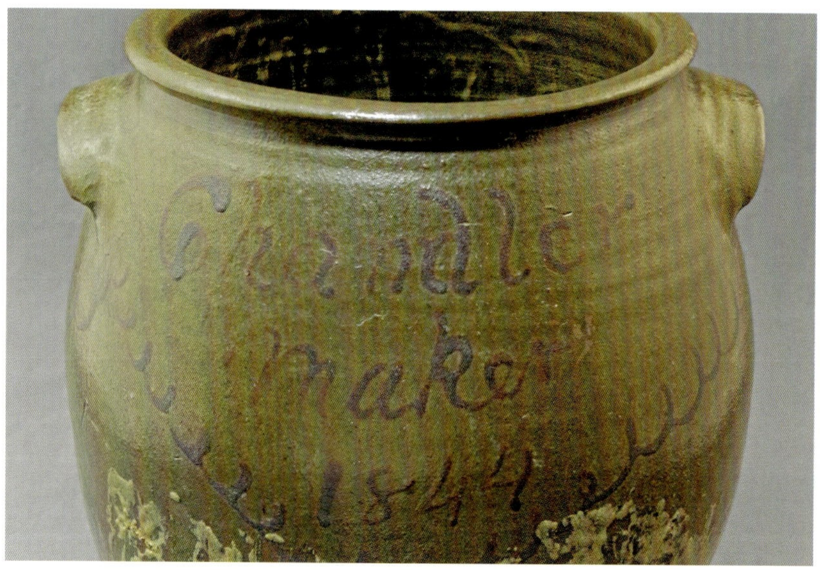

Figure 14 Storage jar, Thomas Chandler Pottery, Edgefield District, South Carolina, 1844. Alkaline-glazed stoneware. H. 17¾". Inscribed in iron slip: "Chandler / Maker / 1844"; on reverse, in iron slip: "Mr. B. Harland" (Courtesy, Steve and Terry Ferrell; photo, Jan Todd.) The handwriting appears to be the same as that on the 1836 jug illustrated in fig. 13.

brought his decorating method and distinctive fluid script to the area (fig. 12).[31] Two presentation pieces dated 1836 have pronouncements written in cobalt blue in what I believe is Chandler's script.[32] The Drake and Rhodes jug (fig. 13) and the Hiram Gibbs jar (not illustrated) are the first known examples of Edgefield stoneware to have any kind of slip applied for use other than glazing. Most ceramics historians accept that Chandler was working in Edgefield by 1838, but on the basis of these two presentation examples, I believe he was there in 1836.[33] For comparison, a signed "Chandler Maker" jar, dated 1844, confirms Chandler's distinctive script (fig. 14).

Thomas Chandler Production Marks in the Edgefield District

A distinctive characteristic of Edgefield District stoneware made during the first and second quarter of the nineteenth century is the production mark. These marks were used by both slave and non-slave potters and they help us

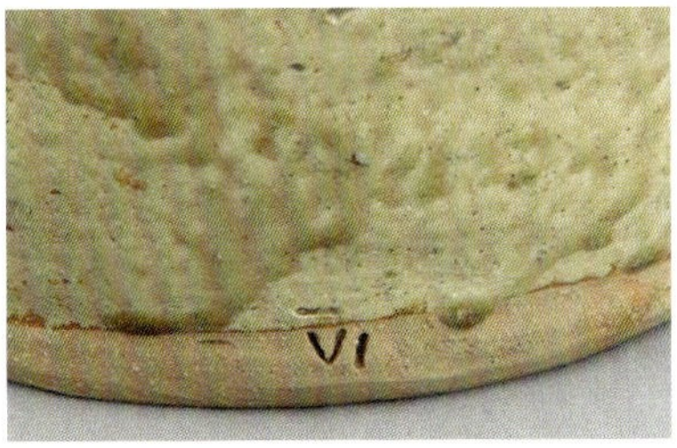

Figure 15 Detail of production/capacity mark, roman numeral VI. (Photo, Jan Todd.) This mark is attributed to Thomas Chandler at the Pottersville stoneware factory and the Trapp and Chandler Pottery location.

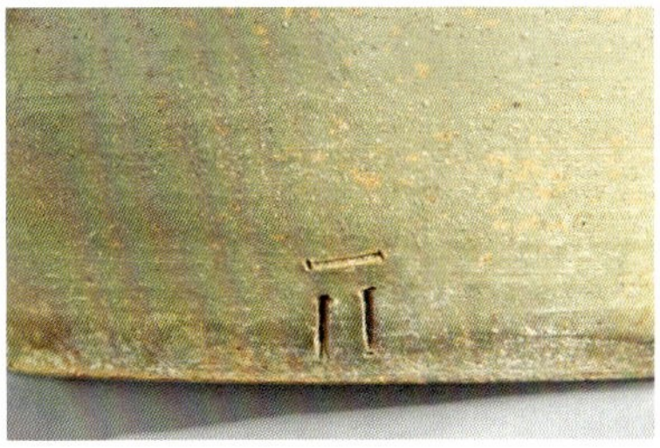

Figure 16 Detail of production/capacity mark roman numeral II. (Photo, Jan Todd.) This mark has been found at the pre–Trapp and Chandler site (38GN169).

identify potters and their movement from site to site. Most itinerant potters impressed their mark near the base edge of the pot, but occasionally marks were placed near the rim.

The marks usually had two functions: to identify ware turned by a specific potter so the shop owner could monitor productivity; and to indicate capacity. Most marks were either stamped letters, symbols, and punctuates, or cursive slashes, letters, dots, and symbols. There were also stamped roman numerals ranging from I to XX, usually with a crossbar at the top or bottom.

While Thomas Chandler was a production potter in Edgefield at Pottersville and the pre–Trapp and Chandler site during the period 1836–1838, he primarily used stamped roman numerals as his production mark (figs. 15, 16). I base this finding on a comparison of surviving ware attributed to him with sherds collected from these sites as shown in the two major archaeological reports on Pottersville and the report by Gerald K. Landreth on the Trapp and Chandler site. I also studied sherds from these sites collected by Steve and Terry Ferrell.[34] Moreover, at the pre–Trapp

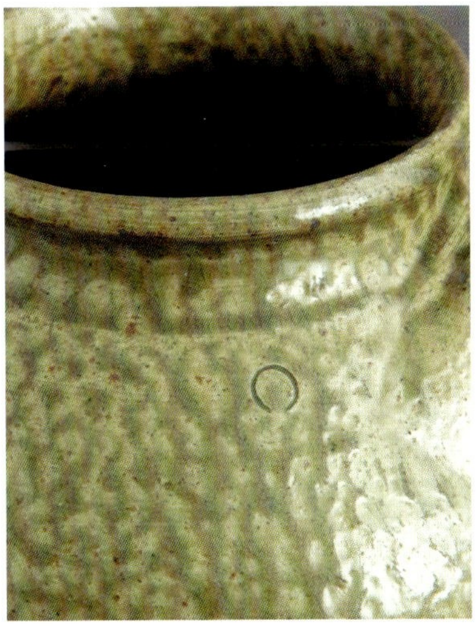

Figure 17 Storage jar, detail of production mark oblate or elongated "C" near the rim, attributed to Thomas Chandler, Edgefield District, South Carolina, 1835–1840 (38GN169). Alkaline-glazed stoneware. H. 13½". Production/capacity mark near base: "II" (see fig. 16). (Author's collection; photo, Jan Todd.)

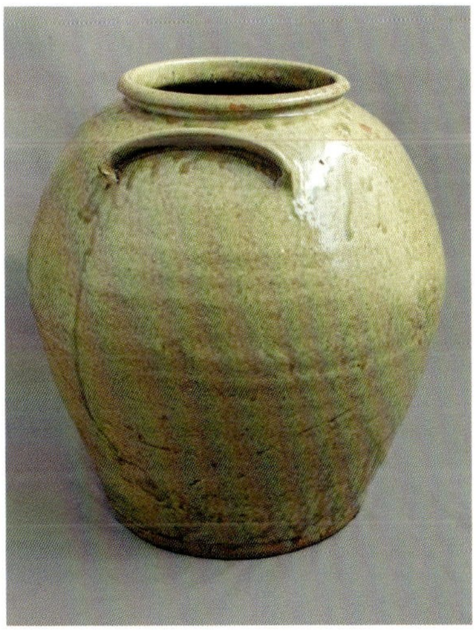

Figure 18 Storage jar, attributed to Thomas Chandler, Edgefield District, South Carolina, ca. 1837. Alkaline-glazed stoneware. H. 16". Production/capacity mark at base: "VI" (see fig. 15). (Author's collection; photo, Jan Todd.)

and Chandler site, circa 1837, he sometimes used an oblate or elongated "C" (figs. 17–22). By the time he was at Shaw's Creek and Phoenix Factory, circa 1838–1840, he was using production marks rarely, with the exception of a deeply impressed lowercase "c" as well as some abstract symbols on the earliest decorated examples (figs. 23, 24). After he left Shaw's Creek, about 1844, he did not use production marks at all, but had begun signing his

Figure 19 Storage jar, attributed to Thomas Chandler or John Isaac Durham, Edgefield District, South Carolina (site 38GN169), ca. 1837. Alkaline-glazed stoneware. H. 12½". Stamped capacity mark near base: "II" (see fig. 16). (Author's collection; photo, Jan Todd.)

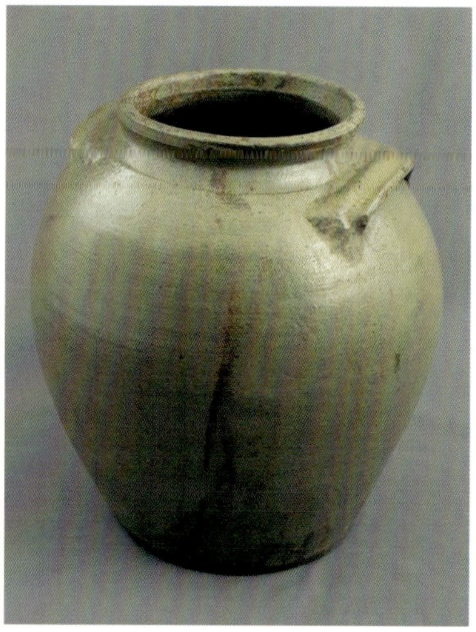

Figure 20 Storage jar, attributed to Thomas Chandler, Edgefield District, South Carolina (site 38GN169), ca. 1837. Alkaline-glazed stoneware. H. 11½". Production mark: oblate or elongated "C" (see fig. 17); production/capacity mark near base: "II" (Author's collection; photo, Jan Todd.)

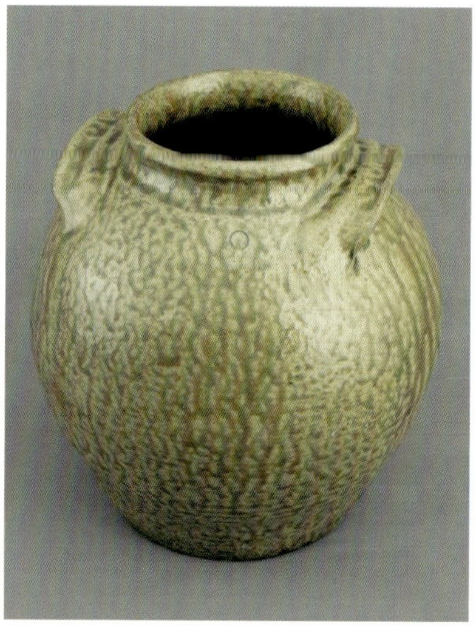

Figure 21 Storage jar, attributed to Thomas Chandler, Edgefield District, South Carolina (site 38GN169), ca. 1837. Alkaline-glazed stoneware. H. 13½". Production/capacity mark: "III" partially filled in with glaze. (Author's collection; photo, Jan Todd.) In this surviving example, Chandler had begun to transition his glaze toward celadon.

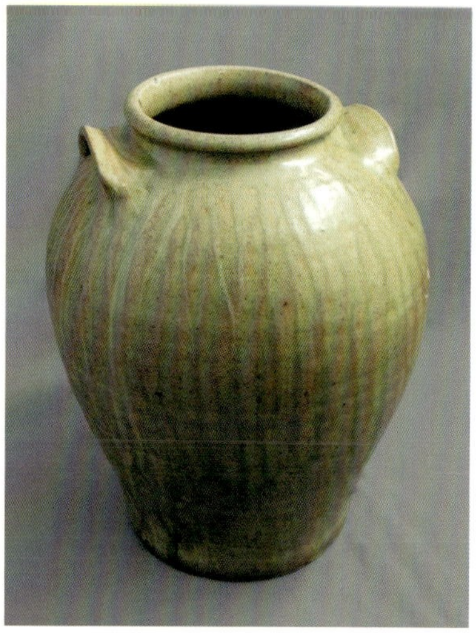

Figure 22 Detail of the production/capacity mark "II" on the jar illustrated in fig. 20.

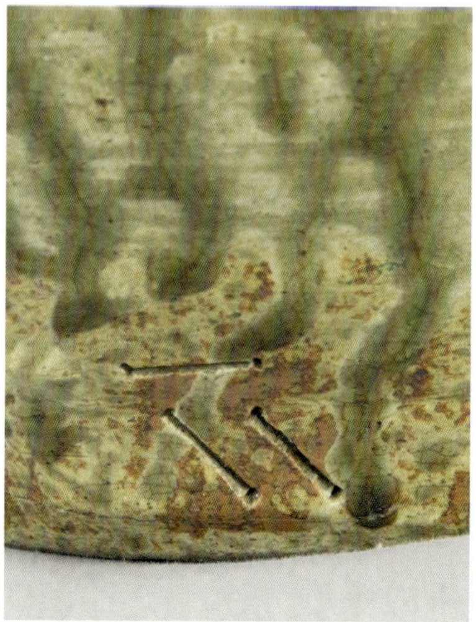

ware "Chandler Maker." Confusing the matter is the fact that not all surviving stoneware made by production potters during this period display production marks. Also, some surviving pottery examples display more than one production mark, sometimes multiples of the same mark.[35]

At least three potteries on the old Martintown Road north of Edgefield were in operation between 1830 and 1850, and Chandler worked at all three. By late 1837 I believe he was working at site 38GN169 with Durham and others (later known as the Trapp and Chandler Pottery) (see fig. 11). From circa 1844 until about 1848, Chandler worked at a pottery a short distance away (Chandler Site I, or site 38GN16), where he first attempted to produce stoneware on his own.[36] In 1848, at site 38GN169, Chandler entered into a partnership with the Reverend John Trapp, which lasted for about two years, until 1850. From circa 1850 to 1853 he operated his own pottery (Chandler Site II, or site 38GN343), which was about a mile south down the Martintown Road.

At site 3GN343, Chandler signed the majority of his work and no production marks have been found archaeologically.[37] Marks representing shop

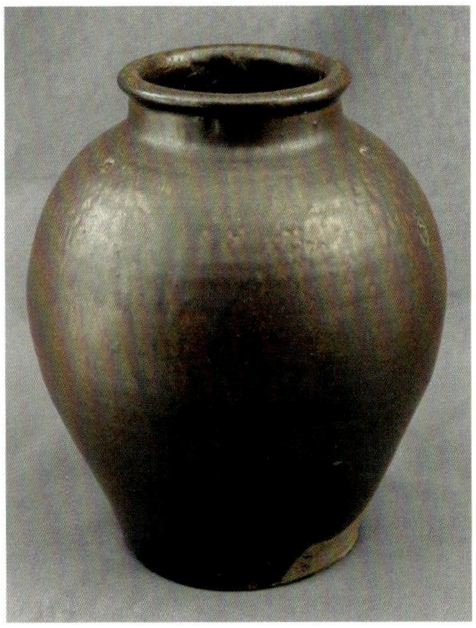

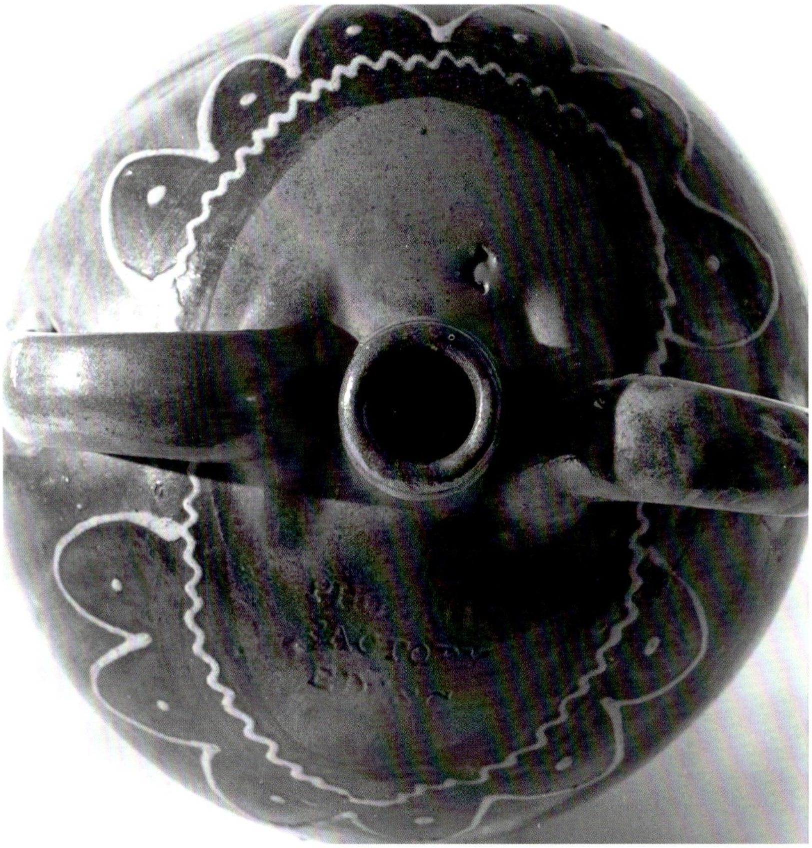

Figure 23 Storage jar, attributed to Thomas Chandler, Edgefield District, South Carolina, ca. 1837. Iron-slip alkaline-glazed stoneware. H. 10". Three small "C"s are impressed equidistant around the rim and three additional roman numeral marks are at the base (see fig. 10); compare with the mark on the jug illustrated in fig. 24.

Figure 24 Spirits jug, attributed to Thomas Chandler, Phoenix Factory, Edgefield District, South Carolina, 1840. Alkaline-glazed stoneware. H. 16¼". Impressed mark: "PHOENIX FACTORY ED: SC / C" (Courtesy, Collection of the Museum of Early Southern Decorative Arts.)

owners are usually found as letters stamped near the rim, spout, or base and are not considered production marks. For example, the following letter marks found at the Trapp and Chandler Pottery and on existing ware represent the potters: "ID" for John Isaac Durham; "M" for Robert Mathis; and "P" (possibly for John Presley).[38]

After a comprehensive study of this material, I conclude that Thomas Chandler used roman numerals as his most common production mark. In most cases the roman numerals were also used to indicate capacity (such markings are also found at Pottersville).[39] Another mark previously mentioned—a deeply impressed "C" stamped in different places and in a variety of patterns—is known from existing ware made at several Edgefield sites. Although it has not yet been found in site archaeology, I consider this to be a Chandler production mark.

Chandler's marriage into the Durham family in 1838 placed him in the midst of the primary potters at Pottersville and the Martintown Road pottery during the 1830s: Rueben Drake, John Isaac Durham, Amos Landrum, Robert Mathis, John Presley, and Collin Rhodes. By 1838 I believe Chandler had left the Martintown Road area where he had been working with Durham and others, and began working at a pottery established on Shaw's Creek south and east of Edgefield. There he most likely was working with Mathis, Durham, and Presley. A copy of the Ramey & Hughes Ledger, Pottersville, South Carolina, shows that by January 1, 1839, Mathis was at Shaw's Creek producing stoneware that the Pottersville operation, in at

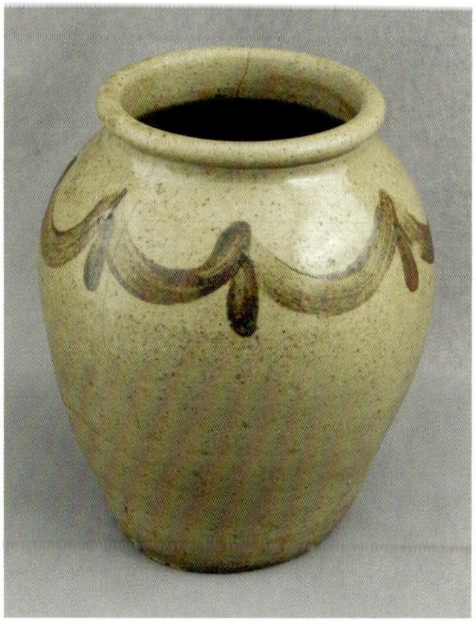

Figure 25 Jar, Shaw's Creek Pottery, Edgefield District, South Carolina, 1837–1839. Alkaline-glazed stoneware. H. 9¼". (Author's collection; photo, Jan Todd.) Chandler adapted and developed the swag-and-tassel decoration from the Remmey family of potters of New York and Baltimore; this is one of his first examples decorated with iron slip.

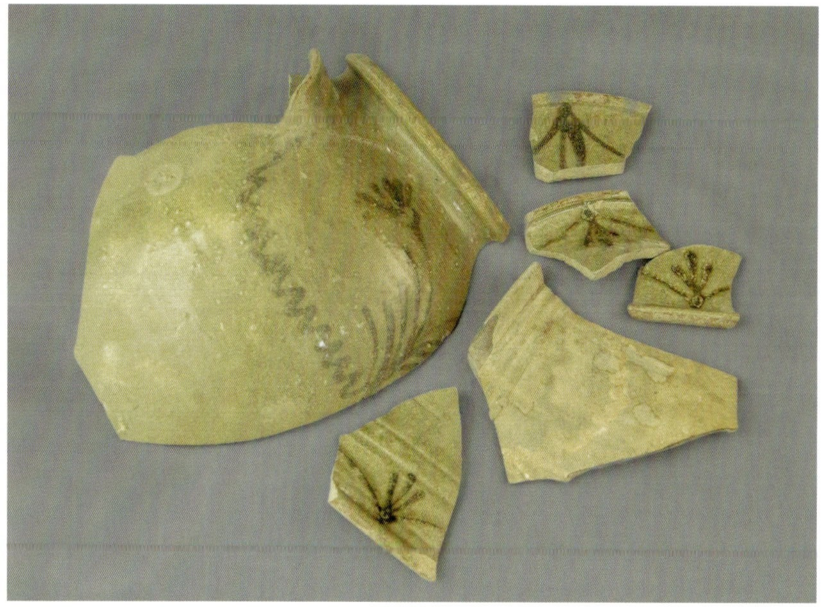

Figure 26 Storage-jar fragments, Shaw's Creek Pottery, Edgefield District, South Carolina, 1837–1839. Alkaline-glazed stoneware. (Courtesy, Steve and Terry Ferrell; photo, Jan Todd.) These fragments, found by ceramics historian Steve Ferrell at Shaw's Creek, exhibit slip-trailed decoration that is remarkably similar to the delicate swag-and-tassel decoration in cobalt on the salt-glazed stoneware jar illustrated in fig. 8. See also fig. 38.

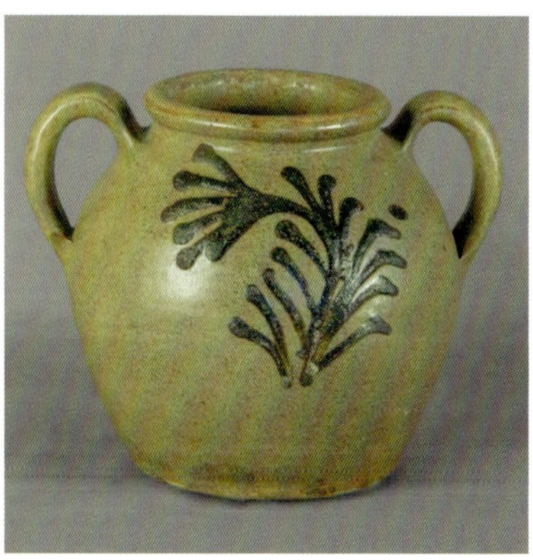

Figure 27 Stew pot, Shaw's Creek Pottery, Edgefield District, South Carolina, 1837–1839. Alkaline-glazed stoneware. H. 8". (Courtesy, Randy Windham; photo, Jan Todd.) This early stew pot presents iron-slip decorative elements that are some of the first developed by Chandler in Edgefield. One can see from this example that Chandler had not yet developed the right liquid consistency for his decorative slip.

least one instance, shipped to Union, South Carolina, for him. Collin Rhodes was not yet at Shaw's Creek, as he is listed as the overseer of the Pottersville operation for the entire year 1839.

At Shaw's Creek, Chandler began the experimentation process needed to master the art of decorating alkaline-glazed stoneware (fig. 25). Fragments found by Steve Ferrell at the site are decorated in early, simple motifs (fig. 26). One of the motifs, the swag-and-tassel, remained in Chandler's repertoire throughout his career as a potter. Some of the early ware from Shaw's Creek looks very much like ware from Pottersville, having a very light or white clay body and a translucent glaze; however, early Shaw's Creek ware is sometimes decorated, and the contrast of dark iron slip against a light background demonstrates a disciplined training and execution of design. It is here at Shaw's Creek that Chandler first implemented a unique decoration that I believe mimics an artillery cannon blast (fig. 27).

Chandler's ability to decorate stoneware at Shaw's Creek earned him significant recognition. By 1840 he was part of a group of potters working at a major stoneware enterprise known as Phoenix Factory (3AK495), literally built on top of the Shaw's Creek site (fig. 28). Presumably the location of the factory was chosen for its access to the newly built railroad that ran from Charleston to Hamburg, South Carolina. For the first time, many of the major Edgefield pottery concerns are working together, and it is at Phoenix Factory that Chandler might have shared, to a limited degree, his knowledge of decorating stoneware with such potters as Collin and Milton

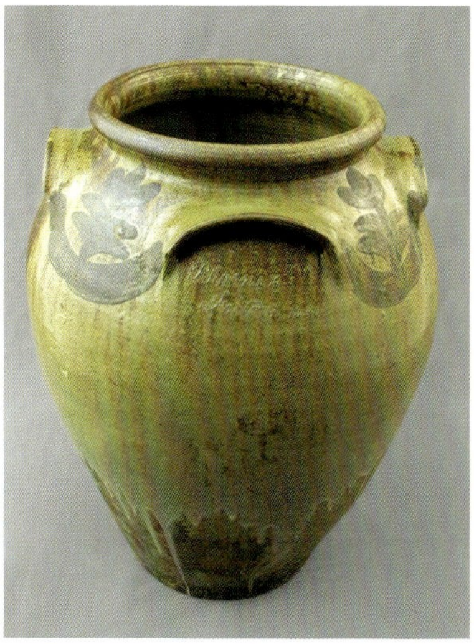

Figure 28 Storage jar, Thomas Chandler, Phoenix Factory, Edgefield District, South Carolina, 1840. Alkaline-glazed stoneware with brushed iron-slip decoration. H. 19¼". Incised: "Phoenix Factory"; "1840" (Courtesy, Steve and Terry Ferrell; photo, Jan Todd.) Compare the iron-slip brushed flowers between the handles on this example to the cobalt trailed flowers on the jar illustrated in fig. 8.

Rhodes, John Isaac Durham, William Durham, Robert Mathis, and Amos Landrum. After a tremendous one-year run of stoneware production that included some of the most beautiful stoneware ever produced in America, the partnership known as Phoenix Factory ended. Whereas only a few surviving decorated stoneware examples in the Edgefield District can be dated pre-1840, many can be attributed to Phoenix Factory during the year it was in operation.

Most collectors and ceramics historians agree that Chandler was the principal turner at Phoenix Factory. Virtually any form Chandler had seen or produced in Baltimore during the 1820s was replicated by him during his time in the Edgefield District. His output included colossal water coolers and huge four-handled storage vessels along with smaller, intricately decorated milk pans, buttermilk pitchers, and bowls. It is here that he likely produced the first face vessel in the district. This varied and dynamic production was in stark contrast to what had been made in Edgefield in the previous twenty-five years.

Based on archaeology and the vast number of fragments found at the Shaw's Creek/Phoenix Factory site, Chandler must have remained there for a few years, making and decorating stoneware most likely for Collin Rhodes (fig. 29). By about 1844 Chandler can be found in the Kirksey Crossroads area again near the Trapp and Chandler site, at a pottery site designated

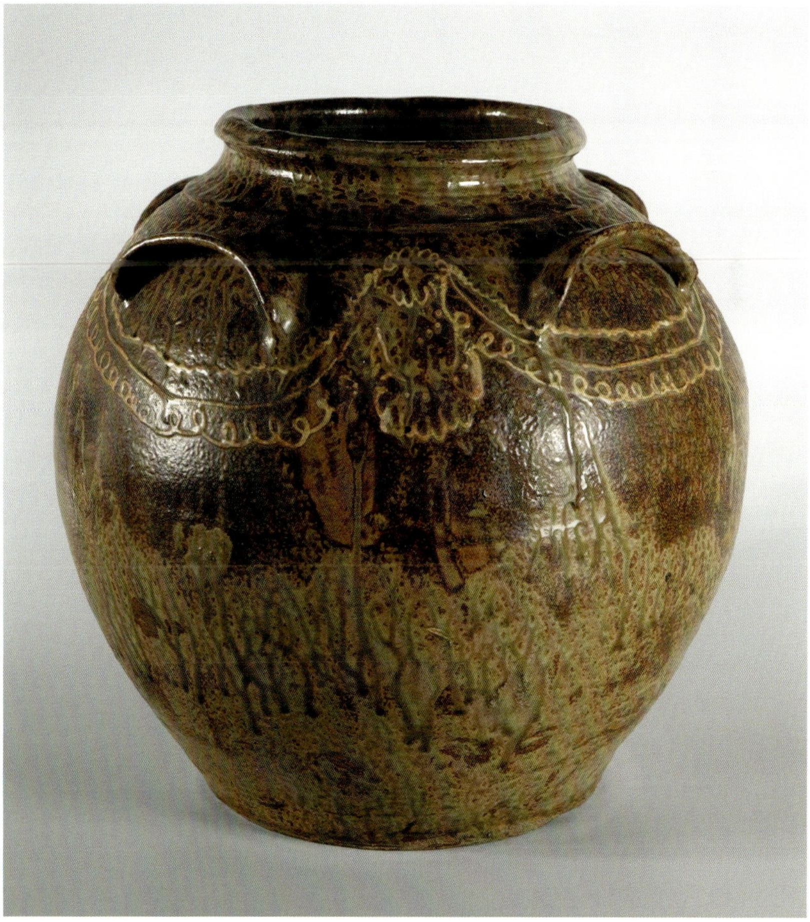

Figure 29 Storage jar, attributed to Thomas Chandler, Shaw's Creek/Phoenix Factory area, Edgefield District, South Carolina, 1841–1845. Alkaline-glazed stoneware. H. 16¾". (Courtesy, Museum of Early Southern Decorative Arts.) With a shape more bulbous than ovoid, this approximately ten-gallon storage jar was likely produced after Phoenix Factory had dissolved. Based on the large volume of waster material at Phoenix Factory, it appears that Chandler continued potting at the site under the auspices of Collin Rhodes.

38GN16 (Chandler Site I). He worked there for about four years, making a variety of vessels (fig. 30). Most of the vessels attributed to this site are decorated with an intermittent wreath or sprig, some with loop and swag, others with a broken-stem flower. In addition, some of the pots are decorated with capacity numbers surrounded by a cartouche, mainly in a dark slip. In 1848

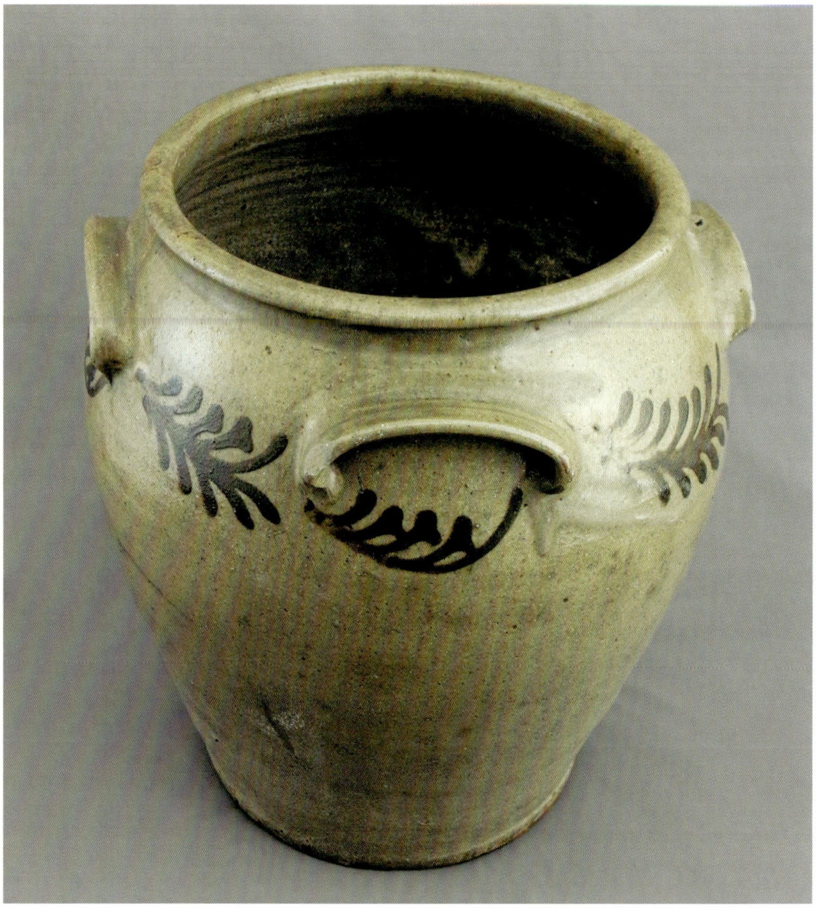

Figure 30 Storage jar, Edgefield District, South Carolina, ca. 1845. Alkaline-glazed stoneware. H. 14½". (Courtesy, Vivian and Gerald Gowitt; photo, Jan Todd.) In my opinion this example was made at the first site Chandler operated independently (38GN16), near Trapp and Chandler (38GN169). The iron-trailed sprigs match potsherds found at site 38GN16.

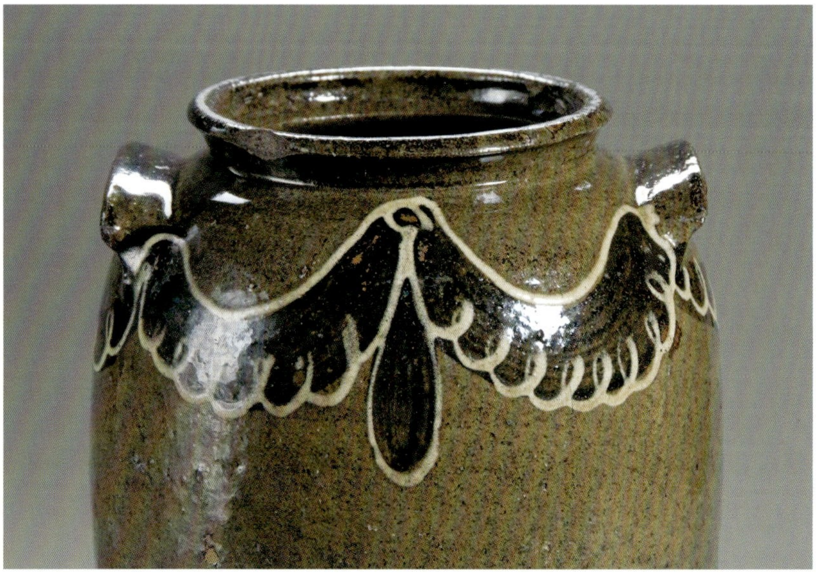

Figure 31 Storage jar, Thomas Chandler Pottery, Edgefield District, South Carolina, 1845–1850. Alkaline-glazed stoneware. H. 12½". (Courtesy, Doug and Brenda Howard; photo, Louis Tonsmeire Studio.) Thomas Chandler began using two-color slip decoration and elaborated on the swag-and-tassel theme by 1840 while at Phoenix Factory. The design on this example is almost identical to the design on the jar illustrated in fig. 25, but the outline in white slip makes a bolder, more neoclassical statement.

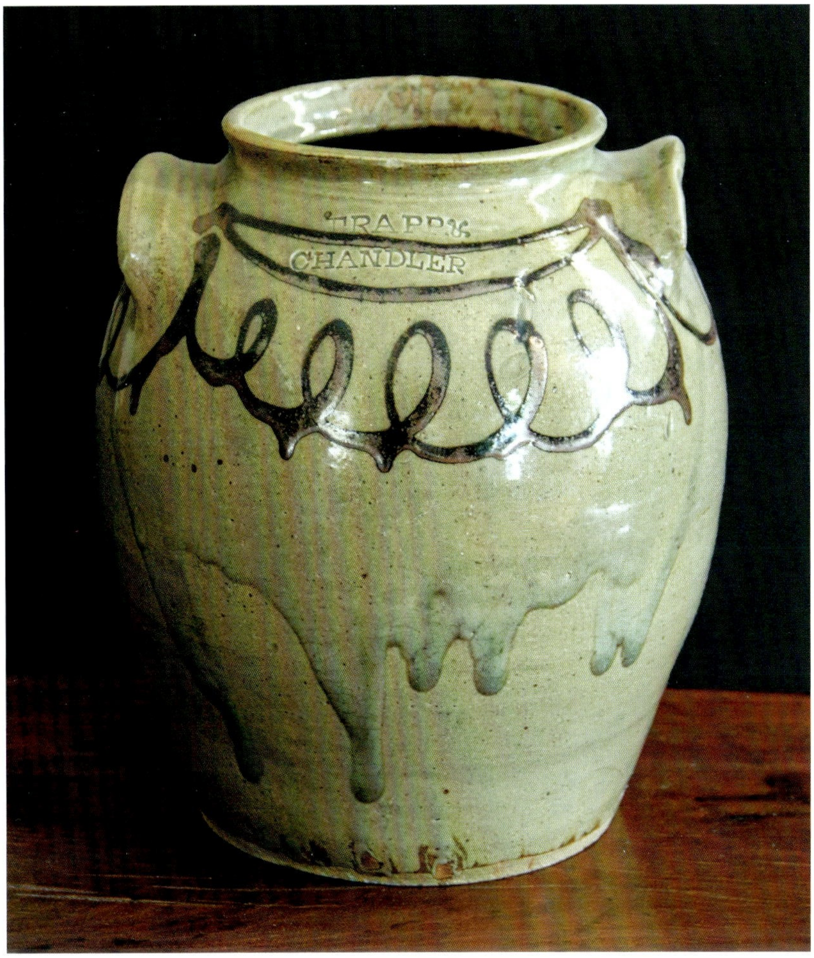

Figure 32 Storage jar, Trapp and Chandler Pottery, Edgefield District, South Carolina, 1848–1850. Alkaline-glazed stoneware. H. 11½". Impressed mark: "TRAPP & CHANDLER" (Courtesy, L. C. Lynch; photo, Jan Todd.) By 1848 Chandler was using draped lines with loops beneath, in iron or kaolin slip, as his primary decorative motif. The thick yet translucent celadon overglaze makes a stunning presentation.

Thomas Chandler moved a short distance to the old Presley/Mathis/Durham pottery on Martintown Road (site 38GN169), where he formed a partnership with the Reverend John Trapp. Chandler found the clays north of Edgefield to be very suitable for stoneware, containing more feldspar and iron and less alumina than the kaolin clays being used in the lower part of the district (figs. 12, 25). Whereas kaolin clays could be fired at a high temperature, they lacked elasticity, which partly explains the dearth of large vessels made during the early period of stoneware manufacture in the Old Edgefield District, when a higher percentage of kaolin clay was being used. At the Trapp and Chandler Pottery, Chandler had a tremendous run of making beautiful functional stoneware from about 1848 until 1850 (figs. 31, 32).

While still operating the Trapp/Chandler site in 1850, Chandler built a new kiln and pottery about a mile south. This site, 38GN343 (Chandler Site II) was built on a high ridge. The placement of his kiln on a ridge may have produced a better draw or draft for the chimney, allowing for a higher firing temperature. This perhaps helped Chandler achieve "Flint Ware" status, his latest experiment in making stoneware. He advertised the business and boldly stated that his ware was warranted from defects, the ware being so marked. For two years at this site, Chandler produced some of his best stoneware, often glazed with true celadon, decorated again in white frilly kaolin and bold, brushed-iron strokes, and fired to a high-pitched ring, becoming the ultimate utilitarian product, beautiful, durable, and useful.

Thomas Chandler and his family left Edgefield in late 1852. Perhaps a connection with the Partlow family brought them to North Carolina's Steele Creek Township, just across the state line from York County and Fort Mill Township near the Catawba River. I believe Chandler's mother-in-law, Nancy Durham, who according to the 1860 South Carolina census was born in York County, was the daughter of Boswell (Bozel) Partlow, who held at the time of his death in 1845 a large estate and several large tracts of land on the Catawba River near the state line and in both townships.[40] It is noteworthy that the last shop Chandler opened and operated (38GN343) was one mile south of the Trapp and Chandler Pottery (38GN169) in the Edgefield District, next to a Partlow homesite and probably on Partlow land.[41] The land that the Chandler family settled on in 1853 in North Carolina was very near the Partlows.[42]

A deed of trust executed in the Edgefield District between Thomas Chandler and his wife, Margaret, in February 1852 indicates that Chandler had become ill or to some extent was incapacitated. In moving his tangible assets to his wife and children, he was limiting any loss that might occur against his estate (his use of the phrase "notwithstanding improvidence or misfortune" suggests that he was expecting God's wrath or bad luck). Four slaves—Simon, Ned, John, and Easter—were included in his estate, as were a large road wagon and four mules, two two-horse wagons with four mules, one mare, and the household furniture.[43] Although it is unclear what prompted Chandler to execute this deed of trust, it might have had something to do with the drowning of his five-year-old son, Thomas H. Chandler, in 1850.

A coroner's inquest into the death of this son occurred on Wednesday, September 11, 1850, "at the Chandler house and the old pottery" three days after the incident.[44] Fourteen neighborhood men came to Chandler's home to view the body and to determine the cause of death. One of them was John Partlow. According to Chandler's account, after realizing his son was missing on Sunday evening, he immediately began a search. He found his son lying face down in a hole of water about five feet deep. Stewart Durham reported that Chandler, carrying his son and both soaking wet, came up to the house where they tried to revive the boy. There was a red mark above the right eye of the child, but otherwise no signs of violence. It is of interest that there is no mention of Chandler's wife or any of the other children at the inquest with the exception of his eldest daughter, Caroline, who was eleven. She witnessed her father pull his son from the small pond of water. Francis Devlin, listed in 1850 as a potter next door to Chandler, is also mentioned in the inquest. The death was ruled accidental.[45] Both Devlin and Stewart Durham, who was a few doors down, likely worked at the Trapp and Chandler Pottery.[46] The death of his son at the Trapp/Chandler site might have prompted Chandler to move to a new location (38GN343), building a kiln next to the Partlow homesite a mile south on Martintown Road.

Thomas Chandler died on July 23, 1854, in the Steele Creek community of Mecklenburg County, North Carolina. Records indicate he had moved there from Edgefield by January 1853.[47] Because he died without a will, Chandler's widow, Margaret, and her brother William Durham were administrators of the estate. After searching through hundreds of records it does not appear that Chandler owned any land in North Carolina or South Carolina. Although his deed of trust showed he owned four slaves, none was listed in his estate at this time, suggesting all were given their freedom. Manumitting slaves was a Chandler family tradition, as well as a common practice in Accomack County, Virginia, in the first half of the nineteenth century.[48]

A free black man by the name of John Chandler was making pottery in Texas at the Guadalupe Pottery, owned and operated by John Wilson from 1857 to 1869 (fig. 33). Although it is not clear when this John Chandler first arrived in Texas, it is believed that he is the slave listed as John in Chandler's

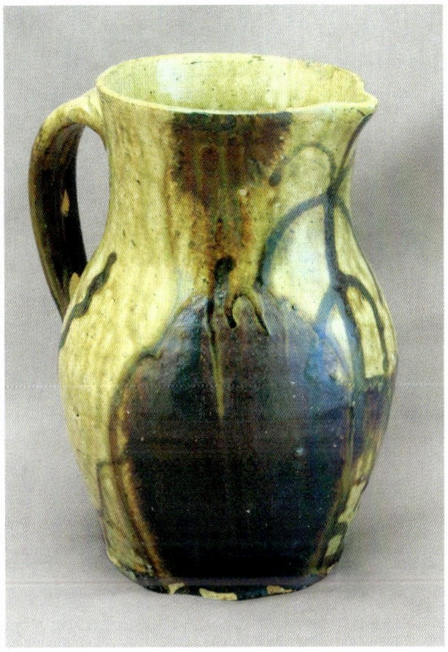

Figure 33 Buttermilk pitcher, possibly eastern Texas, 1860–1870. Alkaline-glazed stoneware. H. 8½". (Courtesy, John Lafoy; photo, Jan Todd.) One of the few slip-decorated stoneware examples surviving from the post-Chandler era, ca. 1860. After Chandler's death, in 1854, only a handful of slip-decorated Southern stoneware examples were produced by potters in South Carolina or points west.

deed of trust executed in 1852. In addition to John Chandler, Marion Jefferson Durham, Thomas Chandler's nephew, who had attended Chandler's estate sale in 1854, was also working at the Guadalupe Pottery and in 1869 bought out John Wilson.[49] John Wilson was from the lower part of Mecklenburg County, North Carolina, but was educated at Washington and Lee in Virginia, as was his father.[50] He was living near the Chandler family in the mid-1850s. There may be a connection between this John Wilson family and the Samuel Wilson who worked for John P. Schermerhorn in Richmond during the first quarter of the nineteenth century.

On September 27, 1854, the personal belongings of Thomas Chandler were sold at auction.[51] The estate sale featured a number of lots of books; among those books listed and sold were a volume of Shakespeare, *The History of the World*, *The History of the United States*, and *Camp Stories of the Revolution*; another lot was simply "books." Clearly, Chandler was a well-read and educated man. Additional items included in the sale ranged from several lots of tobacco, an accordion (perhaps the instrument that Chandler played), two beds and other bedroom furniture, one bureau, one cupboard, one saddle, a half-bushel measure, two wagons, a fine carriage, and three mules. Pottery and related items included a jackscrew, a turning lathe (potter's wheel), and a butter pot made by Isaac Durham, Chandler's father-in-law, and a few pieces of green ware. One hundred gallons of ware (likely decorated) sold for $.32 a gallon, 100 gallons of (likely plain) ware sold for $.08 a gallon, 3,119 gallons of "less good" ware sold for $.0775 a gallon, and 2,000 gallons of "damaged" ware was sold for $.01 a gallon. Chandler's widow bought all of the furniture and the turning lathe; Stewart Durham, his brother-in-law, bought most of the pottery and the jackscrew.[52] Attendees were given six months to pay for their purchases. Based on the amount of pottery sold at the estate sale (5,300 gallons, more or less), plus the hundreds of gallons of stoneware sold from January 1853, when Chandler first arrived in Mecklenburg County, until his death in July 1854, it is possible that Chandler had a pottery operation in Mecklenburg County.[53] Good stoneware clays abound in the area, along with flint, iron ore, and feldspar.

The Edgefield Wares and Baltimore Influences

In Baltimore, Thomas Chandler saw a wide range of wares—personal flasks, water coolers, large storage jars, jugs, stew pots, snuff jars, pitchers, pans, and bowls, many decorated with cobalt sprigs, open bell flowers, or daisies intertwined with wreaths and an occasional incised bird perched on a branch. Many of these forms he brought to Edgefield (figs. 34, 35), where he produced his most significant stoneware circa 1837–1853: thousands of well-executed stoneware pots including water coolers, storage jars, syrup and whiskey jugs, flasks, pans, bowls, butter pots, stew pots, snuff jars, buttermilk pitchers, churns, crocks, and face vessels. Chandler

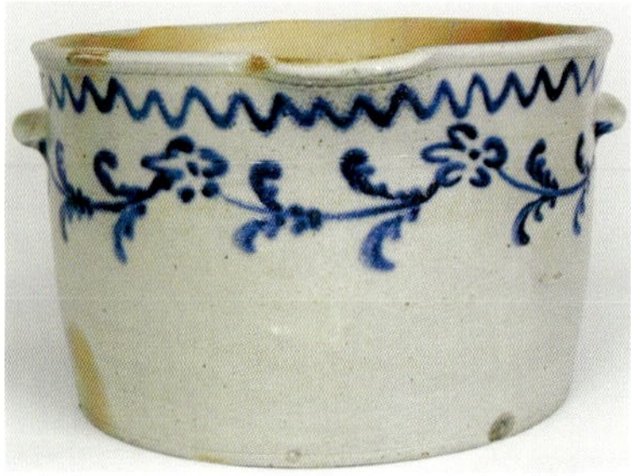

Figure 34 Mixing pan, William Morgan Pottery, Baltimore, Maryland, 1825. Salt-glazed stoneware. H. 6¾". (Courtesy, Crocker Farm Auction.) The squared pouring spout and the slab handles were adapted by Thomas Chandler. Compare the zigzag decoration to that on the jug illustrated in fig. 24.

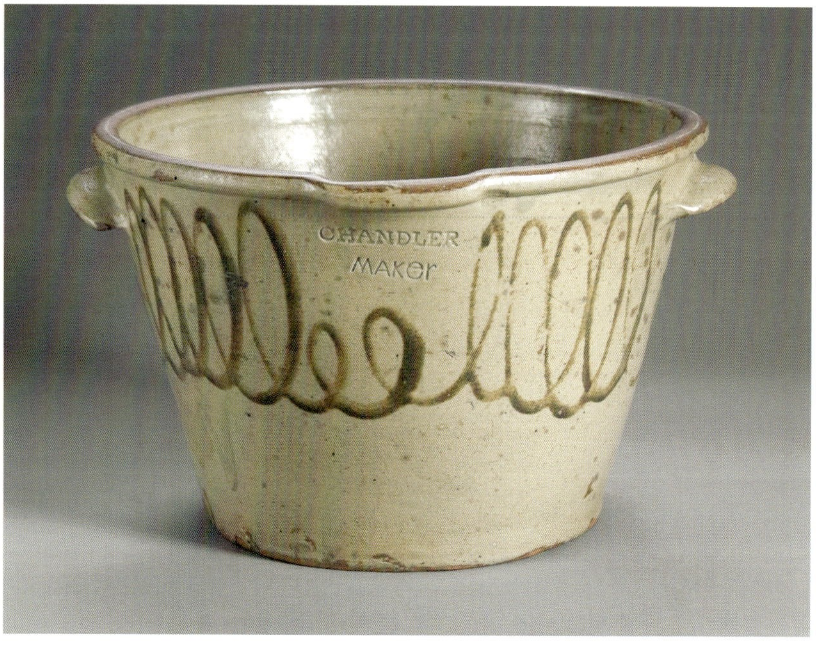

Figure 35 Mixing pan, Trapp and Chandler Pottery, Edgefield District, South Carolina, 1848–1850. Alkaline-glazed stoneware. H. 7¼". Impressed mark: "CHANDLER MAKer"(Collection, Doug and Brenda Howard; photo, Louis Tonsmeire Studio.) Squared pouring spouts are commonly seen on Baltimore stoneware; they were not made in Edgefield prior to Chandler's arrival.

also brought a superior knowledge of stoneware manufacturing that I believe he learned from Henry Remmey. Just as Remmey's arrival in Baltimore had an immediate effect on stoneware manufacturing there, Chandler's arrival in Edgefield, and in particular the back country of South Carolina, effected an immediate improvement to product utility and decorative design for stoneware manufactured in Edgefield.

Chandler was able to produce, on a consistent basis, a high-quality alkaline glaze like that of the ancient Chinese celadon. He used it on stoneware in South Carolina to great effect, and far more examples of it have been found on Chandler-produced ware than on wares produced by any other potter in the area. Chandler was proficient at finding the right kinds of stoneware clays and knowing precisely how to mix them. His firing technique was equally proficient, reaching the highest kiln temperature for best stoneware results—often approaching such a high level of vitrification that the pots would warp and collapse under their own weight (fig. 36).

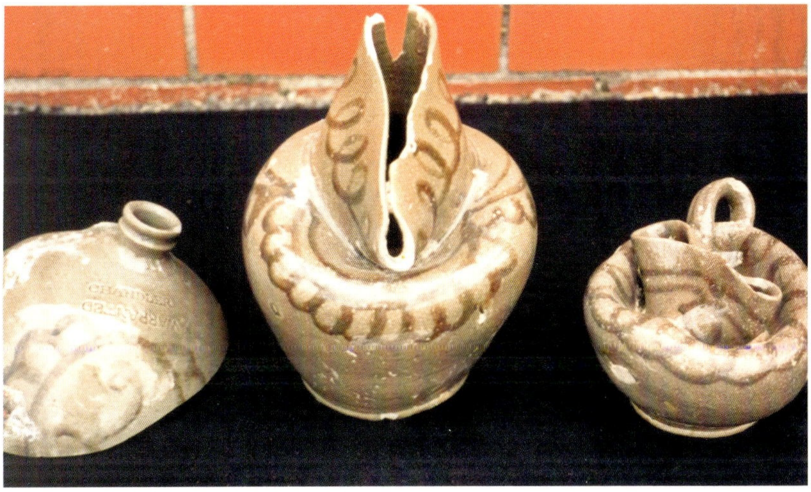

Figure 36 Stoneware wasters, Thomas Chandler Pottery, Edgefield District, South Carolina, 1840. Alkaline-glazed stoneware. (Courtesy, University of South Carolina, McKissick Museum; photo, Carl Steen.) Archeologist Carl Steen found these collapsed vessels on top of the waster at the Thomas Chandler Pottery during a survey in 1988.

The majority of Chandler's stoneware was decorated with iron and/or kaolin slip, trailed by slip-cup or brushed classic designs, including swag and tassels, loops and swags, cartouches, daisy and bell flowers, tulips, and three-leaf clovers, along with some figural as well as abstract creations. His most elaborate decoration was in two colors, with a brushed iron slip outlined in white kaolin. Another important link to Baltimore is Chandler's classic swag and three-leaf or tassel design, which he used on his earliest decorated Edgefield stoneware pots (circa 1837–1839) and then on many of his major four-handled, two-color decorated storage jars made at Phoenix

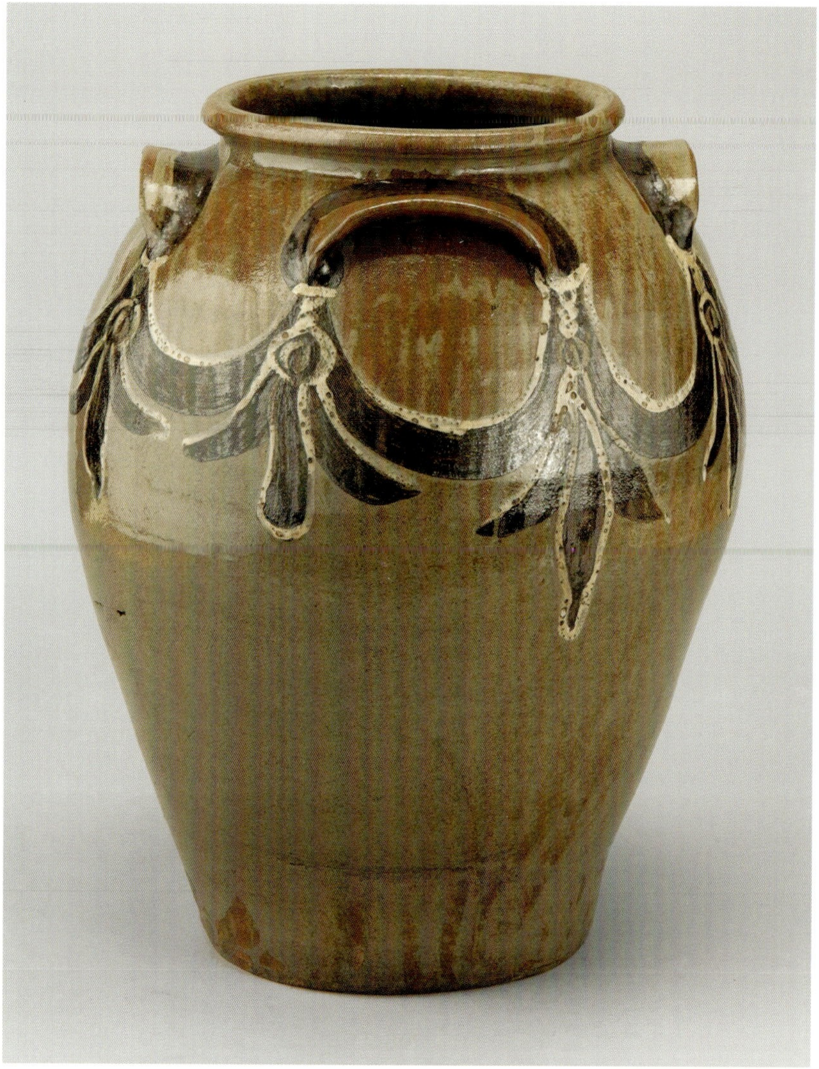

Figure 37 Storage jar, attributed to Thomas Chandler, Phoenix Factory, Edgefield District, South Carolina, 1840. Alkaline-glazed stoneware. (Courtesy, Dr. and Mrs. John Hoar; photo, Brewster Photography). A magnificent surviving example of Chandler's work at Phoenix Factory. Here the kaolin is applied with a brush instead of a slip cup.

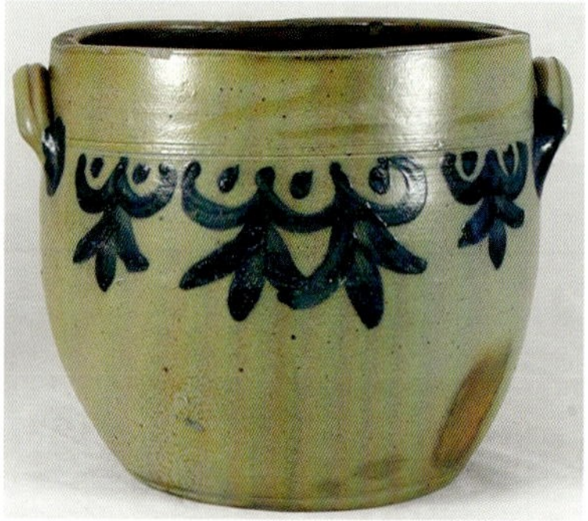

Figure 38 Storage jar, attributed to John Remmey III, New York or New Jersey, 1810–1820. Salt-glazed stoneware. H. 9". (Courtesy, Bruce and Vicki Waasdorp, American Pottery Auction.) This decoration matches the decoration on the Phoenix Factory jar illustrated in fig. 37.

Factory in 1840 (fig. 37). This intricate and distinctive decoration was designed and used by John Remmey and family in New York City in the late eighteenth and early nineteenth century, and has been described as their signature decoration (fig. 38).[54]

Chandler also adopted the flower head on a bent stem, a motif Remmey used with the flowers often placed horizontally in a series (figs. 39, 40). Chandler used the daisy flower extensively in Edgefield, mainly vertically on a bent or broken stem in iron slip but many times between a series of loops and swags (see fig. 51). The flower heads, which are nearly identical to those by Remmey, are seen on Chandler's pots from his time at the Shaw's Creek shops (including the Collin Rhodes pottery) to his time at the Martintown Road shops in Edgefield. By most accounts Henry Remmey was the most accomplished stoneware potter in Baltimore in the 1820s, and from 1821 until 1829 he worked for Henry Myers.[55] Chandler seems to have been greatly influenced by Remmey during this time period. Another influence was William Morgan, who in Baltimore in the 1820s signed his pots "Morgan Maker," phrasing Chandler would use extensively

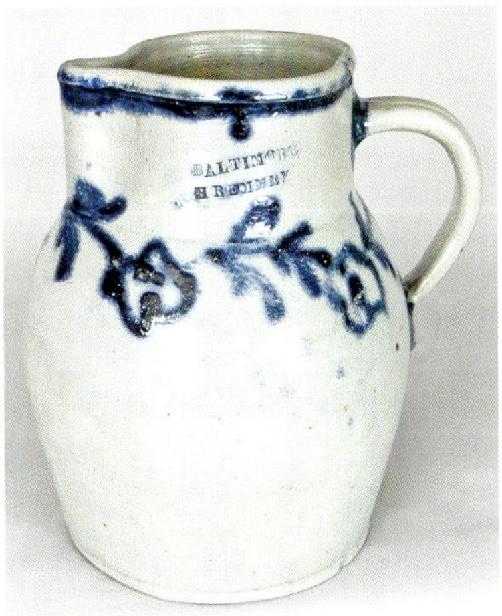

Figure 39 Pitcher, Baltimore Stoneware Manufactory, Baltimore, Maryland, 1825–1828. Salt-glazed stoneware. H. 10". Stamped: "BALTIMORE, / H. REMMEY" (Courtesy, Crocker Farm Auction.) The cobalt-trailed floral decoration is the same basic design that Chandler would use some twenty years later.

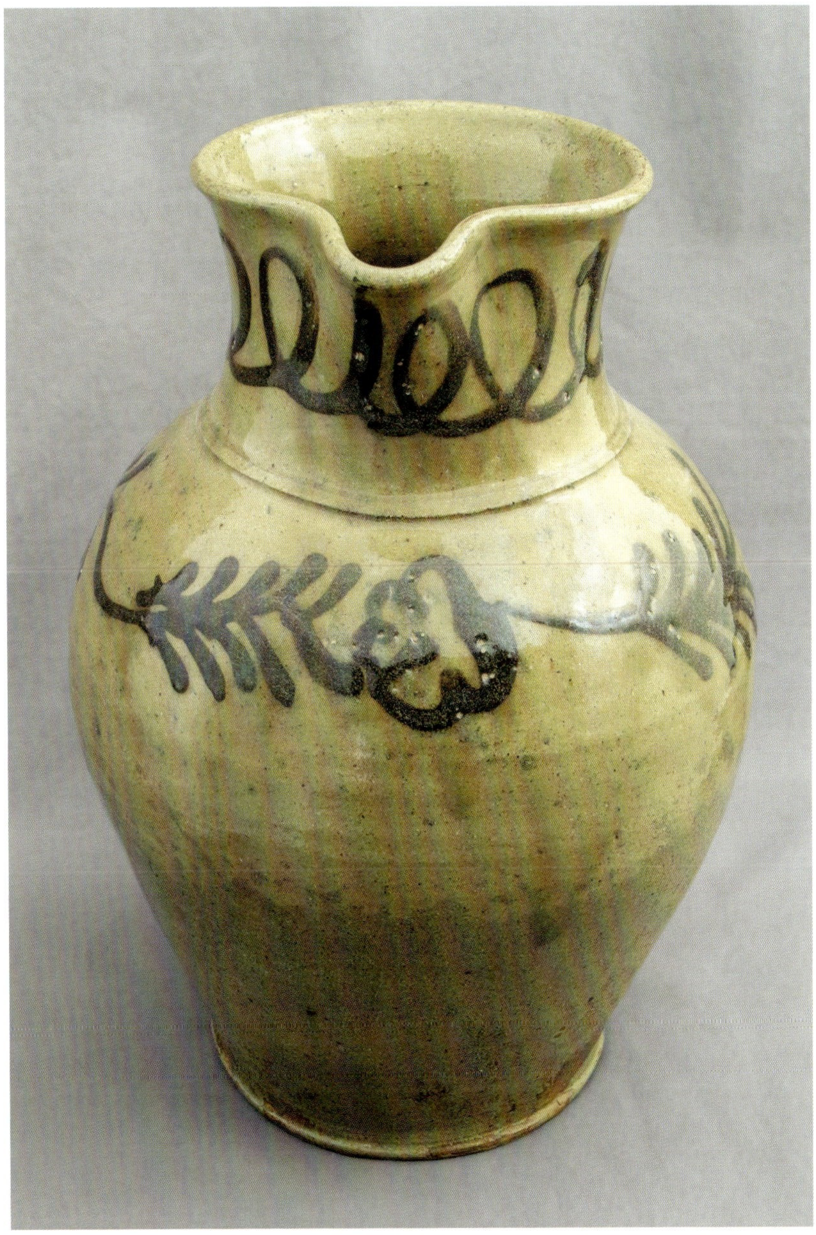

Figure 40 Pitcher, Trapp and Chandler Pottery, Edgefield District, South Carolina, 1848–1850. Alkaline-glazed stoneware. H. 13½". (Courtesy, Donny Taylor; photo, Jan Todd.) This wonderfully executed sprig-and-flower-head decoration is a Chandler adaptation of Henry Remmey's design in Baltimore.

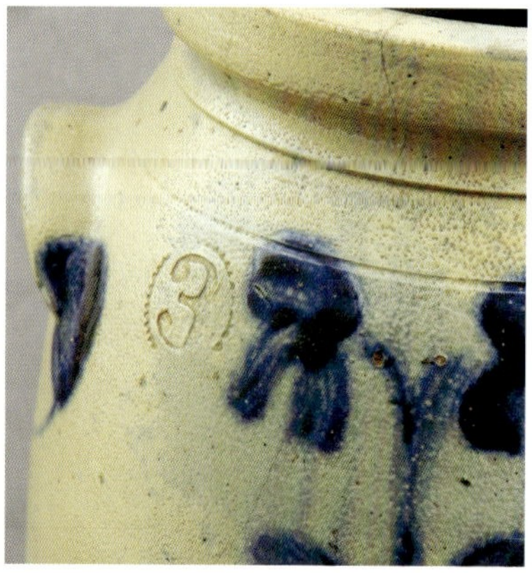

Figure 41 Detail of mark on a storage jar, David or Elisha Parr Pottery, Baltimore, Maryland, 1820–1830. Salt-glazed stoneware. Capacity mark, stamped within a circle of coggle-impressed teeth: "3" (Author's collection; photo, Jan Todd.)

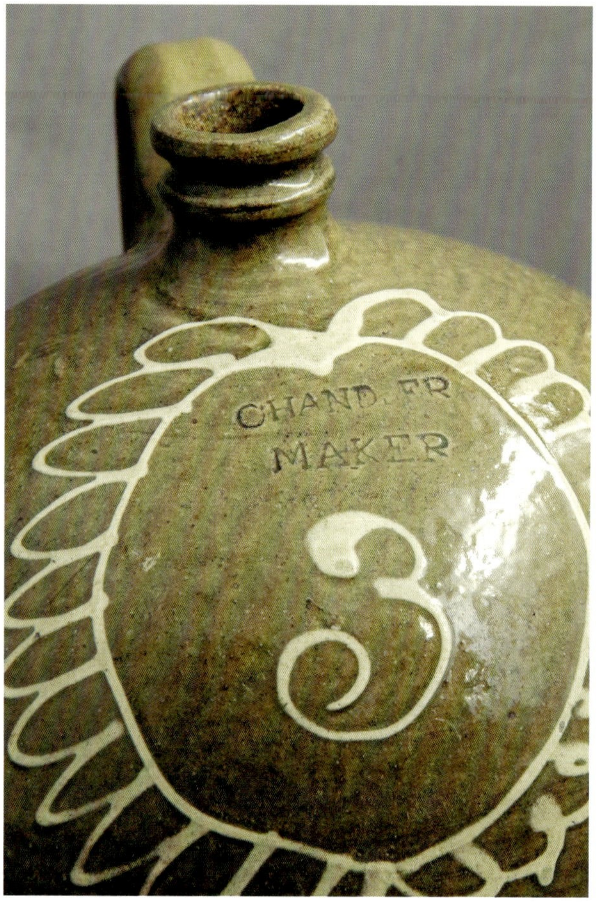

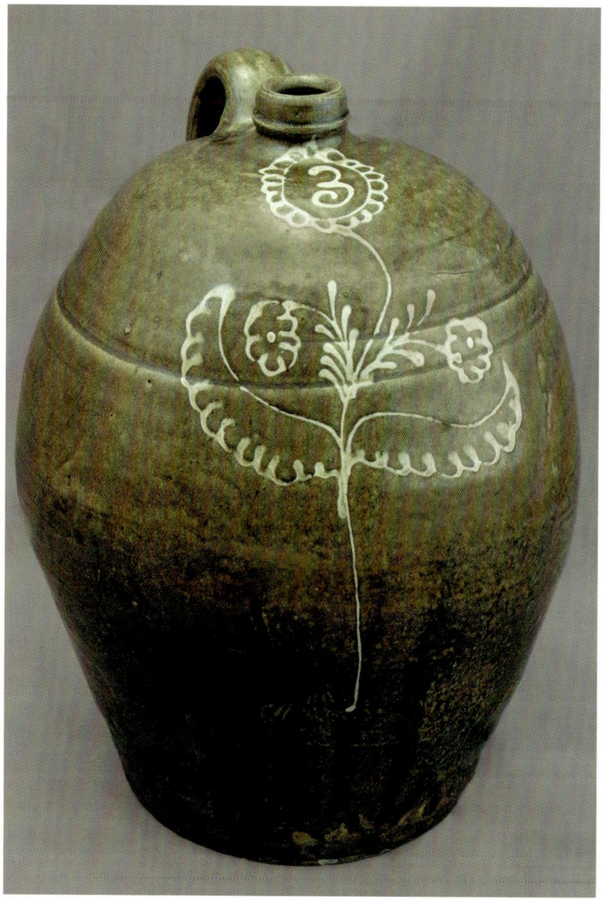

Figure 42 Spirits jug, Thomas Chandler Pottery, Edgefield District, South Carolina, 1850–1852. Alkaline-glazed stoneware. H. 16". Capacity mark: "3"; stamped signature: "CHANDLER / MAKER" (Author's collection; photo, Jan Todd.) Chandler developed his own technique for indicating capacity by evolving the coggled circle on the jar illustrated in fig. 41 into a large, slip-cup-executed cartouche surrounding a slip-drawn freehand arabic numeral. In some instances he framed the capacity with a daisy flower head instead of a cartouche. This was done in a kaolin or iron slip. After 1840 Chandler used this method to indicate gallons, as he no longer marked his ware with roman numerals, slashes, capacity stamps, or incised arabic numerals. This is in stark contrast to his contemporaries, who continued to use slashes, punctuates, crosses, and dots to indicate gallons until the 1860s.

Figure 43 Jug, Collin Rhodes Stoneware Factory, Edgefield District, South Carolina, 1844–1851. Alkaline-glazed stoneware. H. 15½". Capacity mark: "3" (Courtesy, Sam Phillips; photo, Jan Todd.) This jug was decorated by Chandler, who decorated virtually all of the ware produced at this pottery. Note the capacity surrounded by a daisy flower head.

in Edgefield when he signed his own pots.[56] The nearly identical multi-lobed stamped flower-head designs found on Morgan's work in Baltimore and on Chandler's work in Edgefield years later show a strong connection between these two potters.

While in Edgefield, Chandler adapted a decorative technique for marking capacity on his pots. Rather than use a stamped numeral surrounded by a coggle-wheel–executed cartouche (fig. 41), a tool that he and the other potters of Baltimore had used, Chandler executed the number in freehand slip and surrounded it with a scalloped cartouche done in a white kaolin or dark iron color (fig. 42), which could be applied with either a slip cup or a brush. Occasionally Chandler dispensed with the cartouche and drew the capacity mark inside a flower head at the end of a frilly stem, all executed in iron or kaolin slip (fig. 43). This small but significant touch that he added to his stoneware can be seen on many of the decorated pots made in Edgefield during the 1840s and early 1850s. The practice of incorporating the capacity stamp into the cobalt-blue flower-head decoration was used on some Richmond area pots (see, in this volume, Kurt Russ et al., "The Remarkable 19th-Century Stoneware of Virginia's Lower James River Valley," p. 239–241) and may further indicate the presence of Chandler in that location between 1829 and 1832.

Other innovations Chandler brought to Edgefield, perhaps influenced by his urban upbringing, were to advertise stoneware for sale in the local newspaper and to create "presentation" stoneware, used to celebrate an event or individual or to promote the stoneware business (fig. 44). I can find no example of advertising or presentation stoneware made in the Edgefield District before 1836.

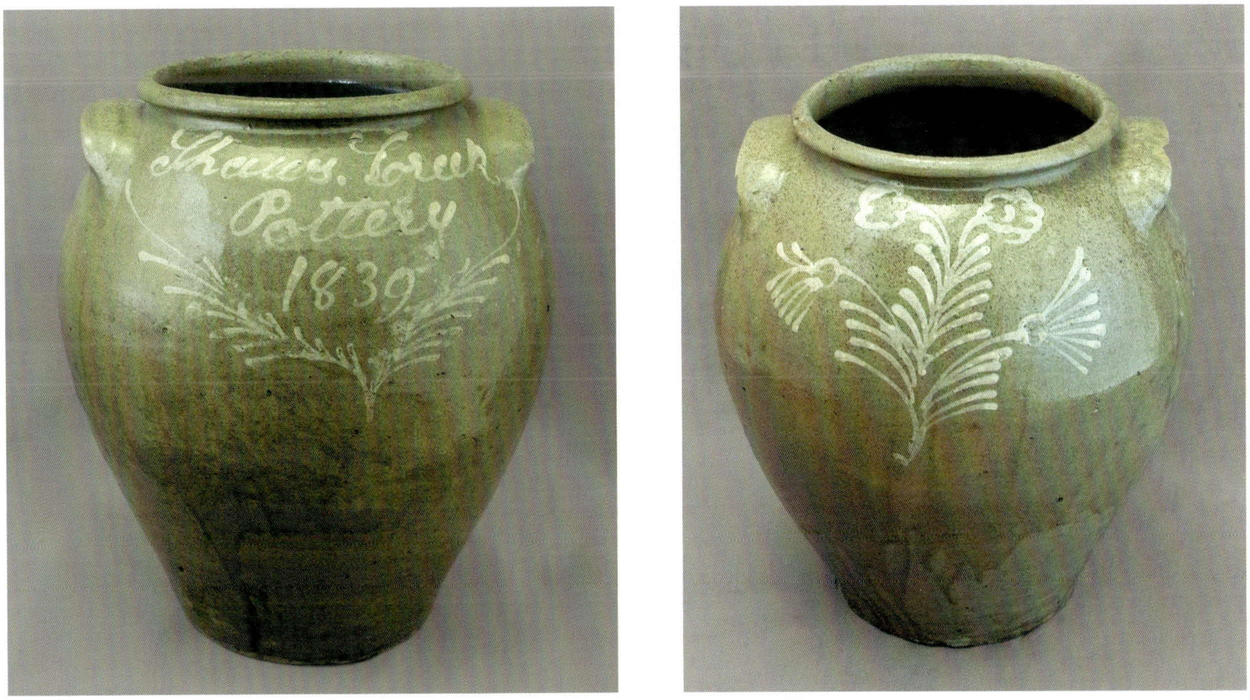

Figure 44 Storage jar, attributed to Thomas Chandler, Shaw's Creek Pottery, Edgefield District, South Carolina, 1839. Alkaline-glazed stoneware. H. 17". Inscribed in slip: "Shaws Creek / Pottery / 1839" (Courtesy, Sam Phillips; photo, Jan Todd.) This is an early example of a Chandler presentation piece, bearing a type of flower decoration he used primarily for Collin Rhodes. The sprigs, bell flowers, and daisy heads are all Baltimore-influenced Chandler designs. The writing is in his hand. This establishes that Chandler had mastered the art of slip decoration on alkaline-glazed stoneware by 1839.

Another important presentation piece that Chandler made is a huge, approximately 25-gallon, alkaline-glazed stoneware water cooler produced at Phoenix Factory.[57] The sophisticated presentation piece (fig. 45) depicts perhaps a slave wedding, complete with wine-goblet toasting and a curly-tailed sow. Notably there are iron slip–brushed three-leaf clovers on either side of the bride and groom. The incising on both the Baltimore churn illustrated in figures 5–7 and the Edgefield water cooler appears to be in Chandler's style (fig. 46). In addition, the clover decorations are reminiscent of those used in Baltimore by the Parr brothers. The reverse of the water cooler has a long-leaf wreath generously brushed in iron slip, a style similar to that of Henry Remmey's work in Baltimore. On the obverse, Chandler again chose iron-slip–brushed inside incised lines and white kaolin-brushed accents on the woman's bonnet, her leggings, and her handkerchief. It is a monumental example of stoneware and arguably Chandler's finest surviving work.

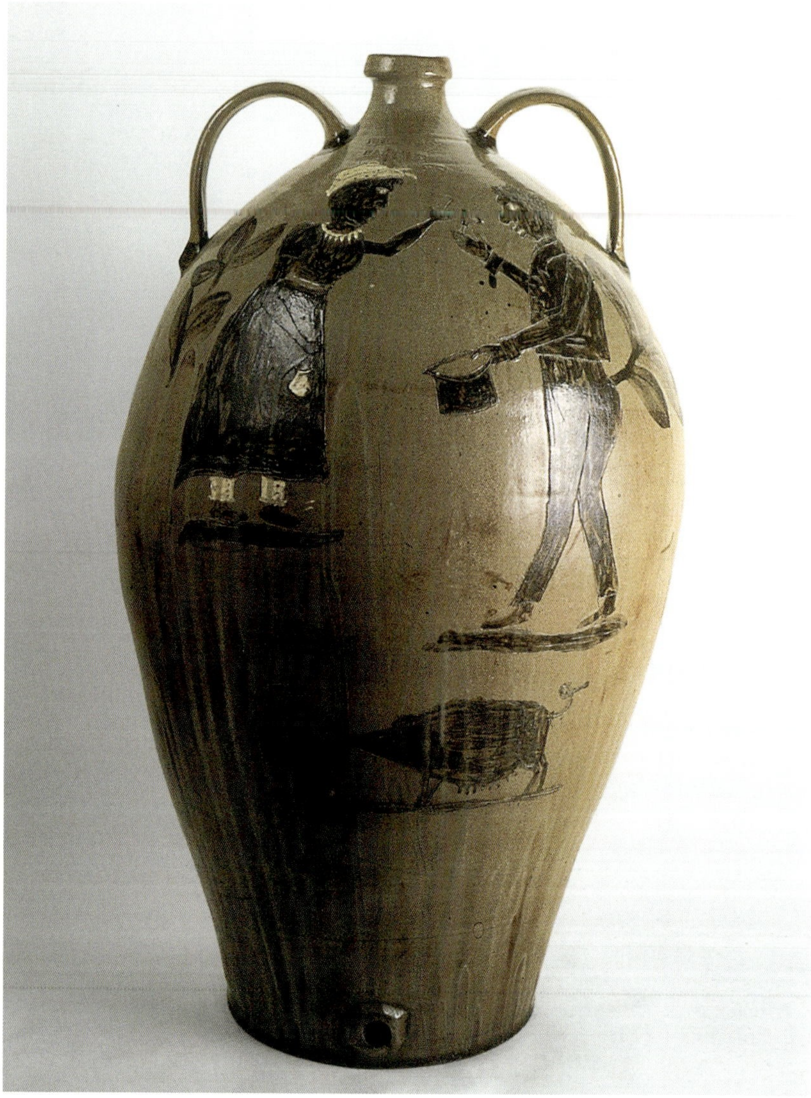

Figure 45 Water cooler, attributed to Thomas Chandler, Phoenix Factory, Edgefield District, South Carolina, 1840. Alkaline-glazed stoneware. H. 31¼". Stamped: "PHOENIX FACTORY ED:SC" on reverse. (Courtesy, High Museum of Art, Purchase in honor of Audrey Shilt, President of the Members Guild, 1996–1997, with funds from the Decorative Arts Endowment and Decorative Arts Acquisition Trust.) This mammoth water cooler was possibly a presentation to slaves in the Chandler household. The drawing on this surviving example is very similar to the drawing on the Chandler Baltimore churn illustrated in figs. 5–7.

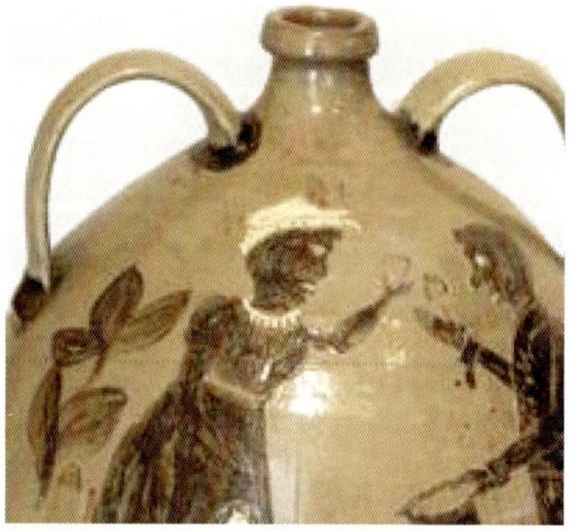

Figure 46 Detail of the water cooler illustrated in fig. 45, showing the scratch decoration of the facing slaves. Compare the hands and facial features with the incising of the Baltimore churn illustrated in fig. 5.

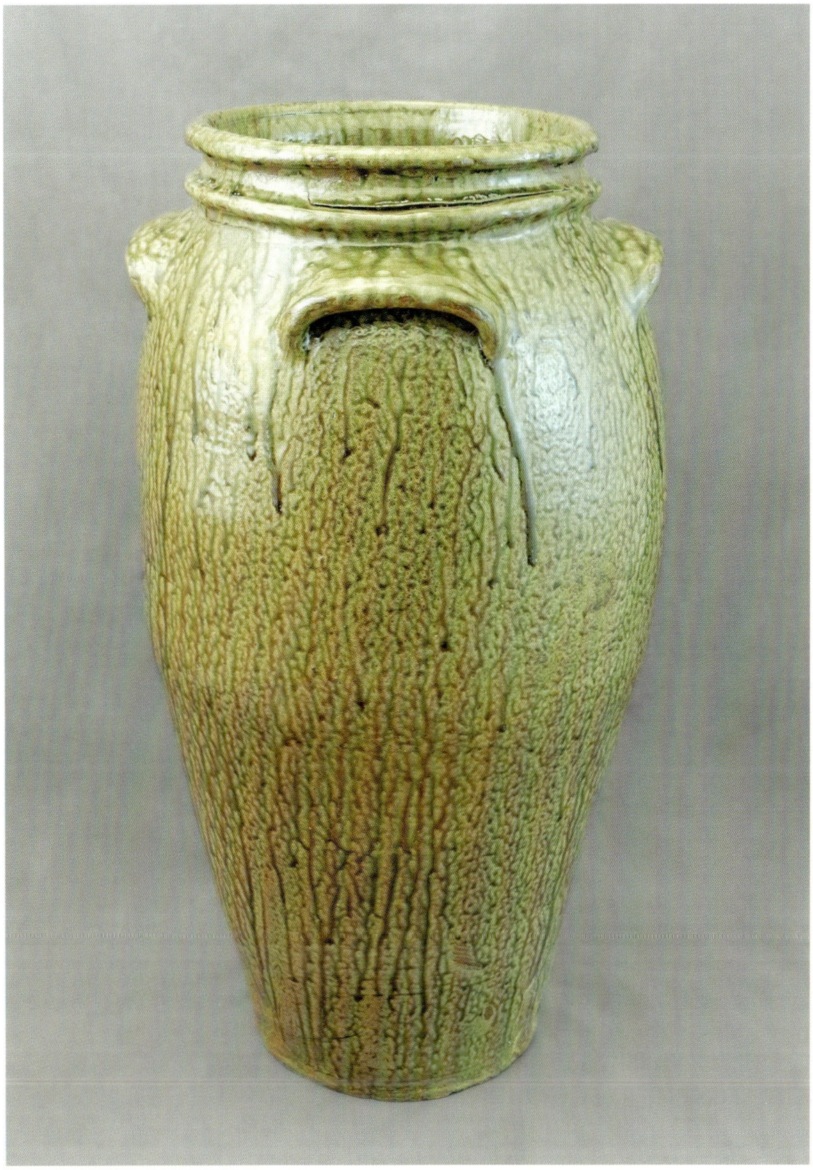

Figure 47 Storage jar, Trapp and Chandler Pottery, Edgefield District, South Carolina, 1845–1850. Alkaline-glazed stoneware. H. 29". (Courtesy, Sam Phillips; photo, Jan Todd.) This massive jar with an agglutinated glaze and four handles has an unusual double-rolled rim, apparently adopted by Chandler from his time working with Remmey in Baltimore.

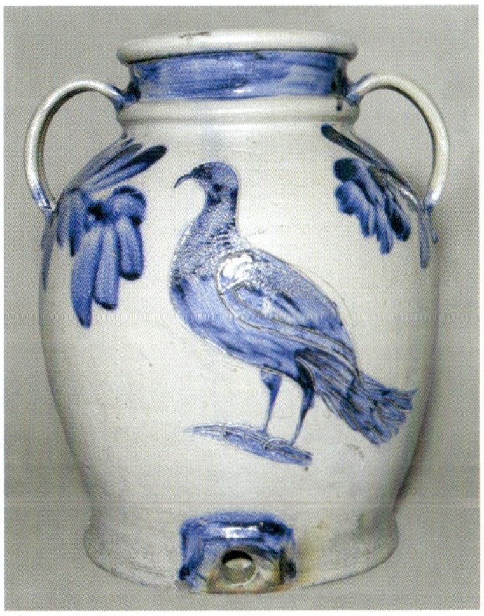

Figure 48 Water cooler, Baltimore Manufactory, Baltimore, Maryland, 1821–1829. Salt-glazed stoneware. H. 15". Impressed mark: "H. Myers" (Courtesy, Crocker Farm Auction.) Attributed to Henry Remmey while he was working for Henry Myers. The unusual double-rolled rim was adapted by Chandler and used by him sporadically in the Edgefield District.

Remmey also used a distinctive double-rolled rim on some of his pots, a technique Chandler used circa 1845 on a huge four-handle, approximately 20-gallon jar he made at one of the Martintown Road potteries (fig. 47). Chandler must have been exposed to this rim form while he was working in Baltimore in the 1820s (fig. 48).

Other Baltimore potters had obvious influences on Chandler. His pan form with two slab handles and a squared pouring spout is a nearly exact copy of those made by Morgan and Amoss in Baltimore (see figs. 34, 35). The cobalt decoration Morgan and Amoss applied to their pans (a series of horizontal sprays or sprigs punctuated by a daisy-rosette pattern) was used by Chandler on a different form, usually four-handle storage jars and large water coolers. However, Chandler chose to decorate his pans primarily with either iron slip–brushed swags or loops, or kaolin slip trailed in the same manner, contrasting boldly against a blue-green celadon glaze. Mor-

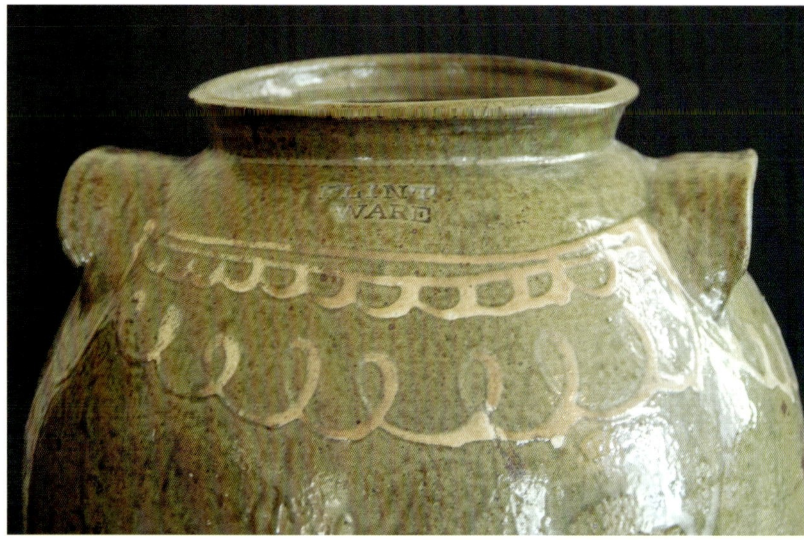

Figure 49 Detail of a storage jar, Thomas Chandler Pottery, Edgefield District, South Carolina, 1850–1852. Alkaline-glazed stoneware. Stamped on the shoulder: "FLINT WARE" (Courtesy, L. C. Lynch; photo, Jan Todd.) Stamped "flint ware" examples are some of the highest-fired stoneware made by Chandler.

gan and Amoss sometimes used a zigzag trailed cobalt line on their ware; this is also seen on Chandler-made Phoenix Factory ware (see figs. 24, 34) and Trapp and Chandler examples, but using kaolin slip or iron slip instead of cobalt.

Anywhere Chandler worked in Edgefield, he produced stoneware that rivaled any produced in America during the nineteenth century (fig. 49). His final great experiment, Flint Ware, though not a commercial success proved he was working toward a high-fired stoneware product that could withstand high- and low-temperature extremes consistent with stovetop use. The ware produced at the Trapp and Chandler site was occasionally signed and, less frequently, marked "Chandler Maker" and sometimes "War[r]anted," but it could also be identified by its bold iron-slip loop-and-swag decoration and occasional flower (fig. 50).

In 1850 Chandler established his own factory about a mile south, closer to Edgefield.[58] There he typically used kaolin slip-trailed decoration instead

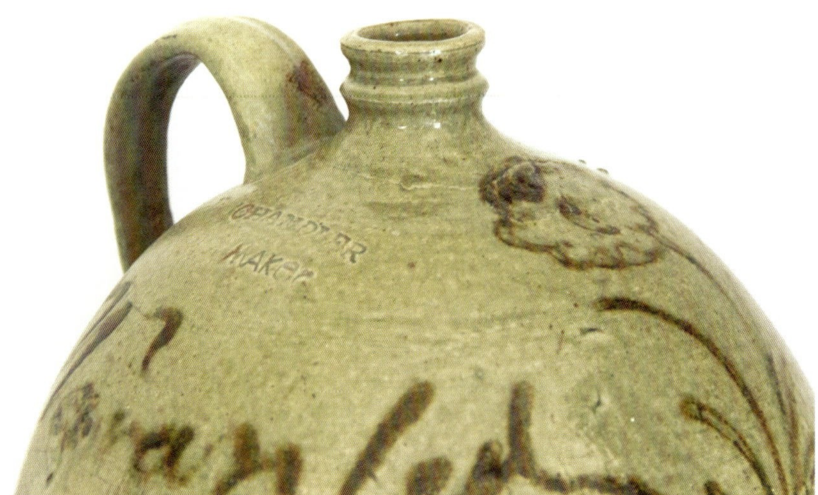

Figure 50 Jug, Thomas Chandler Pottery, Edgefield District, South Carolina, 1850. Alkaline-glazed stoneware. H. 19". Inscribed in dark slip on shoulder: "Waranted"; "1850"; and "5" within a cartouche; stamped: "CHANDLER MAKer" (Courtesy, L. C. Lynch; photo, Jan Todd.) This five-gallon spirits jug is decorated with the Chandler daisy flower head.

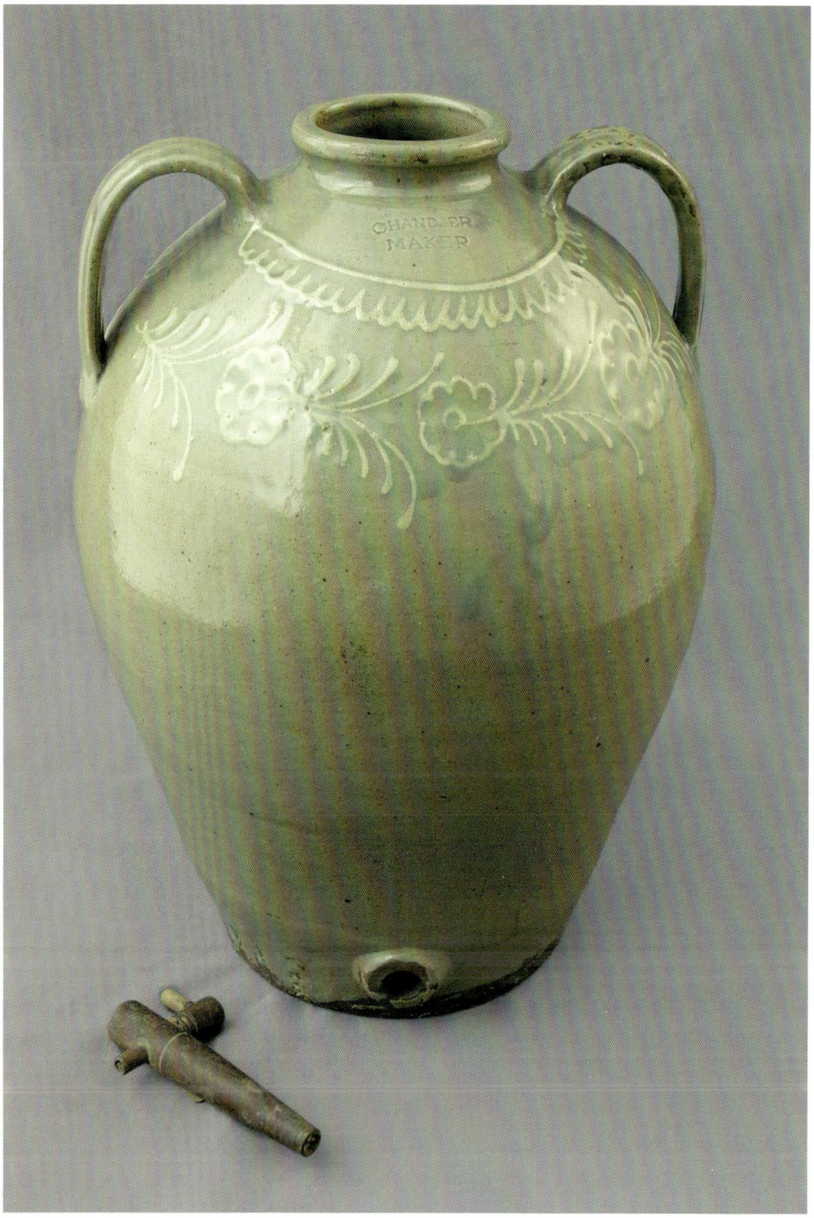

Figure 51 Water cooler, Thomas Chandler Pottery, Edgefield District, South Carolina, 1852. Alkaline-glazed stoneware. H. 20". Stamped: "CHANDLER MAKER" (Courtesy, Steve and Terry Ferrell; photo, Jan Todd.) This approximately fifteen-gallon vessel displays a delicate slip-trailed kaolin decoration as well as a classic celadon glaze. The sprig-and-flower-head decorations on this piece are nearly identical to those on the jar illustrated in fig. 44 right, which were executed some thirteen years earlier.

of iron slip but in a similar format, and he signed much of his work "Chandler Maker," frequently adding the guarantee "War[r]anted." He made large, classic water coolers, perfect in form and glaze, and exquisite four-handled storage jars of ten–fifteen gallons decorated in dark iron slip or white kaolin slip loops, swags, and tassels, continuing a decorative technique used at Phoenix Factory but also developing his own decorative signature style—two to three lines of loops below a series of draped lines (fig. 51).

Face Vessels

One of the most distinctive forms Thomas Chandler introduced to the South was the face vessel (figs. 52, 53). I have identified at least nine that I believe have a strong connection to Chandler. The only known signed

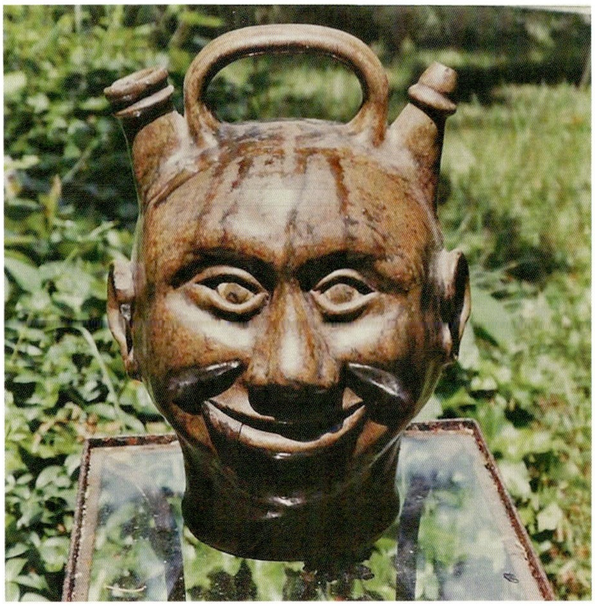

Figure 52 Harvest face vessel, Thomas Chandler, Edgefield District, South Carolina, ca. 1845. Alkaline-glazed stoneware. H. 11". Stamped on reverse, near center top beneath handle: "CHANDLER / MAKER" (Private collection; photo, Philip Wingard.) The bell handle, nipple spout, and double-collar spout, not to mention the facial expression, are distinctive. I believe this was Chandler's personal drinking jug, and remained in his possession until his death.

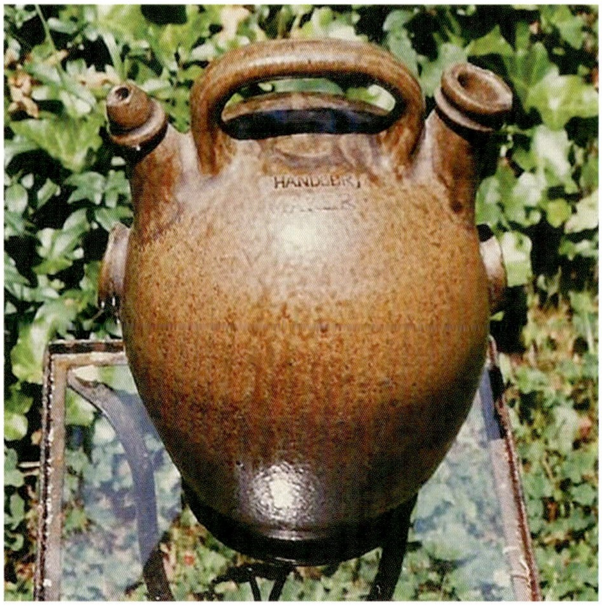

Figure 53 Reverse side of the harvest face vessel illustrated in fig. 52. (Photo, Philip Wingard.)

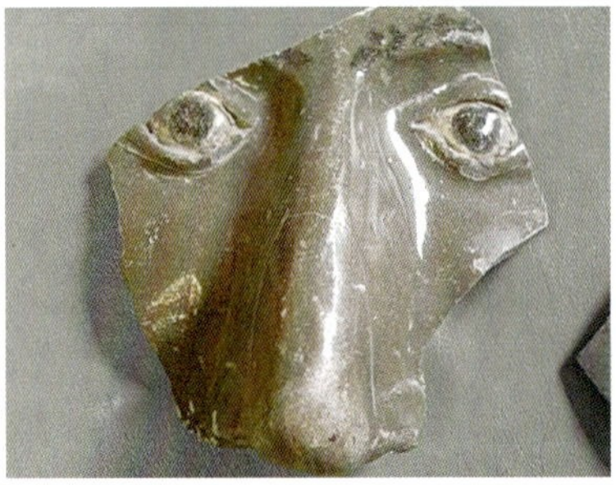

Figure 54 Face-vessel fragment, Phoenix Factory, Edgefield District, South Carolina, ca. 1840. Alkaline-glazed stoneware. (Courtesy, Steve and Terry Ferrell; photo, Dr. John Hoar.) The true-to-life eyes are like those executed on other face vessels of 1838–1852 by Thomas Chandler. In the Edgefield District, no fragments or surviving face vessels from a period prior to 1838 are known to exist. Chandler brought the idea of the face vessel with him, having learned about the form while in Baltimore, most likely from Henry Remmey.

Chandler face jug, made in Edgefield, suggests that Chandler was the first potter in the South and one of the first in America to make a stoneware face vessel. At least four other Chandler-related stoneware examples from Edgefield survive, two from Phoenix Factory, one a fragment found at the site by Stephen Ferrell, and a nearly complete example found in the marsh of the Charleston, South Carolina, harbor. Two face vessels attributed to the Trapp and Chandler Pottery also exist (figs. 54–57).

I believe Chandler learned how to make face vessels while he was in Baltimore. At least three earthenware face vessels exist that I attribute to this Chandler school, and I believe they were made in Baltimore during the 1820s (figs. 58, 59). These vessels are in the shape of what I call the Southern harvest-jug form, with a pouring or drinking spout and a nipple spout for sipping or venting, and with a bell handle placed across the top of the vessel. This form matches the known Chandler-signed Edgefield example, and was also used on several other Edgefield face jugs in this group, which strengthens the connection. The redware examples share the following characteristics with the signed stoneware Chandler face vessel: kaolin inset eyes with colored pupils, kaolin inset shaped teeth, large and distinctive lips and noses, and sunken eyes, to varying degrees, with Toby sculpted cheeks. The three earthenware examples and the stoneware example are near-exact copies, sharing a highly developed expressive face and the distinctive harvest-jug form.

Figure 55 Face vessel, attributed to the Phoenix Factory, Edgefield District, South Carolina, ca. 1840. Alkaline-glazed stoneware. H. 9¼". (Courtesy, Dr. and Mrs. John Hoar; photo, Dr. John Hoar.) This face vessel was recovered from the marsh of the Charleston harbor in the 1920s. Based on the archaeology from site 38GN169, where Chandler worked prior to Shaw's Creek and Phoenix Factory, dipping large sections of vessels in a contrasting iron wash was being done at this Martintown Road site. This decorative technique was first developed here by Chandler and others prior to their arrival at Shaw's Creek.

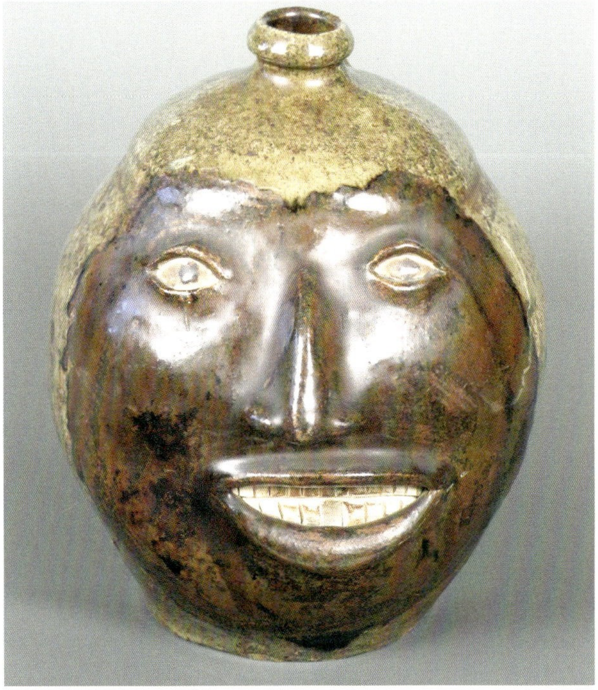

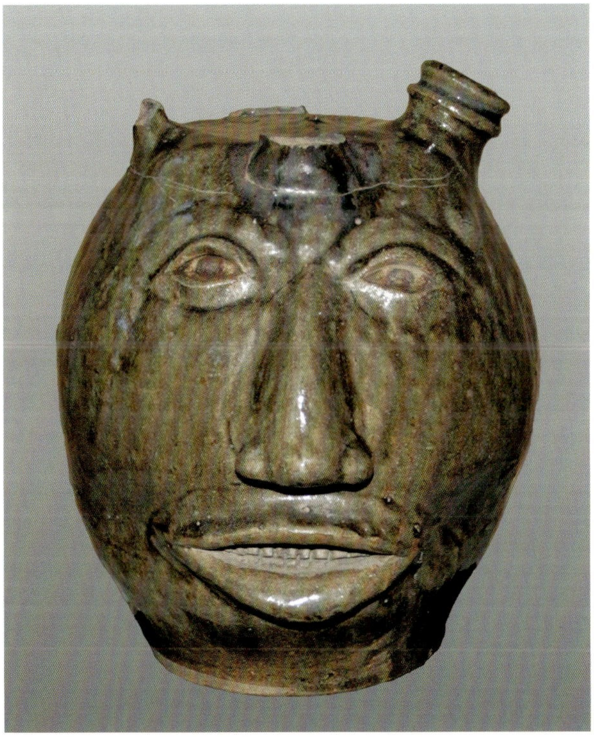

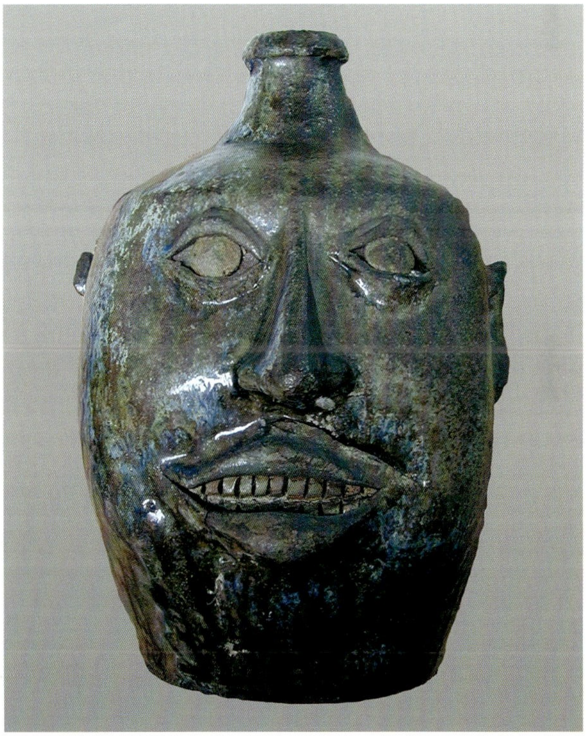

Figure 56 Face harvest vessel, attributed to the Trapp and Chandler Pottery, Edgefield District, South Carolina, 1845–1848. Alkaline-glazed stoneware. H. 8½". (Private collection; photo, Carl Steen.) The nose, mouth, and eyes are very similar to those on the signed Chandler example illustrated in fig. 52. Also shared are the small nipple spout and double-collar spout. There is applied iron wash at the base of the missing bail handle.

Figure 57 Face jug, Trapp and Chandler Pottery or possibly Mecklenburg County, North Carolina, attributed to Thomas Chandler, 1848–1854. Alkaline-glazed stoneware. H. 10¼". (Courtesy, David Good; photo, Carole Wahler.) Traces of blue rutile, a characteristic of alkaline-glazed North Carolina stoneware, and the darker clay body and glaze indicate that this jug might have been made in Mecklenburg County, North Carolina, during Chandler's short tenure there.

A ninth face vessel in the harvest-jug form is associated with this group, a jug that is signed with a "J. H. Stone" stamp (fig. 60). However, it has very different facial features. J. H. Stone was a potter in Buncombe County, North Carolina, from about 1867 until the 1890s. His father, Edward Stone, worked in Edgefield for a time at one of the Martintown Road potteries in the 1840s, possibly associated with Thomas Chandler.[59] Although the J. H. Stone harvest face jug likely was made in the mid- to late 1860s, the similarity of the facial features to those found on several Mid-Atlantic face vessels leads me to conclude that Chandler passed on his knowledge of design and artistry to Edward Stone before the mid-1840s, when Stone left Edgefield for Buncombe County, North Carolina. Edward later passed on this knowledge to his son J. H. Stone.

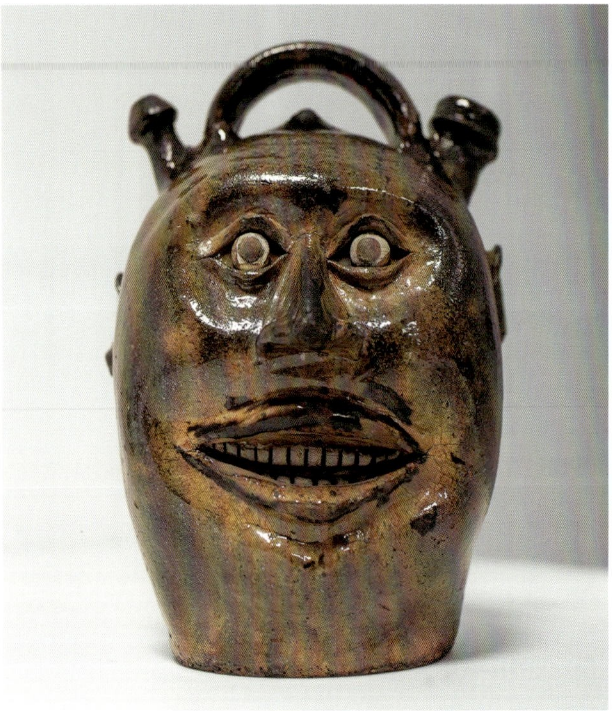

Figure 58 Face harvest vessel, Baltimore, Maryland, 1825–1829, Chandler school. Lead-glazed earthenware. H. 10⅛". (Courtesy, Colonial Williamsburg, Abby Aldrich Folk Art Museum.) There are at least three known lead-glazed earthenware face harvest vessels made by the same potter. That these vessels have facial features very closely related to the known signed Chandler example, as well as the four unsigned stoneware examples attributed to Chandler, suggests that they were likely made in Baltimore and that Chandler either made them or was trained by the potter who did.

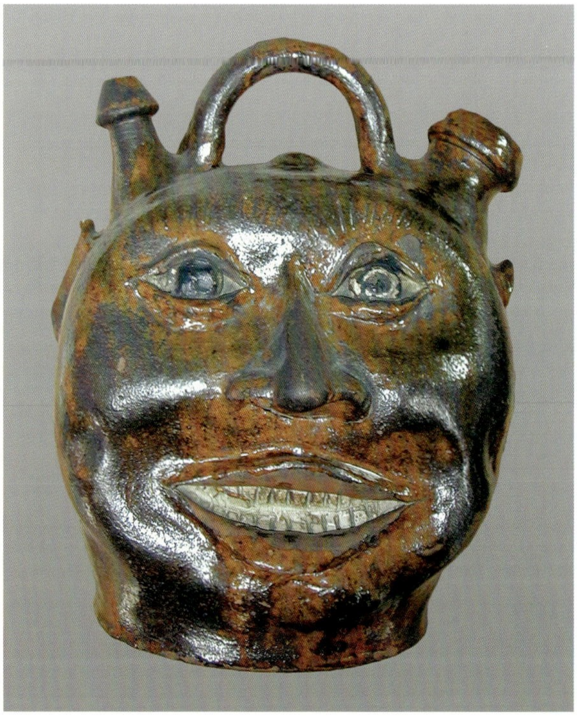

Figure 59 Face harvest vessel, Chandler school, Baltimore, Maryland, 1825–1829. Lead-glazed earthenware. H. 9". (Courtesy, David Good; photo, Carole Wahler.) This example has a strong provenance to Baltimore.

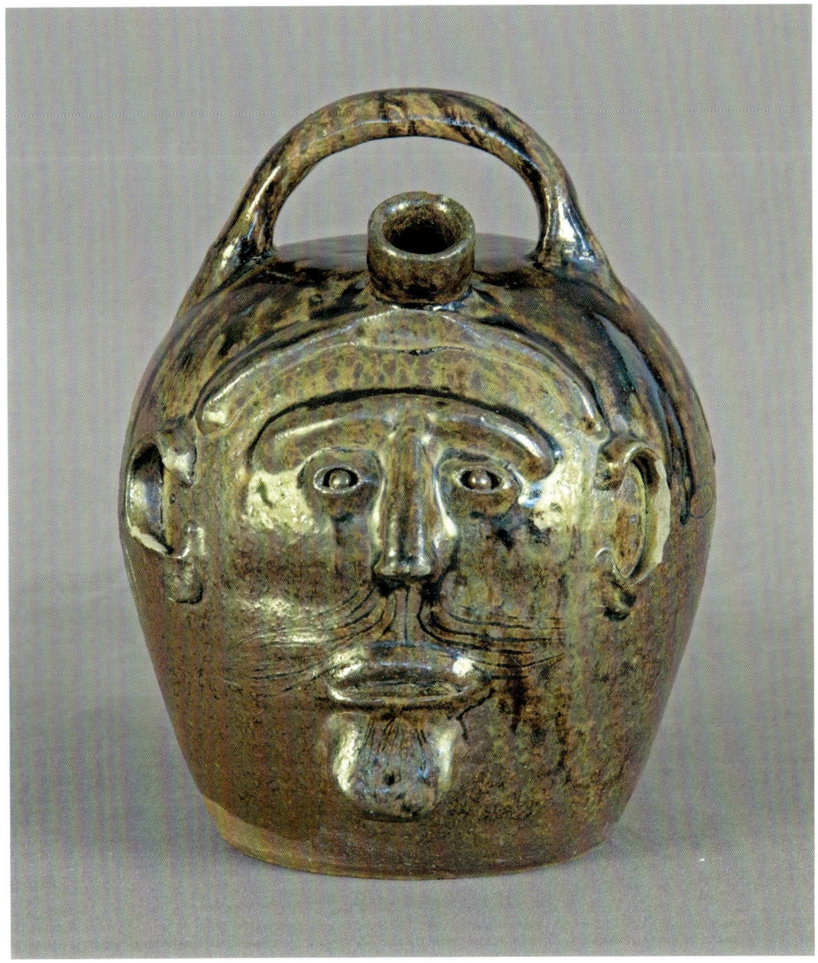

Figure 60 Face harvest vessel, Edward Stone Pottery, Buncombe County, North Carolina, 1860–1870. Alkaline-glazed stoneware. H. 9". Stamped: "J. H. Stone" (Courtesy, Donnie Garrett; photo, Jan Todd.) James Henry Stone, son of Edward Stone, was born in Buncombe County in 1847.

The signed "Chandler Maker" face harvest vessel (see figs. 52, 53) descended to James Hunter from his great-great-grandfather Isaac Faires.[60] According to Hunter, Isaac acquired the vessel from a neighbor who had gotten into trouble with the law. The jug was payment for resolving the issue. Researching the 1860 and 1870 censuses for North Carolina and South Carolina, I found that during the period 1855–1870 Isaac Faires was living very near Chandler's widow, Margaret, and her children in Fort Mill Township, South Carolina.

Concluding Thoughts

After thirty years of studying traditional Southern stoneware manufactured during the nineteenth century, I have come to the following conclusions: Chandler's stoneware is unmistakable, tracing it from his circa 1837 days working with John Isaac Durham at the pre–Trapp and Chandler Martintown Road site (38GN169), to his work and development of slip decoration at Shaw's Creek in the late 1830s, his work at a new site on Marintown Road near the Trapp Chandler site from 1844–1848, his work while in a partnership with John Trapp in the late 1840s, and finally his operation a mile south of the Trapp and Chandler Pottery (38GN343) in 1850–1852. His glazes range from a red iron slip, a honey-colored glossy, an agglutinated or slick gray-blue, a thin translucent, and true celadon, and his decoration is bold and well executed. When comparing his ware today with surviving stoneware made by other potters from the period, Chandler's stoneware is better made, fired to a higher temperature, and coated with a superior glaze. His firing temperature matched the vitrification point of his clay and glaze to perfection.

Chandler's attainment of perfection in stoneware manufacture is evident in the large volume of his work that survives. The greatest testament to a nineteenth-century potter is the utility of his ware and how it provided for an easier way of life. Thomas Chandler produced wares of lasting utility and great aesthetic beauty that represent the highest degree of stoneware art of his time—and perhaps ours.

ACKNOWLEDGMENTS This project could not have come to fruition were it not for the contributions of living and deceased Edgefield stoneware collectors, ceramics researchers, ceramics historians, and archaeologists. Howard Smith, Fred Holcombe, Joe Holcombe, Georgiana Greer, and Gerald K. Landreth all made significant discoveries regarding Chandler's life. Cinda K. Baldwin, whose book *Great and Noble Jar* started many of us down the research road of Southern stoneware. Jane Przybysz, executive director of McKissick Museum in Columbia, South Carolina, along with Jill Koverman, chief curator of collections, gave invaluable advice and encouragement. Steve and Terry Ferrell allowed access to their tremendous Ferrell Pottery Museum, including the archaeology of several of the sites where Chandler potted. Prominent collectors John Hoar and Bill Cox listened to me almost daily as I sorted through data to reach my conclusions. My wife, Debbie, critiqued my many revisions and continually reminded me that I was doing good work. Carl Steen helped with understanding the archaeology and shared anything he had pertaining to Chandler. The Reverend Jerry Robinson shared his collection with me in 1980, exposing me to my first example of alkaline-glazed stoneware. William Robert Jordan's work as an archaeologist in Washington County, Georgia, was very helpful in tracking Chandler. Both John E. Kille's work on the Baltimore stoneware industry and Kurt C. Russ and W. Sterling Schermerhorn's work on Richmond-area potteries were very helpful. Carole Wahler gave me productive leads to follow and shared important photos of face vessels. David Good shared his vast knowledge of Americana and allowed me to study his collection of Edgefield face vessels, which includes several important Chandler examples. My friend Jim Hunter shared his family connection to Chandler and was the inspiration for me to tackle this project. The folks at the Eastern Shore Public Library in Accomack were very helpful in unraveling the Chandler lineage. William Mathis, a direct descendant of Thomas Chandler, shared photos and genealogical data of great importance. Thanks to Gordon Lohr for sharing his monumental find, the signed Chandler Baltimore churn, and to Al Zimmerman, who gave me the tip that such a pot existed. Conversations with Charles Santore about American Windsor chairs, and with Jenny Barker about the Methodist movement in Accomack County, led me to new discoveries about Chandler. Tony Zipp of Crocker Farm Auctions, Gary Albert of MESDA, Bruce and Vicki Waasdorp of American Pottery Auctions, and Whitney Bounty at Freeman's Auctioneers and Appraisers shared important photography. For her great work behind the camera, I thank my friend Jan Todd, who followed me to a half-dozen collections with camera, backdrop, and accessories in tow. Huge thanks to Wade Fairey, Charles Comolli, Steve and Terry Ferrell, Donnie Garrett, David Good, Vivian and Gerald Gowitt, Doug and Brenda Howard, John Lafoy, L. C. Lynch, Sam Phillips, Jason Shull, Mary Smith, Donny Taylor, David Ward, Randy Windham, Jim Witkowski, and Jeremy Wooten for allowing me to photograph their Chandler stoneware collections, and to Bill Cox, John Hoar, Carl Steen, Carole Wahler, and Colonial Williamsburg for sharing photos with me. I am particularly grateful to the Chipstone Foundation for funding *Ceramics in America*, and to Rob Hunter for giving me the opportunity to publish something we both agreed was important. Whether I follow a path or blaze a trail, my friends help me find my way.

1. Records of the Adjutant General's Office, 1798–1914, s.v. "US Army Register of Enlistments"; National Archives Microfilm Publication M233, Washington, D.C. In the 1840 U.S. Census for the Edgefield District, Chandler is listed as a pensioner for service in the military.

2. Baltimore County Indentures, 1794–1799, book 1, folio 189, Eastern Shore Public Library, Accomac, Virginia.

3. Thomas Chandler Sr., "Evidence Explained: Citing the Wise File on Chandler Genealogy," Eastern Shore Public Library, Accomac, Va.

4. Gail Walczyk, *Accomack Indentures, 1798–1835* (Coram, N.Y.: Peter's Row, 2002), pp. 16–17.

5. In 1940 the Virginia General Assembly officially added a "k" to the end of the county's name to arrive at its current spelling of Accomack.

6. Nancy Goyne Evans (*American Windsor Chairs* [New York: Hudson Hills Press, 1996], p. 687) has Thomas Chandler in Baltimore in 1817. See also 1820 U.S. Census Records, Baltimore Ward 2, Baltimore, Md., p. 100; National Archives and Records Administration, Washington, D.C. (hereafter cited as NARA), roll M33, image 62; available at www.archives.gov.

7. Elizabeth Wise and Thomas Chandler married in Baltimore on July 18, 1802. Thomas Chandler was a chairmaker and grocer. *Baltimore City Directories* lists Elizabeth Chandler as a shopkeeper in 1829 and grocer in the early 1830s. The couple had eight children. The painting was handed down by descent from Elizabeth Wise Chandler to her daughter, Ann Elizabeth Chandler (Mrs. Andrew Jackson Bandel), to her eldest daughter Mary Elizabeth Bandel (Mrs. George Hakesley), to her daughter Grace Hakesley Bryn, to Elizabeth Bandel, to the present owner. See also Carolyn J. Weekley and Stiles Tuttle Colwill, eds., *Joshua Johnson: Freeman and Early American Portrait Painter*, exh. cat. (Baltimore: Abby Aldrich Rockefeller Folk Art Center, Maryland Historical Society, 1987), pp. 164–65, no. 82. The people of Accomack were very sympathetic to the plight of African-American slaves due to the Methodist Revivalism that occurred in that county beginning in 1800. See Kirk Mariner, *Revival's Children: A Religious History of Virginia's Eastern Shore* (Salisbury, Md.: Peninsula Press, 1979), pp. 34–52.

8. C. Keenan, *The Baltimore Directory for 1822 and '23* (Baltimore, Md.: Printed by R. J. Matchett, 1822–23).

9. Luke Zipp, "Henry Remmey & Son, Late of New York: A Rediscovery of a Master Potter's Lost Years," in *Ceramics in America*, edited by Robert Hunter (Hanover, N.H.: University Press of New England, 2004), pp. 143–61.

10. Luke Zipp, "Baltimore Stoneware," *Antiques & Fine Art* (Summer 2006): 154–58.

11. Gordon Lohr, collector, telephone interview by Philip Wingard, Fall 2005.

12. On June 26, 1829, Thomas Chandler Jr. paid $11.29 for a Baltimore City grocery license. His mother, Elizabeth, is listed as head of household in Baltimore in the 1830 U.S. Census.

13. George F. Chandler completed his apprenticeship when he reached the age of twenty-one. He is listed in the 1837 *Baltimore City Directory* as a potter on 21 Comet Street.

14. Zipp, "Henry Remmey & Son, Late of New York," p. 146.

15. A review of the 1831 *Baltimore City Directory* published by Matchett shows a reduction in the number of potteries in operation.

16. Records of the Adjutant General's Office, 1798–1914, "US Army Register of Enlistments."

17. *U.S. History: Pre-Columbian to the New Millennium*, "24c. The South Carolina Nullification Controversy," www.ushistory.org/us/24c.asp (accessed on June 15, 2012).

18. Ibid.

19. Returns from Regular Army Infantry Regiments, June 1821–December 1916, microfilm serial M665, roll 42, NARA.

20. Ibid.

21. William Robert Jordan, "Archaeological Investigations of Alkaline-Glazed Stoneware Production Sites in Northern Washington County, Georgia" (master's thesis, Georgia State University, 1996).

22. My Wingard ancestors in the 1820s used the Meriwether Trail to travel west from Lexington, South Carolina—via Augusta, Georgia—to settle in southeastern Alabama, near Fort Mitchell.

23. Jordan, "Archaeological Investigations," p. 38, fig. 4. I researched several deployments to Fort Mitchell from Augusta and found that the march took, on average, nineteen days.

24. Ibid., pp. 174–78, fig. 54.

25. Returns from Regular Army Infantry Regiments, June 1821–December 1916, microfilm serial M665, roll 42, NARA.

26. Court of Pleas and Quarto Sessions, Mecklenburg County, North Carolina, December 21, 1857, transcribed from microfilm, Carolina Room, Charlotte Mecklenburg Library, Charlotte, N.C.

27. A history of this artillery regiment, written by Lieut. W. A. Simpson, Adjutant 2nd U.S. Artillery, gives a concise account of the actions of the 2nd Regiment and all companies associated with it before and during the Second Seminole Indian campaign of 1835. This account is available online at www.history.army.mil/books/R&H/R&H-2Art.htm (accessed August 30, 2013).

28. Returns from Regular Army Infantry Regiments, June 1821–December 1916, microfilm serial M665, roll 42, NARA.

29. Court of Pleas and Quarto Sessions, Mecklenburg County, North Carolina, December 21, 1857.

30. U.S. Army Register of Enlistments, 1798–1914, record for Thomas Chandler, NARA, roll M233; see also the marriage notice found in McClendon's Edgefield Marriage Records, p. 32.

31. Cinda K. Baldwin, *Great and Noble Jar: Traditional Stoneware of South Carolina* (Athens: University of Georgia Press, 1993), pp. 32–57.

32. This conclusion is based on my comparison of the handwriting on the two Pottersville presentation examples made in 1836 with the handwriting on a Chandler-signed presentation jar to B. Harland in 1844.

33. Baldwin, *Great and Noble Jar*, p. 148.

34. See George J. Castille et al., *Archaeological Survey of Alkaline-Glazed Pottery Kiln Sites in Old Edgefield District, South Carolina* (Columbia: McKissick Museum, University of South Carolina, South Carolina Institute of Archaeology and Anthropology, 1988), pp. 15–17, 38; Gerald K. Landreth, "Archaeological Investigations at the Trapp and Chandler Pottery, Kirksey, South Carolina" (master's thesis, University of Idaho, 1986). See also Joe L. Holcombe and Fred E. Holcombe, "South Carolina Potters and Their Wares: The Landrums of Pottersville," *South Carolina Antiquities* 18, nos. 1–2 (1986): 47–62. I also studied the artifacts recovered by Steve and Terry Ferrell at the Trapp/Chandler and the Shaw's Creek sites, which included sherds from Phoenix Factory.

35. Holcombe and Holcombe, "South Carolina Potters and Their Wares." William and John Isaac Durham, Harvey and Rueben Drake, and Thomas and John Presley were all living near Abner Landrum in Edgefield in 1820 (1820 U.S. Census), and all were living very near him in 1830 at Pottersville (1830 U.S. Census). Rueben Drake, John Isaac Durham, and John Presley, along with Collin Rhodes, were there when Chandler arrived in 1836. I believe that later on Chandler, John Isaac Durham, Robert Mathis, and John Presley moved to the early Martintown Road pottery site. John Presley sold this pottery to John Isaac Durham in 1840. When Amos Landrum sold his interest in a stoneware factory on Shaw's Creek to Collin Rhodes in 1835, it set the stage for Chandler to move to Shaw's Creek by late 1838.

36. In the late 1860s, this pottery was operated by the Reverend Jesse P. Bodie and William Durham, son of John Isaac Durham and executor of Chandler's estate.

37. Landreth, "Archaeological Investigations at the Trapp and Chandler Pottery, Kirksey, South Carolina."

38. Robert Mathis was likely one of the first potters to work here. John Isaac Durham, Francis Devilin, and John Presley were here early as well. It is my conclusion that the roman-numeral base markings found at site 38GN169 indicate gallon capacity and an itinerant potter (Chandler), and that the letter markings near the shoulder were signatures of shop owners or operators. Fred Holcomb and I discussed this as well as his work at the site in 1981, which yielded fragments stamped with an "M," representing potter Robert Mathis, and with a "D," for John Isaac Durham. A partial "P" mark was found by South Carolina Institute of Archaeology and Anthropology during its survey of alkaline-glazed sites in the Edgefield District in 1988. The mark was found at the Shaw's Creek/Phoenix Factory site (39AK495) in Test Unit 2 "P" and "M" and level 10 3A and 3J.

39. I concluded that these were Chandler's markings, made circa 1838, when I found a jar with them as well as an oval "C," which was also found on a potsherd with a partial signature.

40. Joseph Hart Sr. genealogical files, McCelvey Center, York, S.C.; York County Probate Records, case 4, files 170 and 171, and case 75, file 3706. See also the data for the two townships in the 1860 and 1870 U.S. Census for North Carolina and South Carolina.

41. Notes of Gerald K. Landreth on file at the McKissick Museum research room, Columbia, South Carolina. I found a hand-drawn map by Landreth showing the Thomas Chandler Pottery and its proximity to the Partlow home site.

42. Joseph Hart Sr. genealogical files, McCelvey Center, York, South Carolina; see also York County Probate Records, case 4, files 170 and 171, and case 75, file 3706.

43. Edgefield County Archives, GGG 383.

44. Edgefield County Archives: s.v. "Coroner's Book of Inquisitions," 1851–1859, Edgefield County Judge of Probate, Edgefield, WPA typescript, 1937, no. 165-33-7172. I interpret "and the Chandler house and old pottery" to mean site 38GN169, the Trapp/Chandler pottery. This information establishes that in September 1850 Chandler was living near the old pottery, known as the Trapp and Chandler Pottery, on Martintown Road, although he was also working at a new site, Chandler Site II (38GN343) a mile down the road toward Edgefield.

45. Edgefield County Archives, "Coroner's Book of Inquisitions," 1851–59.

46. Baldwin, *Great and Noble Jar*, p. 54.

47. Estate of Thomas M. Chandler, Mecklenburg County, N.C. (1854), Record of 1762–1957, Mecklenburg County Estate, volume Cathy Cherry, Microfilm Collection, Robinson-Spangler Carolina Room, Charlotte Mecklenburg County Library, Charlotte, N.C.

48. Kirk Mariner, *Revival's Children: A Religious History of Virginia's Eastern Shore* (Salisbury, Md.: Peninsula Press, 1979), pp. 34–93.

49. Estate of Thomas M. Chandler, Mecklenburg County, N.C. (1854). See also Michael K. Brown, *The Wilson Potters: An African-American Enterprise in 19th-Century Texas*, exh. cat. (Houston: Bayou Bend Collection and Gardens, Museum of Fine Arts, 2002), pp. 11–26.

50. Brown, *The Wilson Potters*, p. 11.

51. Estate of Thomas M. Chandler, Mecklenburg County, N.C. (1854).

52. Ibid.

53. Ibid. On the second page of the estate-sale account given by William Durham, co-administrator of the estate, Durham references a Chandler account book with monies not settled. The first entry is for February 5, 1853, and runs through March 1854. All of the customers who owed Chandler money for pottery purchases were neighbors in the Steele Creek and Fort Mill Township areas, and many of them attended his estate sale. This establishes the likelihood that Thomas Chandler was in Mecklenburg County by February 1853.

54. W. Oakley Raymond, "Remmey Family: American Potters, Parts I and II," in *The Art of the Potter*, edited by Diana Stradling and J. Garrison Stradling (New York: Main Street/Universe Books, 1977), p. 114.

55. Zipp, "Henry Remmey & Son, Late of New York," p. 146.

56. John E. Kille, "Distinguishing Marks and Flowering Designs: Baltimore's Utilitarian Stoneware Industry," in *Ceramics in America*, edited by Robert Hunter (Hanover, N.H.: University Press of New England for the Chipstone Foundation, 2005), p. 100, fig. 13.

57. Baldwin, *Great and Noble Jar*, p. 149.

58. There is the possibility that these two sites operated simultaneously and that Chandler managed both. A coroner's inquest in September 1850 indicates he was living near 38GN169 in late 1850, but I suspect his shop a mile down the road (38GN343) was already in operation because of his advertisement in the *Edgefield Advertiser* on May 15, 1850.

59. Charles G. Zug III, *Turners and Burners: The Folk Potters of North Carolina* (Chapel Hill: University of North Carolina Press, 1986), pp. 96–98.

60. James Hunter, conversation with the author in March 1992. I met Mr. Hunter at an antiques show at Brattonsville, South Carolina, in the fall of 1991. He told me about the Chandler face vessel he had inherited, and I spent an afternoon with him the following spring, during which I took photos of the vessel and he shared with me its history along with his family genealogy.

Barbara H. Magid

"Stone-ware of excellent quality, Alexandria manufacture" Part II: The Pottery of B. C. Milburn

▼ STONEWARE from Alexandria, Virginia, is indeed of excellent quality, as Alexandria merchant Robert H. Miller advertised in 1841.[1] Alexandria's best-known and longest-operating stoneware manufactory was located on the 600 block of Wilkes Street circa 1810–1876. While the business was referred to in some contemporary advertisements as the Alexandria Pottery or the Milburn Pottery, it is now commonly known as the Wilkes Street Pottery.[2]

In Part I of this study, published in the 2012 volume of *Ceramics in America*, I reviewed the period in which Milburn's predecessor, John Swann, worked at Wilkes Street, circa 1810–1831.[3] In Part II, I focus on the period in which Benedict C. Milburn and his sons operated the manufactory, circa 1831–1876. I have examined approximately 150 Alexandria stoneware vessels of this period in museums and private collections, as well as extensive archaeological materials.

The earliest Milburn stoneware was stamped "H.C.SMITH / ALEX^{A} / D.C" for the china merchant from whom he rented the pottery. After Milburn purchased the pottery in 1841, just three months before Miller's advertisement for his stoneware appeared in the *Alexandria Gazette*, much of his stoneware was stamped with his own name.[4] During the earlier Swann period, stoneware was stamped "J. SWANN / ALEX^{A}", "HUGH SMITH & CO.", and "H.SMITH & C^{O}"

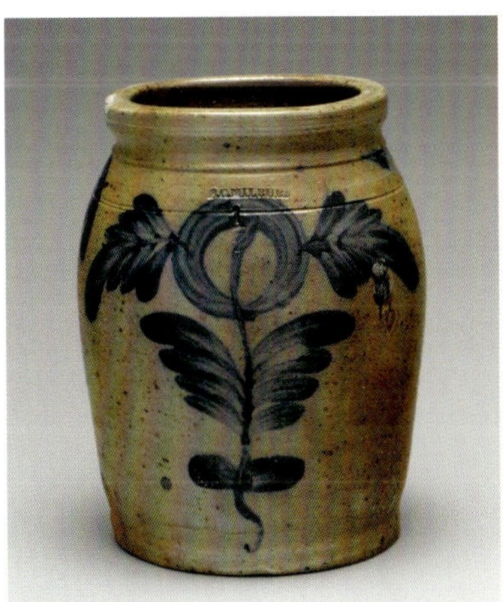

Figure 1 Jar, Alexandria, Virginia, 1847–1861. Salt-glazed stoneware, brushed cobalt. H. 7½". Capacity: ½ gallon. Impressed mark: B.C.MILBURN. (Courtesy, Al Marzorini; photos by Gavin Ashworth unless otherwise noted.) The central round flower is known as the Alexandria motif.

Many of Milburn's stoneware vessels still survive, decorated with elaborate brushed-cobalt or slip-trailed decoration (fig. 1). According to Mary Powell's 1928 history of Alexandria,

> Here all kinds of vessels were made of gray stoneware ornamented with blue leaves and flowers,—jugs, milk crocks, jars, pitchers and even little money jugs for children. Many are still owned and treasured by old families of the town. That kind of ware apparently is not now manufactured in this country, but a returned soldier from the great World War in France said it was frequently seen there, and made him feel almost homesick when he recalled the old potteries of his childhood at Alexandria.[5]

Some Alexandria stoneware is still owned by families in Alexandria and surrounding counties whose ancestors bought and used it in the nineteenth century. One family owned a house on Prince Street from 1817 until 2000, and pieces of Milburn's stoneware remained in the outbuildings. Alexandria stoneware is now highly sought after by collectors and can also be found in a number of museums, including the National Museum of American History, Smithsonian Institution; the Museum of Early Southern

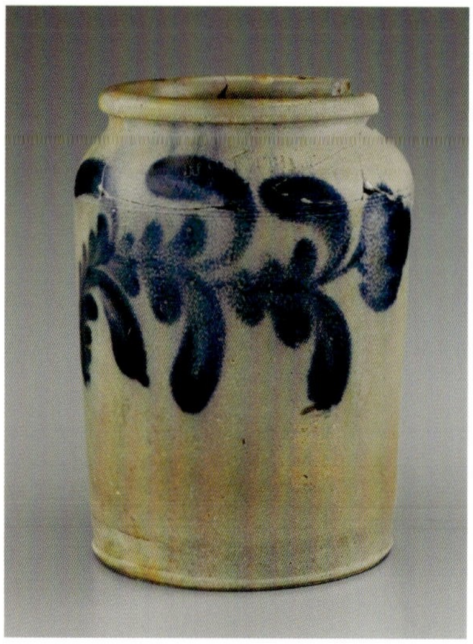

Figure 2 Jar, Alexandria, Virginia, 1847–1851. Salt-glazed stoneware, brushed cobalt. H. 9½". Capacity: 1 gallon. Impressed mark: H.C.SMITH / ALEX^A^ (Courtesy, Alexandria Archaeology Museum.) Encircling foliage is another common motif on Alexandria stoneware. Brushed-cobalt decoration was common until the Civil War. This piece was found in archaeological excavations on a residential property at 524 King Street.

Figure 3 Pitcher, B.C. Milburn, Alexandria, Virginia, 1847–1861. Salt-glazed stoneware, slip-trailed cobalt. H. 11". Capacity: 2 gallons. Impressed mark: B.C.MILBURN / ALEX^A^ (Courtesy, Alexandria Archaeology Museum.) Slip-trailed cobalt decoration is typical of the later Milburn period. The pitcher was found in a privy at 112 South Royal Street, in the midst of Alexandria's commercial district.

Decorative Arts; the DAR Museum; the Brooklyn Museum; and The Lyceum: Alexandria's History Museum.

Vessels marked "H.C.SMITH / ALEX^A^ / D.C" (fig. 2), made circa 1847–1851, are common among Alexandria stoneware, although the most commonly collected bears the mark "B.C.MILBURN / ALEX^A^" and was made circa 1847 to 1867 (fig. 3). Other Milburn marks are seen less often, and rare pieces include those made at Wilkes Street with the marks of Alexandria merchants Dubois & Reddick and James P. Smith. While most of the extant pieces of Alexandria stoneware were made at Wilkes Street, a few marked examples made at the potteries of James Black, Tildon Easton, and James Miller also survive.[6]

Much of our knowledge of Alexandria stoneware comes from archaeological work conducted in the city over the past forty years. In 1977 a brief rescue excavation at the Wilkes Street Pottery yielded about 14,500 stoneware sherds, 1,500 of which had cobalt decoration. Fifteen stamped sherds from the Milburn period were among the finds.[7] Milburn's stoneware was so widely used that it has been recognized at all of the Alexandria archaeological sites relating to his time period, and the distinctive slip-trailed decoration he employed has also been noted at archaeological sites in Fairfax County, Virginia, in Washington, D.C., and elsewhere in the region (see fig. 3).

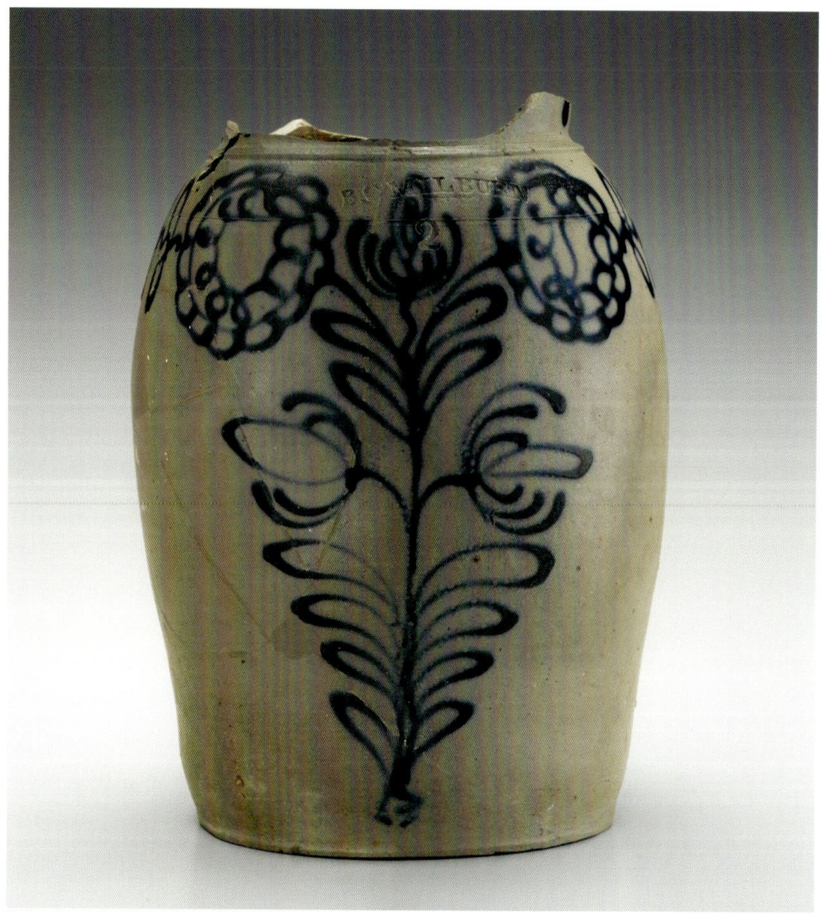

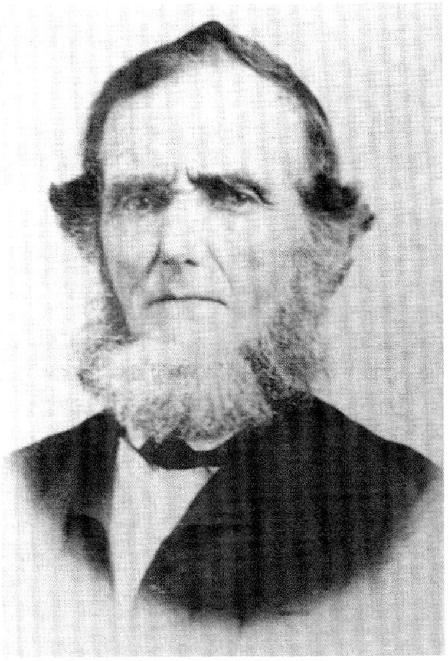

Figure 4 Benedict C. Milburn (1805–1867). (Courtesy, Smithsonian Institution, National Museum of American History.)

B.C. Milburn and the Wilkes Street Pottery

Benedict Cuthbert Milburn came to Alexandria from St. Mary's County, Maryland, in 1822 (fig. 4). During his first years in Alexandria, he may have worked with earlier Wilkes Street potter John Swann. The Swann and Milburn families both came from St. Mary's County and had close ties there. By 1833 Milburn was running the stoneware manufactory, and purchased it from china merchant Hugh Charles Smith in 1841. After Milburn's death in 1867, his sons continued to operate the pottery until 1876.

According to family records, the seventeen-year-old Milburn came to Alexandria to apprentice himself with a potter.[8] Milburn probably apprenticed at Wilkes Street with John Swann, but in 1822 there was at least one other pottery in town: Evans and Griggs, successors to Lewis Plum at Wolfe and South Columbus streets. Milburn could also have worked with potter James Black before Black's first pottery failed in 1832.[9] According to the *Alexandria Gazette*, Milburn and Black were partners in a grocery business in 1828, close to the time Milburn would have finished an apprenticeship. Milburn purchased Black's interest in the grocery in June of that year.[10] We do not know if he worked as a potter and grocer at the same time, nor do we know if his grocery business developed into the "family grocery store" advertised in 1865 by one of his sons, J. Clinton Milburn.[11]

Milburn was probably working for H. C. Smith at Wilkes Street at least by 1831, when he rented Swann's old house. He was certainly there by 1833 when his name first appeared on tax rolls for the pottery. In an 1869 advertisement by his son, "Milburn's Pottery" is said to be "Established 1833."[12] B. C. Milburn was most likely managing the pottery from that date, either as an employee of H. C. Smith or renting the premises from him.

In 1839 the editor of the *Alexandria Gazette* wrote:

> As soon as our Canal is completed we shall be abundantly supplied with water for manufacturing purposes; nor, in truth, do we see the reason, why this should not become, at once, a MANUFACTURING TOWN, having the means already of a flourishing commerce to rid us in the transportation of goods, and Ship Yards where the finest merchant vessels are yearly turned off the stocks. . . . Let us, then, by all means, commence early with our MANUFACTORIES.[13]

It was in this hopeful economic climate that Milburn purchased the pottery in 1841. He appears to have been an immediate success, with his beautifully formed and decorated stoneware.

Milburn faced only brief competition from other Alexandria potters. Milburn's one-time partner in the grocery business, James Black, worked at Wilkes Street under Milburn in 1834. Before then, Black operated his own stoneware manufactory in Alexandria, which ended in failure. He was confined briefly in a debtors' prison before declaring bankruptcy in 1832 and, according to the *Alexandria Gazette*, endured a court-ordered sale of "burned ware, clay, wheelbarrow etc."[14] In 1836 Black was again working on his own as a potter, at the northwest corner of Wolfe and Patrick streets. This site has not been excavated, but a few extant pots are known, with the stamps "J. A. BLACK / ALEX$^{\underline{A}}$ / D. C." and "J. A. BLACK / ALEX".

Tildon Easton opened a manufactory on the 1400 block of King Street in 1841, the same year that Milburn purchased the Wilkes Street Pottery. Only one extant piece is known—marked "TILDON EASTON"—but fragments of more than 650 vessels were found in the archaeological excavation of his kiln site. Easton produced salt-glazed stoneware and earthenware for a short period of time, before declaring bankruptcy in 1843.[15] Perhaps Black and Easton were unable to compete with the well-established Wilkes Street Pottery, which had been operating since 1810.

Additional competition known from the 1840 Census of Manufactures included two potteries in neighboring Washington, D.C., and nine in nearby Baltimore. No stoneware sherds with Washington or Baltimore marks have been found on Alexandria's archaeological sites, though unmarked pieces would be difficult to distinguish from Alexandria wares. A few early-nineteenth-century unmarked vessels from Baltimore have been identified in the archaeological record. Many potters worked in Virginia's Shenandoah Valley in the second half of the nineteenth century, and some of their wares were used in Alexandria and are recognized on archaeological sites in the city.

A few of Milburn's employees are known from tax records. The tax assessor recorded names of workers who were employed on the day of his yearly visits, but others were most likely employed at other times. B.C. Milburn himself was first listed in 1833. Michael Morris was first listed in 1828 and in 1833 was listed alongside Milburn. Kitty Marshall, the first known female employee, was listed in 1832 and again in 1842. Two additional women worked with Milburn, perhaps as decorators: Milly Jackson (1834–1838) and Sylvia Rogers (1834–1847). John Simms was listed on the tax rolls for 1839, and Alfred Merrick in 1840 and 1841. After Milburn purchased the pottery in 1841, a few additional names appear: Chas. Richardson (1842), Wm. Nickens (listed as Colored, 1842–1848), Milburn's twelve-year-old son, Stephen (1845), Thos. Grant (1849), Dennis Hasket (also listed as Colored, 1847 and 1850), and John Payne and James Shemick (both in 1850).[16]

Milburn's business continued to expand during the late 1840s and 1850s. The Alexandria Canal opened at the end of 1843, connecting the Alexandria waterfront with Georgetown and the C&O Canal, which by 1850 reached all the way to Cumberland, Maryland. The canal opened trade to the west, enabling better transport of both raw materials and manufactured goods. The period following Alexandria's retrocession to Virginia in 1847 brought economic growth. The city fought long and hard for retrocession, as the town's tenure as a part of the federal district had stifled economic growth. In particular, banking restrictions imposed by Congress on the District of Columbia had affected the town's economy. Following retrocession, the town ranked ninth in the U.S. in overall trade, following New York, New Orleans, Philadelphia, Baltimore, Boston, Richmond, Petersburg, and Cleveland.[17] Milburn and Hugh Smith were among those Alexandrians who voted for retrocession, and presumably they were happy to discontinue the use of "D.C." in their pottery marks.

Baltimore's early adoption of railroads helped that city eclipse Alexandria as a center of manufacturing and trade. In 1847 Baltimore potter Edwin Bennett built a more industrialized pottery, producing yellow wares and Rockingham wares using molds and jigger/jolly machines in place of the potter's wheel. These wares are common on Alexandria's archaeological sites, along with Milburn's stoneware.

During the 1840s and early 1850s, Milburn continued to make stoneware with the stamp of his former employer, H. C. Smith. Hugh Charles Smith was the senior partner in his family's retail business until his death in 1850, and the partnership was dissolved the next year.[18] His brother James P. Smith then ran the retail shop, and the Milburn Pottery made stoneware stamped with his name until September 1854, when the business was sold out of the Smith family.[19]

Stoneware with the impressed marks of another Alexandria merchant, Dubois and Reddick, was also made at Wilkes Street sometime between 1841 and 1847. After the Civil War, Milburn and his sons made undecorated stoneware with the mark of prominent china merchant Elisha Janney (E. J.) Miller, with vessel forms identical to those with the Milburn's own stamps. From the early 1820s Robert H. Miller had been a prominent china merchant and a rival to Hugh Smith, with a store on the 200 block of King Street. Although Robert H. Miller and his son Elisha Janney Miller both advertised Alexandria stoneware,[20] the ware was not stamped with the company name until after Elisha took over the business in 1865, upon his father's death. The marked pieces may date from E. J. Miller's advertisement in the *Alexandria Gazette* in 1869: "Alexandria Pottery being again in full operation, I am prepared to furnish my customers and the public generally with stoneware of all kinds and of first quality."[21] After the Wilkes Street Pottery closed, E. J. Miller imported stenciled stoneware made by James Hamilton in Greensboro, Pennsylvania.

Milburn also supplied stoneware without merchant's marks to druggist William Stabler. An invoice in the collection of the Stabler-Leadbeater Apothecary Museum, dated July 10, 1848, shows a sale from Milburn to Stabler of 100 two-inch pots and 100 three-inch pots, possibly for Stabler's wholesale seed business.

Milburn's first advertisement, for flowerpots, appeared in the *Alexandria Gazette* on September 1, 1845:

> ALEXANDRIA POTTERY
>
> Those who have never witnessed the operations of shaping and finishing Earthenware will be gratified by a visit to the manufactory of Mr. Milburn, on Wilkes Street, of this city. The material employed is a species of bluish white clay, found in various parts of the country, and composed of such proportions of alumina and other ingredients as to make it very tenacious and plastic when moistened. The clay used at Mr. Milburn's factory is brought from the vicinity of Baltimore City. After the clay is thoroughly kneaded and prepared, a certain portion, according to the size of the vessel to be made, is placed upon a circular board fixed horizontally and connected with a treadle by which a rotary motion is given to it. While the clay is revolving in common with the board on which it lies, the operator shapes it with his hands, into whatever vessel it is designed to make.

> The judgment shown in choosing just the proper quantity for the vessel designed, and the skill and regularity with which it is brought to the shape and size desired, by the aid of machinery so simple, excite the admiration of the beholder. The vessels thus prepared are dried a while in the sun; after which they are placed in the kiln where the process of burning and glazing complete the work.[22]

In 1855 there was enough interest in the Wilkes Street Pottery that a reporter for the *Alexandria Gazette* toured the premises and offered a description of the process of making stoneware. The author mentions turning, drying, and firing, but not decorating, although this practice most likely continued until the Civil War. The most surprising statement from this article is that Milburn used clay brought in from the Baltimore area. At least some Alexandria stoneware is chemically consistent with clay samples taken locally, within a few blocks of the kiln site. In 1851 a rail line opened in Alexandria that ran right past Milburn's Pottery, and this may have made it cost effective to import clay. Baltimore clays were also exported to industrial potteries in New Jersey and Ohio into the early twentieth century.[23]

Business remained strong throughout the 1850s, but like all Alexandria businesses, the Wilkes Street Pottery was severely affected by the Civil War. The day after the first shots were fired, federal troops occupied Alexandria. The town housed a major supply depot and other support services. Anthony Trollope wrote:

> We landed also on this occasion at Alexandria, and saw as melancholy and miserable a town as the mind of man can conceive. . . . All trade was at an end. In no town at that time was trade very flourishing; but here it was killed altogether—except that absolutely necessary trade of bread.[24]

The pottery may have closed during the war, as did many of Alexandria's businesses. If Milburn did supply stoneware to the U. S. military, it would have been under duress. Three of Milburn's sons enlisted in the Confederate Army. A 1912 obituary for his son James Clinton Milburn refers to his father, "who previous to the Civil War conducted a pottery on the north side of Wilkes Street."[25] The pottery was in full operation once again by November 1866, when the *Alexandria Gazette* reported:

> The old established Pottery of Mr. Milburn, in this place, is, we are glad to see, in successful operation, turning out the best ware from the best materials. This establishment has obtained a well-merited reputation for earthenware, which it keeps up to this day.[26]

Following Milburn's death in 1867, a notice in the *Alexandria Gazette* referred to the success and prominence of the pottery:

> THE POTTERY of the late Mr. Milburn, which he carried on, with so much credit to himself, for many years, will be continued under the control of his son. This is another of the old and successful manufacturing establishments of this place. Its wares are well known throughout the country, and considered the very best of their kind.[27]

Milburn's Sons

At least three of Milburn's seven sons were trained as potters, and two of them managed the pottery in its later years.[28] Stephen Calvert Milburn, who first worked at the pottery in 1845 at the age of twelve, already had a major role there before his father's death in 1867. His name appears beneath his father's in the 1860 Census; both men are listed as potters.[29] Stephen advertised in the local paper in 1869, stating that he was the successor to B.C. Milburn and was continuing the manufacture of stoneware and of earthenware flower and chimney pots.[30] The 1870 Census shows S.C. Milburn as having four male employees, one clay mill, three lathes (potter's wheels), and one kiln.

Although Stephen's brothers John and W. Lewis both trained in their father's business, they were both listed as apothecaries in the 1860 Census and W. Lewis Milburn was listed in an 1870 city directory as "druggist."[31] An 1873 city directory lists W. Lewis Milburn as proprietor of the pottery,[32] and Stephen had moved away from Alexandria by 1875.

Court records from 1874 provide some insight into the pottery's financial difficulties in its last years. Henry D. Wolff, a journeyman potter, filed a mechanic's lien against W. Lewis Milburn for twenty-three days' work. The lien was for $46 in back wages, with $16.30 deducted for three weeks' board, cracked pots, and cash. The lien lists an inventory of 904 stoneware vessels: 356 jugs, 169 pots, 336 pans, and 43 churns.[33] The sizes produced are also enumerated, with one-half, one-, and two-gallon jugs; three- to six-gallon pots; one- to three-gallon pans; and three- and four-gallon churns.

The pottery closed sometime before April 1876, at which time the land was sold to the Smoot tannery across Wilkes Street and a bark shed was constructed on the site.

Makers' and Merchants' Marks

At least ten distinct marks were used during the Milburn period. Milburn and his sons made stoneware with their own marks at the same time they were producing stoneware with the stamps of Alexandria merchants. Dates attributed to each mark are based on historical records, the 1847 date of Alexandria's retrocession to Virginia, and formal characteristics of the stoneware.

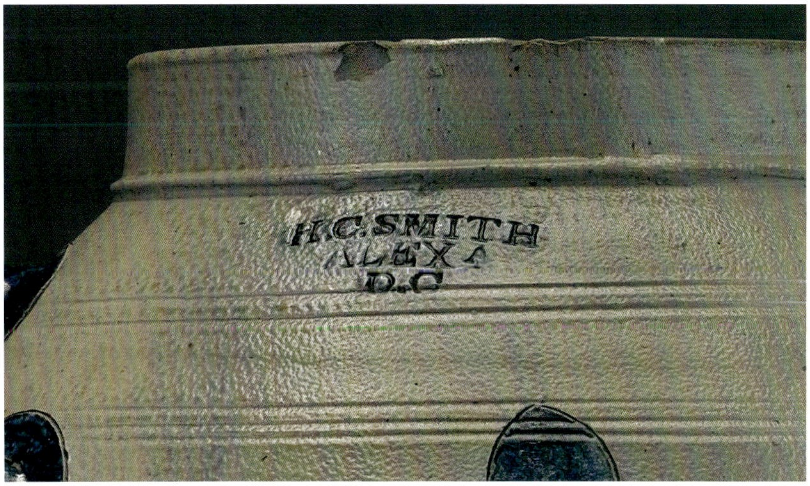

Figure 5 Detail of the "H.C.SMITH / ALEX^A / D.C" mark on the cooler illustrated in fig. 12. Although the stamp is found on many vessels of this period, this is the only known example in which it is highlighted with cobalt.

Marks were impressed into the clay vessel using a stamp consisting of lead type, probably set into a wood frame. The lettering ranges from 14- to 28-point type. In one example, a large cooler made for J. W. Smith, the lettering was highlighted with cobalt (figs. 5, 12).

Capacity marks were usually made with impressed lead type in larger type than the maker's mark. On some of the five- and six-gallon vessels, however, marks were incised into the leather-hard surface (fig. 6). Some 1- and 1½-gallon capacity marks on "B. C. MILBURN" and "B.C.MILBURN / ALEX^A^" vessels were stamped within a dashed circle, similar to marks on some of the earliest Swann vessels (fig. 7).

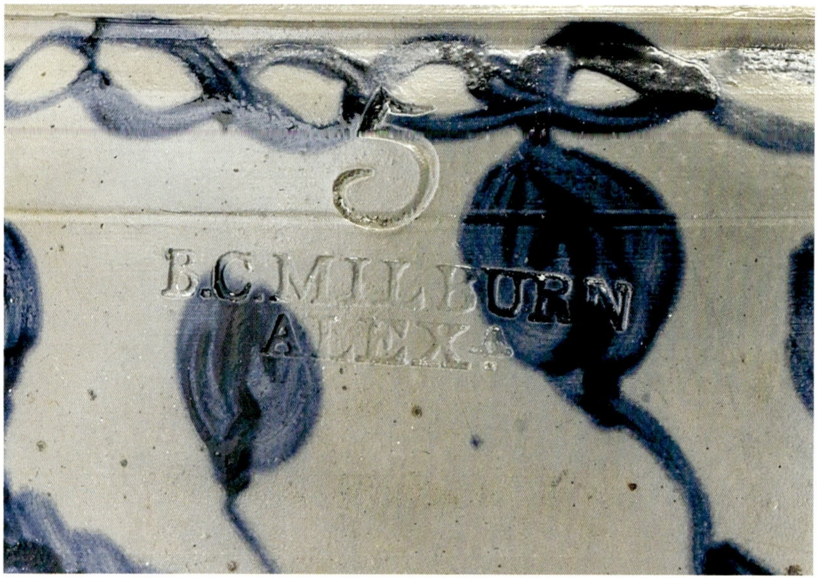

Figure 6 Detail of the cooler illustrated in fig. 68 showing capacity mark with "B.C.MILBURN / ALEX^A^" potter's mark, 1847–1861. Whereas most capacity marks were made with impressed lead type, some 5- and 6-gallon capacity marks were drawn freehand on the leather-hard surface.

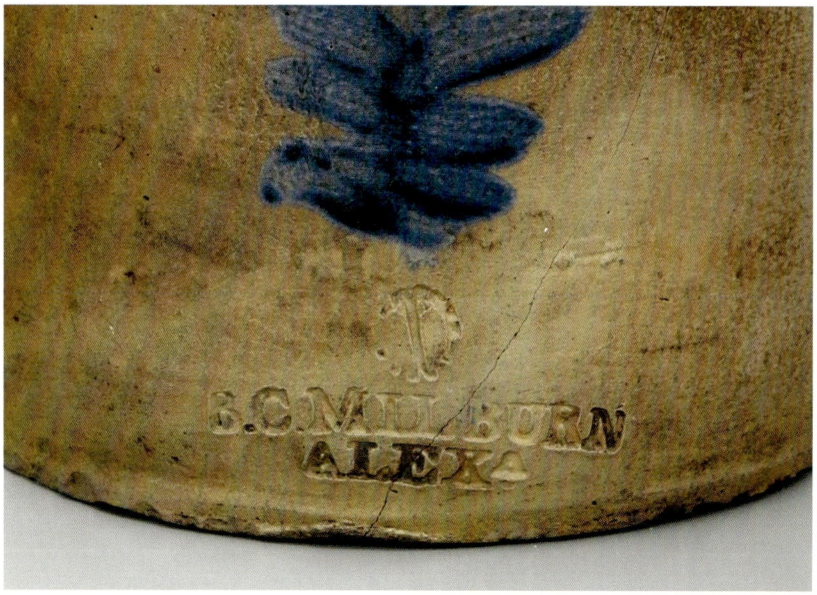

Figure 7 Detail of the cylindrical jar illustrated in fig. 27 showing capacity mark with "B.C.MILBURN / ALEX^A^" potter's mark, 1847–1861. Some of Milburn's 1- and 1½-gallon capacity marks include a dashed circle, similar to marks used on some of the earliest Swann vessels.

The marked Milburn stoneware can be divided into three periods. The first (1841–1847) begins with Milburn's purchase of the pottery, and includes the time in which Alexandria is part of the District of Columbia. The second period (1847–1861) begins when Alexandria was retroceded to Virginia, and includes stoneware decorated with slip-trailed cobalt. Most of the stoneware produced by Milburn's sons after the Civil War, in the third period (1865–1876), is undecorated.

Early B.C. Milburn Marks

Mark	Size	Date
H.C.SMITH / ALEX^{A} / D. C	28 pt.	ca. 1831–1847
B. C. MILBURN, / *ALEXANDRIA D. C.*	12 pt. / 8 pt.	ca. 1841–1847
DUBOIS & REDDICK / ALEXANDRIA D C	18 pt.	ca. 1841–1847
H.C.SMITH / ALEX^{A}	26 pt.	ca. 1847–1851

Four marks were stamped on Milburn pottery prior to his use of slip-decorated cobalt decoration. Three of these marks were used during the period when Alexandria was part of the District of Columbia, prior to the city's retrocession to Virginia in 1847.

The mark "H.C.SMITH / ALEX^{A} / D.C" came into use sometime between 1831, when Hugh Smith dissolved an earlier partnership and left his son Hugh Charles Smith to manage the company, and 1833, when Milburn leased the pottery. After Milburn became owner of the pottery in 1841, vessels with B.C. Milburn and H.C. Smith marks were probably made concurrently. The "B. C. MILBURN, / *ALEXANDRIA D. C.*" mark is unusual in that the place name is executed in smaller italic type (fig. 8).

While the history of Alexandria merchant Dubois & Reddick is not known, the mark probably dates to the same era as the Milburn mark in which the city name is also spelled out. The only documented jar with the Dubois & Reddick mark is similar in appearance to others from the Wilkes Street Pottery dating from 1825 to 1847 (see fig. 17).

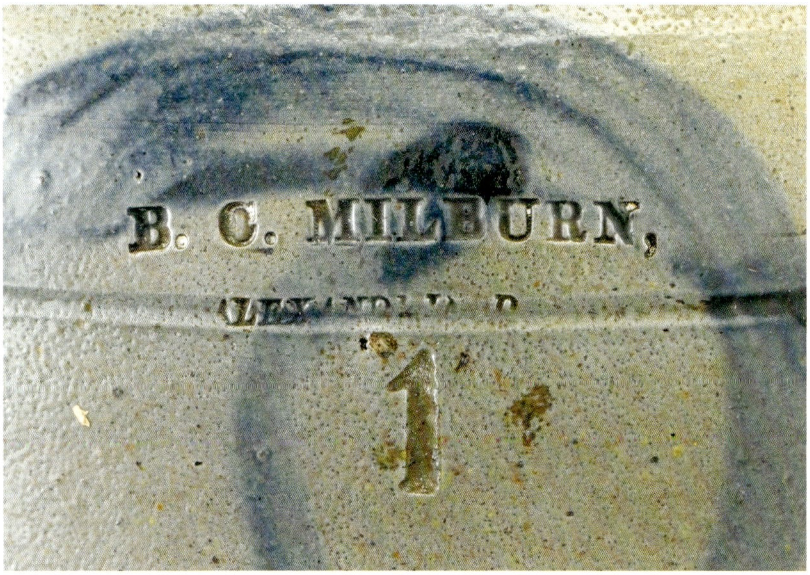

Figure 8 Detail showing the potter's mark of the cylindrical jar illustrated in fig. 19: "B. C. MILBURN, / *ALEXANDRIA D. C.*" This is the only mark in which the place name is executed in smaller, italic type. The city's name is spelled out only on this mark and that made for Dubois & Reddick, illustrated in fig. 17.

The "H.C.SMITH / ALEX^{A}" mark was used after Alexandria's 1847 retrocession until 1850, when Hugh Charles Smith passed away, or perhaps 1851, when company ownership passed to his brother James P. Smith.

Later B.C. Milburn Marks

J.P.SMITH	26 pt.	ca. 1851–1854
B. C. MILBURN.	16 pt.	ca. 1847–1867
B C MILBURN	14 pt.	ca. 1847–1867
B.C.MILBURN / ALEX^{A}	26 pt.	ca. 1847–1867

B. C. Milburn and J. P. Smith marks have been found on vessels with slip-trailed cobalt and with brushed-cobalt decoration.

The mark "J.P.SMITH" was used for a few short years, from July 1851 to September 1854. During this period, the company was owned by James P. Smith, Hugh Smith's son and former partner.

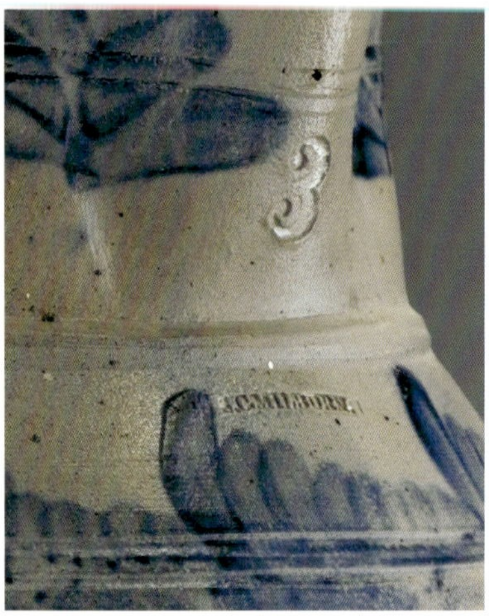

Figure 9 Detail of the pitcher illustrated in fig. 51 showing the "S. C. MILBURN." potter's mark. This late mark is usually found on undecorated stoneware.

Most of Milburn's vessels of the post-retrocession period are stamped "B.C.MILBURN / ALEX^{A}", but the mark was probably used concurrently, for at least a part of this period, with the marks "B. C. MILBURN." and "B C MILBURN". The "B. C. MILBURN." mark includes a period after the name, as do the marks of Milburn's sons, but it is executed in 16-point type, whereas the unpunctuated mark is stamped with 14-point type, like that of Milburn's sons. These marks were documented on both decorated and plain vessels, and were most likely in use before the Civil War.

Marks Used by Milburn's Sons

S.C.MILBURN.	14 pt.	ca. 1867–1873
W. LEWIS MILBURN.	14 pt.	ca. 1873–1876
E.J.MILLER / ALEX^{A}	24 pt.	ca. 1865–1876

Stephen C. Milburn took over the business after B.C. Milburn's death in 1867, and by 1873 his brother W. Lewis Milburn had succeeded him. The S.C. Milburn mark is found on a few vessels with brushed-cobalt decoration, but both marks appear mostly on plain, undecorated wares (figs. 9, 10). On one plain, two-gallon storage jar the small stamped mark "S.C.MILBURN." appears above his father's larger "B.C.MILBURN / ALEX^{A}" stamp. The "S.C.MILBURN." stamp was also used together with that of china merchant E. J. Miller and with the stamped initials "SDS"—

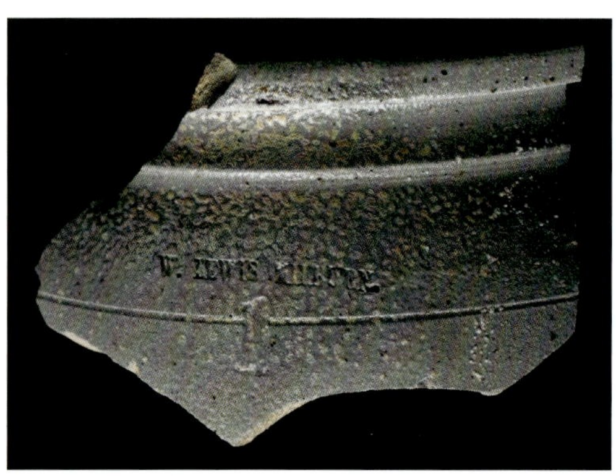

Figure 10 "W. LEWIS MILBURN." potter's mark, 1873–1876. D. of rim 7". Capacity: 1 gallon. (Courtesy, Alexandria Archaeology Museum.) This is the only piece found at the Milburn kiln site with this mark.

possibly the mark of another local merchant. Stoneware stamped with Miller's name is dated after 1865, when he took over his father's King Street store.

Techniques of Decoration

Most pieces of early Milburn stoneware made between circa 1831 and 1851 were decorated with brushed cobalt. Somewhat surprising, only one piece with incised decoration is known. Slip-trailed decoration was added to the pottery's repertoire sometime between 1847 and 1854. After the Civil War, the pottery made mostly undecorated wares.

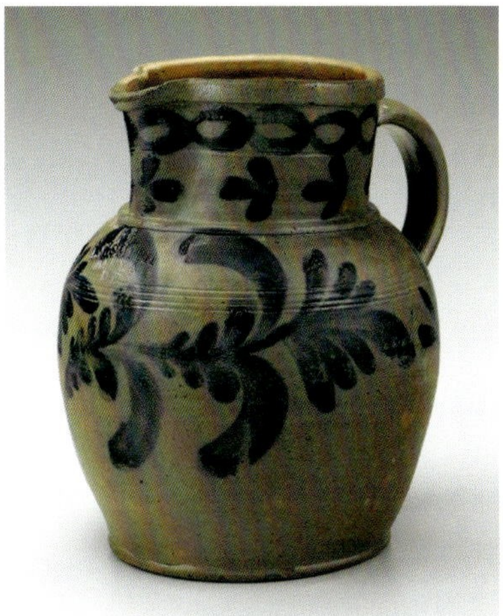

Figure 11 Pitcher, Alexandria, Virginia, 1831–1847. Salt-glazed stoneware, brushed cobalt. H. 11". Impressed mark: H.C.SMITH / ALEX^{A} / D.C (Courtesy, Smithsonian Institution, National Museum of American History.) Brushed cobalt is the most common form of decoration on Alexandria stoneware.

Brushed Cobalt. Brushed cobalt is the most common form of decoration on American stoneware, widely used after the late 1780s when cobalt pigment was commercially produced in this country. Milburn's predecessor, John Swann, first used brushed-cobalt decoration around 1819. By the early 1830s, this method of decoration had already reached its mature form in Alexandria, with exuberant flowers and foliage covering a large portion of the pot (fig. 11). Milburn continued to use Swann's earlier motifs, including forward-facing designs with a round flower (known as the Alexandria motif), and foliage encircling the vessel, often combined with tulips. Brushed-cobalt decoration was common on Alexandria stoneware until the Civil War.

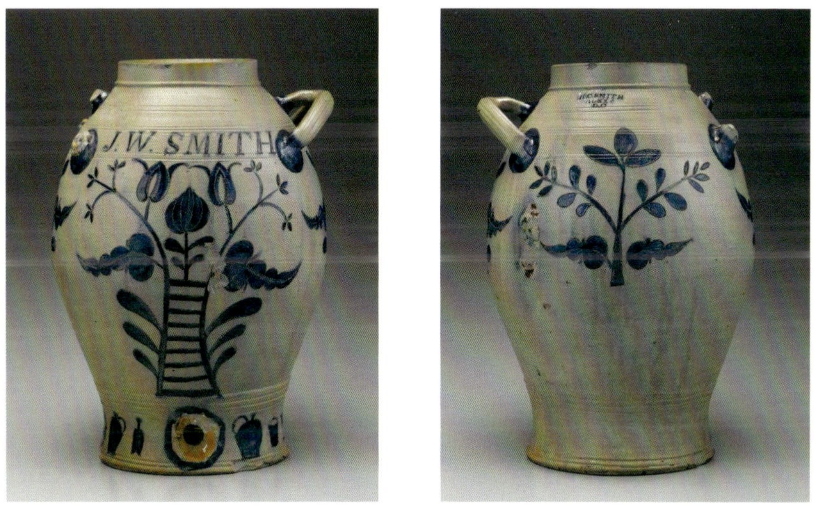

Figure 12 Cooler, Alexandria, Virginia, 1831–1847. Salt-glazed stoneware, incised decoration. H. 18½". Inscribed: J. W. SMITH; impressed mark: H.C.SMITH / ALEX^{A} / D.C (Courtesy, Smithsonian Institution, National Museum of American History.) This is the only known piece of Alexandria stoneware with incised and cobalt-filled decoration. This unusual cooler appears to have been made for a tavern, with images of bottles and jugs around its base. Leafy branches appear on the reverse and under each handle. The most likely owner of this piece was James W. Smith, listed in an 1834 business directory as proprietor of the Globe Tavern on Cameron Street, opposite the Market House.

Incised Decoration. On one cooler stamped "H.C.SMITH / ALEX^{A} / D.C" the potter incised the outline of the letters and decorative elements and then filled the designs with cobalt (fig. 12). This technique was used on early-nineteenth-century American stoneware from Connecticut, Massachusetts, and New York. Incised decoration is not known on other Alexandria stoneware, but the maker might have been influenced by stoneware from the Baltimore pottery of Morgan and Amoss (1812–1822)[34] or by Richmond, Virginia, potter John Poole Schermerhorn. Schermerhorn was originally from New York State, where this technique is more common, and he might have introduced the filled incised decoration at Richmond's

DuVal Pottery circa 1813. The technique continued to be used in Richmond as late as 1856.[35] The Alexandria cooler's extruded, horizontal, open-loop handles and straight neck are also reminiscent of New York stoneware from the early nineteenth century.[36] While Benedict C. Milburn and some of the other known employees of the Wilkes Street Pottery came to Alexandria from St. Mary's County, Maryland, perhaps one of the potters had worked with Schermerhorn or another potter in the New York area—or Milburn might have been influenced directly by examples of their wares or by German stoneware, which is found on early Alexandria sites.

Slip-Trailed Cobalt. Slip trailing was done concurrently with brushed decoration, but the two techniques are not found together on the same vessel. Instead of using a brush, the potter blew through a slip-cup, dripping a thin line of blue cobalt from a quill onto the surface of the vessel (fig. 13). This produced a finer line, but similar motifs and arrangements of decorative elements were used with both brushed and slip-trailed cobalt. Both techniques are found with the marks "B. C. MILBURN." and "B.C.MILBURN / ALEX^{A}" (both dated circa 1847–1867), and "J.P.SMITH" (1851–1854), indicating a date between 1847 and 1854 for the introduction of slip trailing.

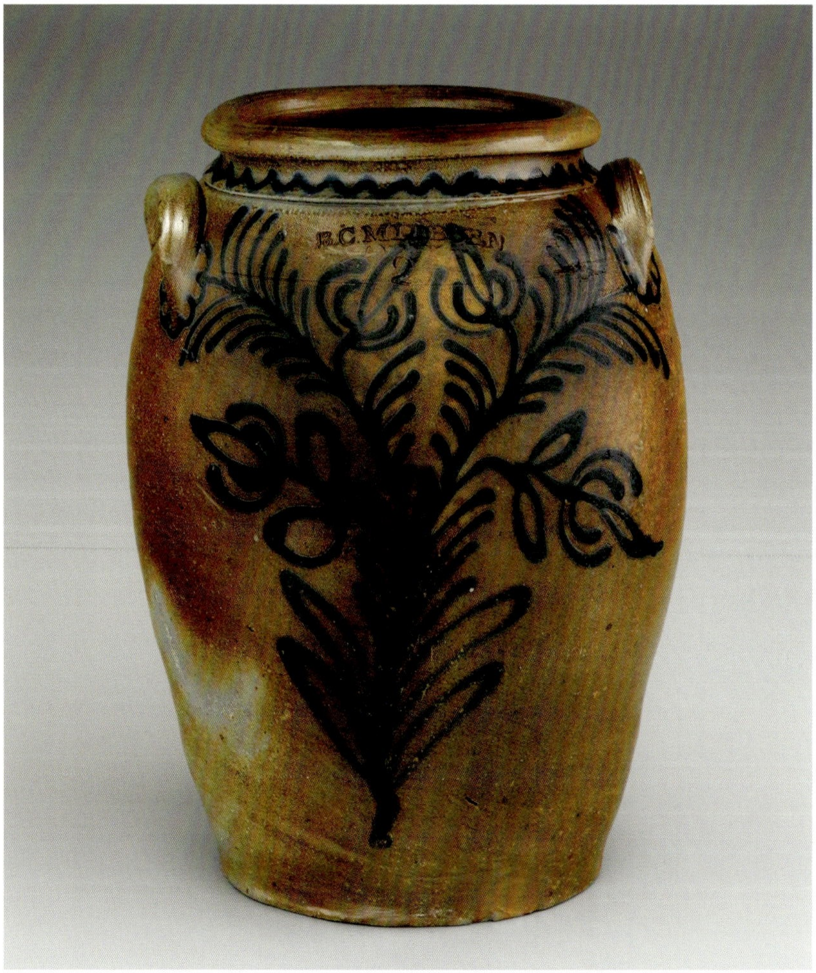

Figure 13 Jar, Alexandria, Virginia, 1847–1861. Salt-glazed stoneware, slip-trailed cobalt. H. 12½". Capacity: 2 gallons. Impressed mark: B. C. MILBURN / ALEX^{A} (Courtesy, Al Marzorini.) Slip trailing, a technique seen much earlier on Baltimore, New Jersey, and New York stoneware, was not introduced in Alexandria until sometime between 1847 and 1854.

This technique was used by Baltimore potters James Burland, William H. Morgan, Morgan and Amoss, and David Parr in the early 1820s, but their designs are less elaborate than Milburn's work. Their small dabbed leaves are reminiscent of Swann's brushwork of that period.[37] There is also one known example of slip-trailed cobalt decoration on a jar with a simple floral design, made circa 1824–1826 by another Alexandria potter, James Miller.[38]

Undecorated Stoneware. Plain, utilitarian wares, such as jugs and flowerpots, were made prior to the Civil War alongside decorative wares. In the period following the war until 1876, when the Wilkes Street Pottery closed, Milburn and his sons mostly made undecorated utilitarian wares (fig. 14).

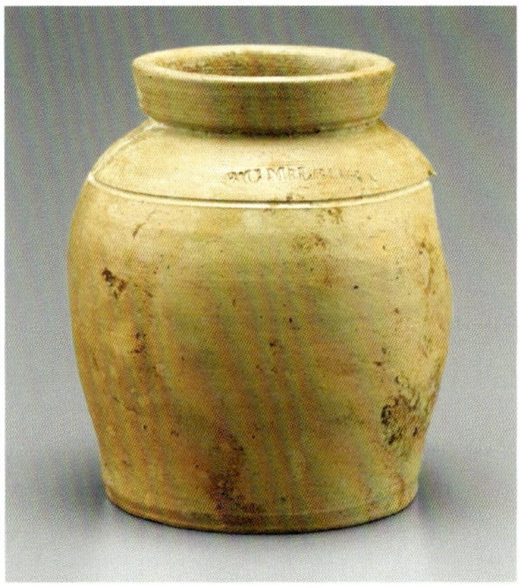

Figure 14 Jar, Alexandria, Virginia, 1847–1867. Salt-glazed stoneware. H. 5". Impressed mark: B C MILBURN (Courtesy, Alexandria Archaeology Museum.) Plain, utilitarian wares were made throughout the history of the Wilkes Street Pottery, but predominated after the Civil War. This jar was found in a privy behind 307 South Saint Asaph Street, an elegant house that was occupied, as were many Alexandria homes, by Union troops during the war.

Decoration on Early Milburn Stoneware

The "H.C.SMITH / ALEX^A^ / D.C" mark is the most common from this period, documented on thirty vessels, followed by "H.C.SMITH / ALEX^A^" on eight vessels. The "B. C. MILBURN, / *ALEXANDRIA D. C.*" mark was documented on only five vessels, as most stoneware of this period was probably marked with the Smith Company name. Only one "DUBOIS & REDDICK / ALEXANDRIA D C" pot is known. Vessels with all of these marks bear similar decorative motifs.

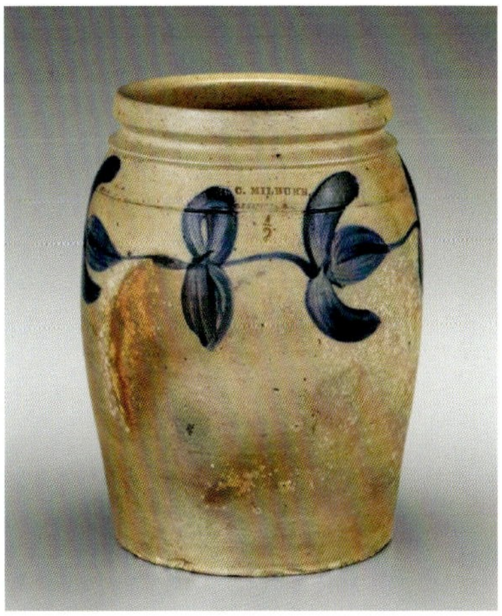

Figure 15 Jar, Alexandria, Virginia, 1841–1847. Salt-glazed stoneware, brushed cobalt. H. 8". Capacity: ½ gallon. Impressed mark: B. C. MILBURN, / *ALEXANDRIA D. C.* (Courtesy, Toby and Oscar Fitzgerald.) Encircling foliage is the most common decorative motif on early Milburn stoneware.

Encircling foliage (fig. 15) is the most frequent decoration in this period, followed by the so-called Alexandria motif, consisting of a central round flower on a stem with graduated leaves (fig. 16). Some vessels of this period closely resemble those made by Swann before 1831. For instance, the execu-

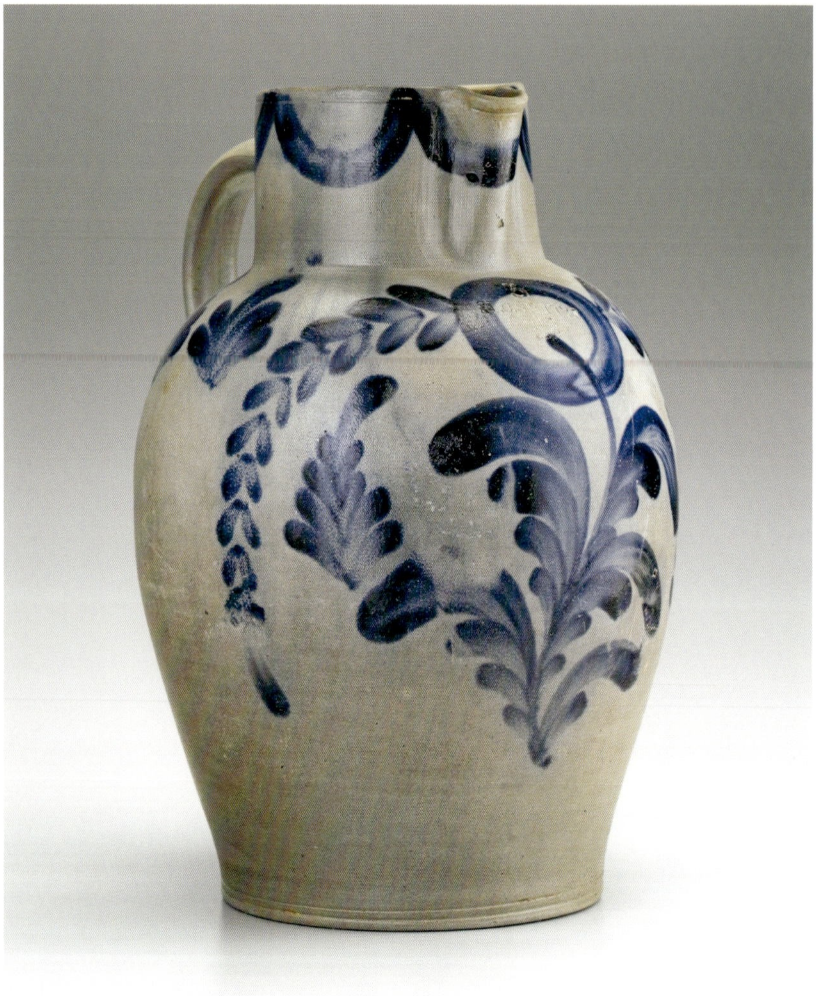

Figure 16 Pitcher, Alexandria, Virginia, 1831–1847. Salt-glazed stoneware, brushed cobalt. H. 15¼". Capacity: 3 gallons. Impressed mark: H.C.SMITH / ALEX^{A} / D.C (Private collection.) The Alexandria motif, consisting of a central round flower on a stem of graduated leaves, was introduced about 1820 by earlier Wilkes Street potter John Swann.

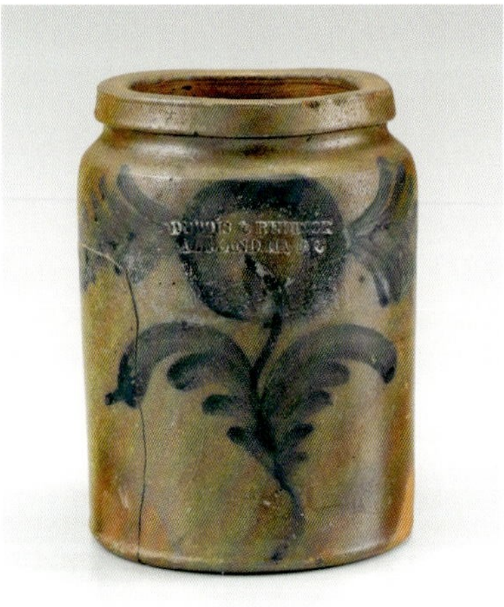

Figure 17 Jar, Alexandria, Virginia, 1841–1847. Salt-glazed stoneware, brushed cobalt. H. 7¼". Impressed mark: DUBOIS & REDDICK / ALEXANDRIA D C (Private collection.) This jar, made at Milburn's pottery for an Alexandria merchant, is the only documented example with this mark.

tion of the Alexandria motif on the Dubois & Reddick jar (fig. 17) is similar to that on Swann-period vessels marked "H.SMITH & C^{O}" (1825–1831) and on Milburn-period vessels marked "H.C.SMITH / ALEX^{A} / D.C".

In some vessels of this period, longer straight leaves appear near the base of the leafy stem. The decorators used from one to three pairs of leaves (figs. 18, 19). The same design is found with later marks, executed in brushed and slip-trailed cobalt.

In another variation, the encircling foliage is much more crowded than on earlier and later jars. Yet another variation, likely the work of one particular decorator, is a thick stem, seen on a few jars with each of the H. C. Smith marks (fig. 20).

A new design of pendant leafy fronds, sometimes called "swag and tassel," was first used on stoneware with the "H.C.SMITH / ALEX^{A} / D.C"

Figure 18 Jars, Alexandria, Virginia, 1831–1847. Salt-glazed stoneware, brushed cobalt. *Left:* H. 8". Capacity: ½ gallon. *Right:* H. 9½". Capacity: 1 gallon. Impressed mark: H.C.SMITH / ALEX^{A} / D.C (Private collection.) A modification of the Alexandria motif introduced on early Milburn stoneware includes pairs of longer, straight leaves near the base of the leafy stems.

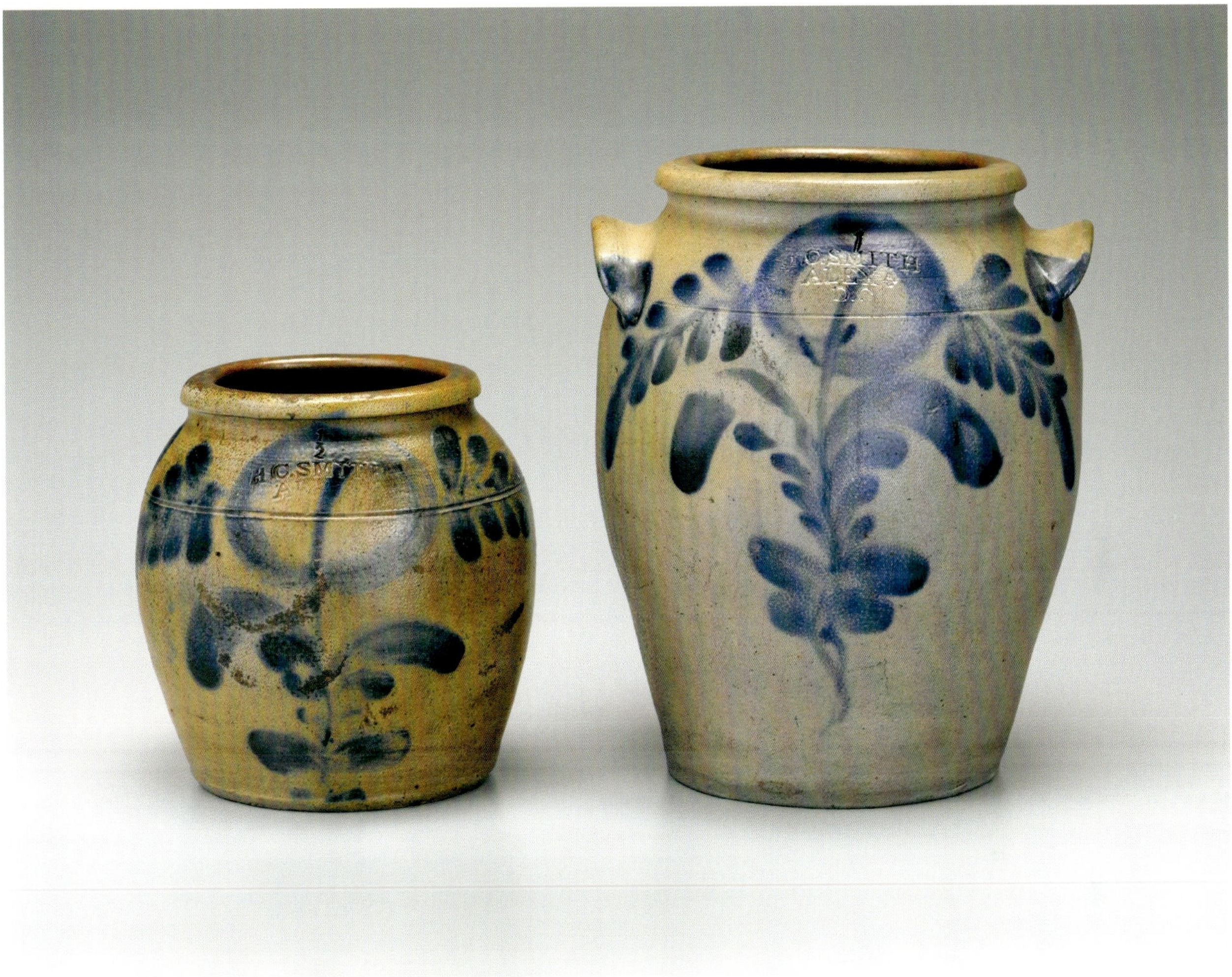

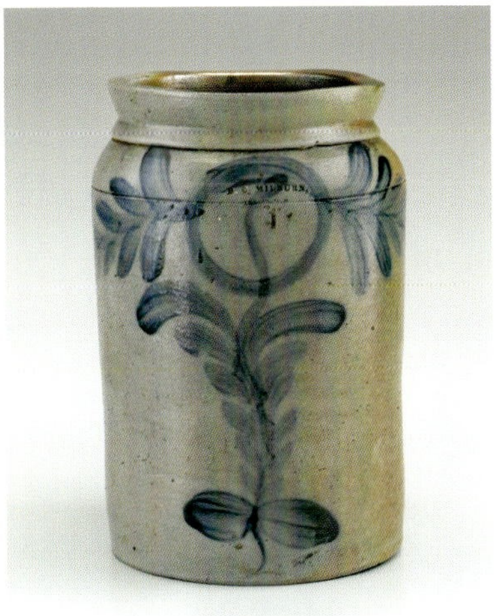

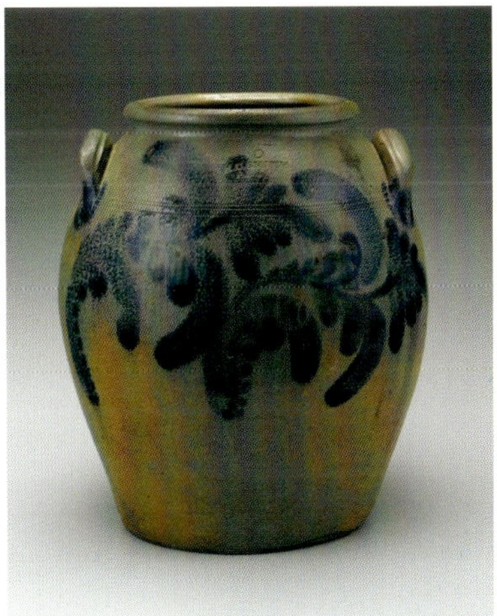

Figure 19 Jar, Alexandria, Virginia, 1841–1847. Salt-glazed stoneware, brushed cobalt. H. 9½". Capacity: 1 gallon. Impressed mark: B. C. MILBURN, / *ALEXANDRIA D. C.* (Private collection.)

Figure 20 Jar, Alexandria, Virginia, 1831–1847. Salt-glazed stoneware, brushed cobalt. H. 15". Impressed mark: H.C.SMITH / ALEX^{A} / D.C (Courtesy, The Lyceum: Alexandria's History Museum.) The thick leafy stem, found with both of the H. C. Smith marks, appears to be the work of one decorator. The more crowded, encircling foliage seen here is only on jars from the early Milburn period.

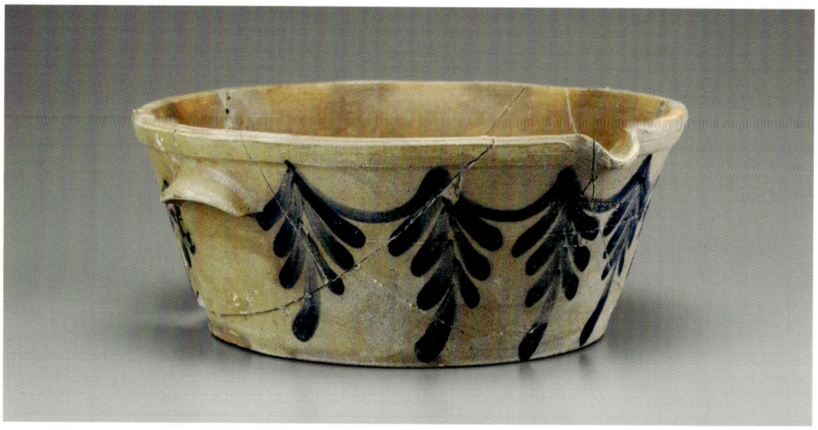

Figure 21 Milk pan, Alexandria, Virginia, 1831–1847. Salt-glazed stoneware, brushed cobalt. D. 10". Capacity: 1 gallon. Impressed mark: H.C.SMITH / ALEX^A / D.C (Courtesy, Alexandria Archaeology Museum.) A design of pendant leafy fronds, sometimes called "swag and tassel," first appeared with this mark. It is more common on the back of Alexandria stoneware, but on this piece, separate, shorter fronds appear on the reverse. The milk pan was discarded in a privy at 106 South Saint Asaph Street in the latter half of the nineteenth century.

mark, and continued to be used in the later Milburn period. The fronds are used as the main design on pitchers and a milk pan (fig. 21), but are more commonly found on the back of Alexandria stoneware vessels.

Decoration on Later Milburn Stoneware

The later period of B. C. Milburn's stoneware is marked by the introduction of the technique of slip-trailed decoration. Slip trailing is found only with specific marks—"B.C.MILBURN / ALEX^A"; "B. C. MILBURN."; and "J.P.SMITH"—and so was used from circa 1847/54 until about 1861, the start of the Civil War. Slip trailing was used concurrently with brushed-cobalt decoration, and similar patterns were executed with both techniques (figs. 22–25). Decorators may have specialized, but it is possible that some workers were proficient in both techniques. Although slip-trailed stoneware was made in Alexandria for only about ten years, extant examples are

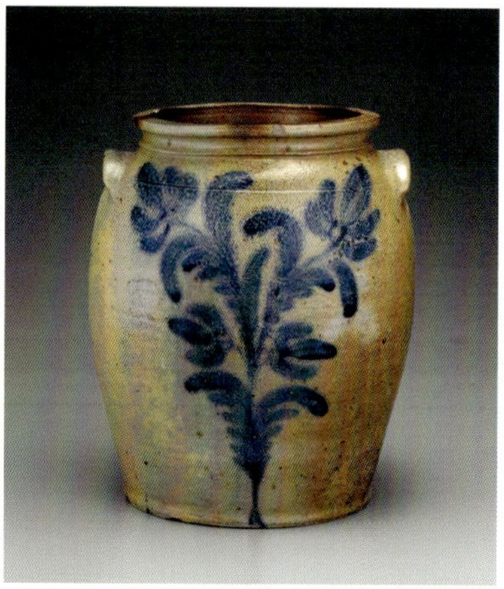

Figure 22 Jar, Alexandria, Virginia, 1847–1861. Salt-glazed stoneware, brushed cobalt. H. 14¼". Capacity: 5 gallons. Impressed mark: B.C.MILBURN / ALEX^A (Courtesy, Al Marzorini.) Brushed and slip-trailed designs were used concurrently, and similar designs were executed with both techniques.

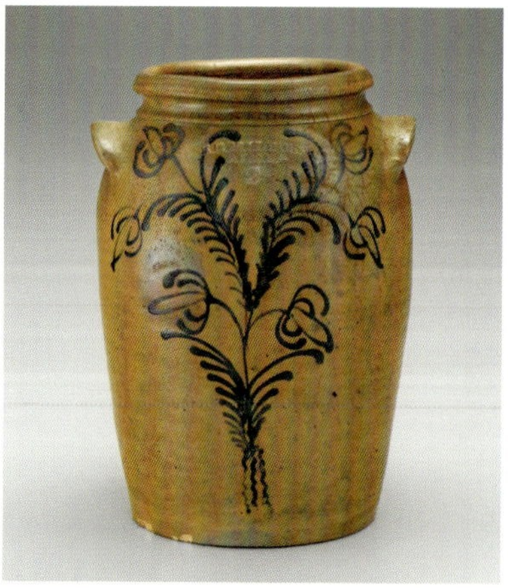

Figure 23 Jar, Alexandria, Virginia, 1847–1861. Salt-glazed stoneware, slip-trailed cobalt. Dimensions not provided. Capacity: 2 gallons. Impressed mark: B.C.MILBURN / ALEX^A (Private collection.)

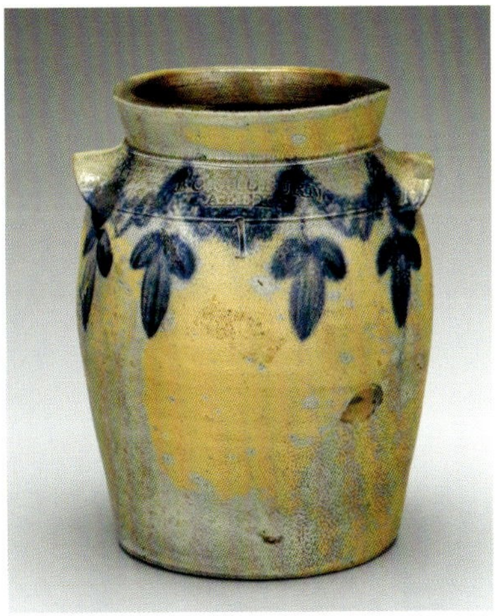

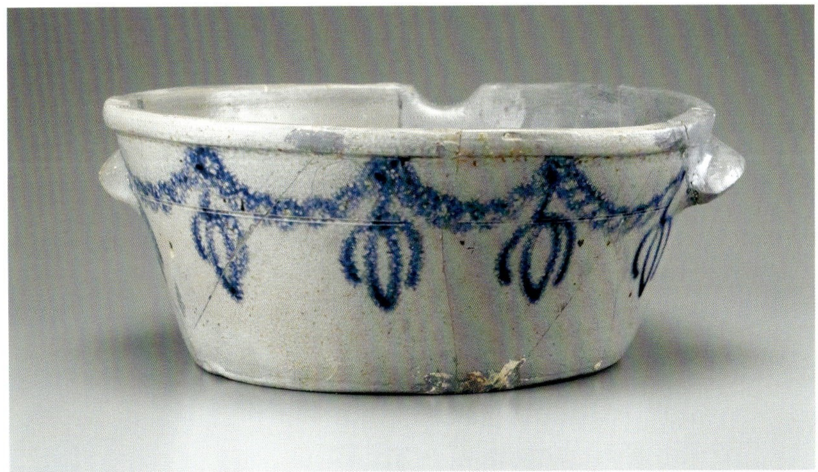

Figure 24 Jar, Alexandria, Virginia, 1847–1861. Salt-glazed stoneware, brushed cobalt. H. 9½". Capacity: 1 gallon. Impressed mark: B.C.MILBURN / ALEX^{A} (Private collection.) Chains and pendant flowers encircle this jar. Similar designs were also executed in slip-trailed cobalt.

Figure 25 Milk pan, Alexandria, Virginia, 1847–1861. Salt-glazed stoneware, slip-trailed cobalt. D. 11¼". Impressed mark: B.C.MILBURN / ALEX^{A} (Courtesy, Alexandria Archaeology Museum.) Chains and pendant flowers are more commonly seen on the back of vessels, as on this milk pan found in an excavation at 112 South Royal Street.

Figure 26 Jar, Alexandria, Virginia, 1847–1861. Salt-glazed stoneware, slip-trailed cobalt. H. 14½". Capacity: 4 gallons. Impressed mark: B.C.MILBURN / ALEX^{A} (Courtesy, Alexandria Archaeology Museum.) This elaborately decorated jar was found in excavations at Alfred and South Columbus streets, in a nineteenth-century free black neighborhood.

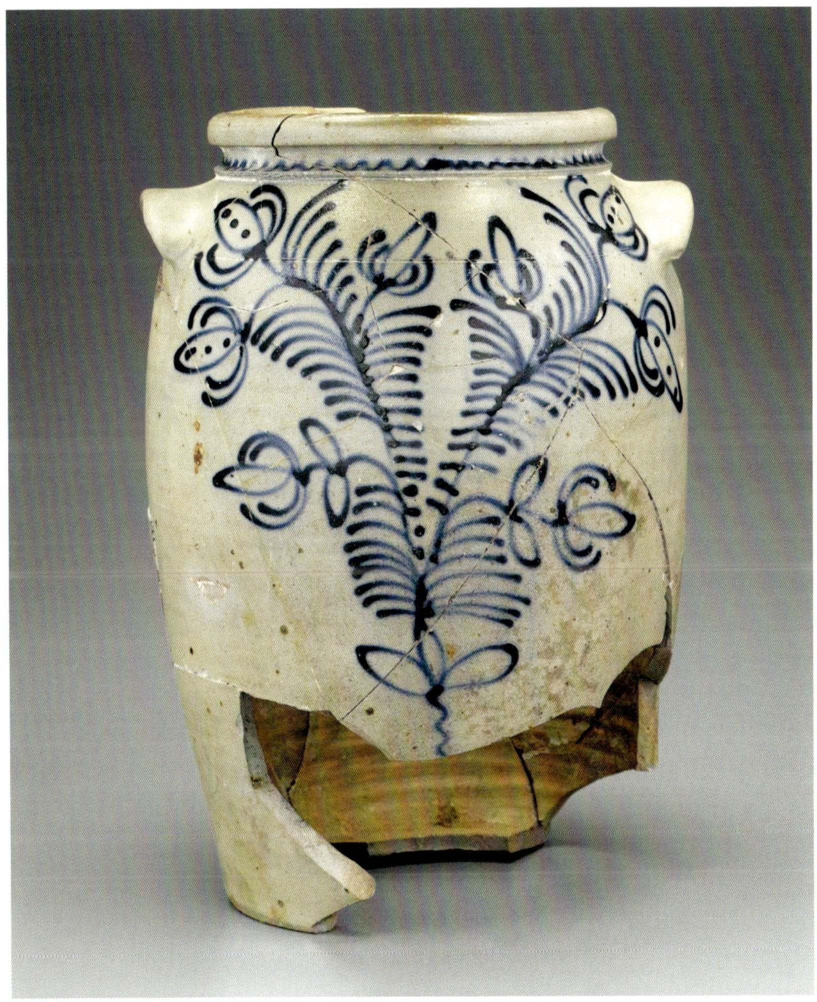

fairly common because by the 1850s Milburn had increased production at the pottery.

While many vessels marked "B.C.MILBURN / ALEX^{A}" have beautifully executed and elaborate artwork (fig. 26), others have designs executed in a surprisingly primitive style (fig. 27). The Smith Company merchant marks are not found on the more primitive artwork, which might indicate that it was produced after the company closed, in 1854.

In this period, the central round flower or Alexandria motif was the most common design (figs. 28, 29), followed by encircling foliage (figs. 30, 31). Foliage branching from a central stem was also common, and pendant fronds or foliate sprigs continued to be used.

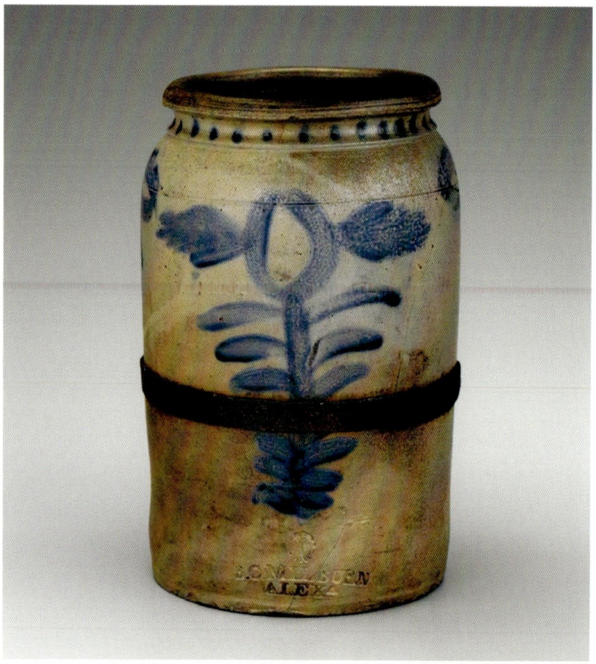

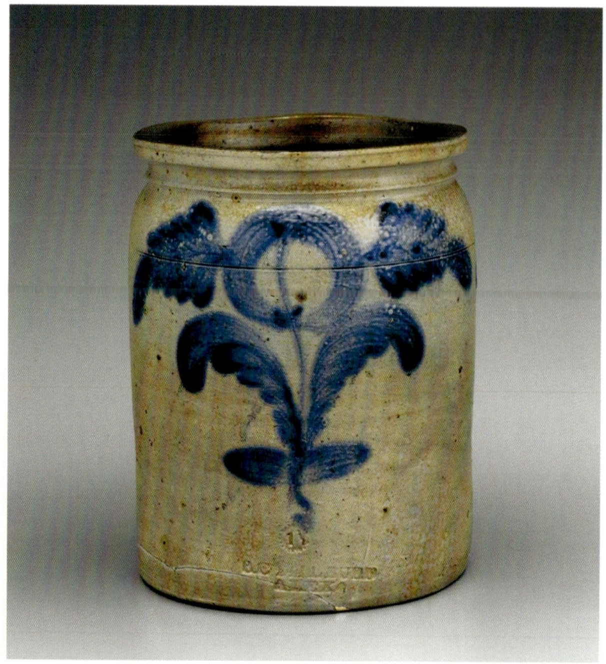

Figure 27 Jar, Alexandria, Virginia, 1847–1861. Salt-glazed stoneware, brushed cobalt. H. 10½". Capacity: 1 gallon. Impressed mark: B.C.MILBURN / ALEX^A (Courtesy, Al Marzorini.) This jar has a very primitive version of the central round flower. Large chain links with pendant branches appear on the reverse. The iron band is an old "make do" repair, probably from the nineteenth century.

Figure 28 Jar, Alexandria, Virginia, 1847–1861. Salt-glazed stoneware, brushed cobalt. H. 9½". Capacity: 1½ gallons. Impressed mark: B.C.MILBURN / ALEX^A (Courtesy, Al Marzorini.) This mark was documented on nearly one hundred vessels, split evenly between those with brushed and slip-trailed designs. The Alexandria motif is the most common design on later Milburn stoneware.

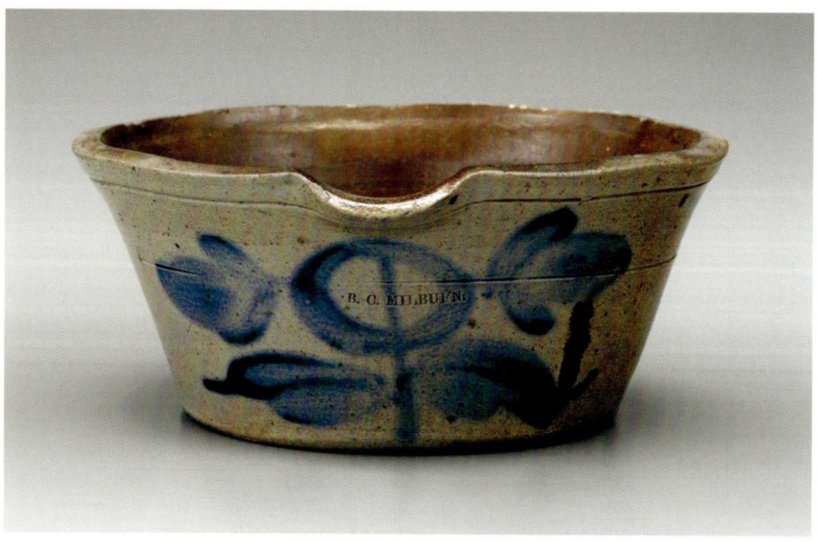

Figure 29 Milk pan, Alexandria, Virginia, 1847–1861. Salt-glazed stoneware, brushed cobalt. D. 9". Capacity: 1½ gallons. Impressed mark: B. C. MILBURN. (Courtesy, Alan Darby.) Vessels with this mark have similar form and decoration as those stamped "B. C. MILBURN / ALEX^A"

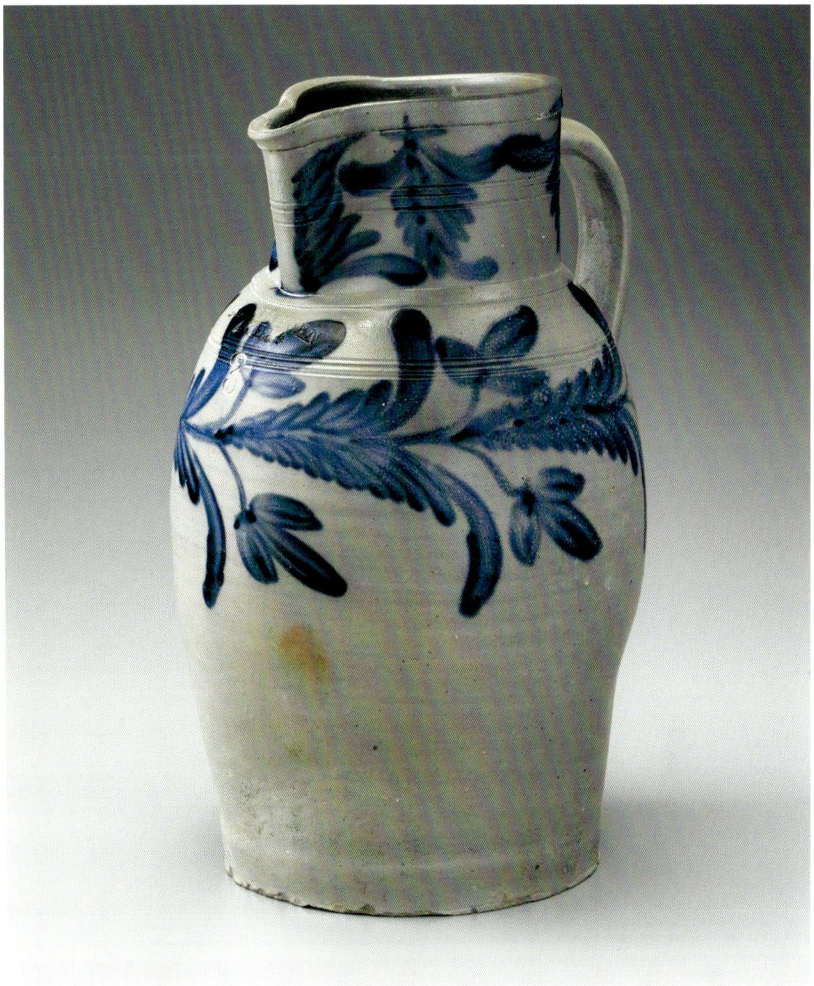

Figure 30 Pitcher, Alexandria, Virginia, 1847–1861. Salt-glazed stoneware, brushed cobalt. H. 14¾". Capacity: 3 gallons. Impressed mark: B.C.MILBURN / ALEX[A] (Courtesy, Alan Darby.) Encircling foliage remains popular in the later Milburn period.

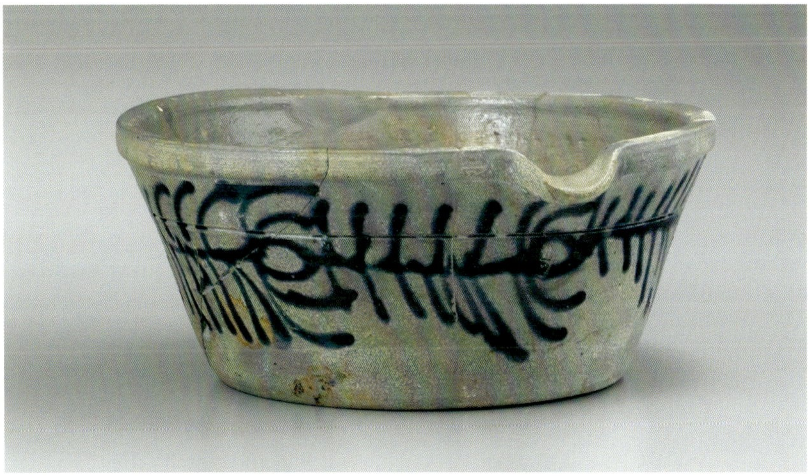

Figure 31 Milk pan, Alexandria, Virginia, 1847–1861. Salt-glazed stoneware, slip-trailed cobalt. D. 8". Capacity: ½ gallon. Impressed mark: B.C.MILBURN / ALEX[A] (Courtesy, Alexandria Archaeology Museum.) Encircling foliage is also common in slip-trailed designs. This milk pan was found in excavations on the 400 block of King Street, Alexandria's main commercial district.

Several new design elements were used in this period, in both brushed and slip-trailed cobalt. While the forward-facing flower continued to be common, flowers appeared sideways on the backs of vessels or in a variant of the encircling foliage motif (fig. 32). Milburn used a variety of flowers in addition to the simple round flower introduced by Swann. The round flower was sometimes embellished with seeds (fig. 33). The sunflower was common in this period (figs. 34, 35).

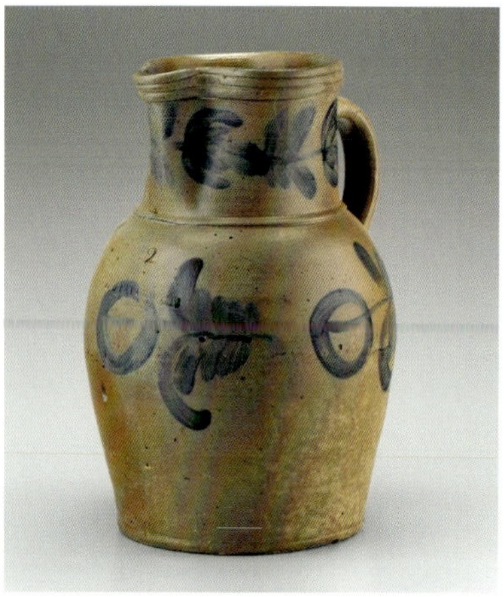

Figure 32 Pitcher, Alexandria, Virginia, 1847–1861. Salt-glazed stoneware, brushed cobalt. H. 13¼". Capacity: 3 gallons. Impressed mark: B.C.MILBURN / ALEX^{A} (Private collection.) A sideways version of the round flower first appears in the later Milburn period, and is more commonly found on the back of vessels.

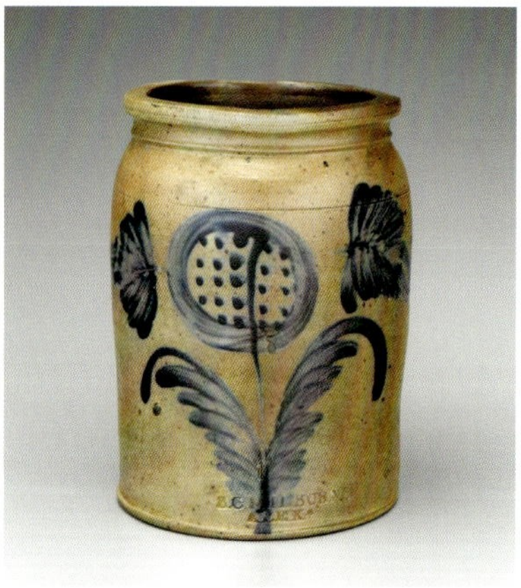

Figure 33 Jar, Alexandria, Virginia, 1847–1861. Salt-glazed stoneware, brushed cobalt. H. 9". Capacity: 1 gallon. Impressed mark: B.C.MILBURN / ALEX^{A} (Private collection.) The round flower is dotted to represent seeds.

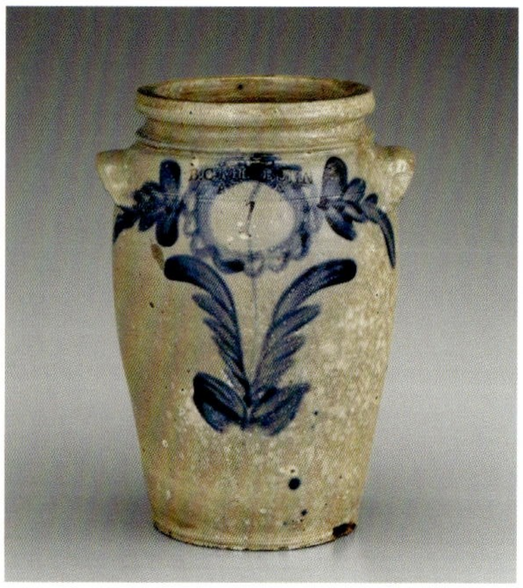

Figure 34 Jar, Alexandria, Virginia, 1847–1861. Salt-glazed stoneware, brushed cobalt. H. 10½". Capacity: 1 gallon. Impressed mark: B.C.MILBURN / ALEX^{A} (Courtesy, Al Marzorini.) The sunflower was first depicted by Swann but was not used on early Milburn stoneware.

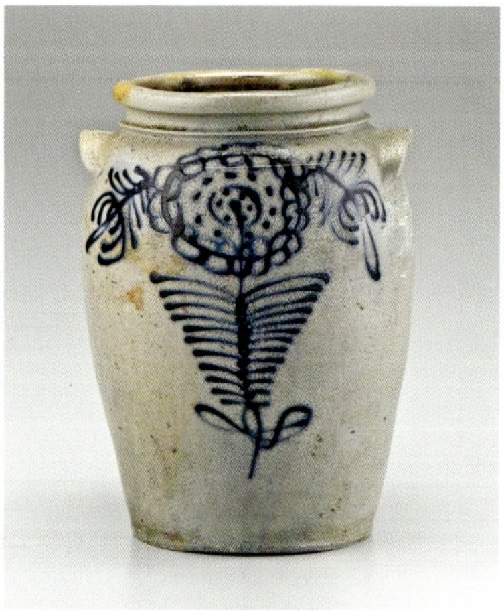

Figure 35 Jar, Alexandria, Virginia, 1847–1861. Salt-glazed stoneware, slip-trailed cobalt. H. 12". Capacity: 2 gallons. Impressed mark: B.C.MILBURN / ALEX^{A} (Private collection.) This design includes a sunflower with seeds and petals, flanked by tulips.

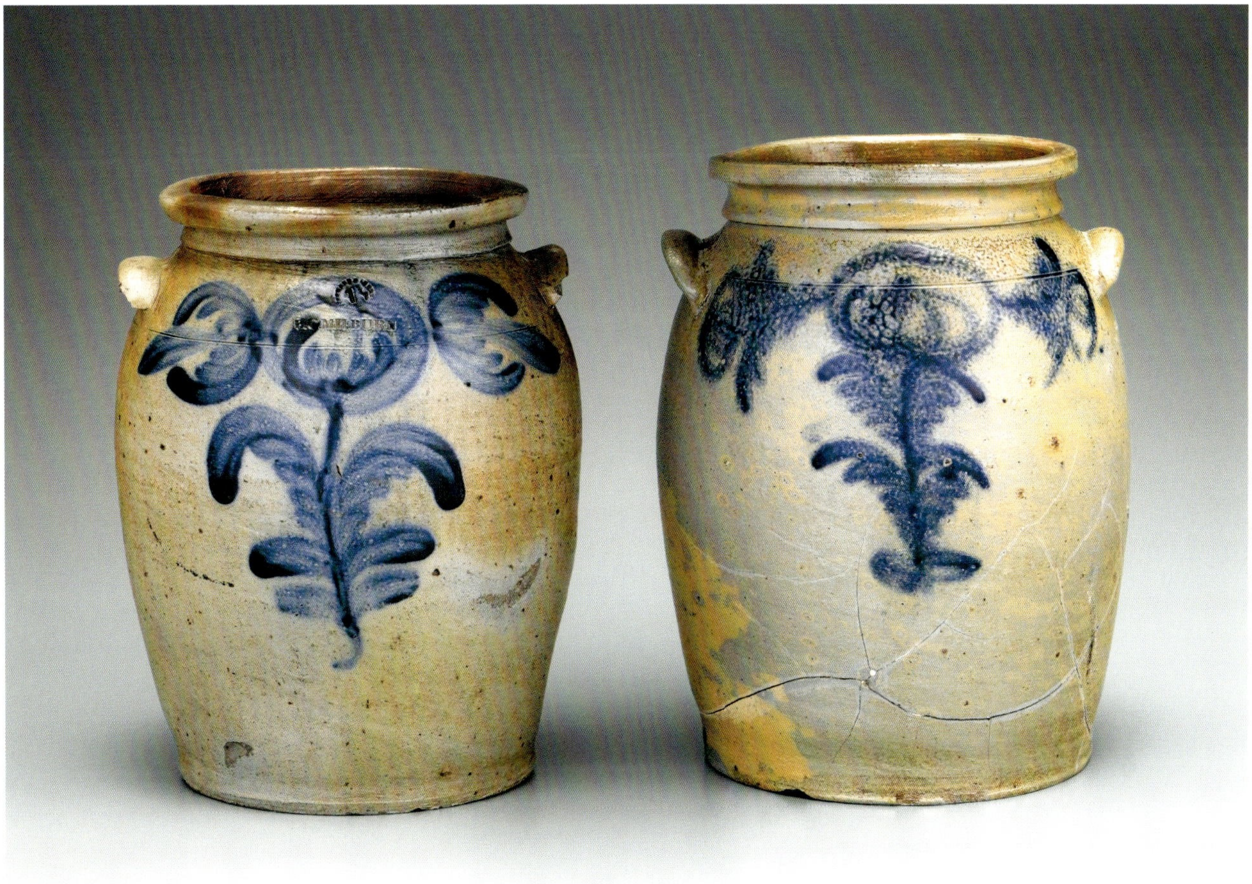

Figure 36 Jars, Alexandria, Virginia, 1847–1861. Salt-glazed stoneware, brushed cobalt. *Left:* H. 10¾". Capacity: 1½ gallons. *Right:* H. 11". Capacity: 2 gallons. Impressed mark: B. C. MILBURN. (Courtesy, Al Marzorini.) This flower, with petals within a circle, appears for the first time with this mark.

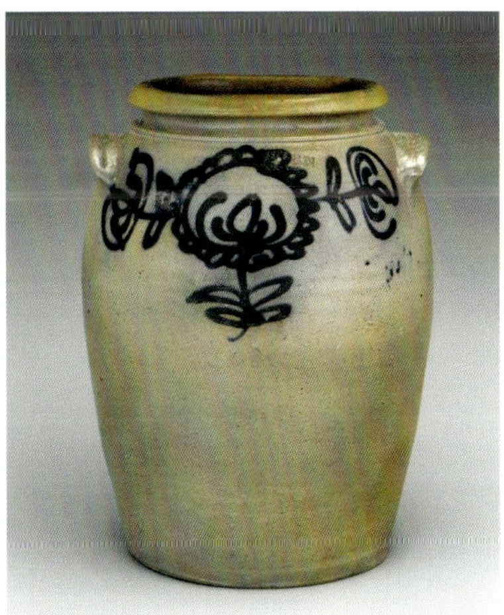

Figure 37 Jar, Alexandria, Virginia, 1847–1861. Salt-glazed stoneware, slip-trailed cobalt. H. 12". Capacity: 1½ gallons. Impressed mark: B.C.MILBURN / ALEX[A] (Courtesy, Al Marzorini.) The decorator combined two common elements in slip-trailed cobalt, the sunflower and the tulip.

Although sunflowers were depicted on Swann's pottery, they were not documented in the early Milburn period. A new flower, with petals inside an open circle, may be a depiction of a cabbage rose or simply the artist's fancy (fig. 36). In odd juxtapositions, the round flower with petals encircles a tulip-like flower on a slip-trailed jar and milk pan (figs. 37, 38), and tulips sprout from the edges of a large sunflower on a jar (fig. 39). Tulips, first used by Swann, become more common in this period and are used in

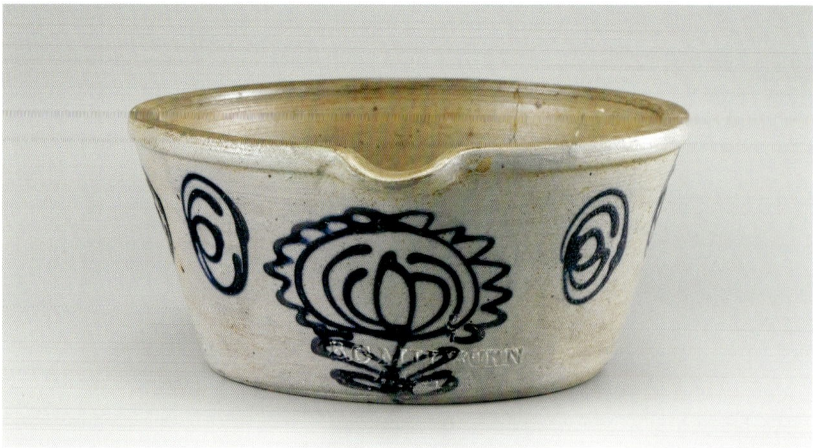

Figure 38 Milk pan, Alexandria, Virginia, 1847–1861. Salt-glazed stoneware, slip-trailed cobalt. D. 8". Capacity: ½ gallon. Impressed mark: B.C.MILBURN / ALEX^A (Private collection.)

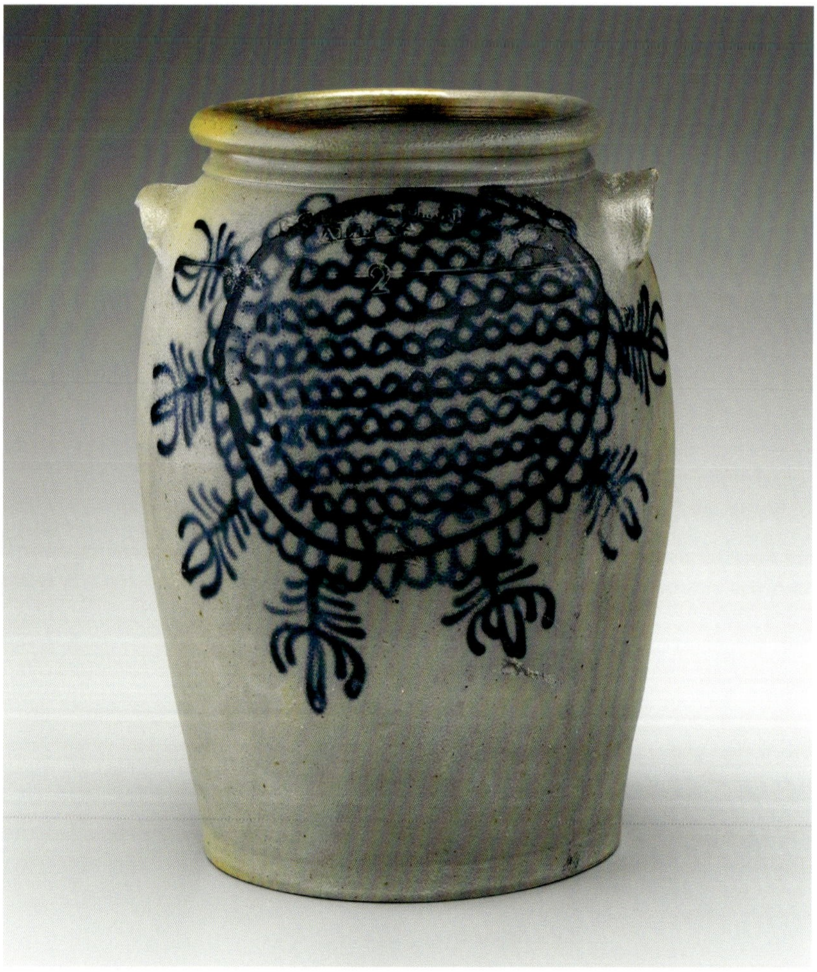

Figure 39 Jar, Alexandria, Virginia, 1847–1861. Salt-glazed stoneware, slip-trailed cobalt. H. 12½". Capacity: 2 gallons. Impressed mark: B.C.MILBURN / ALEX^A (Courtesy, Al Marzorini.) Tulips sprout from the edges of a large sunflower. The seeds are made up of a series of chain links.

forward-facing designs (figs. 40–42) and also in ones that encircle the pot (figs. 43). Encircling foliage continues to be popular in this period and is similar to that seen on earlier Alexandria pots.

Pendant leafy fronds (or swags and tassels) were first documented in the early Milburn period. In the later period, pendant fronds or tulips are usually combined with chains instead of plain swags (figs. 44, 45). New

Figure 40 Jars, Alexandria, Virginia, 1847–1861. Salt-glazed stoneware, brushed cobalt. H. 8¼" and 7". Capacity: ½ gallon. Impressed mark: B.C.MILBURN / ALEX^{A} [on both]. (Courtesy, Toby and Oscar Fitzgerald.) Tulips, first used by Swann, are more common in the later Milburn period.

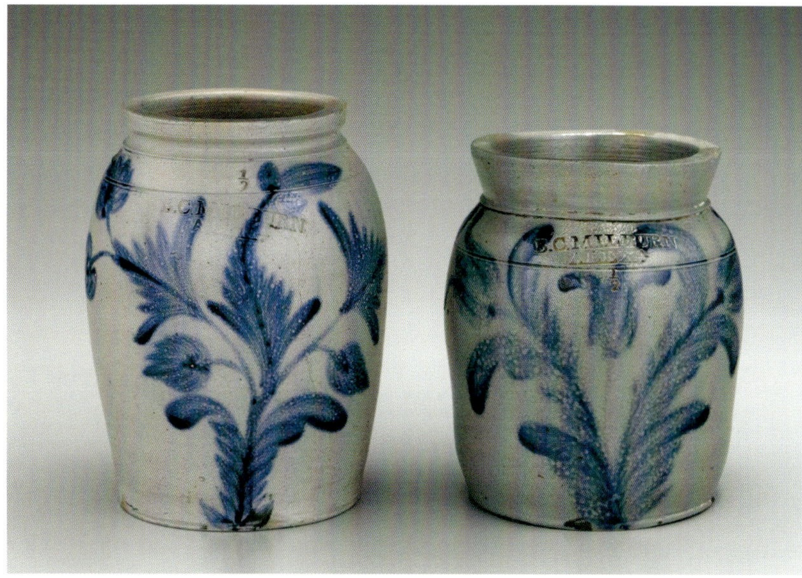

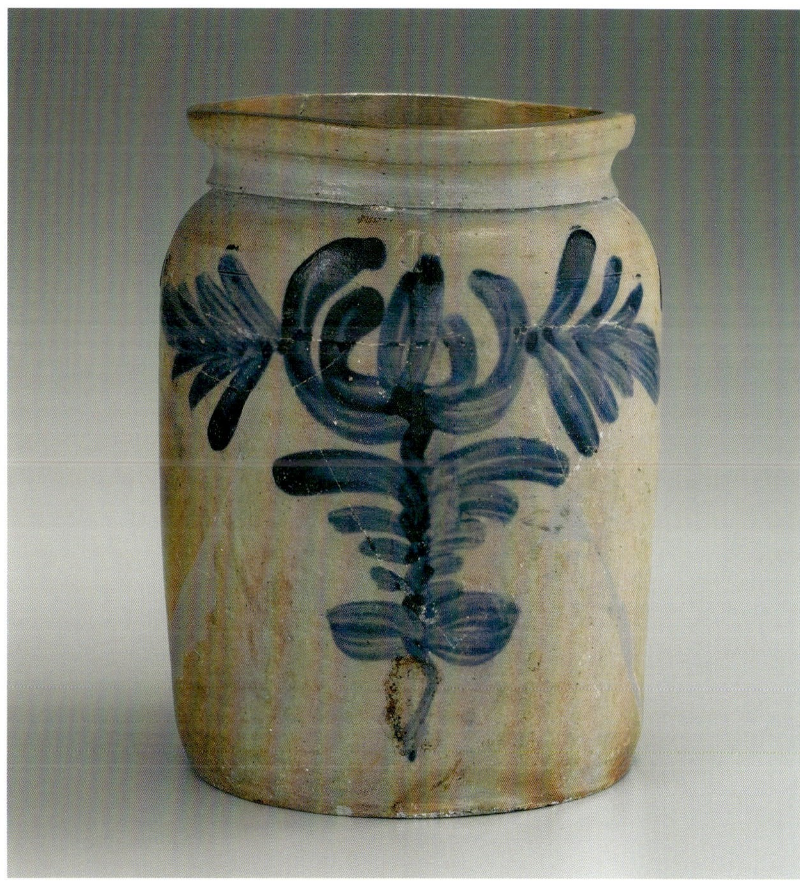

Figure 41 Jar, attributed to B.C. Milburn, Alexandria, Virginia, 1847–1861. Salt-glazed stoneware, brushed cobalt. H. 9¼". Capacity: 1 gallon. (Courtesy, Alexandria Archaeology Museum.) Unlike most of the decorated vessels from the Wilkes Street Pottery, this jar is not marked. The tulip was commonly used as the central flower on vessels with the "B.C.MILBURN / ALEX^{A}" mark, and the primitive brushwork matches other vessels from this period. The jar was found in archaeological excavations on Alexandria's Market Square.

Figure 42 Jars, Alexandria, Virginia, 1847–1861. Salt-glazed stoneware, slip-trailed cobalt. *Left:* H. 10¼". Capacity: 1 gallon. *Center:* H. 11". Capacity: 1½ gallons. *Right:* H. 12½". Capacity: 2 gallons. Impressed mark: B.C.MILBURN / ALEX^{A} (Private collection.) Milburn employed several decorators, who executed similar motifs in their own style.

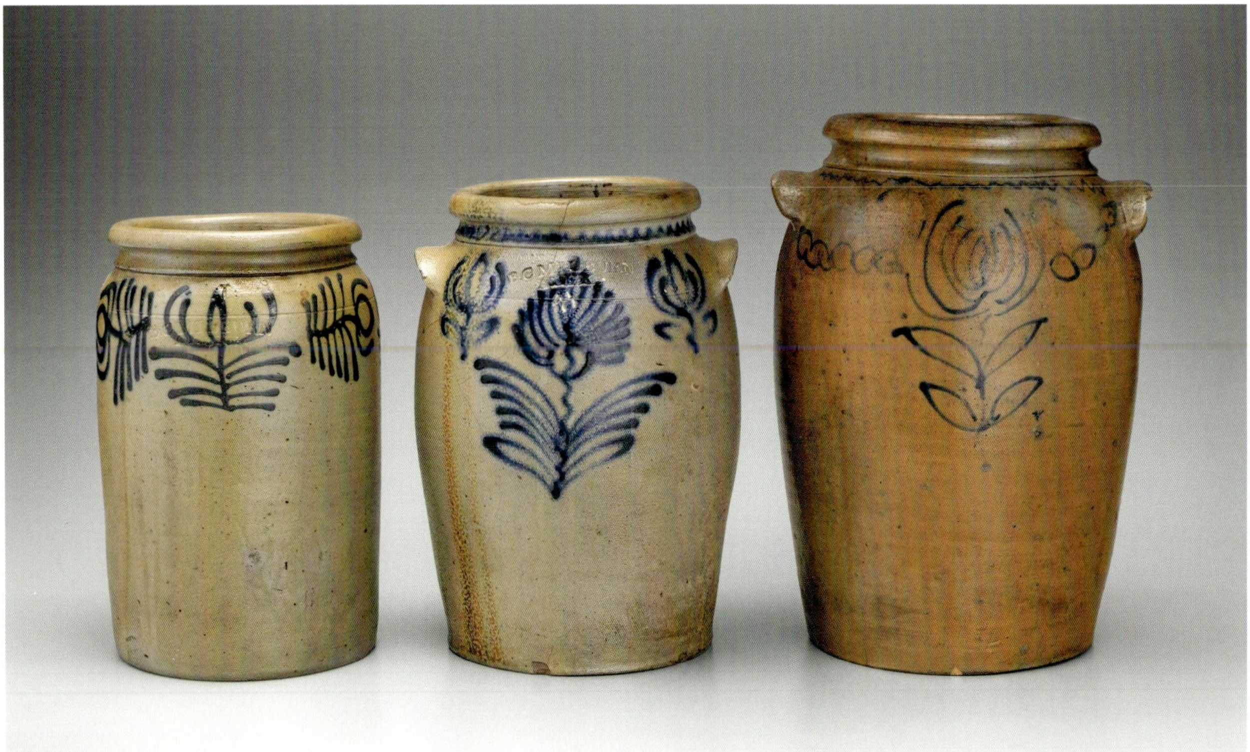

Figure 43 Jar, Alexandria, Virginia, 1847–1861. Salt-glazed stoneware, slip-trailed cobalt. H. 15¾". Capacity: 4 gallons. Impressed mark: B.C.MILBURN / ALEX^{A} (Courtesy, John and Lil Palmer.) The tulip motif is also used with encircling foliage.

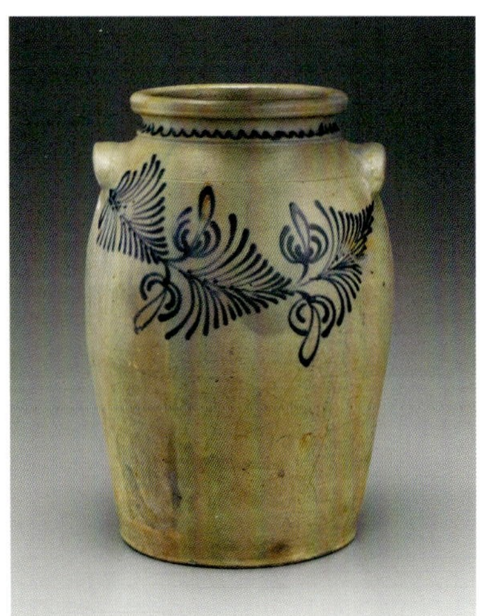

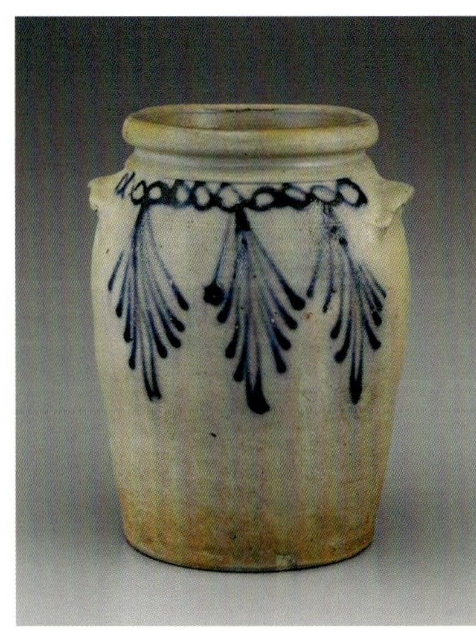

Figure 44 Jar, Alexandria, Virginia, 1847–1861. Salt-glazed stoneware, slip-trailed cobalt. H. 10½". Capacity: 1½ gallons. Impressed mark: B.C.MILBURN / ALEX^{A} (Courtesy, William and Jane Yeingst.) In the later Milburn period, pendant leafy fronds were usually combined with chains instead of plain swags. The reverse is decorated with two sideways tulip-like flowers.

designs of chains and scalloped or wavy lines also commonly appear on the reverse of vessels but sometimes surround the rim (figs. 46, 47).

Some vessels of this period resemble Shenandoah Valley stoneware, with graduated foliage at the rim (fig. 48). Often, the graduated foliage is combined with pendant tulips. The exchange of ideas between potters from

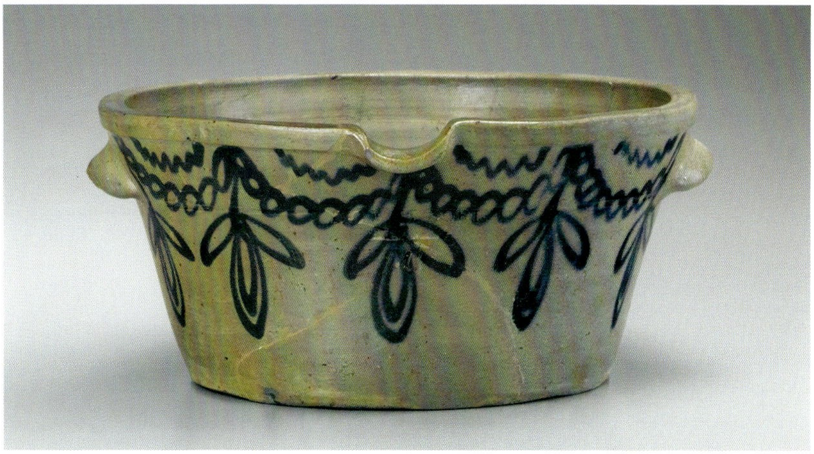

Figure 45 Milk pan, Alexandria, Virginia, 1847–1861. Salt-glazed stoneware, slip-trailed cobalt. D. 12". Capacity: 1 gallon. Impressed mark: B.C.MILBURN / ALEX^{A} (Courtesy, Alexandria Archaeology Museum.) The chains and pendant flowers continue on the reverse. This milk pan was found in excavations at 112 South Royal Street.

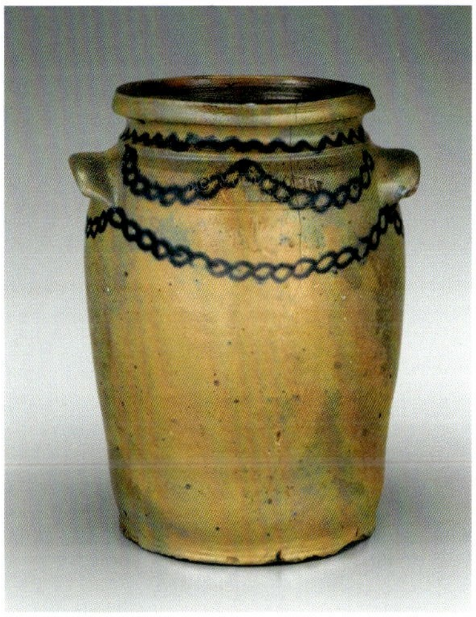

Figure 46 Jar, Alexandria, Virginia, 1847–1861. Salt-glazed stoneware, slip-trailed cobalt. H. 10". Capacity: 1 gallon: Impressed mark: B.C.MILBURN/ALEX^{A} (Courtesy, Al Marzorini.) Chains or wavy lines usually appear on the reverse of Milburn's vessels, but they decorate both sides of this jar.

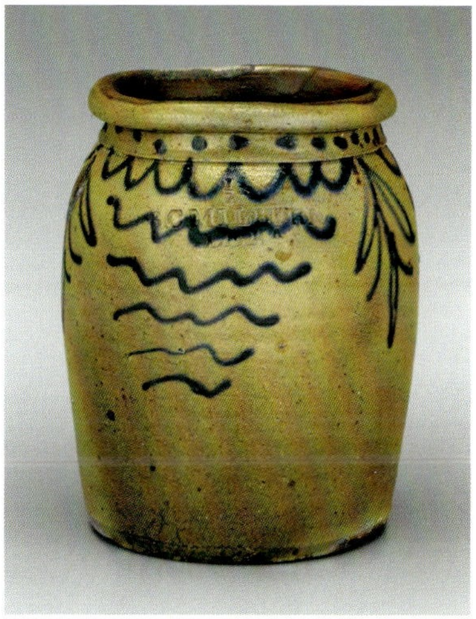

Figure 47 Jar, Alexandria, Virginia, 1847–1861. Salt-glazed stoneware, slip-trailed cobalt. H. 7½". Capacity: ½ gallon. Impressed mark: B.C.MILBURN / ALEX^{A} (Courtesy, John and Lil Palmer.) A similar pattern of dots and wavy lines appears on the reverse of this jar, with the front and back separated by pendant fronds.

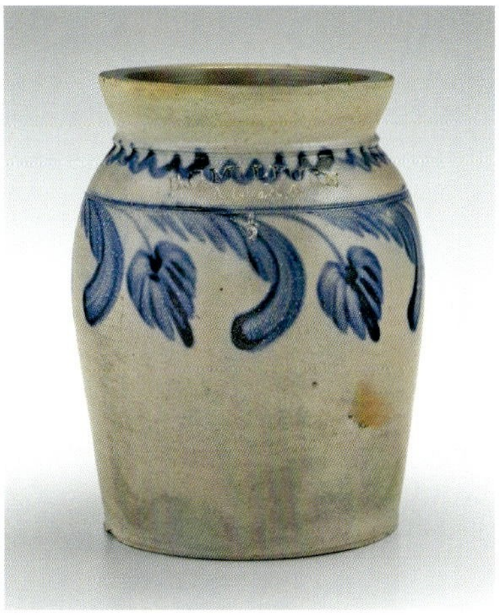

Figure 48 Jar, Alexandria, Virginia, 1847–1861. Salt-glazed stoneware, brushed cobalt. H. 8". Capacity: ½ gallon. Impressed mark: B.C.MILBURN / ALEX^{A} (Private collection.) The graduated foliage at the shoulder is similar to that used by Shenandoah Valley potters.

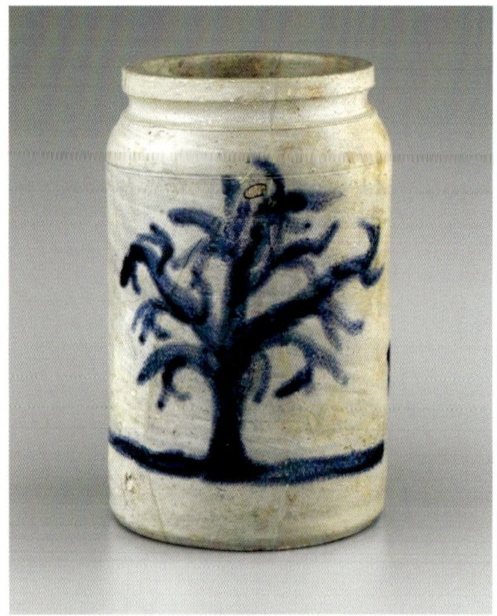

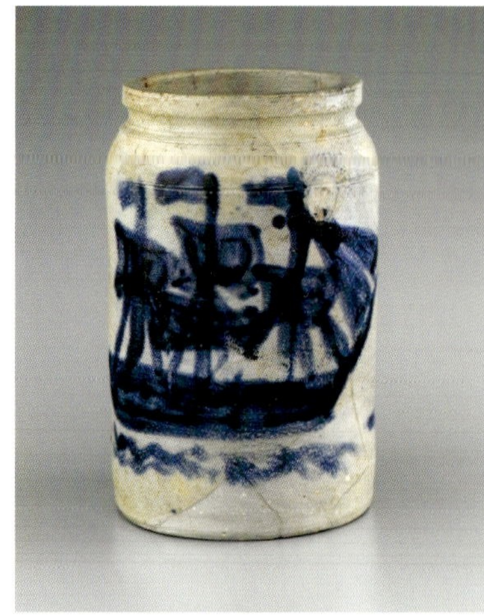

Figure 49 Jar, attributed to B.C. Milburn, Alexandria, Virginia, 1847–1861. Salt-glazed stoneware, brushed cobalt. H. 8¼". (Courtesy, Alexandria Archaeology Museum.) The representative subject matter is unusual, but the clay, the shape of the vessel, and the use of slip trailing are similar to other Milburn pieces. This jar was found in excavations in Alexandria, at the site of a nineteenth-century restaurant and residence on the 400 block of King Street.

Alexandria and the Shenandoah Valley may have gone in both directions. The decoration on a salt-glazed stoneware pitcher produced circa 1866–1867 and marked "IRELAND / & / DUEY. / MT. CRAWFORD / VA." bears a striking resemblance to the work of the Milburn pottery made between 1847 and 1861, with chains, pendant flowers, and graduated leaves.[39]

The marked Alexandria stoneware vessels show no representational motifs apart from flowers and foliage. However, a wonderful jar depicting a sailing ship and tree has long been attributed to Milburn (fig. 49). This intriguing find is from archaeological excavations undertaken on Alexandria's King Street in the early 1970s, at the site of a nineteenth-century restaurant. As with most unmarked pieces, we cannot make a definite attribution, but this jar does bear a significant resemblance to other Milburn pieces. The clay is similar in appearance to Alexandria stoneware, and the shape of the straight-sided jar and its square rim with collar are typical of jars of the Milburn period. The ship and the tree are both common nineteenth-century motifs.

Stoneware decoration was common at least through the J. P. Smith period (1851–1854), and probably until the Civil War. As Alexandria recovered from its long occupation by Union troops, businesses struggled to survive. The use of cobalt may have been discontinued due to the costs of materials and labor. Undecorated jars, jugs, and other forms with the "B.C.MILBURN / ALEX^{A}" mark are similar to those with the marks of Milburn's sons.

Stoneware by Milburn's Sons

S. C. Milburn, 1867–1873

The "S.C.MILBURN." stamp appears primarily on undecorated utilitarian wares, including jars, jugs, and milk pans with forms similar to those with his father's mark (fig. 50). A spittoon (not illustrated) and a pitcher have sparse cobalt decoration (fig. 51).[40]

W. Lewis Milburn, 1871–1876

The mark "W. LEWIS MILBURN." is found on just one rim sherd from the kiln site, from a one-gallon straight-sided jar with a square rim and

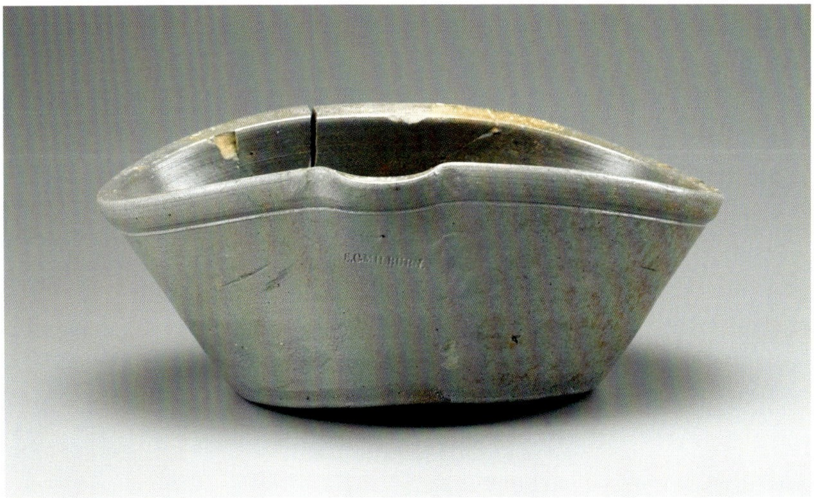

Figure 50 Milk pan, Alexandria, Virginia, 1867–1873. Salt-glazed stoneware. H. 4½". Impressed mark: S.C.MILBURN. (Courtesy, Alexandria Archaeology Museum.) Despite the crack, this badly warped "second" was used at a residence at 307 South Saint Asaph Street.

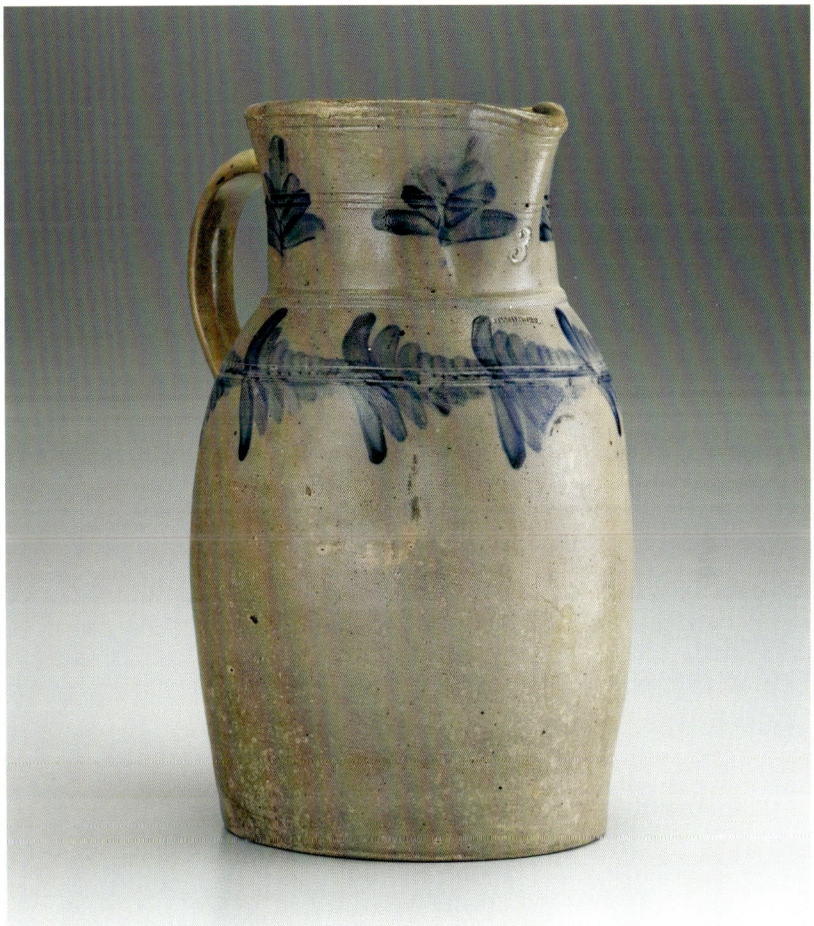

Figure 51 Pitcher, Alexandria, Virginia, 1867–1873. Salt-glazed stoneware, brushed cobalt. H. 15¾". Capacity: 3 gallons. Impressed mark: S.C.MILBURN. (Private collection.) This is a rare example of a decorated S. C. Milburn vessel.

collar (see fig. 10). Other surviving examples with this mark include similar jars and a squat, straight-sided, half-gallon jug with a straight rim (see fig. 76), which is similar to others marked "E.J.MILLER/ALEX^{A}." Milburn descendants were known to have had a pitcher marked in script "Mrs. W. L. Milburn / 1871,"[41] but even this special presentation piece was undecorated.[42]

E. J. Miller, 1865–1876

The Miller mark appears on five undecorated sherds from the kiln site, from vessels that include milk pans and jugs or bottles. While some of the Miller stoneware may have been made before B.C. Milburn's death in 1867, S.C. Milburn probably manufactured most of the Miller pots between 1867 and 1873. Identical jars bear marks for B.C. Milburn, S.C. Milburn, and E. J. Miller (fig. 52).

E. J. Miller & Co. (also called E. J. Miller & Son) survived until sometime between 1907 and 1917. After the Wilkes Street Pottery closed, James Hamilton in Greensboro, Pennsylvania, supplied stoneware for E. J. Miller (fig. 53). Hamilton's stoneware was sold to merchants in several states and

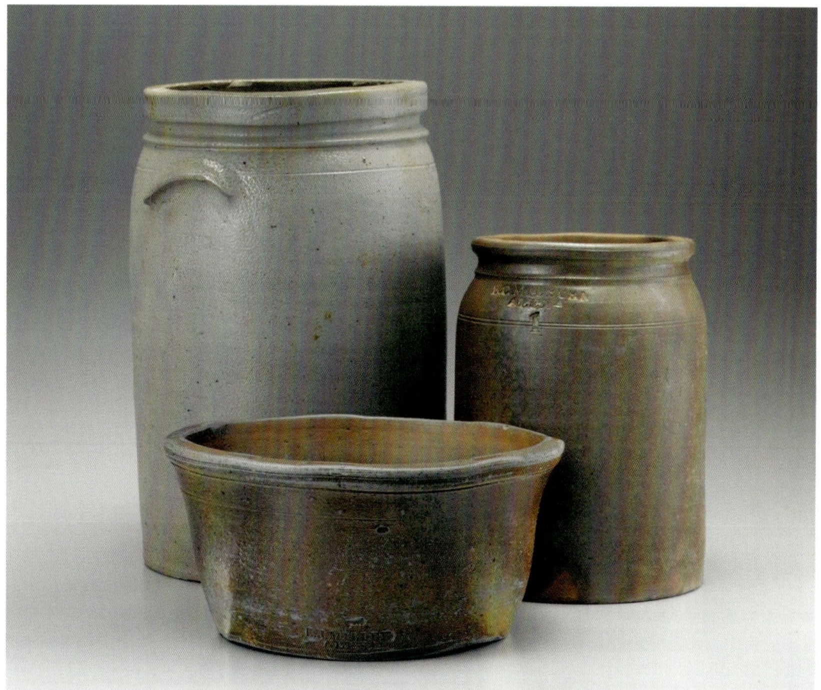

Figure 52 Jars and bowl, Alexandria, Virginia. Salt-glazed stoneware. *Jar, back left*, 1865–1876. H. 13¼". Impressed mark: E.J.MILLER / ALEX^{A}. *Jar, back right*, 1847–1867. H. 9½". Impressed mark: B.C.MILBURN/ ALEX^{A}. (Courtesy, The Lyceum: Alexandria's History Museum.) *Bowl, front*, 1865–1876. D. 10". Impressed mark: E.J.MILLER / ALEX^{A} (Courtesy, Smithsonian Institution, National Museum of American History.) E. J. Miller took over his father's china shop in 1865, and jars with his mark are identical to ones with the marks of B.C. and S.C. Milburn. Five sherds marked E. J. Miller were found at the Milburn kiln site.

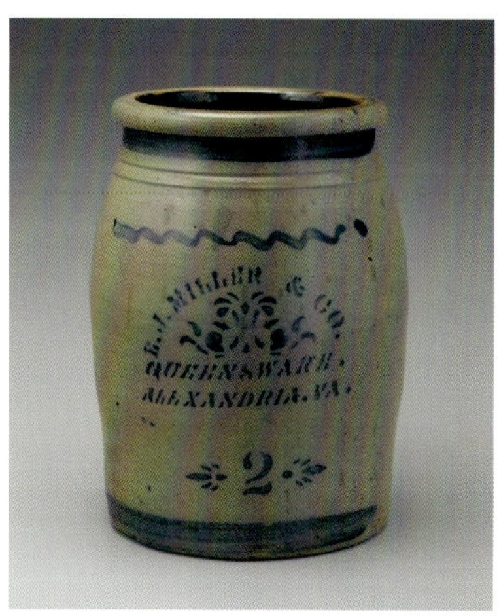

Figure 53 Jar, James Hamilton, Greensboro, Pennsylvania, after ca. 1876. Salt-glazed stoneware with Albany-slip interior, brushed and stenciled cobalt. H. 11½". Capacity: 2 gallons. Stenciling: E.J.MILLER & CO. / QUEENSWARE. / *ALEXANDRIA.VA.* (Courtesy, The Lyceum: Alexandria's History Museum.) Stenciled advertising is not found on stoneware manufactured in Alexandria. After the Alexandria Pottery closed, E. J. Miller & Co. purchased stoneware from this large Pennsylvania manufactory.

is fairly common in Alexandria. It is easily distinguished from that made at the Wilkes Street Pottery, with a shiny brown Albany-slip interior and elaborate stenciled-cobalt advertising. Stenciling was never done at the Alexandria potteries, but is typical of the wares of southwestern Pennsylvania. The pots made for Miller were stenciled with various advertising for E. J. Miller & Co. or E. J. Miller & Son. The business was located at 65 King Street (now 317 King Street) until sometime between 1900 and 1907, and some of the stenciling includes this address. The design sometimes includes stenciled foliage, or may have wavy lines or a capacity mark and leaves in brushed cobalt.

Handles

A simple, straight lug handle with cut ends is found on pots with all of the marks used at Wilkes Street, beginning with Swann, and is found on ovoid jars, churns, cake pots, milk pans, and some coolers (fig. 54).

Extruded lug handles were introduced before 1847 on vessels marked "H.C.SMITH / ALEX^{A} / D.C". They are also found on ovoid jars with the later marks "B. C. MILBURN." and "B.C.MILBURN / ALEX^{A}" (fig. 55). Extruded handles were made by putting clay into a tube and using a plunger to force it through a template, in a way similar to a cookie press. This technique became more common on American stoneware, including that of the Shenandoah Valley, in the mid-nineteenth century.

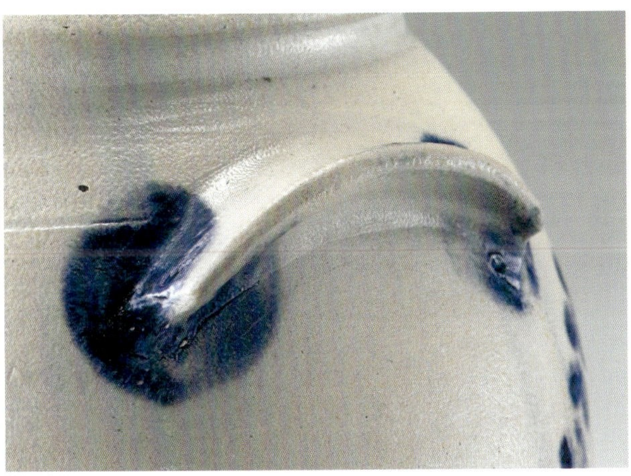

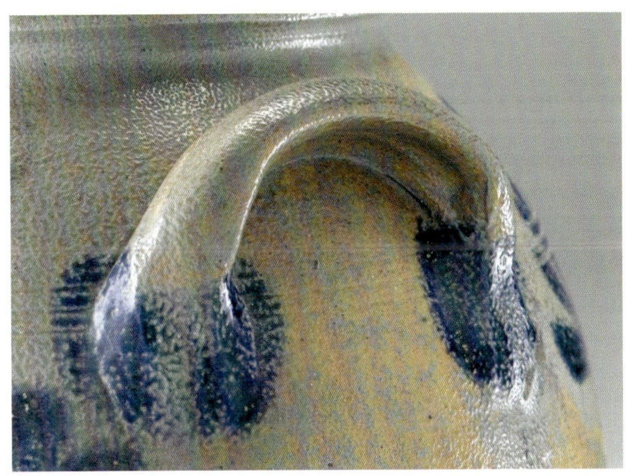

Figure 54 Lug handle. L. 4". Detail of 1819–1821 jar bearing "J.SWANN / ALEX^{A}" mark. A simple, straight handle with cut ends was first used by John Swann and continued to be used on jars, churns, cake pots, and milk pans throughout the history of the Wilkes Street Pottery.

Figure 55 Extruded lug handle. L. 5". Detail of the jar illustrated in fig. 20. Extruded handles, in which clay was forced through a template, first appeared on Alexandria jars between 1831 and 1847; they became common on American stoneware in the mid-nineteenth century.

Extruded strap handles were used in all periods. Vertical strap handles are found on pitchers, chamber pots, and coolers (fig. 56), while horizontal open-loop handles are found on larger coolers (fig. 57).

Throughout the history of the pottery, cylindrical jars and some smaller ovoid jars and milk pans were made with no handles.

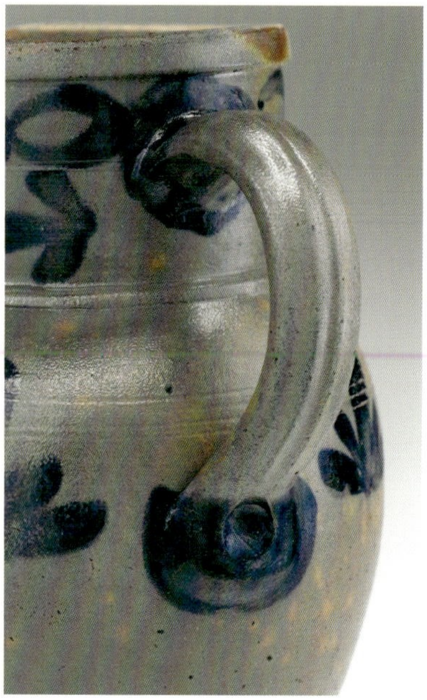

Figure 56 Strap handle. L. 5½". Detail of the pitcher illustrated in fig. 11. Vertical strap handles were used on pitchers, chamber pots, and coolers of all periods.

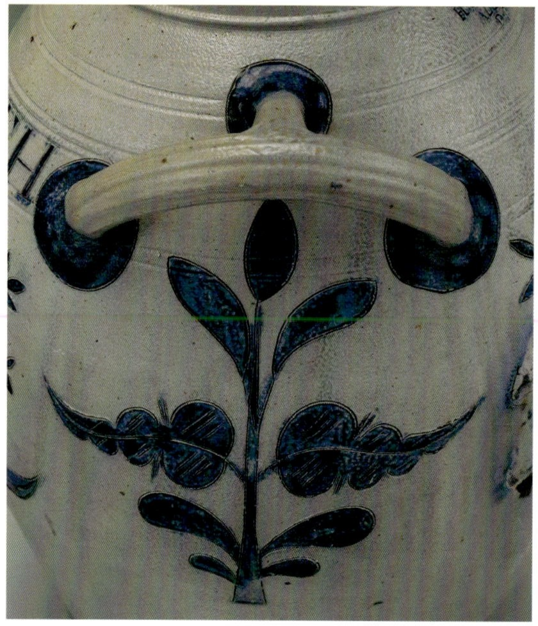

Figure 57 Extruded horizontal open-loop handle. L. 7". Detail of the cooler illustrated in fig. 12.

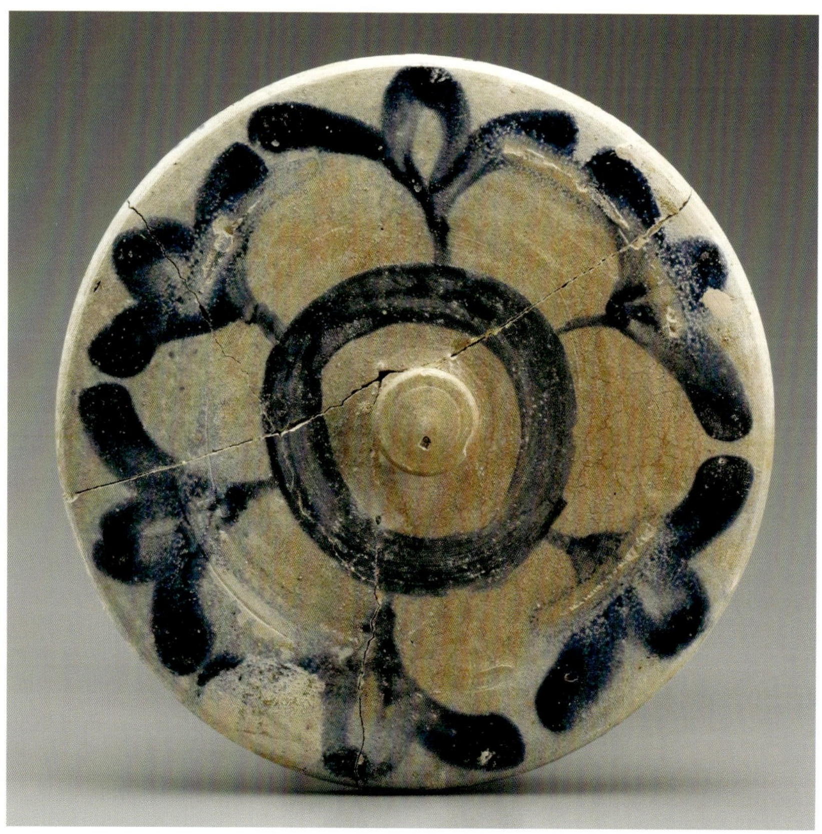

Figure 58 Lid, Alexandria, Virginia, 1825–1861. Salt-glazed stoneware, brushed cobalt. D. 7½". (Courtesy, Alexandria Archaeology Museum.) Some lids are flat on the underside, but others, like this lid from excavations at 524 King Street, have a shallow flange one inch from the edge.

Rims

At least six distinct rim forms were used on stoneware from the Milburn period. Within these forms are numerous minor variations, such as small differences in size and angle. The most variation is seen in jar rims. Certain rim shapes were specifically designed to be used with stoneware or wooden lids (fig. 58). These include flared rims used on some jars and coolers (fig. 59), flat rims used on butter or cake pots (fig. 60), and tall, flared churn rims that allow a wooden or stoneware guide for the dasher to sit down inside the rim (fig. 61).

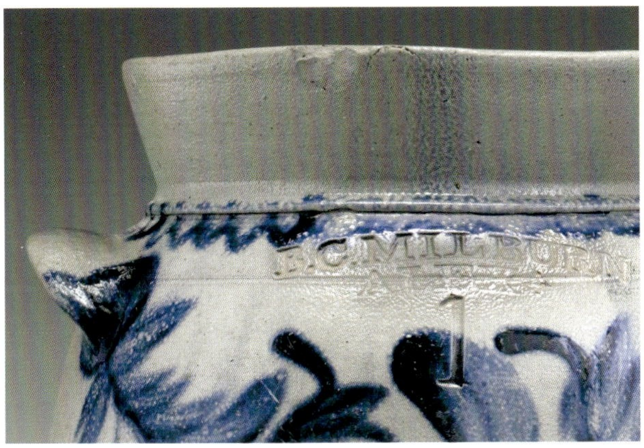

Figure 59 Flared rim. Detail of the jar illustrated in fig. 73. Flared rims accommodated flat stoneware or wooden lids.

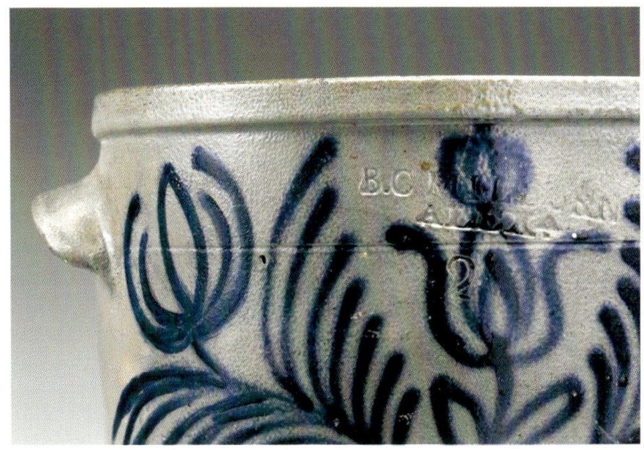

Figure 60 Flat rim. Detail of the butter pot illustrated in fig. 64. Flat rims supported a flanged lid, as shown in fig. 58.

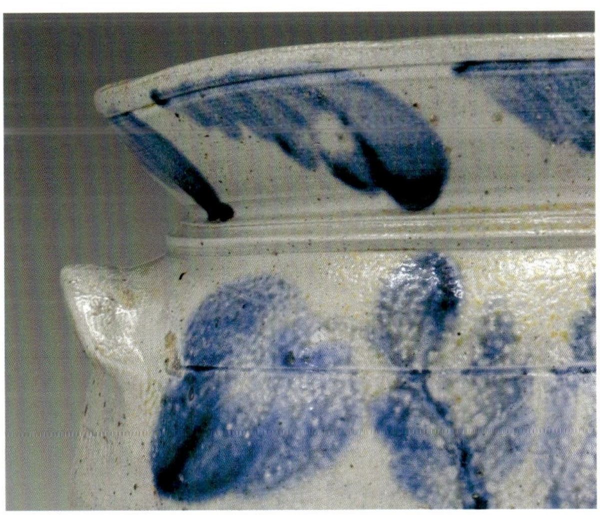

Figure 61 Flared rim. Detail of the churn illustrated in fig. 66. Tall, flared churn rims allow a flat wooden or stoneware disk with a hole for the dasher to sit down inside the rim.

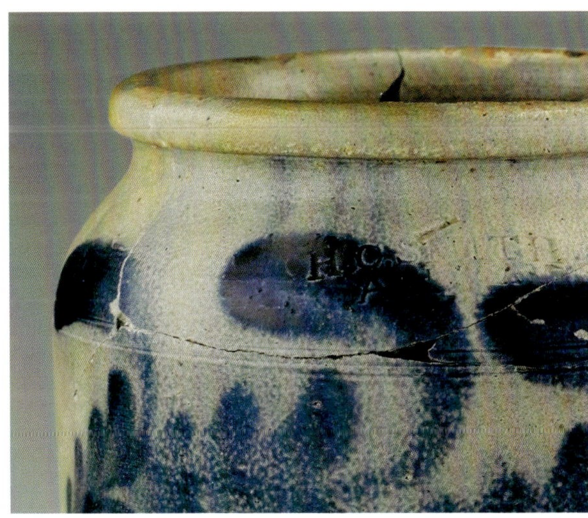

Figure 62 Round rim. Detail of the jar illustrated in fig. 2. The round rim was used on jars throughout the history of the pottery.

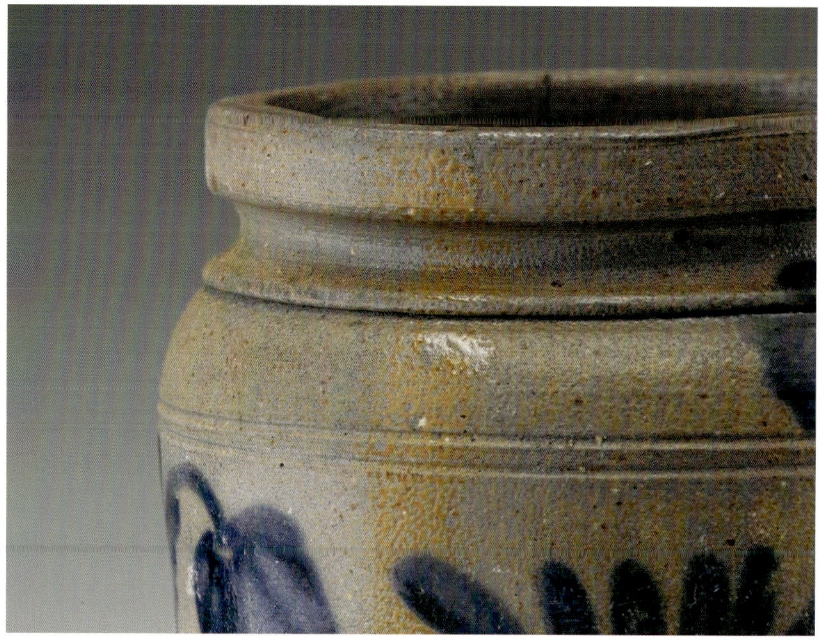

Figure 63 Collared rim. (Detail; jar not illustrated.) (Courtesy, William and Jane Yeingst.) A collar below the rim was first used with the stamp "H. C. SMITH / ALEX^{A} / D. C" between 1831 and 1847.

The round rim is the most common form on jars made throughout the history of the pottery (fig. 62). A new collared-rim form was introduced by Milburn sometime between 1831 and 1847 (fig. 63). This innovation was first seen, along with extruded lug handles, with the "H.C.SMITH / ALEX^{A} / D.C" mark, and the square rim with collar was documented with all of the marks from the Milburn period. The round rim with collar may have been limited to jars with post-retrocession B.C. Milburn marks. Collared rims are common on both ovoid and cylindrical jars. They appear on some jars with brushed-cobalt decoration and on all of the jars recorded with trailed-slip decoration. Collared rims were also common on Shenandoah Valley stoneware by the mid-nineteenth century.

Vessel Forms

Milburn expanded the repertoire of vessel forms produced at Wilkes Street. While eight vessel forms are known from the Swann period, fifteen forms were recorded for the Milburn period. Milburn continued to make butter pots, chamber pots, churns, jars, jugs, milk pans, and pitchers, but there is no evidence that he continued to make the portable earthen furnaces advertised by H. Smith & Co. in 1829. Additions in the Milburn period include banks, chimney pots, coolers, cups or bowls, flowerpots, spittoons, and stovepipe collars.

Banks. In her 1928 history of Alexandria, Mary C. Powell wrote that the Milburn Pottery made "little money jugs for children."[43] No bank fragments were found at the kiln site, but Suzita Myers recorded a bank with a squat bottle form and tapered finial, made for B.C. Milburn's son James Clinton Milburn (b. 1844), which was decorated with encircling leaves and buds.[44]

Bowls A deep, straight-sided bowl was made for E. J. Miller after 1865 (see fig. 52).

Butter Pots Butter pots are tall, straight-sided jars with lug handles and plain rims that are flat on top to support a lid.[45] Milburn's 1859 trade card lists "butter jars" as covered wares.[46] A butter pot marked "B.C.MILBURN / ALEX^A" has slip-trailed decoration (fig. 64). Rim fragments of twenty-eight butter or cake pots were found at the kiln site, but most if not all were from the earlier Swann period.

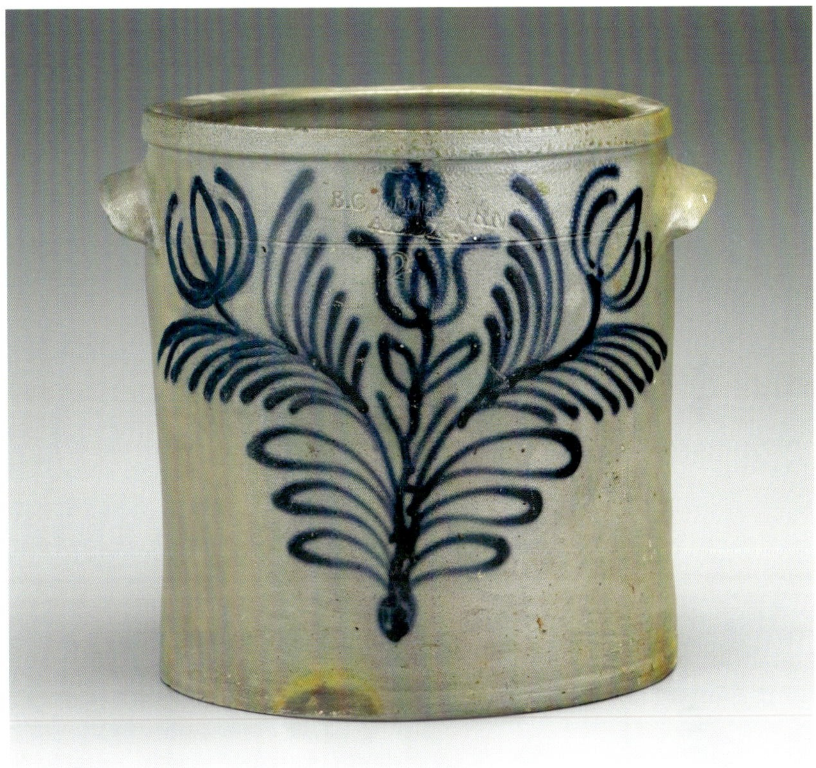

Figure 64 Butter pot, Alexandria, Virginia, 1847–1861. Salt-glazed stoneware, slip-trailed cobalt. H. 9½". Capacity: 2 gallons. Impressed mark: B.C.MILBURN / ALEX^A (Courtesy, Alan Darby.) Milburn's 1859 trade card lists "butter jars" as covered wares. These would have had flat lids like that shown in fig. 58.

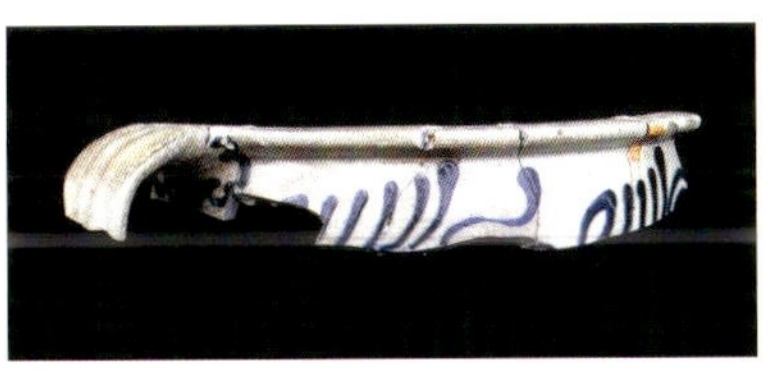

Figure 65 Chamber pot fragment, Alexandria, Virginia, 1847–1861. Salt-glazed stoneware, slip-trailed cobalt. D. 7". (Courtesy, Alexandria Archaeology Museum.) This fragment, found in excavations on the 200 block of North Union Street, is attributed to Milburn based on the familiar slip-trailed design.

Chamber Pots. Only one chamber pot is known to have an Alexandria mark, from the earlier Swann period. However, several unmarked chamber-pot fragments, with flat everted rims and brushed or slip-trailed decoration typical of Wilkes Street, have been found on Alexandria archaeological sites (fig. 65).

Chimney Pots. An 1869 advertisement by S. C. Milburn mentions earthen chimney pots.[47] No fragments were found at the kiln site. Chimney pots would most likely be unmarked. If small fragments of cylindrical unglazed earthenware chimney pots have been found archaeologically in Alexandria, they have probably been misidentified as flowerpots or drainpipes. Some Milburn chimney pots probably still sit atop Alexandria houses.

Churns. Churns are tall and cylindrical with lug handles and flared or cup-shaped rims to support the stoneware or wooden disk holding the handle

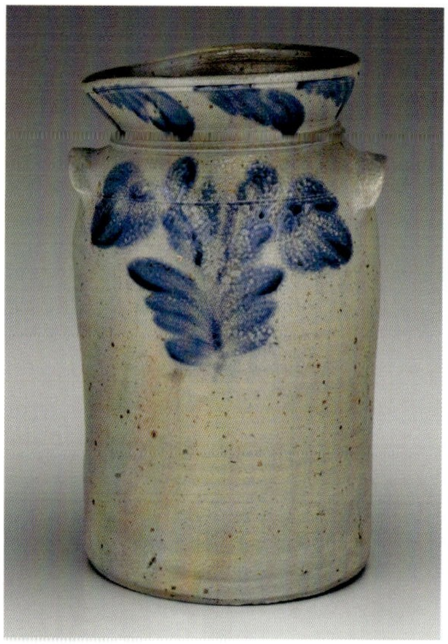

Figure 66 Churn, Alexandria, Virginia, 1847–1861. Salt-glazed stoneware, brushed cobalt. H. 12". Capacity: 2 gallons. Impressed mark: B.C.MILBURN / ALEX^{A} (Courtesy, John and Lil Palmer.)

of the dasher. One churn recorded with the mark "H.C.SMITH / ALEX^{A}" and two with the mark "B.C.MILBURN / ALEX^{A}" have brushed-cobalt decoration and flared rims (fig. 66). Milburn's 1859 trade card listed churns under the heading of covered ware. Henry Wolff 's 1874 mechanic's lien against W. Lewis Milburn mentions that he made three- and four-gallon churns.[48]

Coolers. The four coolers recorded from Alexandria are from the Milburn period. Coolers have an ovoid shape, a bunghole near the bottom, and a variety of rims and handles. Also known as fountains or kegs, these vessels could have held water, cider, or liquor. Two coolers are marked "H.C.SMITH /ALEX^{A} / D.C" (ca. 1831–1847), and two are marked "B.C.MILBURN / ALEX^{A}" (after 1847).

The Smith mark is found on the simplest cooler, shaped like a pitcher but with two strap handles and a bunghole (fig. 67). The mark is also found on the most elaborate cooler, with extruded horizontal handles, a straight rim, and unusual decoration with filled incised lines.

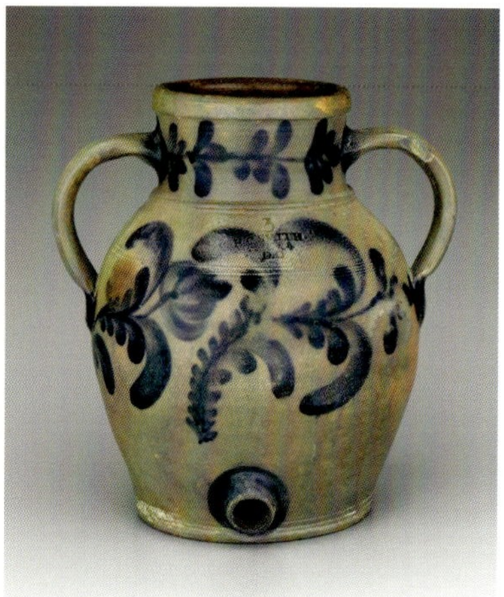

Figure 67 Cooler, Alexandria, Virginia, 1831–1847. Salt-glazed stoneware, brushed cobalt. H. 14". Capacity: 3 gallons. Impressed mark: H.C.SMITH / ALEX^{A} / D.C (Courtesy, John and Lil Palmer.) The design and shape of this cooler are similar to pitchers of this period, but with two handles and a bunghole at the base instead of a pouring spout at the rim. On the reverse is a large chain on the neck and five-leaf clusters on the shoulder.

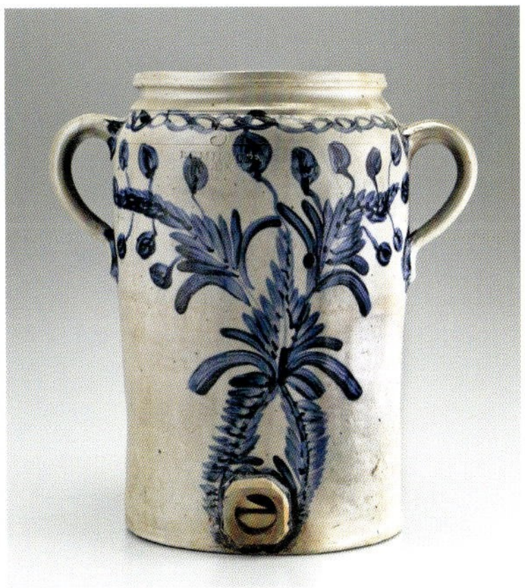

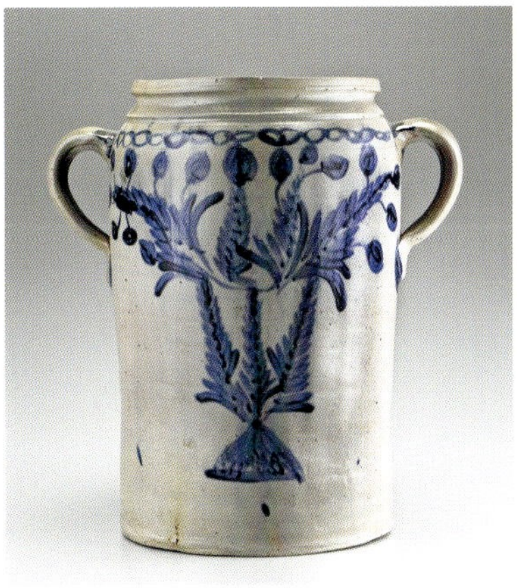

Figure 68 Cooler, Alexandria, Virginia, 1847–1861. Salt-glazed stoneware, brushed cobalt. H. 17". Capacity: 5 gallons. Impressed mark: B.C.MILBURN / ALEX^{A} (Private collection.) The round flower buds are unusual in Alexandria but similar to ones on jugs attributed to Emanuel Suter (ca. 1866–1880), a later Rockingham County, Virginia, potter.

One of the Milburn coolers with slip-trailed decoration (not illustrated) is shaped like a jar, with a slightly tapered body, lug handles, and a flared rim. The bunghole identifies it as a cooler. The other, with a brushed-cobalt design, has a square rim with collar like that used on jars, and strap handles of the type usually seen on pitchers (fig. 68).

Cups. A 1945 article in *Antiques* illustrates a "Brown Stoneware Whimsey," a small mug made at the Milburn Pottery for James McCormick Eaches and bearing his initials; it is dated 1867 on the reverse. The cup was an heirloom of the author, Nancy Lee Tackett. The small brown stoneware cup has a flat bottom, no handle, three incised lines below a slightly flaring rim, and the initials JME written in script.[49]

Flowerpots. Milburn advertised in 1845 that he had "just finished burning a kiln of flower pots of all sizes—from 11 inches to 2 inches."[50] Flowerpots are also listed on his 1859 trade card and in his son S. C. Milburn's 1869 advertisement. Unglazed earthenware flowerpot fragments were found at the kiln site, in four- to eight-inch sizes. Flowerpot trays were also found. The pots have either a slight bulge at the rim or a double flange. Most of the trays are plain with a flat or rounded rim, but some have an incised line just below the rim. One stoneware flowerpot was found archaeologically with the mark "B.C.MILBURN / ALEX^{A}"; it is undecorated, light brown

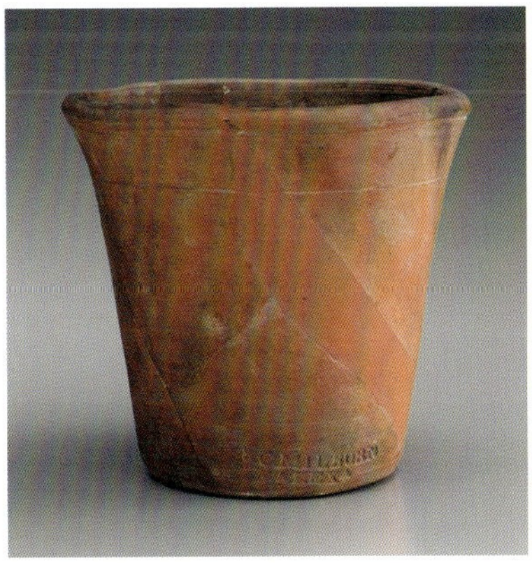

Figure 69 Flowerpot, Alexandria, Virginia, 1841–1867. H. 7¼". Salt-glazed stoneware. Stamped near base: B.C.MILBURN / ALEX^{A} (Courtesy, Alexandria Archaeology Museum.)

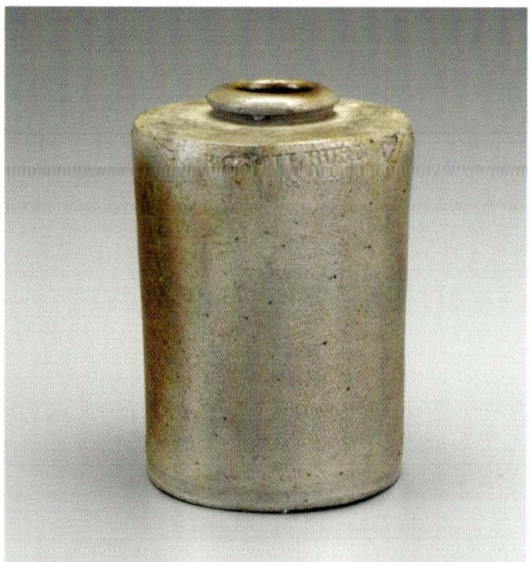

Figure 70 Master ink bottle, Alexandria, Virginia, 1847–1867. Salt-glazed stoneware. H. 6¾". Impressed mark: B.C.MILBURN / ALEX^{A} (Courtesy, Alan Darby.)

in color, and has a plain rounded rim (fig. 69). Swann does not appear to have made flowerpots, but they were made by earlier earthenware potters Henry Piercy and Lewis Plum, and by Tildon Easton, who made them in both earthenware and stoneware.

Ink bottles. One master ink bottle was documented, with the "B.C.MILBURN / ALEX^{A}" mark. This cylindrical bottle has a flat top and rounded rim (fig. 70).

Jars

Jars, the most common form, were made in many shapes and sizes. Jars have been found with each of the impressed marks from the Wilkes Street Pottery. The stamps of merchants Dubois & Reddick and J. P. Smith were found only on jars. Jars usually have capacity marks, ranging from one-half gallon to six gallons. Jars are one of the forms itemized in the 1874 lien against W. Lewis Milburn.

Ovoid jars. There is considerable variation in the shape of ovoid jars, from rounder to more tapered forms, but there are many gradations and no clear trends over time (fig. 71). These are wide-mouthed jars with an opening at least as large as the base. Half-gallon jars have no handles, whereas larger ones have lug handles.

Figure 71 Ovoid jars, Alexandria, Virginia, 1831–1847. Salt-glazed stoneware, brushed cobalt. *Left:* H. 9". Capacity: 1 gallon. Impressed mark: H.C.SMITH / ALEX^{A} / D.C (Courtesy, The Lyceum: Alexandria's History Museum.) *Right:* H. 9½". Capacity: 1 gallon. Impressed mark: H.C.SMITH / ALEX^{A} (Courtesy, Smithsonian Institution, National Museum of American History.)

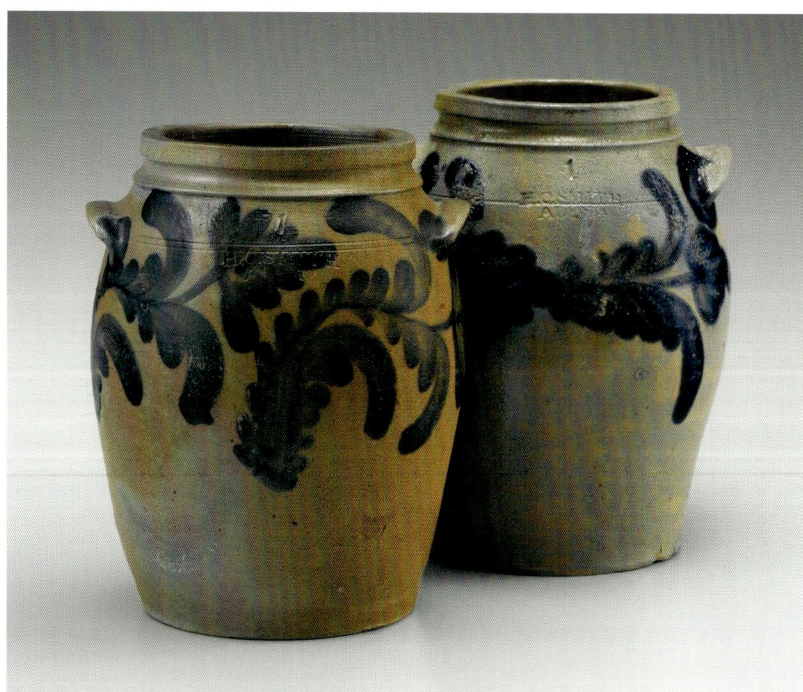

Cylindrical jars. Cylindrical jars have no handles and range from one-half gallon to three-gallons. They are straight sided, and may be truly cylindrical or slightly wider at the shoulder. These jars mostly have rounded shoulders, with rims nearly the same size as the base (fig. 72).

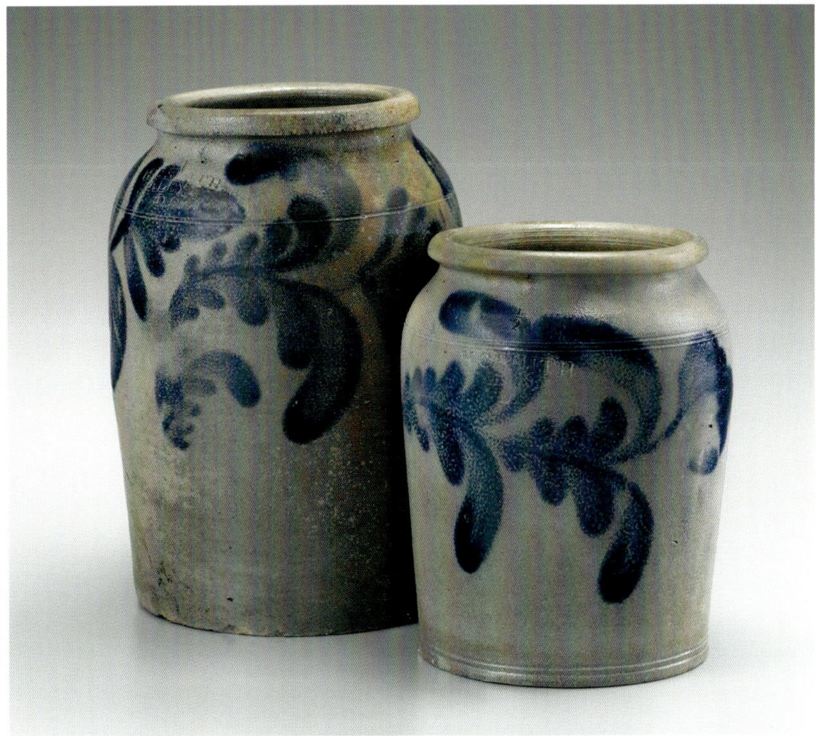

Figure 72 Straight-sided jars, Alexandria, Virginia, 1831–1847. Salt-glazed stoneware, brushed cobalt. *Left:* H. 10". Capacity: 1 gallon. (Courtesy, The Lyceum: Alexandria's History Museum.). *Right:* H. 8". Capacity: ½ gallon. Impressed mark: H.C.SMITH / ALEX^{A} / D.C (Courtesy, Smithsonian Institution, National Museum of American History.) Jars with rounded rims are not quite true cylinders, with sloping shoulders and a narrower rim. This form was first made by John Swann.

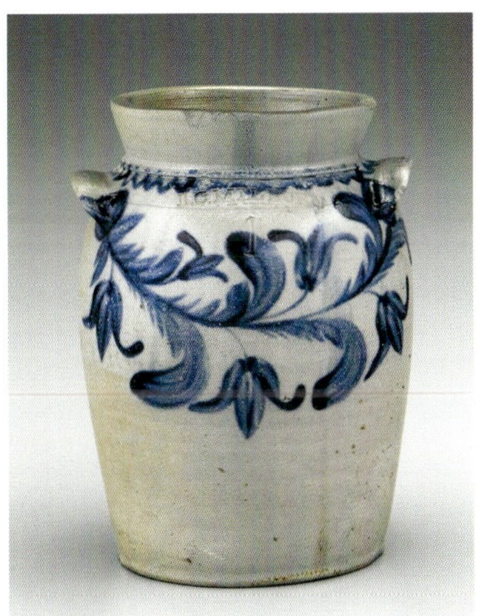

Figure 73 Lidded jar with flared rim, Alexandria, Virginia, 1847–1861. Salt-glazed stoneware, brushed cobalt. H. 9½". Capacity: 1 gallon. Impressed mark: B.C.MILBURN / ALEX^{A} (Courtesy, Alan Darby.)

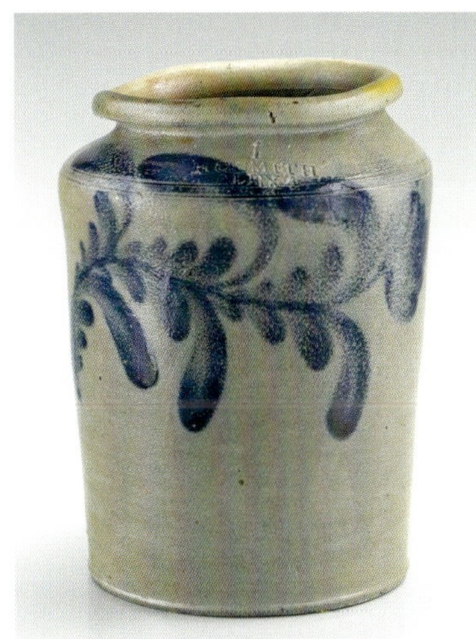

Figure 74 Small-mouthed jar, Alexandria, Virginia, 1831–1847. Salt-glazed stoneware, brushed cobalt. H. 9". Capacity: 1 gallon. Impressed mark: H.C.SMITH / ALEX^{A} / D.C (Private collection.) The shape of this jar, with a sharply angled shoulder and narrow mouth with rounded rim, is similar to that of earthenware jars made in Alexandria and Philadelphia in the late eighteenth century.

Lidded jars with flared rims. Squat, ovoid, lidded jars with tall, flared rims come in sizes from one-half gallon to one-and-one-half gallons, and even the smallest jars of this shape were made with lug handles (fig. 73). The earliest jars of this form have Swann's mark, but they are more common with the "B. C. MILBURN, / *ALEXANDRIA D. C.*" and "B.C.MILBURN /ALEX^{A}" marks. A small ridge is sometimes found at the neck, beneath the flared rim.

Small-mouthed jars. A few small-mouthed jars have more sharply angled shoulders (fig. 74), similar in shape to glazed earthenware preserve jars made in the 1790s in Alexandria and earlier in Philadelphia.

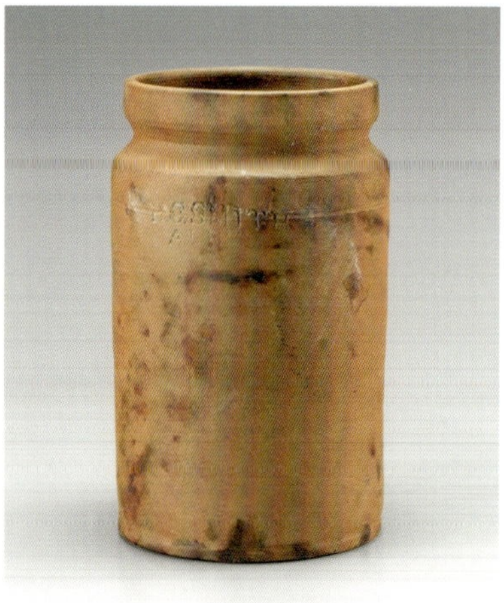

Figure 75 Cylindrical jar, Alexandria, Virginia, 1841–1851. Salt-glazed stoneware. H. 6½". Impressed mark: H.C.SMITH / ALEX^A (Private collection.) It is likely that an 1859 Milburn trade card listing "stone fruit jars filled with cork and cement" referred to jars of this shape. The jar was probably sealed with a small stoneware lid; a cone-shaped section protrudes down through a cork ring.

Fruit jars with cork and cement. Milburn's 1859 trade card lists "fruit jars, with cork and cement" (fig. 75). The "cork and cement" may refer to small, flat stoneware lids found on an Alexandria archaeological site. These have a pointed protrusion that would fit down into a cork ring. In 1868 china merchant E. J. Miller advertised glass fruit jars and "Also, stone jars, of all sizes, pint to 6 gal. Stone fruit jars, quart and half gallon, just as good for putting up tomatoes as the best glass jars, and about one half the cost of glass."[51]

Jugs

Jugs, made to hold liquids, have a narrow neck and a strap handle springing from the shoulder. Milburn's jugs were straight-sided or slightly ovoid, while Swann's were ovoid. The rims also changed over time, from Swann's ringed necks to Milburn's straight narrow necks with collars, which look like the tooled lips with string rings used on nineteenth-century glass liquor bottles (fig. 76). In a variation, a plain square rim sits above a thin, slightly wider ridge. Jugs are often marked with their capacity, with the smallest size a half-gallon. Jugs are listed in advertisements and on the 1859 Milburn trade card. Wolff's 1874 lien lists half-, one-, and two-gallon sizes. No decorated jugs were recorded from the Milburn period.

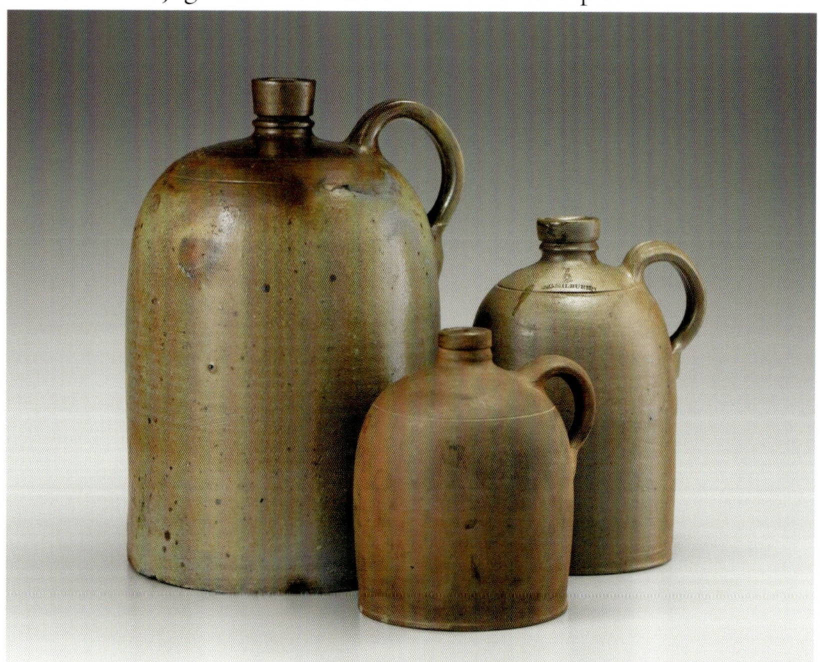

Figure 76 Jugs, Alexandria, Virginia. Salt-glazed stoneware. *Left*, 1847–1867. H. 12¼". Impressed mark: B.C.MILBURN / ALEX^A (Courtesy, The Lyceum: Alexandria's History Museum.) *Center*, 1873–1876. H. 6½". Impressed mark: W. LEWIS MILBURN. (Courtesy, The Lyceum: Alexandria's History Museum.) *Right*, 1867–1873. H. 8¾". Impressed mark: S.C.MILBURN. (Courtesy, Smithsonian Institution, National Museum of American History.)

Milk Pans

Milk pans were made throughout the years of production at Wilkes Street, though none has been recorded with the "B. C. MILBURN, / *ALEXANDRIA D. C.*" or "H.C.SMITH / ALEX^A" marks. Milk pans are mentioned on Milburn's 1859 trade card and in Wolff's 1874 lien.

The Alexandria milk pans range from deep and narrow to wide with more steeply sloping sides, but there appears to be no significant change in shape over time. They were made both with and without lug handles. Milk

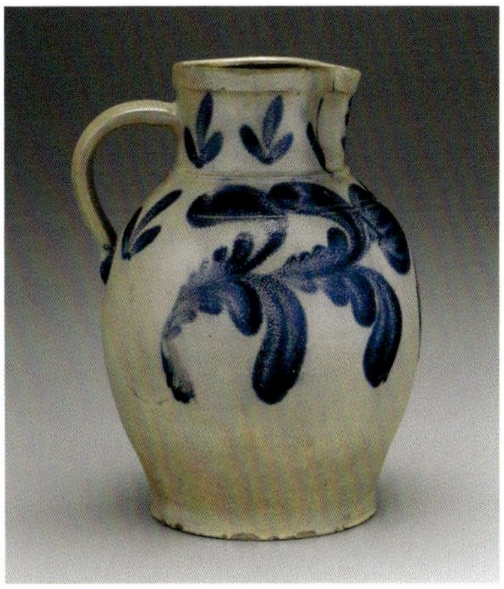

Figure 77 Pitcher, Alexandria, Virginia, 1831–1847. Salt-glazed stoneware, brushed cobalt. H. 14½". Capacity: 3 gallons. Impressed mark: H.C.SMITH / ALEX^A^ / D.C (Courtesy, William and Jane Yeingst.)

pans have flat rims with a square band, or just an incised line to separate rim from body (see figs. 45, 50). The Milburn marks were used on milk pans with brushed-cobalt and slip-trailed decoration. Undecorated milk pans were also made with this mark and with the stamps of S.C. Milburn and E. J. Miller.

Pitchers

Pitchers have been documented with all of the various marks. Most Alexandria pitchers are a baluster shape with an ovoid bottom and a cylindrical or slightly flaring neck (fig. 77).[52] Most of the pitchers have a plain rim, or a small rounded ridge at the top. A pitcher marked "H.C.SMITH / ALEX^A^/ D.C" has a flat, square rim, similar to one on jars and milk pans. Another, marked "B.C.MILBURN / ALEX^A^" has a rim with three ridges. A pitcher in the collection of the Brooklyn Museum bearing the same mark, has a cylindrical body with a sharply defined shoulder (fig. 78). In the second half of the nineteenth century, the baluster and cylindrical shapes were both used in the Shenandoah Valley as well.

Figure 78 Pitcher, Alexandria, Virginia, 1847–1861. Salt-glazed stoneware, brushed cobalt. H. 10¼". Capacity: 1 gallon. Impressed mark: B.C.MILBURN / ALEX^A^ (Courtesy, Brooklyn Museum.) The cylindrical shape of this pitcher is unusual in Alexandria, but was used in the Shenandoah Valley in the late nineteenth century.

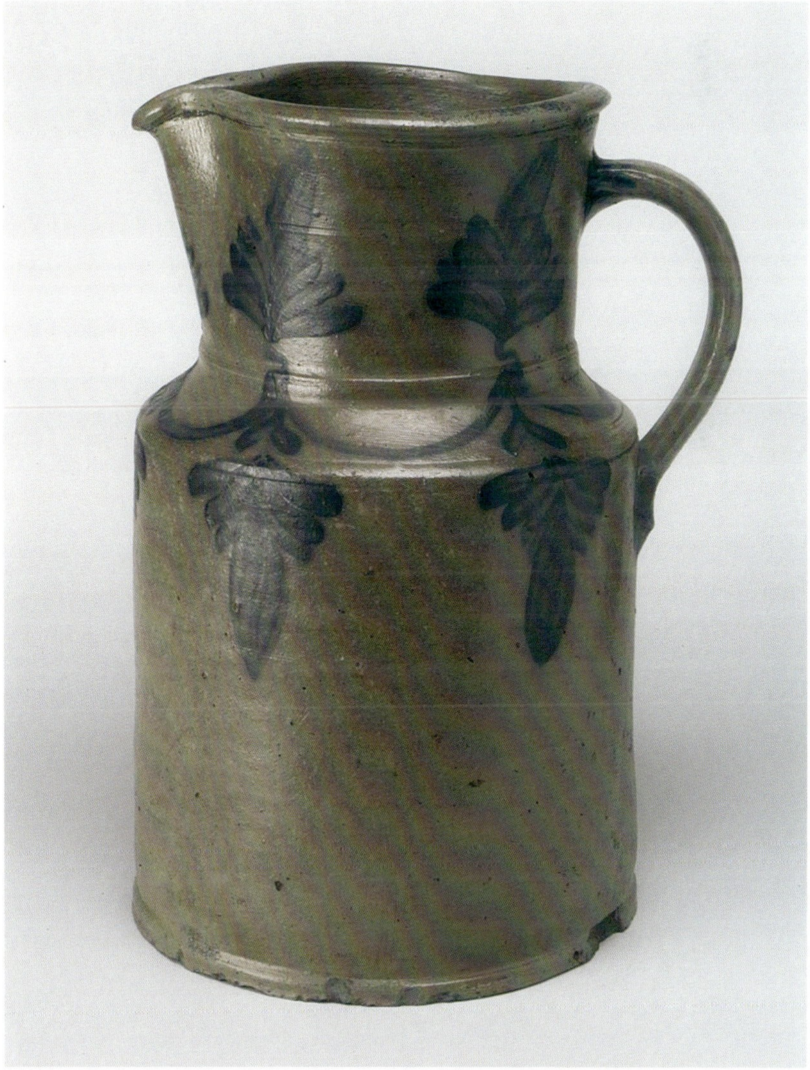

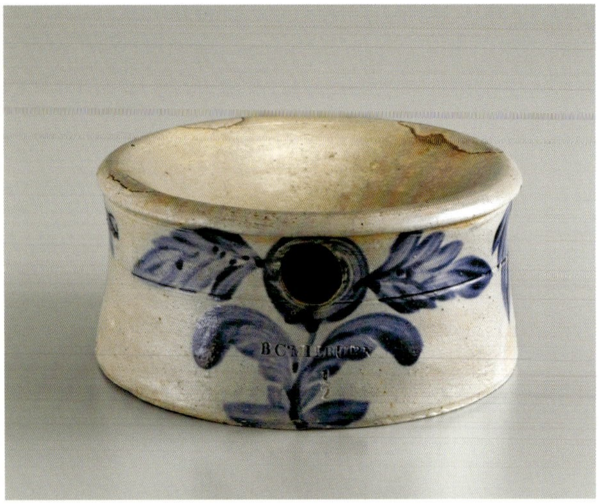

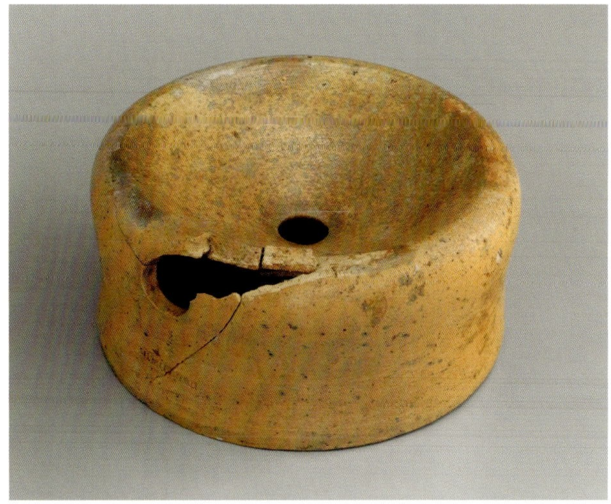

Figure 79 Spittoon, Alexandria, Virginia, 1847–1861. Salt-glazed stoneware, brushed cobalt. D. 7". Capacity: ½ gallon. Impressed mark: B C MILBURN (Private collection.)

Figure 80 Spittoon, Alexandria, Virginia, 1841–1867. D. 7". Salt-glazed stoneware. Capacity: ½ gallon. Impressed mark: B. C. MILBURN. (Courtesy, Alexandria Archaeology Museum.) This spittoon was found in a privy at 524 King Street.

Spittoons

All of the recorded spittoons were manufactured after 1847. One decorated example has the typical round flower in brushed cobalt and the mark "B C MILBURN" (fig. 79). This vessel has an incised line below the angled shoulder and another at the waist. Another decorated spittoon, originally used at St. Paul's Episcopal Church and now in a private collection, is most unusual in that it is marked "S.C.MILBURN". Decorated with three-leaf sprigs, it has a grayish brown glaze, a double incised line below a rounded shoulder, and another line above the base.[53]

Alexandria's Beth El Hebrew Congregation had at least eight Milburn spittoons in the nineteenth century, and still owns two plain ones.[54] One plain, "B.C. MILBURN." spittoon found on an archaeological site in Alexandria is of light-brown stoneware with a rounder shoulder and no incised lines (fig. 80). A plain, gray stoneware spittoon marked "B.C.MILBURN / ALEX^{A}" has incised lines at the shoulder and just above the base. One stamped for merchant E. J. Miller is similar, but with a double-incised line at the shoulder.

Stovepipe Collars

Stovepipe collars are listed on Milburn's 1859 trade card, but no fragments were found at the kiln site; such an object would not have had a maker's mark. As with chimney pots, fragments might not be identifiable in the archaeological record.

Summary

B.C. Milburn established his place in the history of American stoneware—and in the hearts of collectors—with his business acumen and his beautifully decorated wares. He built on the work of his struggling predecessor, John Swann, putting the business on a firm footing so that it served a regional market. Archeological excavations indicate that while Milburn's pottery was in operation very little nonlocal stoneware was used in Alexandria.

By the time the Wilkes Street Pottery closed, nearly a decade after B. C. Milburn's death, small manufactories were being replaced by larger ones and by more industrialized technologies. Railroads transformed the landscape in the 1850s and 1860s and made it profitable for larger industrial potteries to develop in the Ohio Valley and to transport yellow ware and ironstone to eastern destinations such as Alexandria. Yellow ware and ironstone replaced old-fashioned stoneware chamber pots, pans, and baking dishes. Glass screw-top canning jars, first manufactured in 1858, were sold alongside Milburn's stoneware canning jars in an 1868 ad, and had replaced them a decade later. The preference for lower-priced and more uniform industrial goods increased as the century passed, dooming the smaller stoneware manufacturers. After the local manufactories closed, larger stoneware manufacturers such as James Hamilton in Greensboro, Pennsylvania, supplied the more limited needs of Alexandria and other towns for stoneware jars and jugs.

Milburn's manufactory closed in 1876, but many examples of his beautiful stoneware live on, more than 130 years later. Today the Tannery House Condominium sits on the property, and a plaque marks the site of the old Wilkes Street Pottery.

ACKNOWLEDGMENTS I am grateful to collectors of Alexandria stoneware who shared their expertise and were so willing to allow their precious pieces to be examined and photographed for this article. I also thank Kristen Lloyd of The Lyceum: Alexandria's History Museum and Bonnie Lilienfeld of the National Museum of American History, Smithsonian Institution, for the loan of stoneware for this project. I am indebted to the late Suzita C. Myers as well as C. Malcolm Watkins, and John K. Pickens for their research on the history of the Alexandria potters, and to Alain Outlaw, who organized rescue excavations at the Wilkes Street Pottery in 1977. Finally, this research would not have been possible without the assistance and cooperation of my colleagues and volunteers at the Alexandria Archaeology Museum.

The author would appreciate learning about additional collections of Alexandria stoneware and, in particular, about pieces that are different in appearance from those described in this article.

1. The quotation is from an advertisement in the *Alexandria Gazette*, September 15, 1841, for merchant Robert H. Miller, whose King Street shop sold "China, Queensware and Glass." Alexandria stoneware manufactured by Benedict C. Milburn, and Miller's imported goods, were marketed to a wide area. The advertisement indicates that the notice was also to be published in the *National Intelligencer* (Washington, D.C.), *Georgetown Advertiser*, *Warrenton Times*, *Winchester Republican*, *Leesburg Genius of Liberty*, and *Williamsport Banner*.

2. In the nineteenth century, Alexandria stoneware was advertised widely in Virginia newspapers—in Fredericksburg, Falmouth, Leesburg, Winchester, Warrenton, and Williamsport—as well as in Georgetown and Washington, D.C.

3. Barbara H. Magid, "'Stoneware of excellent quality, Alexandria manufacture,' Part 1: The Pottery of John Swann," in *Ceramics in America*, edited by Robert Hunter (Hanover, N.H.: University Press of New England for the Chipstone Foundation, 2012), pp. 111–45.

4. Alexandria Deed Book B-3, pp. 228–29, Alexandria Circuit Court Record Room.

5. Mary G. Powell, *The History of Old Alexandria, Virginia, from July 13, 1749 to May 24, 1861* (Richmond, Va.: William Byrd Press, 1928), p. 305.

6. Brandt Zipp and Mark Zipp, "James Miller, Lost Potter of Alexandria, Virginia," in *Ceramics in America*, edited by Robert Hunter (Hanover, N.H.: University Press of New England for the Chipstone Foundation, 2004), pp. 253–61; Barbara H. Magid, "A New Look at Old Stoneware: The Pottery of Tildon Easton," in *Ceramics in America*, edited by Robert Hunter (Hanover, N.H.: University Press of New England for the Chipstone Foundation, 2004), pp. 249–52.

7. Magid, "'Stoneware of excellent quality,'" pp. 118–19.

8. Suzita Cecil Myers, *The Potters' Art: Salt-glazed Stoneware of 19th-Century Alexandria*, Alexandria Archaeology Publications 91 (1983; 2nd ed., Alexandria, Va.: Alexandria Archaeology, 2003), p. 23, citing Ann Hebb Wintzer family papers.

9. *Alexandria Gazette*, July 16, 1832.

10. *Alexandria Gazette*, June 16, 1828. According to the notice, Milburn purchased Black's interest on June 1, 1828, and continued to operate the grocery business.

11. J. Clinton Milburn advertised a family grocery store on the northeast corner of Royal and Cameron streets in 1865 and remained in the grocery business until his death in 1912. John K. Pickens, "B.C. Milburn and the Alexandria Pottery" (n. d.), files of the Alexandria Archaeology Museum and the Pickens Papers box 57, Alexandria Library, Special Collections.

12. *Alexandria Gazette*, February 19, 1869.

13. *Alexandria Gazette*, December 20, 1839.

14. *Alexandria Gazette*, July 16, 1832. Court documents list Black's personal property as "about two tons of clay for making stoneware and a small quantity of stoneware"; see Arlington County Insolvent Debtors: 1826–1833, pp. 379–82, Alexandria Library, Special Collections, microfilm reel 00548.

15. Magid, "A New Look at Old Stoneware," pp. 249–52. Bankruptcy announcement in the *Alexandria Gazette*, February 20, 1843.

16. Papers of Suzita C. Myers, in the files of the Alexandria Archaeology Museum.

17. Thomas Duffy, "Decline of the Port of Alexandria, 1800–1861," master's thesis, Georgetown University, 1965.

18. *Alexandria Gazette*, July 19, 1851.

19. John K. Pickens, "Master Potters or Merchant Princes," files of the Alexandria Archaeology Museum and the Pickens Papers (box 57), Alexandria Library, Special Collections.

20. Barbara H. Magid, "Robert H. Miller, Importer: Alexandria and St. Louis," in *Ceramics in America*, edited by Robert Hunter (Easthampton, Mass.: Antique Collectors' Club for the Chipstone Foundation, 2008), pp. 143–61. R. H. Miller advertised in the *Alexandria Gazette*, October 24, 1843: "I am constantly receiving from the Pottery supplies of fresh made stoneware of superior quality." Other advertisements appeared in the *Alexandria Gazette* on October 11, 1825; September 15, 1841; and May 3, 1847.

21. *Alexandria Gazette*, February 19, 1869.

22. *Alexandria Gazette*, January 10, 1855.

23. John N. Pearce, "The Early Baltimore Potters and Their Wares, 1763–1850," master's thesis, University of Delaware, 1959, pp. 5–7.

24. Anthony Trollope, *North America* (1862; repr. Adelaide, Australia: University of Adelaide Library [eBooks@Adelaide], 2004), chap. 21 ("Washington").

25. Suzita Cecil Myers, "Alexandria Salt-glazed Stoneware: A Study in Material Culture, 1813–1876," master's thesis, University of Maryland, 1982, p. 44, citing Ann Hebb Wintzer family records.

26. *Alexandria Gazette*, November 4, 1866.

27. *Alexandria Gazette*, April 10, 1867.

28. Myers, "Alexandria Salt-glazed Stoneware," p. 46.

29. *Alexandria Virginia City and County Census 1860*, transcribed by T. Michael Miller, (Alexandria, Va.: Alexandria Library, Lloyd House, Heritage Books, 1986), p. 85.

30. *Alexandria Gazette*, March 1, 1869.

31. *Alexandria Census 1860*, p. 85.

32. Boyd, *Washington & Georgetown Directory* for 1873. Another 1873 directory from Alexandria does not list Stephen C. or W. Lewis Milburn or the pottery, but it does list another druggist at the site of W. Lewis Milburn's former business at the corner of King and Washington streets; see *Alexandria Business Directory*, 1873, on file at the Alexandria Archaeology Museum.

33. Alexandria Deed Book: Mechanic's Lien Book 4, p. 147, June 12, 1874, Alexandria Circuit Court Record Room.

34. John E. Kille, "Distinguishing Marks and Flowering Designs: Baltimore's Utilitarian Stoneware Industry," in *Ceramics in America*, edited by Robert Hunter (Hanover, N.H.: University Press of New England for the Chipstone Foundation, 2005), pp. 104–7.

35. Robert Hunter, Kurt C. Russ, and Marshall Goodman, "Stoneware of Eastern Virginia," *Antiques* 167, no. 4 (April 2005): 129; Robert Hunter and Marshall Goodman, "The Destruction of the Benjamin DuVal Stoneware Manufactory, Richmond, Virginia," in *Ceramics in America* edited by Robert Hunter (Hanover, N.H.: University Press of New England for the Chipstone Foundation, 2005), pp. 37–60. Stoneware fragments with incised decoration (ibid., p. 56, fig. 51), have leaves similar to those on the Alexandria cooler.

36. Howard F. Guilland, *Early American Folk Pottery* (Philadelphia: Chilton Book Co., 1971), pl. following p. 98.

37. Kille, "Distinguishing Marks and Flowering Designs," pp. 95–113.

38. Zipp and Zipp, "James Miller, Lost Potter of Alexandria," pp. 257–58.

39. Jeffery S. Evans and Scott Hamilton Suter, *"A great deal of stone & earthen ware": The Rockingham County, Virginia School of Folk Pottery* (Dayton, Va.: Harrisonburg-Rockingham Historical Society, 2004), fig. 63.

40. Kristin B. Lloyd, *From Potter to Pantry: 19th Century Stoneware. Catalog of the Exhibition at The Lyceum, Alexandria's History Museum* (Alexandria, Va., 1991), no. 35, on file at The Lyceum: Alexandria's History Museum.

41. Myers, "Alexandria Salt-glazed Stoneware," p. 48.

42. Ibid.

43. Powell, *History of Old Alexandria*, p. 305.

44. Myers, *The Potters' Art*, p. 5.

45. Guilland, *Early American Folk Pottery*, p. 53. Covered butter and cake pots are illustrated on an 1862 price list from O. L. & A. K. Ballard in the collection of the Shelburne Museum. According to this illustration, cake pots are wider than they are tall, whereas butter pots are taller.

46. Myers, "Alexandria Salt-glazed Stoneware," p. 116.

47. *Alexandria Gazette*, February 24, 1869.

48. Alexandria Deed Book: Mechanic's Lien Book 4, p. 147, June 12, 1874. Alexandria Circuit Court Record Room.

49. Nancy Lee Tackett, "From Alexandria Cupboards . . ." *Antiques* (February 1945): 118–19. Tackett illustrates three vessels, owned at the time by Milburn descendents, including an elaborately decorated "H. SMITH & Co." jug and two Milburn jars, one with brushed and one with slip-trailed decoration. Another illustrated jar, stamped "H. C. SMITH/ ALEX$^{\underline{A}}$ / D. C", is now in the collection of The Lyceum: Alexandria's History Museum.

50. *Alexandria Gazette*, September 3, 1845.

51. *Alexandria Gazette*, July 21, 1868.

52. Georgeanna H. Greer, *American Stonewares: The Art and Craft of Utilitarian Potters* (Exton, Pa.: Schiffer Publishing, 1981), p. 103.

53. Lloyd, *From Potter to Pantry*, no. 35.

54. C. Malcolm Watkins, "The Pots and Potteries of Alexandria, Virginia: 1792–1876; Early American Craftsmen: B.C. Milburn and the Alexandria Pottery," p. 3, unpublished manuscript in the files of the Alexandria Archaeology Museum.

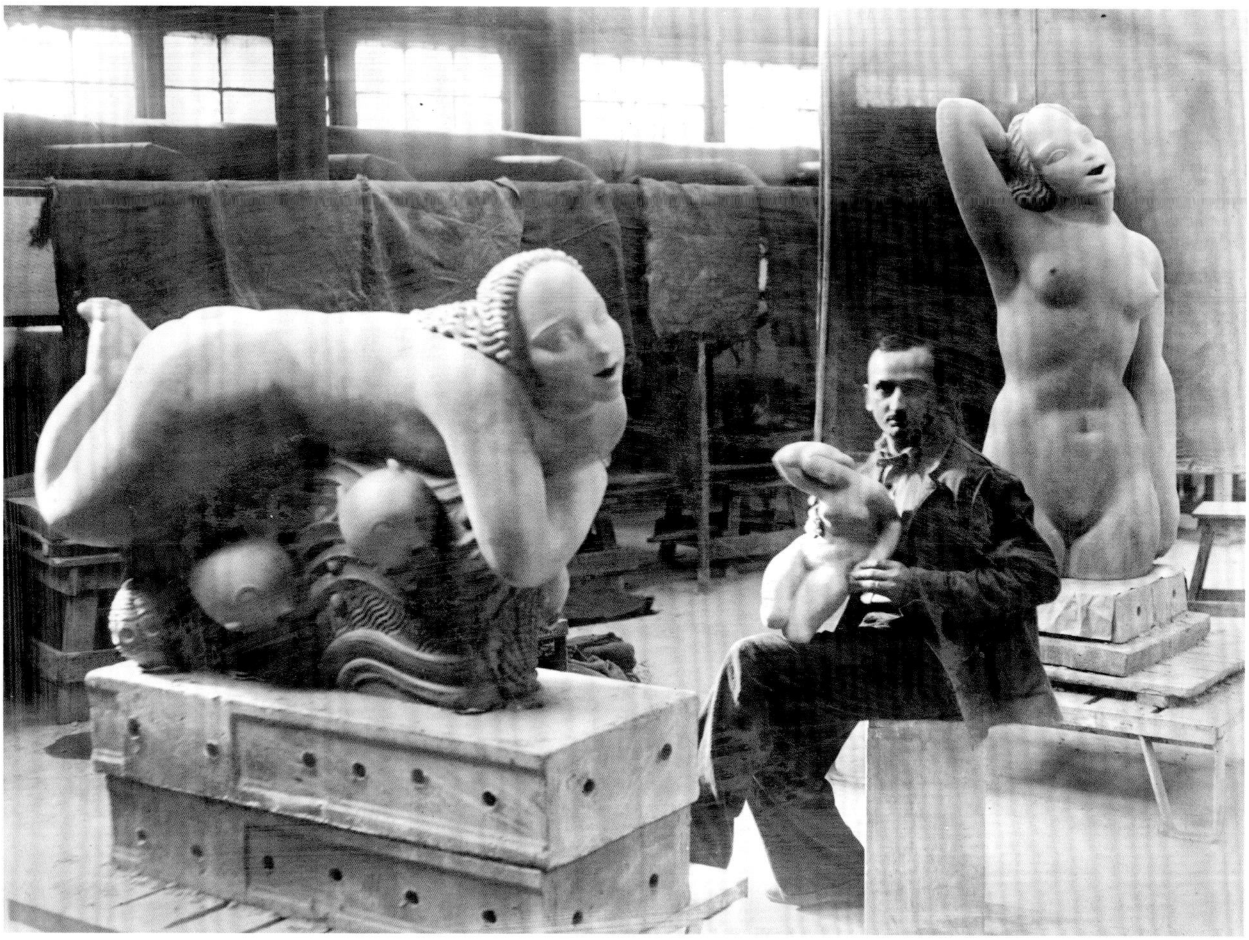

Figure 1 Waylande Gregory, *The Swimmer*, ca. 1933. (Photo, courtesy Waylande Gregory Archives.) Gregory is holding a female nude sculpture; behind him is *The Bather*, ca. 1933.

Thomas C. Folk

Waylande Gregory: Science and Ceramics

▼ WAYLANDE GREGORY is regarded by many as a leading figure in twentieth-century American ceramics. A retrospective exhibition, sponsored by the University of Richmond Museums, Virginia, will be at the Cranbrook Art Museum in Bloomfield Hills, Michigan, from November 16, 2013, to March 23, 2014. The presentation and an accompanying biographical book by the author shed new light on this revolutionary, innovative and, at times, controversial artist (fig. 1).

Gregory was born in 1905 in Baxter Springs, Kansas, and grew up on a ranch near a Cherokee reservation. He was something of a prodigy, and became interested in ceramics when as a child he witnessed their use during a Native American burial. He was also musically inclined and took lessons from his mother, who had been a concert pianist. At eleven, he was enrolled at Kansas State Teacher's College, where he studied carpentry and crafts, including ceramics.

Gregory's early development as a ceramic sculptor was shaped by the encouragement and instruction of Lorado Taft, a leading American sculptor who worked primarily in bronze and marble, and by Guy Cowan, a leading American ceramist. Gregory had caught the attention of Taft after moving to Chicago in 1924. Taft was known to mentor important American sculptors, among them Janet Scudder, and he invited Gregory to study with him privately for eighteen months and to live and work with him at his famed Midway Studios, a complex of thirteen rooms overlooking a courtyard. Taft may have been responsible for getting Gregory interested in creating large-scale sculpture.

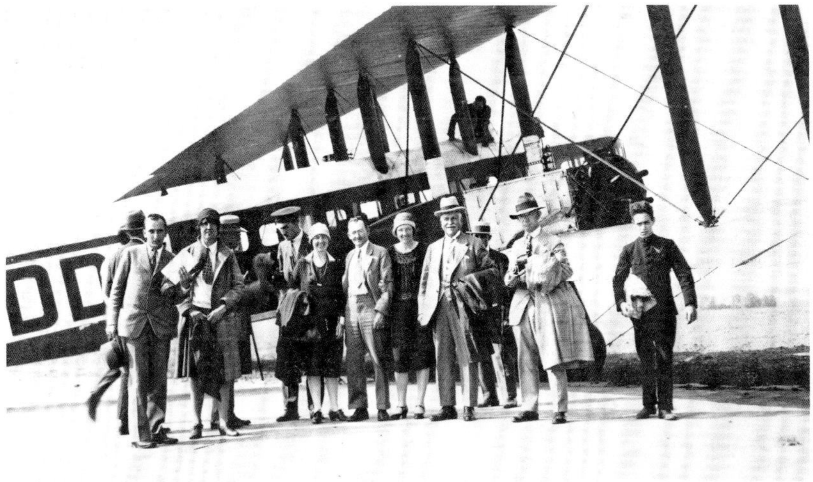

Figure 2 Lorado Taft and students boarding a plane from London to Paris, ca. 1928. Gregory is on the far left; Taft is fourth from the right. (Photo, courtesy Waylande Gregory Archives.)

By the 1920s, however, Taft's brand of academic sculpture was no longer considered progressive. Gregory, who visited Europe with Taft and other students late in the decade (fig. 2), was increasingly drawn to the latest trends of the United States and Europe, and his career in the 1930s mirrored the changing focus of American ceramics from art pottery to studio pottery.

"Kid Gregory," as he was known, was hired circa 1928 by Guy Cowan, the founder of the Cowan Pottery in Cleveland, Ohio. Cowan's family had been in the pottery business for generations, and he was one of the first graduates of the New York State School of Clay-working and Ceramics (now the New York College of Ceramics at Alfred University), which was founded in 1900 and had one of the first programs in production and studio pottery. Cowan, who may have known about pottery production but lacked training as a sculptor, hired Gregory to be the pottery's sole full-time employee, and from 1928 to 1931 Gregory served as its chief designer and sculptor. The focus of the Cowan Pottery was on limited-edition, tabletop ceramic sculptures, the majority of which Gregory designed, including the well-known *Nautch Dancer* and *Burlesque Dancer*. These sculptures referenced Gilda Grey, a popular entertainer of the 1930s who posed for Gregory in the creation of these works.

The shift away from art pottery is reflected in Gregory's next career move when, in 1932, he left the Cowan Pottery, which had fallen victim to the Great Depression, and became the instructor of ceramics at Cranbrook, in Bloomfield Hills, Michigan. Colleges with emerging ceramics programs were instrumental in creating a new role for ceramists. At Cowan, Gregory had enjoyed an able staff to assist him in all phases of production, which is typical of an art pottery. As a studio potter at Cranbrook, however, Gregory directly handled all phases of production—from modeling to glazing to firing. Although he remained at Cranbrook for only a year and a half, it was an extremely productive period for him. He was part of the school's illustrious faculty, which included the architect Eliel Saarinen and sculptor

Figure 3 Waylande Gregory, *Light Dispelling Darkness*, 1936–1938, installed in Roosevelt Park, in Menlo Park, New Jersey. Concrete. H. approx. 20'. (Photo, Randl Bye.)

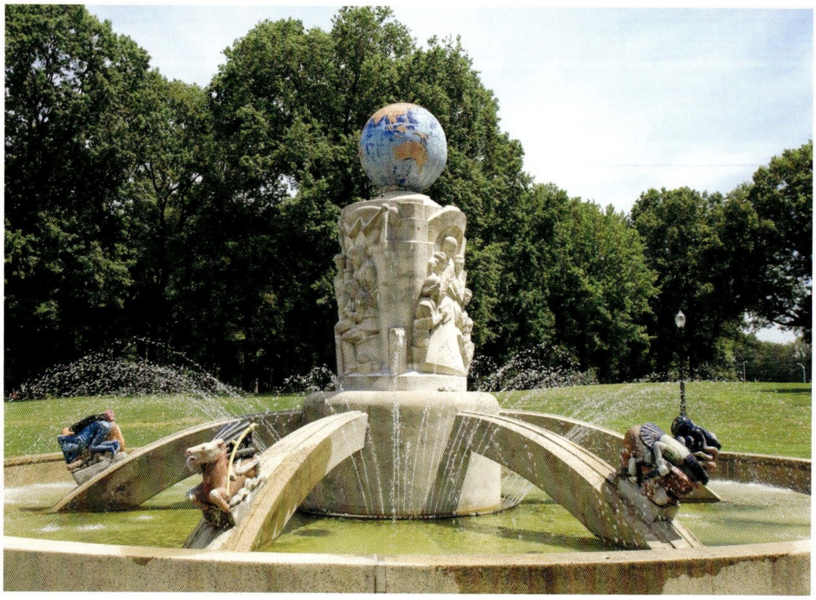

Carl Milles, and he created two of his finest ceramic sculptures, *Kansas Madonna* and *Girl with an Olive*.

In the summer of 1933, Gregory moved to Metuchen, New Jersey, and became very involved in the government-sponsored Works Progress Administration (renamed Work Projects Administration in 1939), better known as the WPA. His mentor, Lorado Taft, had created several fountains and may have inspired Gregory to work on fountain themes. In 1936 Gregory began to plan *Light Dispelling Darkness* (fig. 3), a fountain to be constructed in Menlo Park in part as a tribute to Thomas Edison, who had developed the incandescent lightbulb there. The installation, which is still in place today, was financed by the WPA and the Middlesex County Board of Freeholders.

By 1937 Gregory had been appointed the state head of the sculpture division of the WPA. In an effort to revive interest in commercial terracotta production, which by this time was starting to lose public interest, he selected ten men to assist him in the construction of *Light Dispelling Darkness* at the Atlantic Terra Cotta Company in nearby Perth Amboy; the terracotta elements were sculpted from local Amboy clay. Although the assistants had no previous knowledge of ceramics, it was hoped that the experience they gained could help them find future employment in the ceramics industry.[1]

Figure 4 Waylande Gregory, Universal Peace, from *Light Dispelling Darkness*. (Photo, Randl Bye.)

Figure 5 Waylande Gregory, Agriculture and Industry, from *Light Dispelling Darkness*. (Photo, Randl Bye.)

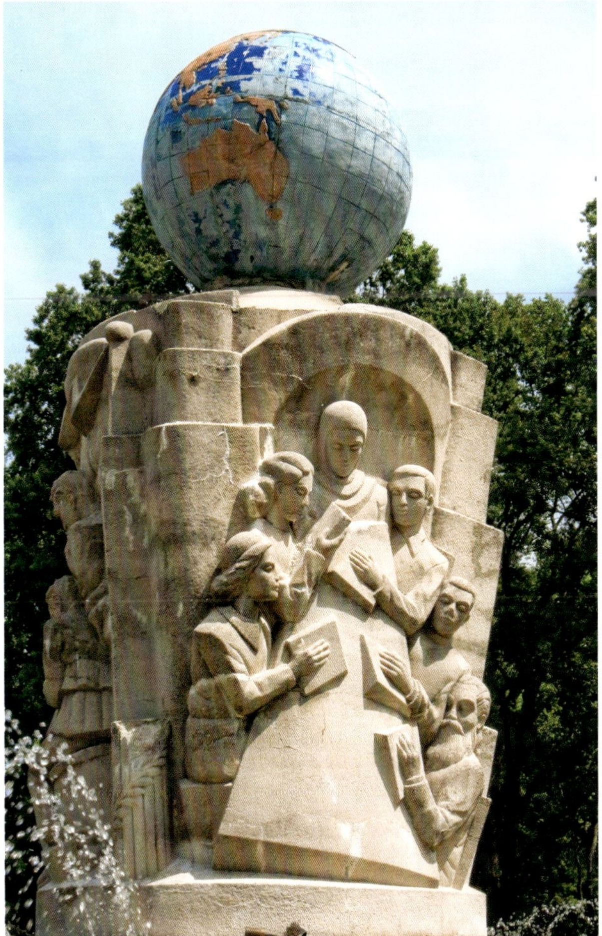

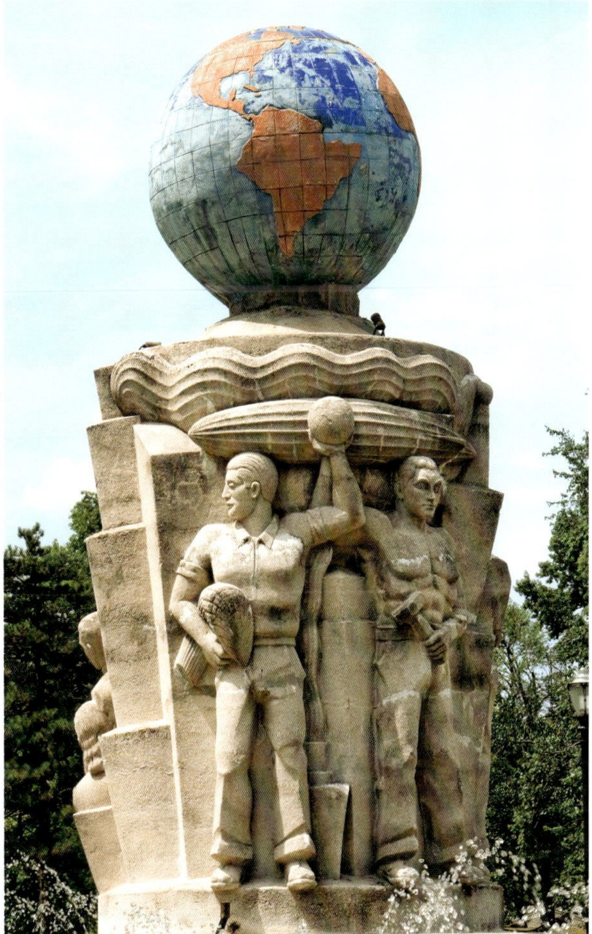

The overall theme of *Light Dispelling Darkness* was the triumph of mankind over evil, achieving universal peace through science, industry, and education (fig. 4). Although the fountain is dedicated to Edison, the inventor is not directly represented or suggested in any of the figures. One of the three reliefs on the central pillar depicts Science and Achievement as a group of professionals at work in their fields of electrical and medical sciences. A figure with a dynamo symbolizes the power available to man through knowledge and perseverance; a man looking through a microscope symbolizes scientific research.

The fountain's circular basin, constructed of concrete aggregate, is forty feet in diameter; its central, bas-relief pillar, also of concrete, is ten feet in diameter and fifteen feet high. The pillar, which is surmounted by a brightly glazed terracotta globe, bears relief figures representing man's achievements in science, industry, medicine, the arts, and other areas. In the relief depicting Agriculture and Industry (fig. 5), for example, Gregory shows some of the marvels of technology of the time: skyscrapers, zeppelins, and ocean liners. In the basin surrounding the pillar, fleeing the

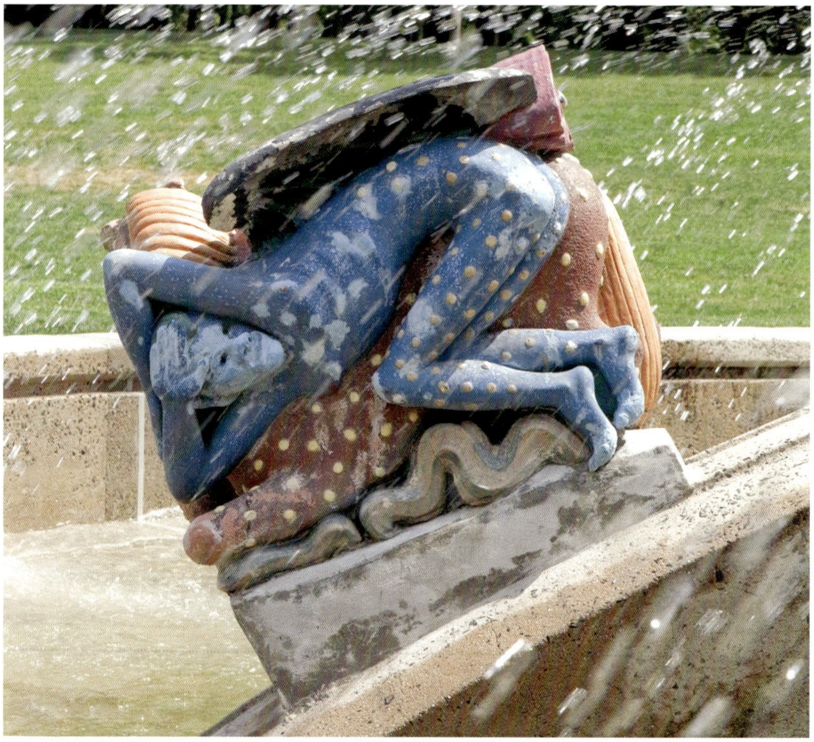

Figure 6 Waylande Gregory, Pestilence, from *Light Dispelling Darkness*. Glazed terracotta. H. 3'. (Photo, Randl Bye.)

"light" of science and knowledge, are six three-foot-tall glazed terracotta figures depicting the evil forces that prey on mankind—Greed (intertwined octopi), Materialism (spewing a ribbon of stock-market ticker tape), and Famine, Pestilence, War, and Death (presented as the Four Horsemen of the Apocalypse).[2]

Pestilence (fig. 6), the first of the Four Horsemen, is shown as a diseased figure with a flowing cape, its body colored blue with bright yellow dots suggesting open sores. The figure almost has a Pop Art quality.

Famine is shown as an emaciated old figure with flowing blond hair riding an equally emaciated horse. The figure is nude, and carries an unbalanced scale in its left hand, symbolizing inequality.

War is depicted as a Roman centurion or soldier clinging to his fallen horse. He wears a contemporary gas mask and a shield adorned with a skull.

Death, a skeleton wearing a flowing black cape and emitting a jagged lightning bolt, is astride a frantic horse.

These figures, arguably the most interesting part of the fountain, are among Gregory's first large-scale ceramic works.

In 1938 Gregory began construction of his new house at 16 Mountain Trail, in the Watchung Mountains of Bound Brook (now Warren), New Jersey. The house, completed in 1939, was constructed of concrete and almost had a California, rather than a New Jersey sensibility, reflecting the influence of Gregory's architectural hero, Frank Lloyd Wright. It does not relate to Wright's Prairie School houses of Chicago, however, but rather to his concrete California houses of the 1920s. The enormous custom kiln was probably included in the initial construction of the house.

The year 1939 also marked a major point in Gregory's professional life, with continued public recognition of his work, specifically his *Fountain of the Atom*, commissioned by the World's Fair Corporation and his most important work. The 1939 World's Fair, situated historically between the Great Depression and World War II, proved to be a microcosm of the rapid changes occurring in the larger world. The theme of the fair was "Building the World of Tomorrow" and it was opened by President Franklin D. Roosevelt on April 30, 1939 (and closed two years later), offering its 44

Figure 7 New York World's Fair, 1939, with the *Perisphere* and *Trylon* by Wallace Harrison rising behind the General Motors Building, designed by Norman Bel Geddes.

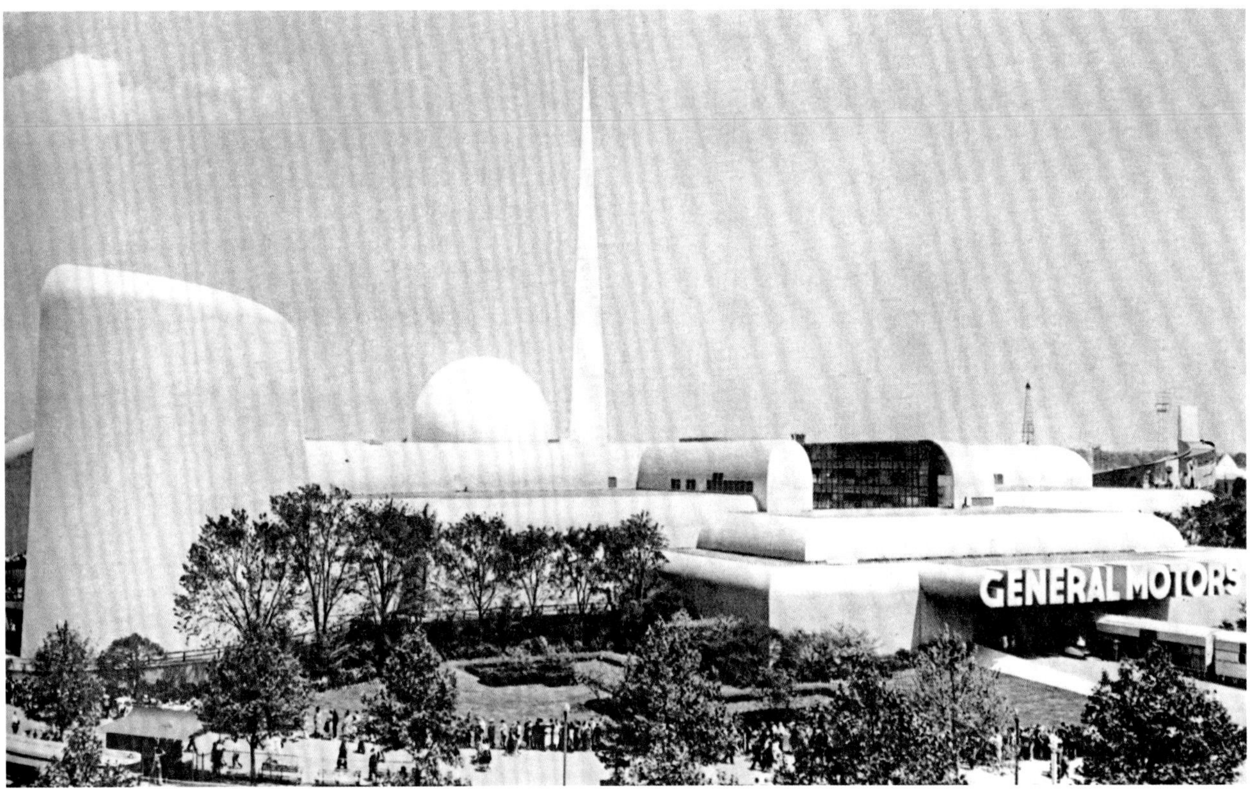

million visitors an optimistic view about new technology. The two iconic white buildings most associated with the fair were the *Perisphere* (200 feet in diameter) and the *Trylon* (more than 700 feet tall), both designed by Wallace Harrison (fig. 7). Within the *Perisphere*, visitors witnessed a diorama of the city of tomorrow called "Democracity."

The most popular exhibit was "Futurama," designed by Norman Bel Geddes and installed in the General Motors Building (also designed by Bel Geddes). This is where Gregory exhibited two additional groupings of ceramic sculptures, *Imports* and *Exports*. New means of transportation were also very evident at the fair, including the futuristic Chrysler Airflow automobile, a transport plane, and the world's fastest train—the 6100—a sleek steam locomotive designed by Raymond Lowey.

The World's Fair, for which Gregory served as an adviser, proved to be the most significant display of art and design in America since the opening of Rockefeller Center in the early 1930s. In addition to the design contributions of Harrison, Bel Geddes, and Lowey, the fair displayed a significant number of works by American sculptors—Malvina Hoffman, Paul Manship, Carl Milles, William Zorach, and others—many of whom worked in bronze; Gregory was the only sculptor to produce monumental outdoor ceramic sculptures.

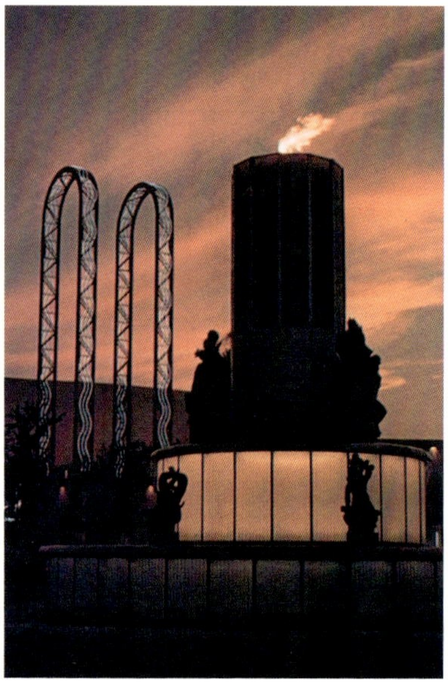

Figure 8 Waylande Gregory, *Fountain of the Atom*, 1939, installed at the World's Fair, Queens, New York. Reproduced from Ross Anderson and Barbara Perry, *The Diversions of Keramos: American Clay Sculpture, 1925–1950*, exh. cat. (Syracuse: Everson Museum of Art, 1984), p. 3.

Gregory's interest in atomic energy was vividly demonstrated by his *Fountain of the Atom* (fig. 8), on display at the entrance of the newly constructed railway entrance at the 1939 New York World's Fair. Composed of twelve large-scale nude and seminude ceramic figures—eight electrons (illustrated in the March 13, 1939, issue of *Life* magazine) and four elements—this major work was fanciful, artistic, and compelling. The installation visualized the growing international interest in atomic energy and may be seen as Gregory's crowning achievement, as well as the culmination of his interest in monumental ceramic sculpture.[3] To help him move these huge sculptures, Gregory hired an African-American father-and-son team, Tyson and Ralph respectively, who resided on the third floor of Gregory's new house. Unfortunately, their last name has been lost to time.

Fountain of the Atom was located on Bowling Green Plaza, very close to the subway that had been renovated for the fair. Partly because of its location, the fountain received considerable attention, including the photographs in *Life* magazine and even a review in the international press.[4] The fountain itself had a steel frame and was constructed of glass bricks. It consisted of a bluish green pool, sixty-five feet in diameter, beneath two concentric, circular tiers, or "terraces" as Gregory called them, the first of which was wider than the second. In the first terrace, relating to the valence shell of an atom, were eight electrons: four male and four female terracotta figures, each approximately 48 inches tall. Above them, on a narrower terrace representing the nucleus of an atom, were four much larger and heavier terracotta figures depicting the Elements, each averaging about 78 inches in height and weighing about one and a half tons: Water and Air (fig. 9) are portrayed as male in gender; and Earth and Fire are portrayed as female. All of the figures were fired in Gregory's new kiln in Warren.

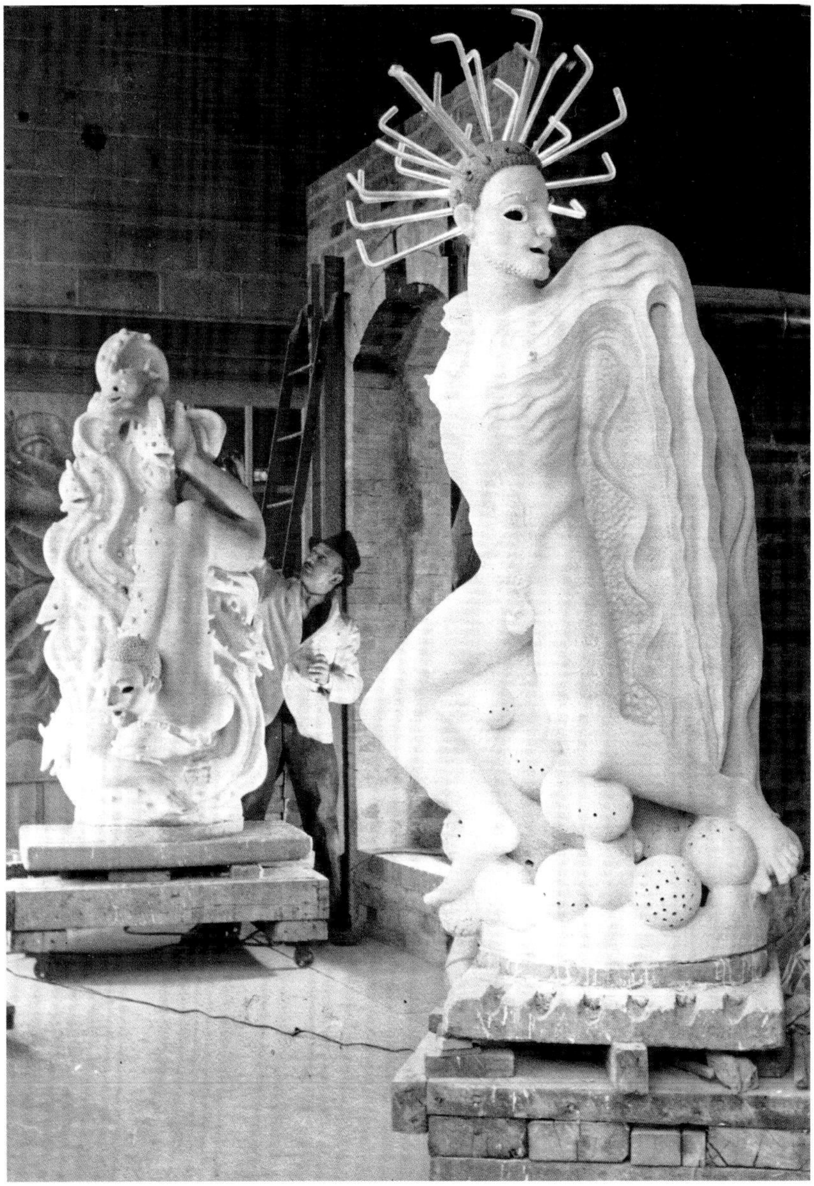

Figure 9 Waylande Gregory, Water (background) and Air (foreground), ca. 1938, from *Fountain of the Atom*. Glazed earthenware. H. 72". (Courtesy, Cranbrook Academy [Water] and private collection [Air]; photo, courtesy Waylande Gregory Archives.) Installed at the 1939 World's Fair; note the artist alongside Water in this studio photograph.

In the center of the fountain, above the elements, was a central shaft of sixteen glass tubes from which water tumbled down, from tier to tier. At the top, a bright flame burned constantly. The glass-block tiers were lit from within, and this, combined with the flowing waters, created a glowing, gurgling effect. Since the fair was temporary, the figures could be removed after its closing. Credit for the design of the structure of the fountain belongs to collaborator Nembhard Culin, who was responsible for several other structures on the fairgrounds as well.[5]

Since so much of the fair focused on new technology, Gregory was very much on target with his design showing the new science of the atom. According to what is known as the "octet rule," atoms tend to combine in such a way that they each have eight electrons in their valence shells, typical of carbon, nitrogen, and oxygen. Molecules tend to be most stable when the outermost electron shells contain eight electrons. But Gregory interpreted this theme artistically using nude figures rather than atomic models, with the elements and the electrons depicted as allegorical figures that represent scientific ideas. The use of allegorical subjects relates to the sculpture of Lorado Taft and the Beaux Arts use of such figures in art, though Gregory's figures are very modern, not Beaux Arts. Paul Manship's

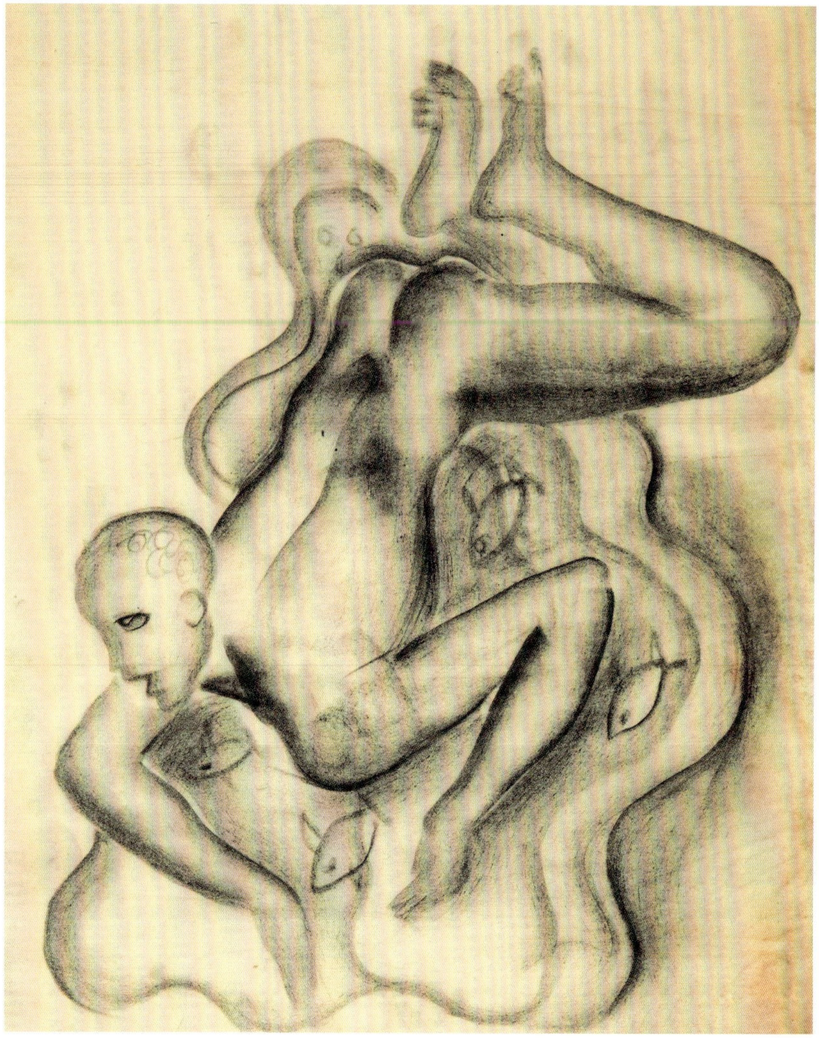

Figure 10 Waylande Gregory, preliminary drawing for Water, ca. 1938. Charcoal on paper. 15" x 13". (Courtesy, private collection; photo, Randl Bye.)

bronze figures for the fair, equally allegorical although Art Deco in style, offered fairgoers his *Time and the Fates Sundial* (now at Brookgreen Gardens, South Carolina) and *Four Models of Time*. At the time Gregory sculpted *Fountain of the Atom*, he had gone beyond Art Deco stylization and was working in a more personal style. The rounded, fleshy nudes had already appeared in some of his monumental works, for example *The Swimmer* and *The Bather*, both of which were created circa 1935. This trend culminated in the female nudes for *Fountain of the Atom*.

Perhaps the most popular of the four elements was Water, ca. 1938 (see fig. 9). In a preliminary drawing for it (fig. 10), Gregory captured the movement of the nude male figure diving down surrounded by fish. Although Gregory never specifically mentioned the ethnicity of Water's swimmer, he

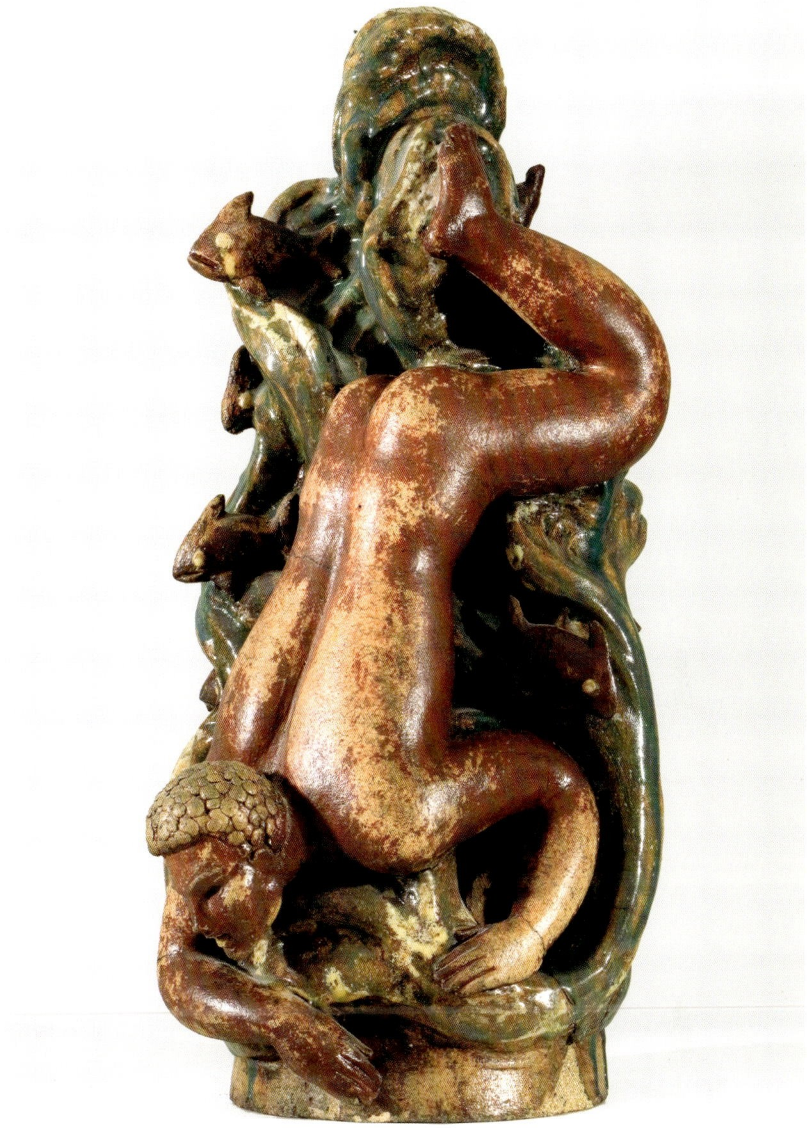

Figure 11 Waylande Gregory, maquette for Water, ca. 1938. Glazed earthenware. H. 22". (Private collection; photo, Randl Bye.)

appears to be an African-American, and studio photographs suggest that Gregory's assistant Ralph might have been the model. The pose is untraditional and unclassical, as the diver seems weightless and is depicted upside down. The finished version of Water was the most colorful of the four elements, with yellow bubbles and pink fish. The glaze is particularly glassy and watery, as Gregory was attempting to interpret the transparent nature of water. The maquette for Water (fig. 11) may be one of the artist's most successful; indeed, he had exhibited it independently of the others.

Another popular representation was Earth, for which Gregory modeled a highly detailed and beautifully glazed maquette (fig. 12); however, the finished version (fig. 13) has harsher features and a more primitive quality. Gregory admitted that Earth, conceived as a "proud earth woman with a

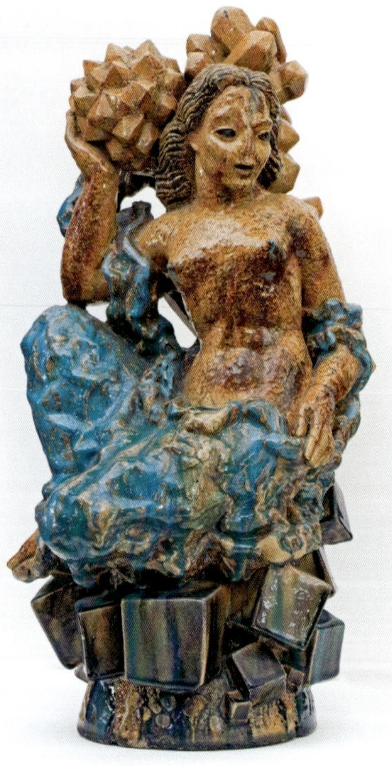

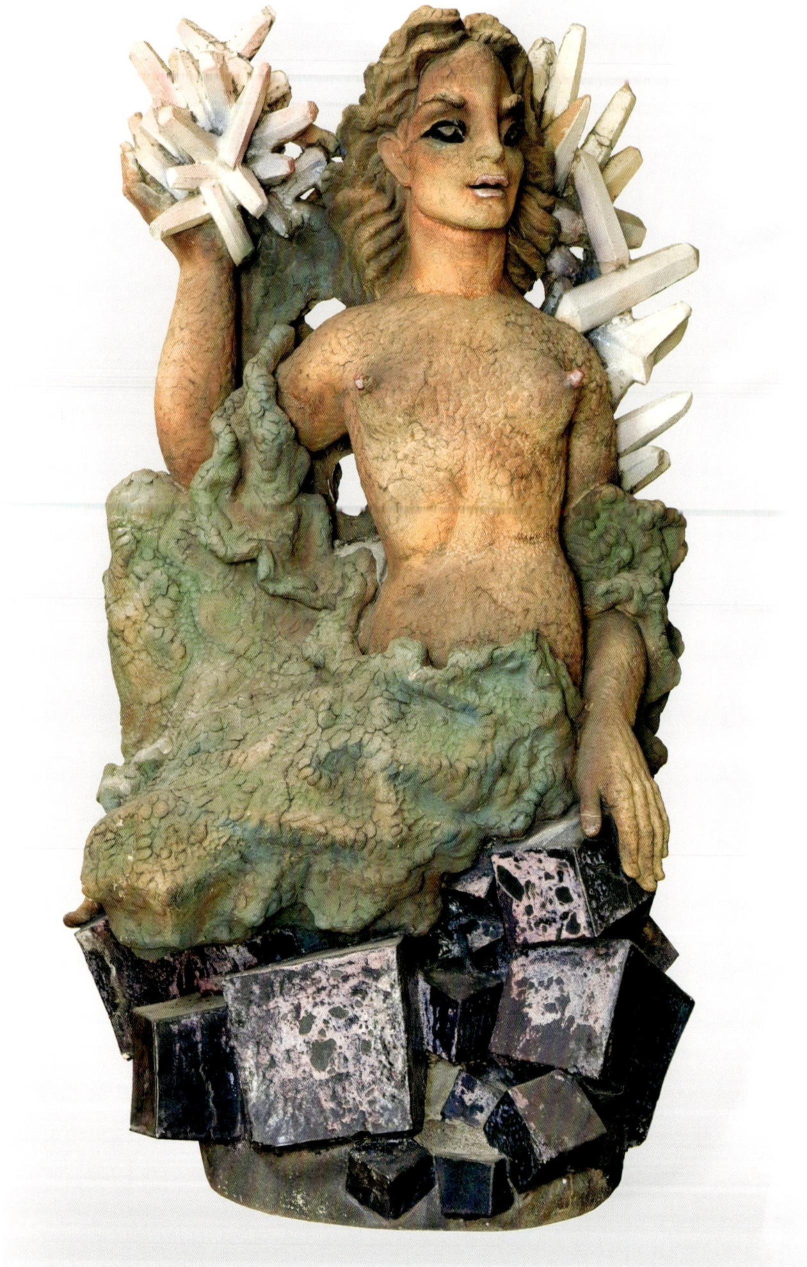

Figure 12 Waylande Gregory, maquette for Earth, ca. 1938. Glazed earthenware. H. 22". (Courtesy, Lea Allen Gebauer collection; photo, Randl Bye.)

Figure 13 Waylande Gregory, Earth, ca. 1938, from *Fountain of the Atom*. Glazed earthenware. H. 77". (Philadelphia Museum of Art; photo, Randl Bye.)

great shock of hair and wrapped in a mantle of green," was his favorite.[6] Her torso is exposed and she sits, goddess-like, on a throne of minerals. Gregory said that he had visualized Earth as "agelessly young, fecund, and virginal, engrossed in the mysteries of evolving."[7] Traditionally, however, earth goddesses in Western art are represented with maternal, not virginal, qualities.

Perhaps the strangest of the four elements was Air (fig. 14), represented by an alien, somewhat like Milton's Lucifer, "a heroic, angel-like demon male figure, naked, winged, swift in movement and evanescent in coloring."[8] The figure, standing on bubble-like clouds, had a mischievous smile and many rods protruded from his head, creating a bizarre halo.[9] Gregory created a self-portrait bust of himself as the devil when he was residing in Cleveland; Air might have been a self-portrait as well.

In *Fountain of the Atom*, atomic energy became a metaphor for sexual energy. Fire (fig. 15) was all about sexual passion; the nude female figure had exaggerated, upward-pointing nipples that underscored her sexuality. Gregory described Fire as "an aquiline female figure being consumed in endless arabesques of flame."[10] As with Water, openings in the sculpture enabled the viewer to see inside the flames.

In the end it was the electrons, rather than the elements, that received the most attention. Also sculpted from Amboy clay and constructed in the same manner as the elements, these brightly colored figures had a charged

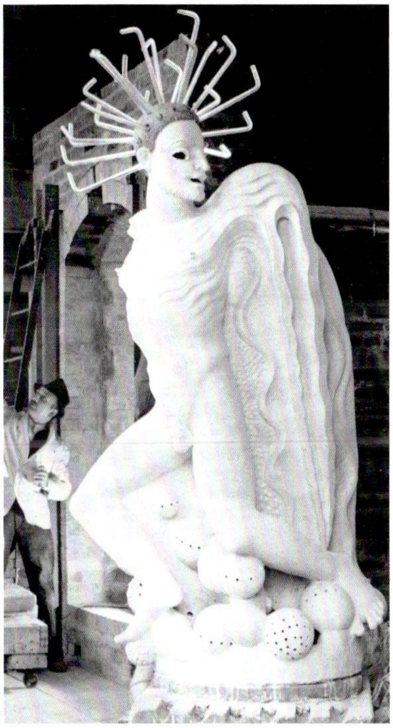

Figure 14 Waylande Gregory, Air, ca. 1938, from *Fountain of the Atom*. Glazed earthenware. 65" x 35" x 39". (Courtesy, private collection; photo, Waylande Gregory Archives.)

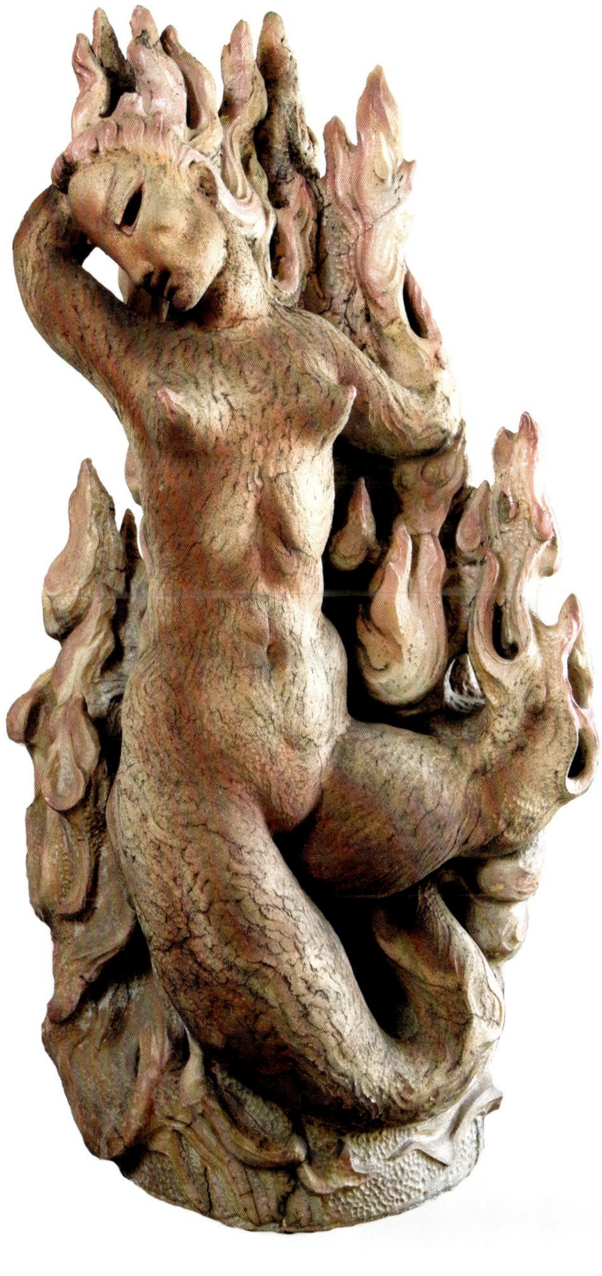

Figure 15 Waylande Gregory, Fire, ca. 1938, from *Fountain of the Atom*. Glazed earthenware. 76" x 41" x 41". (Courtesy, private collection; photo, Randl Bye.)

sensibility. Gregory described them as "elemental little savages of boundless energy."[11] Two of the four female electrons were fierce and played with bolts of electrical energy; one had strange prominent eyes, mounted forward of their orbits (fig. 16). Two of the four male electrons were like Water in that they were depicted upside down. Whereas Water focused on the side and rear of the male figure, these two focused the viewer's attention on the front and genital area. Surprisingly, the figures were publically displayed yet apparently caused no controversy. One of the electrons had fins and a wore a finned crown (fig. 17). The male electrons seemed to be modern interpretations of *putti* from Renaissance art, boy angels symbolic of erotic love.

Figure 16 Waylande Gregory, Female Electron, ca. 1938, from *Fountain of the Atom*. Glazed earthenware. 50" x 25" x 25". (Courtesy, Marty and Judy Stogniew Collection; photo, Randl Bye.)

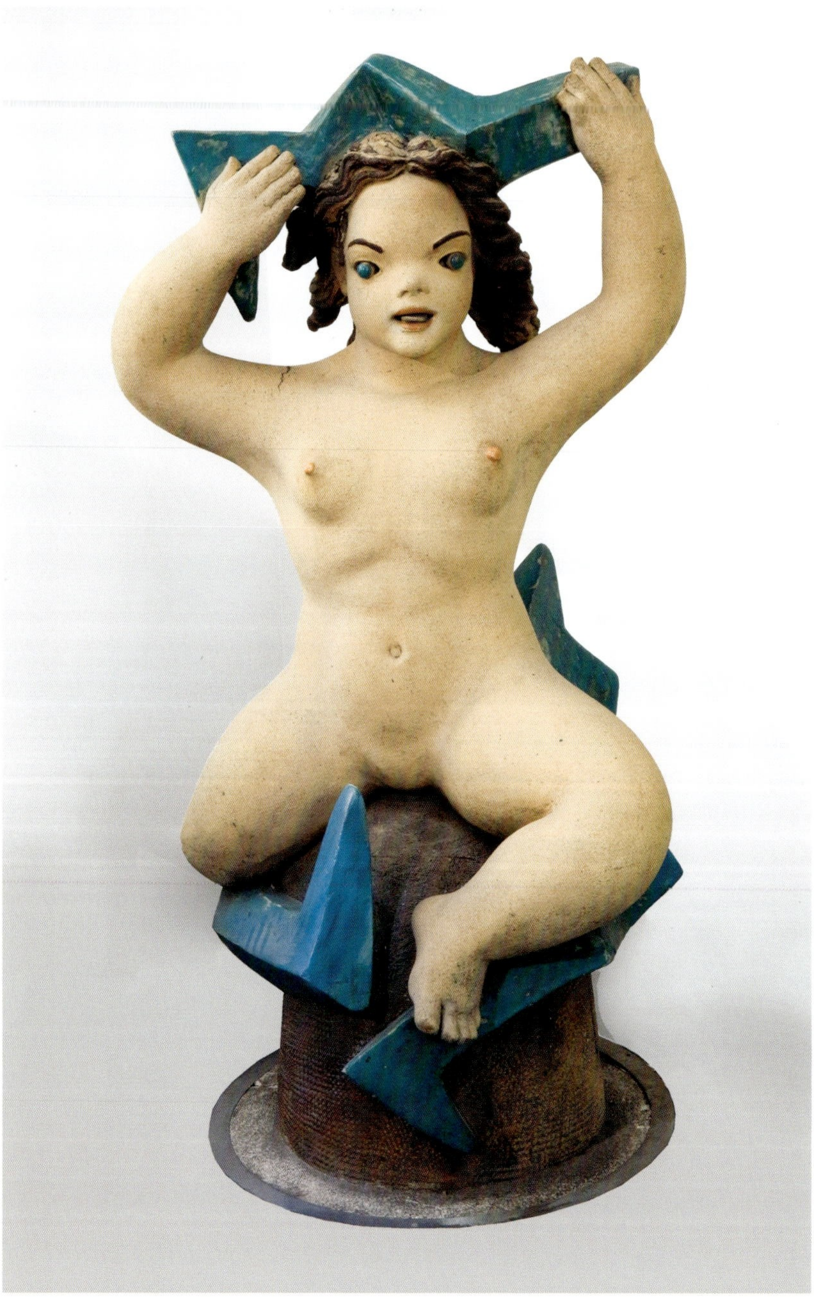

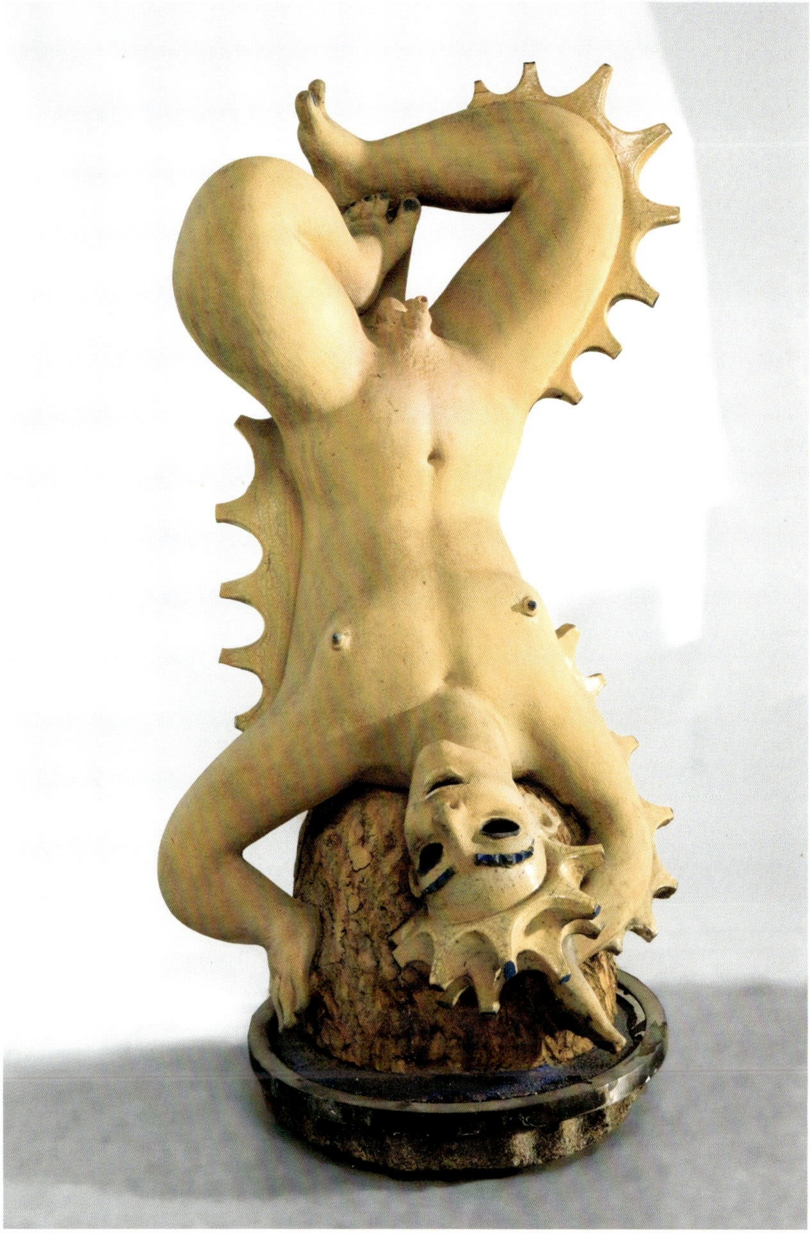

Figure 17 Waylande Gregory, Male Electron, ca. 1938, from *Fountain of the Atom*. Glazed earthenware, 48" x 26" x 26". (Private collection; photo, Randl Bye.)

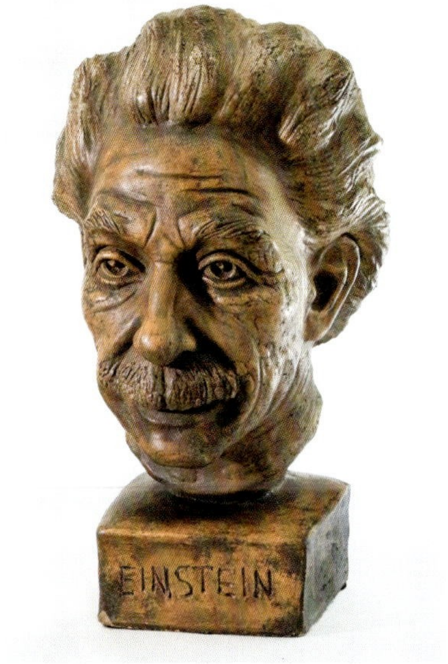

Figure 18 Waylande Gregory, *Portrait of Einstein*, ca. 1940. Glazed terracotta, 15" x 5" x 6". (Courtesy, Marty and Judy Stogniew Collection; photo, Randl Bye.)

Albert Einstein and his entourage had visited the fair and viewed *Fountain of the Atom*. Einstein was very much impressed and is said to have told Gregory, "Young man, I wanted to meet the artist who gave honor to the atom."[12] The two men struck up a friendship, and Gregory went to Princeton to make studies so that he could sculpt Einstein's portrait in terracotta (fig. 18). Gregory had sculpted the terracotta portraits of several celebrities by then, including two of actor Henry Fonda, a personal friend. Gregory made studies of Einstein from the back of his classroom while the scientist was lecturing to his students, and the final sculpture is an accurate likeness. It is significant that Gregory's *Fountain of the Atom* was somewhat playful in nature; by the end of World War II, many people would look at technology and the machine in a much more negative light.

Always interested in helping other Cleveland artists, in October 1938, as an advisor to the fair, Gregory asked Cleveland ceramist Viktor Schreckengost to submit a proposal for four ceramic heads also depicting the four elements of Earth, Water, Fire, and Air. After some initial confusion, the noted industrial designer Walter Dorwin Teague accepted Schreckengost's heads for his interior design in the United States Building.[13] Another Cleveland ceramic artist, Edris Eckhardt, exhibited a single head, representing Earth, at the fair. It is clear that Gregory continued to exert a strong influence on the Cleveland School for several years after he left the Cowan Pottery.

Gregory was the first modern ceramist to create large-scale ceramic sculptures. Similar to the methods developed by the ancient Etruscans, he fired his monumental ceramic sculptures only once. But the new technologies in

Figure 19 Gregory's use of his "honeycomb technique" on *Mother and Child*, ca. 1936. (Photo, courtesy Waylande Gregory Archives.)

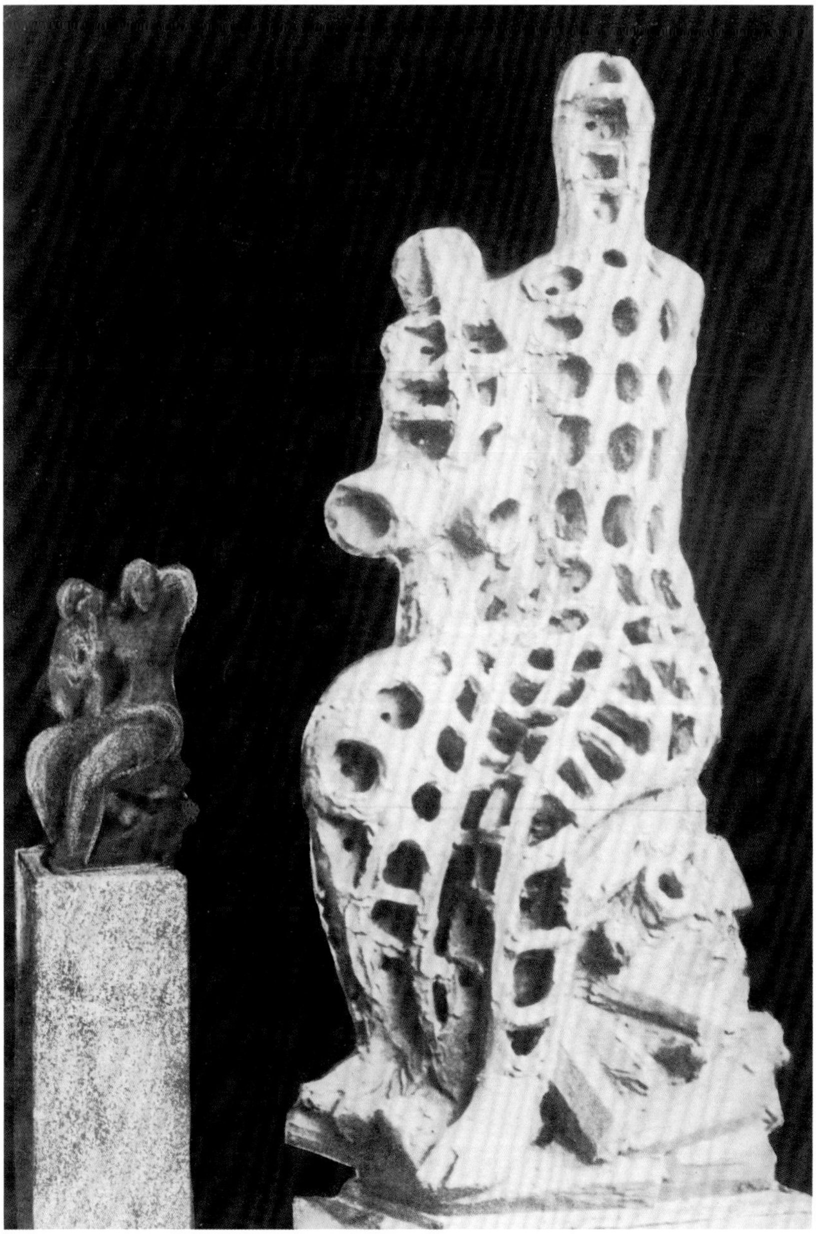

ceramics of the twentieth century were rapidly changing the industry, and Gregory was at the cutting edge of new developments. To create his monumental works (some measure more than seventy inches in height and weigh well more than a ton) the artist developed a "honeycomb technique" (fig. 19), in which an infrastructure of interlaced compartments was covered by a ceramic "skin." These works were fired by Gregory himself in the huge kiln he had constructed at his home and studio. Gregory described his process in 1944:

> Because great thickness of clay cannot be dried and fired successfully, and for other reasons as well, the ceramic work must be constructed hollow, somewhat in the spirit of architectural engineering. Due to shrinkage, no support other than clay can be used. The intense heat of 2,400 degrees Fahrenheit, and more, would burn away all metal or wooden supports or armatures, even if they could be used. Therefore, the structure of the sculpture itself must be entirely of clay, self-supporting from within the sculpture and plan of the sculpture. The sculptural form sequence must be so conceived that it is possible to construct and engineer the work to enable it to survive the stress and strains of drying shrinkages, fire shrinkage, and chemical changes without collapsing or developing cracks. This is not a thing that can be achieved by short cuts. It has taken me many years to acquire this knowledge and experience that it requires to build the massive monumental scaled sculptures that I am now creating. . . . Such sculptures are highly permanent, retain all the texture and veracity of the direct modeling, have great color and distinction over other sculptural mediums.[14]

Gregory fired his monumental sculptures in cone 7 (a level of kiln firing). While working on them, the inner core had to remain moist enough so new wet clay could be added to the sculptures but dry enough so they would not sag. The figures then had to dry before being glazed and fired. After they were placed in the kiln, the entrance to the kiln was walled in with bricks, and the sculptures were fired for two to three weeks. This brick wall was removed after firing, at which time the artist would examine his finished work.[15]

About 1950 Gregory returned to the subject of the atom, although this time in an oil painting rather than a ceramic sculpture. *Atomic Holocaust* (fig. 20) shows a city on fire, evaporating in a blaze of color. Although the painting shows the negative effects of nuclear energy—by the end of World

Figure 20 Waylande Gregory, *Atomic Holocaust*, ca. 1950. Oil on wooden panel. 17" x 47". (Courtesy, Ron Wei; photo, Randl Bye.)

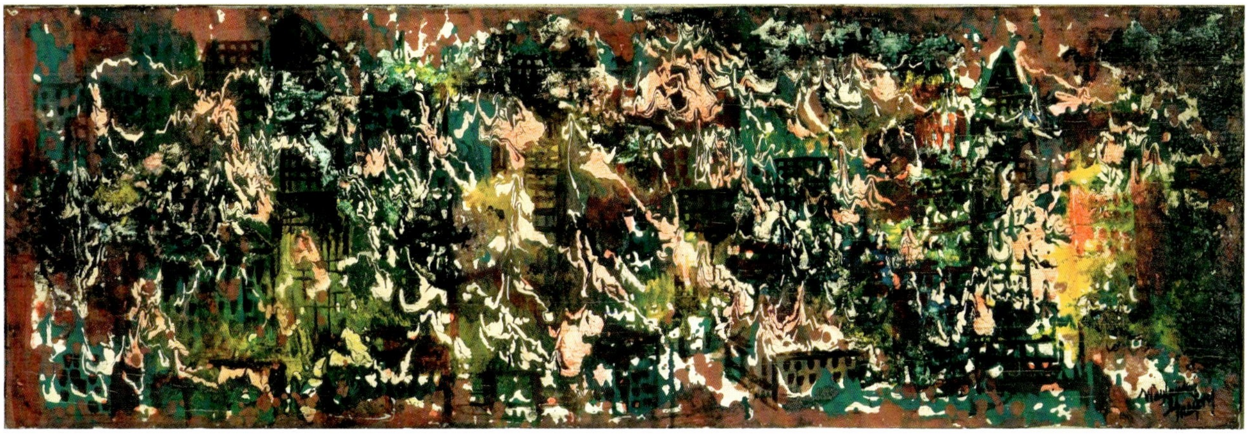

War II most people shared a fear of nuclear annihilation—Gregory focused on the expression of color rather than the human tragedy.

Clearly, the 1939 World's Fair was Gregory's day in the sun. Although he continued to produce a great variety of artwork in ceramics, metal, glass, and paintings, he never again approached the success he had enjoyed with *Fountain of the Atom*. By the same token, even though other sculptors—Robert Arneson, Viola Frey, and Peter Voulkos, for example—tried their hand at monumental ceramic sculpture, no one has attempted Gregory's very personal, sophisticated technique.

The exhibition "Waylande Gregory: Art Deco Ceramics and the Atomic Impulse" was organized by the University of Richmond Museums, Virginia, and travels to the Cranbrook Art Museum, Bloomfield Hills, Michigan, November 16, 2013–March 23, 2014, then to the Canton Museum of Art, Ohio, May 1–July 27, 2014.

1. The men who worked on *Light Dispelling Darkness* included John Corollo, William Deuel, Joseph Kalik, Dominick di Lorenzo, Amadeo Lovi, Henry Lovi, Adolph Maffei, Ned Monti, Andrew Petrovits, and Alfred Rossi.

2. Gregory's description of the iconography he used in *Light Dispelling Darkness* is recorded in an undated letter to the Newark Museum, Newark Museum Archives. I would like to thank Jeffrey Moy, the museum's librarian, for his assistance.

3. For more on Gregory, see his first monograph, *Waylande Gregory: Art Deco Ceramics and the Atomic Impulse*, exh. cat. (Richmond, Va.: University of Richmond Museums, 2013). See also Thomas Folk, "Waylande Gregory," *Ceramics Monthly* 42, no. 9 (November 1994): 27–32; Gene DeGrusin, "No Greater Ecstasy: Waylande Gregory and the Art of Ceramic Sculpture," *Kansas Quarterly* 14, no. 4 (Fall 1982): 65–82; and Elaine Levin, "Monumental Ambitions: Waylande Gregory," *American Ceramics* 5, no. 4 (1987): 40–49.

4. See "World's Fair, Surprise! Surprise! Surprise! They Will Lurk at Every Turn. These Little Savages are Electrons," *Life* 6, no. 11 (March 13, 1939): 37; and "The Octet Theory of the Atom in Terra Cotta: The Fountain of the Atom at the New York World's Fair," *The Illustrated London News*, April 29, 1939, p. 719.

5. Waylande Gregory, cited in Margaret Lindsay, "Unique Sculpture for World's Fair Created in Bound Brook Studio of Waylande Gregory, Ceramic Artist," *The Sunday Times* (New Brunswick, N.J.), March 5, 1939, p. 1.

6. Ibid.

7. Ibid.

8. Ibid.

9. Ibid.

10. Ibid.

11. Ibid.

12. Bianca Brown, a personal friend of Waylande Gregory and his wife, Yolande, related Gregory's experiences with Einstein and Gregory's subsequent portrait bust of him in a telephone conversation with the author, November 4, 1999.

13. Gregory's interest in Schreckengost's submission of his *The Four Elements* in the 1939 World's Fair is related in Henry Adams, *Viktor Schreckengost and 20th-Century Design*, exh. cat. (Cleveland, Ohio: Cleveland Museum of Art, 2000), pp. 72–73.

14. Waylande Gregory, cited in Ernest W. Watson, "Waylande Gregory's Ceramic Art: Interview," *American Artist* 8, no. 7 (September 1944): 14.

15. Lorraine Hoogs discusses the techniques Gregory used in the monumental sculptures he fired in his New Jersey home, in Ross Anderson and Barbara Perry, *The Diversions of Keramos: American Clay Sculpture, 1925–1950*, exh. cat. (Syracuse, N.Y.: Everson Museum of Art, 1983), pp. 19–22.

Ivor Noël Hume

A Hole in One? or, In Search of Piggy Banks and Christmas Boxes

▼ EDITOR'S NOTE This study of early money boxes is the product of three individuals, each approaching an obscure subject from different directions: the potter, the collector, and the museum ceramics specialist (fig. 1). Michelle Erickson, the internationally respected potter, put her contribution into clay rather than on paper. Ivor Noël Hume, the collector, and Jacqui Pearce, ceramics specialist with the Museum of London Archaeology, put theirs in writing, albeit in entirely different styles. One might argue that as a collaborative work of scholarship the texts should blend seamlessly into a single entity, which was certainly possible. The authors agreed, however, that beyond money boxes this contribution demonstrates how different elements of ceramic research can come together. The apparent superficiality of collecting had posed questions that opened the door to scholarly research of enduring value.

The thrill of research at any level is that we never know where it will end. Who, for example would have thought that examining a pinhole in a pot could lead to a nut-cracking, apple-chewing audience in London's Elizabethan theatre The Rose?

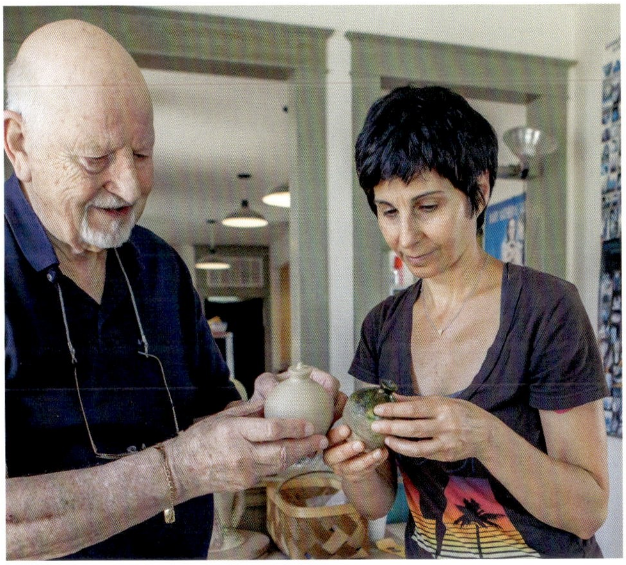

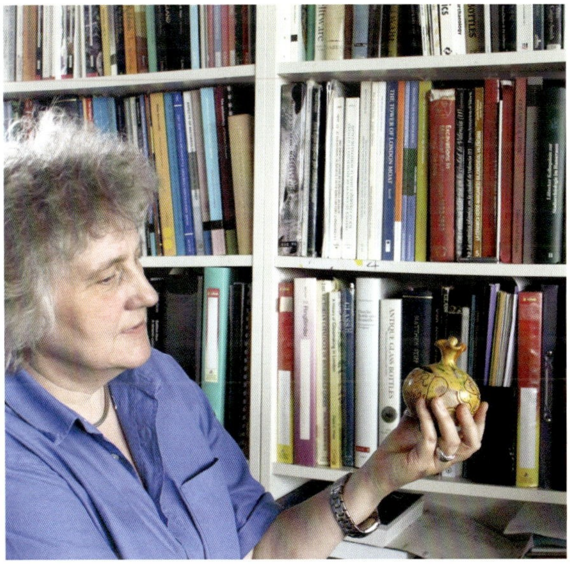

Figure 1 *Left*: Ivor Noël Hume and Michelle Erickson. (Unless otherwise noted, all photos by Robert Hunter.) *Right*: Jacqui Pearce (photo, courtesy of the author).

▼SO WHY THE HOLE? My wife and I had been collecting money boxes for about thirty years before I asked that question and came up with the wrong answer (fig. 2). It became, however, the genesis of this study, and converted the old adage "it takes one to know one" into "it takes one to throw one."

In England, ceramic containers for saved coins have long been known as "money boxes" or "Christmas boxes," whereas in America they are called "piggy banks." The latter term is included in *Webster's Ninth New Collegiate Dictionary* but, surprisingly, cites its first use as 1941, far later than one might have supposed. Equally surprising (at least to an Englishman) is the American dictionary's omission of the term "money box." Perhaps in retaliation, the *Oxford English Dictionary* includes no "piggy bank." It does note that in northern England in the mid-sixteenth century a pig could be a small earthen pot, but fails to associate it with money.

What may be the earliest bank shaped like a pig was in the collection of London's Guildhall Museum (now Museum of London) and cataloged as "in the form of a Sussex pig, reddish glaze with yellow patches, imperfect, XVII cent., perhaps made at Rye."[1] However, a standard British ceramics encyclopedia attributes the beginning of Rye pottery making to the late eighteenth century, but adds that in such factories jugs were made in the form of a pig.[2] It seems, however, that this pig is more than likely a red herring and dates from the nineteenth century.[3]

There is, however, contributing evidence from Amsterdam, where two green-glazed earthenware money boxes in pig form have been found in

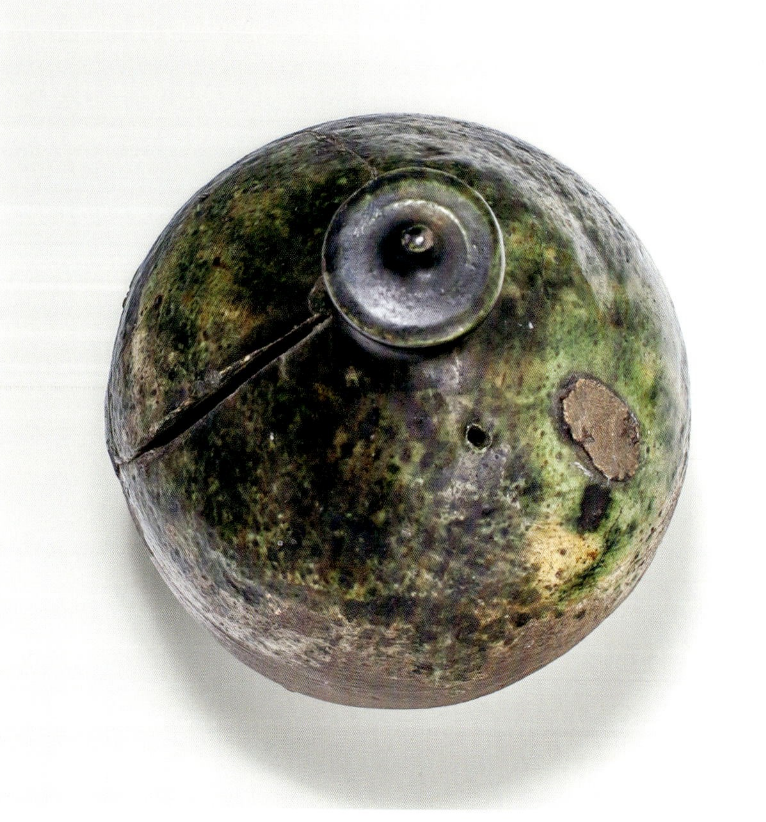

Figure 2 Money box, Surrey/Sussex, England, 1560–1620. Lead-glazed earthenware. H. 3½". (Courtesy, Noël Hume Collection.) This shows the enigmatic hole in the shoulder of the green-glazed Border Ware money box. The lower side of the body has been restored.

contexts attributed to 1550–1575 and 1650–1675.[4] Nevertheless, I am leaving English pigs in their sty and focusing, instead, on globular clay banks or boxes. According to the *Oxford English Dictionary*, they were known as long ago as 1585 and described as "A box for money; especially a closed box into which coin is dropped through a slit." But was that a pottery box? It could have been describing another type, mentioned in a Tudor chronicle, which recorded in 1538 that "the boxes that they have to gather the devotions of the people were taken away first."[5] Those were being removed from religious houses at the time of the Reformation and almost certainly were made of wood.

There are, of course, numerous kinds of boxes, from iron nave boxes on the hubs of wheels to prestigious cubicles in theaters, but as a rule one thinks of boxes being cardboard or, in previous generations, wood. Alms boxes, poor boxes, and ballot boxes were all of wood and their contents received through a narrow slot and carefully locked. In short, they had nothing to do with pottery banks, whose contents were far less secure. A ceramic money box in the collection of London's old Guildhall Museum was cataloged as being "of green glazed ware with slit on shoulder; fractured beneath to extract the money. XVth century." Another of allegedly similar date was described as "red and yellow glaze in the form of a toad (?)." The catalog entry having been penned prior to 1908, there is no assurance that the frog was not a poorly rendered pig.[6]

One has to remember that early museum catalog dating was based on curatorial guesswork, there having been no money-box fragments recovered from sealed archeological contexts. Those that are intact were dug up by London construction workers who had learned that "old stuff" taken to the museum could be worth a beer. Thus, each acquisition has had to speak only for itself and not for the unrecorded context in which it was found. Those shown in figure 3 are typical of the type, but serve to illustrate that they came in sizes large and small.

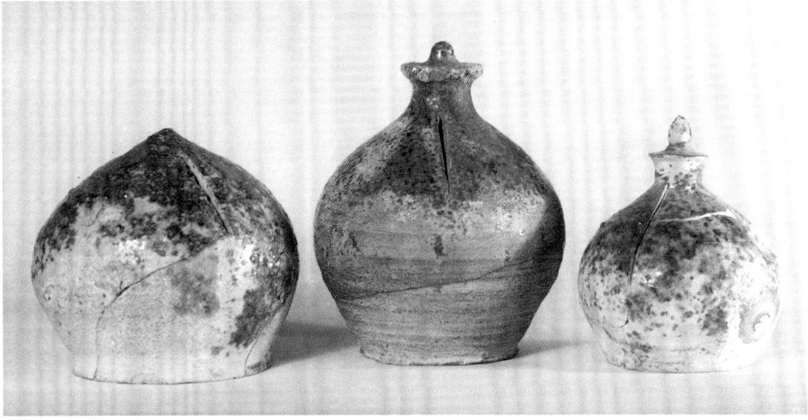

Figure 3 Money boxes from London's old Guildhall Museum Collection. *Left*: miter-shaped, probably from Cheam, Surrey, England, late medieval, lead-glazed earthenware; *center, right*: green-glazed Border Ware, Surrey/Sussex, England, 1560–1620. H. of money box on right $3^{15}/_{16}$". (Courtesy, Museum of London; photo, Ivor Noël Hume.)

In Britain, the first day after Christmas has long been known as Boxing Day, but few of us now remember or care that it relates to the tradition of masters giving money in pottery boxes to their apprentices—hence the variant term *Christmas box*. In 1615 a man of the cloth wrote, "It is a shame for a rich Christian to be like a Christmas-box, that receives all, and nothing can be got out until it is broken in pieces."[7] Clearly, therefore, the box was not of wood but of smashable pottery. But why, one might ask, were Christmas gifts of money not openly handed out to employees with a smile and an assurance of a year well spent?

One of the common traits of apprentices was a lack of politeness or gratitude. While they would look with ill-concealed scorn at a handful of farthings, the same coins bestowed in a clay box deferred the expletives until it was broken and the miserly master unmasked. In 1642 essayist Humphrey Browne wrote of such a master that he "doth exceed in receiving, but is very deficient in giving; like the Christmas earthen boxes of apprentices, apt to take in money, but he restores none until he be broken like a potter's vessel into many shares." Another essayist had a similar thought: "Like a swine, he never doth good till his death: as an apprentice's box of earth, apt he is to take all, but to restore none till hee be broken."[8] Doubtless, it was a fortuitous choice of words that likened the miser to a pig.

In 1714 the eccentric historian John Aubrey noted the discovery of a Roman pot containing silver coins (denarii) and declared that "it resembles in appearance an apprentice's earthen Christmas-box."[9] It is true that pigs (wild boar) frequently appeared on red, molded Roman pottery, but there is no evidence that their potters made clay pigs or that any known type of Roman pot looked like the kinds of earthen Christmas boxes known to Aubrey. He was born in 1626 and would have been most familiar with round pots, green or yellow glazed, tapering to a disc on top and having a narrow slot in one side.

Though no two are exactly alike, the button-knopped form became common in London during the reign of Elizabeth I (1558–1603). Unfortunately the Guildhall Museum's catalog illustrated only one example and was lackadaisical in its descriptions of others, for example Number 400, which is described as "buff ware, ordinary type; XVI century." However, the cataloger did think it worth mentioning that the box had been found in 1897 while building the Central London Railway's Bank Station. That was a reference to the Bank of England, which could well be described as the British nation's ultimate piggy bank.[10]

It is assumed that all the button-knopped money boxes unearthed in London were made in southeastern English kilns along the borders of Surrey and Hampshire counties. Known today as "Border Wares," the industry was in production by the mid-sixteenth century, as Jacqui Pearce discusses in her essay in this volume. The oldest example in my collection (fig. 4) was attributed by a previous collector to a date in the sixteenth century. How his (or her) dating was arrived at has not been ascertained. The type is likened to a bishop's miter and is said to have its origins in the thirteenth century. The shape is spheroid, rising to a point, with a vertically aligned

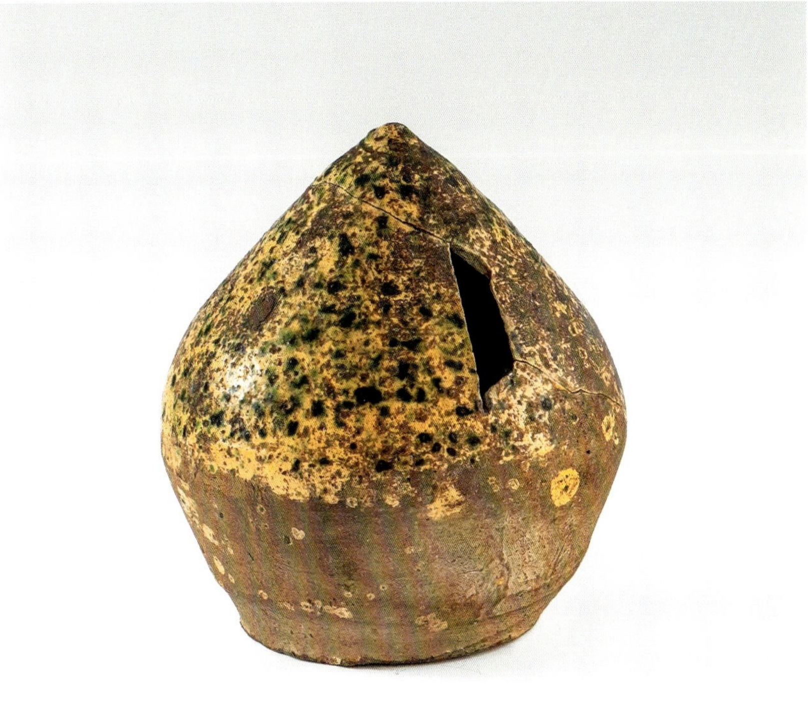

Figure 4 Money box, probably Cheam, Surrey, England, ca. 1500. Lead-glazed earthenware. H. 3½". (Courtesy, Noël Hume Collection.) The coin slot of this miter-shaped money box measured 1¼ x 1/16" before it was widened.

coin slot in one side, barely one millimeter in width. As is true of many of the early surviving boxes, the slot had been roughly widened to extract the contents. The box's yellow lead glaze dappled with copper is undeniably akin to that of jugs made at Cheam in Surrey in the second half of the fifteenth century. However, *looking like* is no more authoritative than descriptions in the old Guildhall Museum catalog. Not having access to any archeological ceramic groups from the fifteenth century, it seemed as though I would have to leave the bishop's box undocumented—until along came one of those coincidences to be found in bad novels.

In May 2013 I received a request from Ian Blair at the Museum of London to remember a Roman wall I had excavated in the city sixty-three years earlier. My memory being what it has become, I had to dig out a booklet on the Roman remains that I had written in 1950. For no apparent reason I turned to its closing paragraph and found this summery of later finds: "One such pit, dating from the late fifteenth century, yielded cooking pots, jugs, and a pottery money box. With them were the bones of two large dogs, three cats, and three mice."[11] This could be the evidence I needed, but I could not recall the pit. I knew I had excavated it before I had established a numerical system of recording, and I knew that the Guildhall Museum where I worked had long since been absorbed into the new Museum of London, meaning that in all likelihood those long-ago discoveries had been

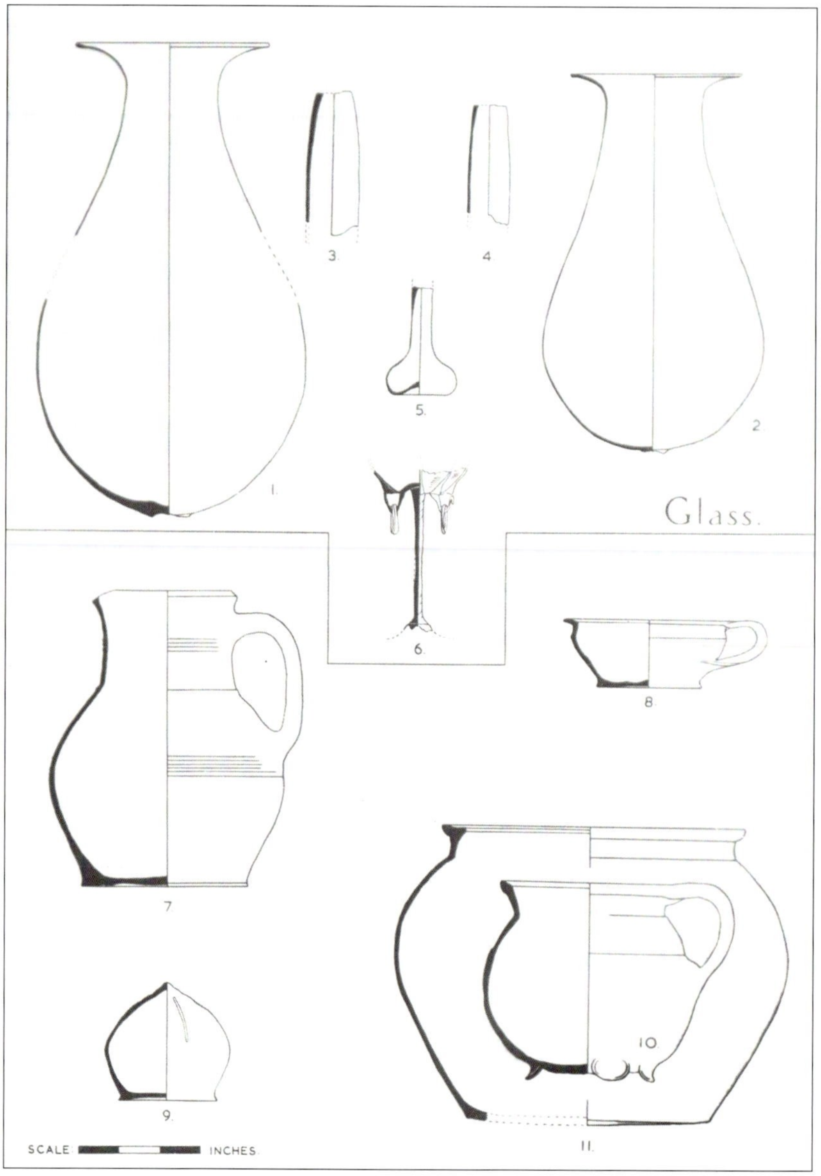

Figure 5 Drawing of a Border Ware pitcher and miter-shaped money box found in a closely datable London refuse pit, 1480–1510. (Courtesy, Guildhall Museum Collection; artwork by Ivor Noël Hume.)

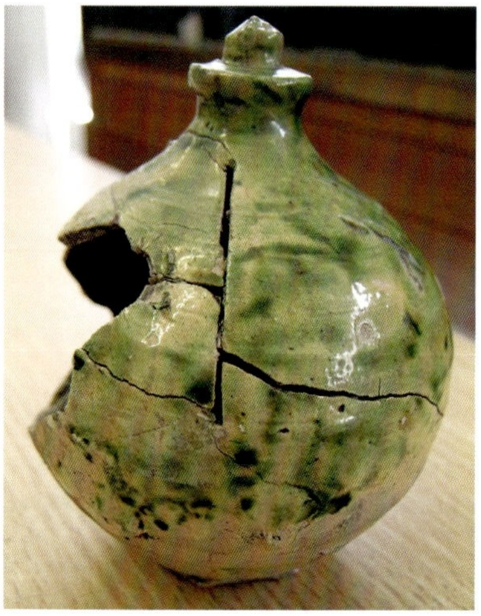

Figure 6 Money box, Farnborough, Hampshire, England, 1565–1580. Lead-glazed earthenware. H. 3½". (Courtesy, Guildford Museum; photo, Jacqui Pearce.) This Border Ware money box is a waster found in a kiln flue (No. 3) on the Farnborough Hill site.

scattered and their evidence lost. But I was wrong again. In my files of 1950s negatives, I found a page of my drawings of that pit group along with a note attributing it to circa 1480–1510 (fig. 5).

The long-forgotten group included a rare German wine-glass stem and two glass urinals.[12] I suspect that there is food there for another bad novel that puts the money box amid dead dogs, cats, and mice, but this is not the place to pursue it.

In the second half of the sixteenth century, the pointed top of the miter-shaped boxes had been replaced by a decorative button with a central nipple (fig. 6). A primary source for the new style was found to be the kilns of Farnborough in Hampshire, where the whiteware date range proved to be 1550–1580. The cracked waster excavated from one of the kiln flues has a knop typical of the form. As in most Elizabethan country potting, however, style changes had a practical purpose. In his *The Collector's Book of English*

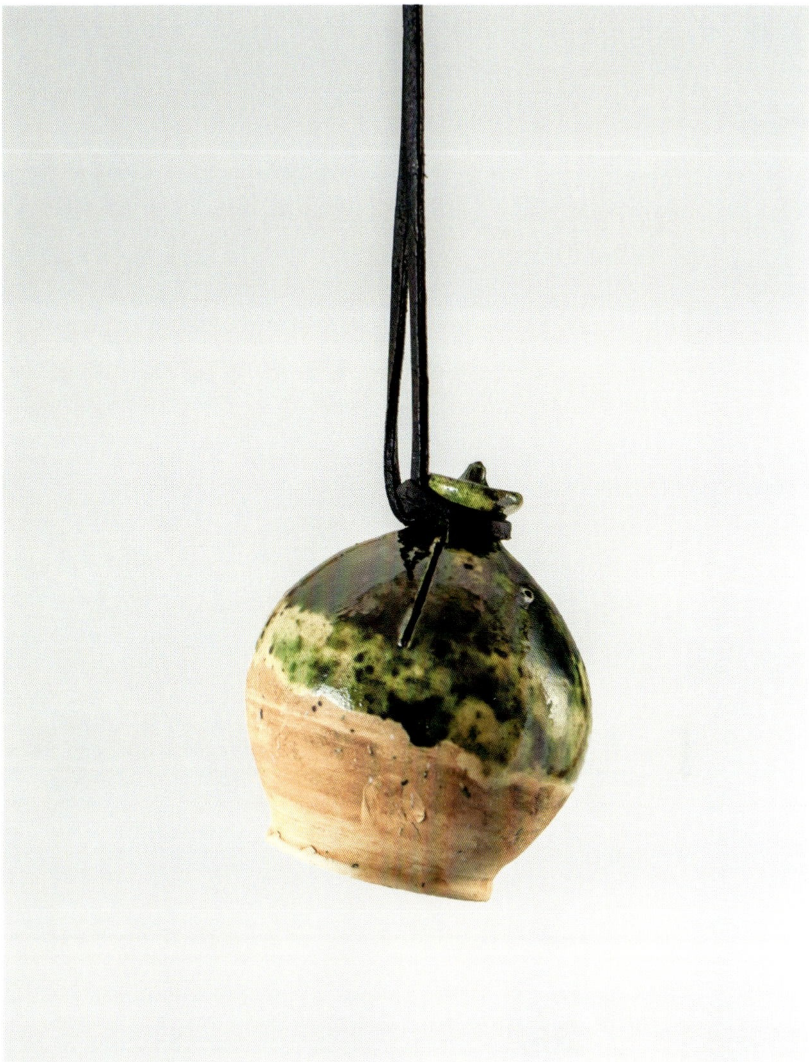

Figure 7 Button-knopped reproduction money box showing its documented method of suspension.

Country Pottery, Peter Brears quotes a reference of 1585 that described a "mony box made of potters clay, wherein boyes put their money to keepe, such as they hang in shops etc., towards Christmas."[13] That explains the need for the collared finial, but does not confirm that the boxes were given to the apprentices only on Boxing Day. Instead, it would appear that they were given earlier, to be suspended in stores, perhaps by the boys themselves, and in locations where customers could be talked into a donation (fig. 7).

The button knop, therefore, allowed a cord to be tied around it just as string-rims secured corks around the necks of glass wine bottles. The button is a small detail but one with significant social implications, as it pointed to the treatment and status of apprentice boys in the reign of Elizabeth I. The shape's demise also heralded the apparent end of the box-breaking tradition, indicative, perhaps, of a yet-to-be-explained societal change—perhaps prompted by the rise of Puritanism that took the fun out of boys' Christmas. Be that as it may, it was not the finial that intrigued this collector, it was that tiny hole (fig. 8).

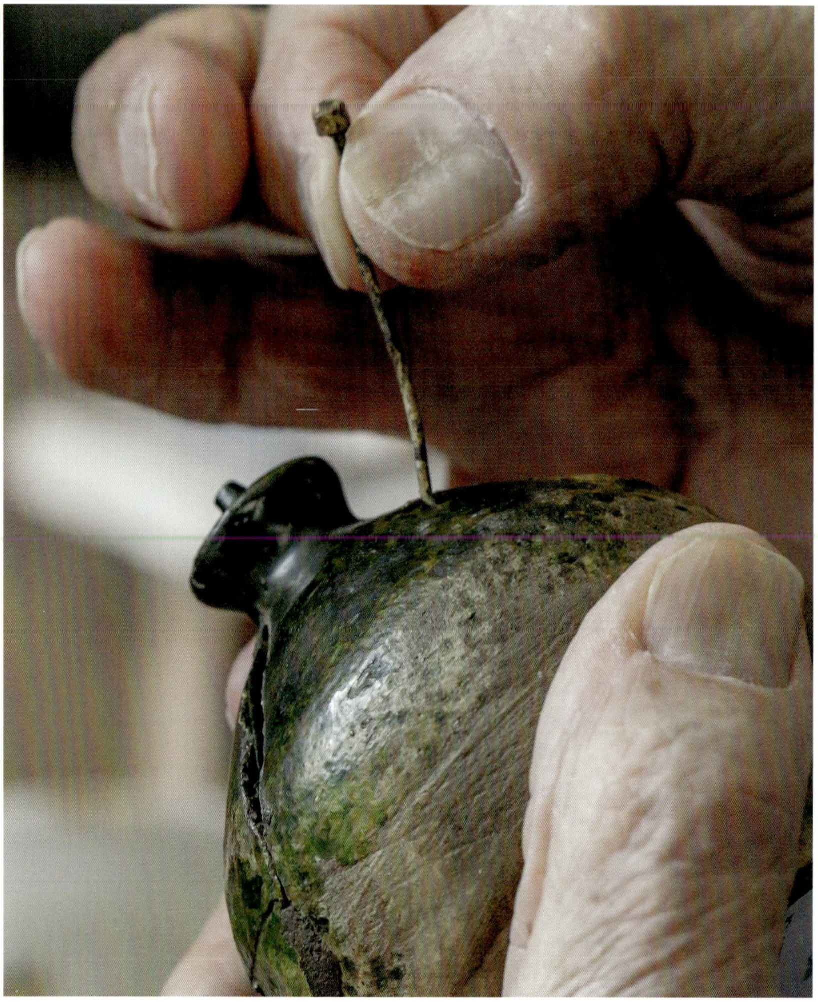

Figure 8 The shaft of an Elizabethan pin snuggly fits the mystery hole on the money box illustrated in fig. 2.

About the diameter of a Tudor pin, it pierced the shoulder of the collection's box (see fig. 2). A study of similar boxes published in archeological reports yielded no parallels. A quick check with Jacqui Pearce caused her to reply that she had never seen pinholes in money pots—which was not surprising, as I had owned my example for decades without noticing the hole. A few days later Jacqui reported that she had subsequently located six pinholed examples in the Museum of London collection, all similarly green-glazed, all of approximately the same dates, and all from the same potting area of southeast English counties. The hole, therefore, was not a "one off" anomaly. But what was its purpose?

Knowing that pots shrink about ten percent between throwing and firing, I deduced that the narrow coin slot might close up during the shrinking and that a spacer had to be inserted into the slot to keep it open; but having done that, air could not escape from the ball unless there was a vent—hence the pinhole.[14]

Michelle Erickson, a highly skilled practitioner of reproductive historical ceramic techniques, agreed to go back to square one and reproduce Tudor money boxes with and without pinholes. Her preliminary conclusion was that I was wrong about the slit closing but half right about the vent.

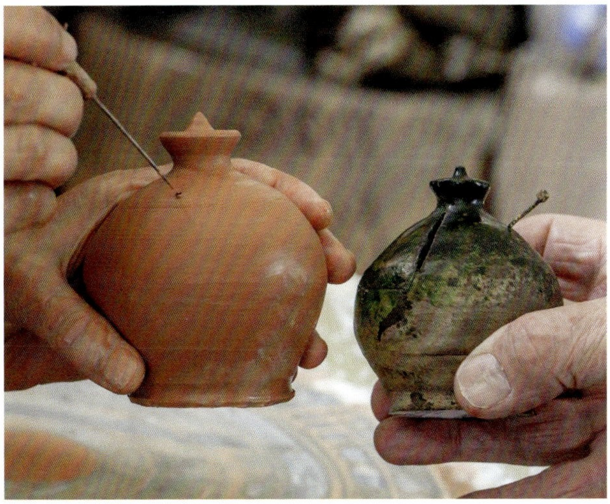

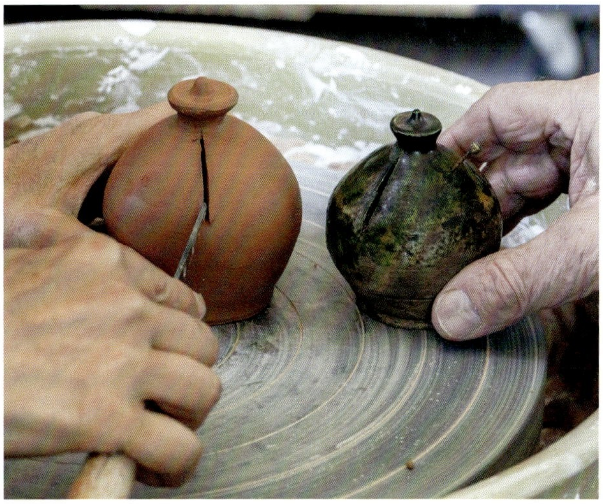

Figure 9 Piercing the hole into the wet clay and cutting the coin slot when leather hard.

Michelle Erickson's early experiments showed that when first thrown, the wet clay ball contains moisture that has to be released during the drying process, requiring the tiny vent. The coin slot cannot be cut until the box is in its leather-hard, prefiring condition—by which time the pinhole served no further purpose (fig. 9). In some cases the pinhole was left open, being in no way detrimental to the box's appearance; in others, potters would smear the hole shut before glazing, causing it to disappear. The same production sequence would be applied to any air-sealed vessel, but when ready for firing the hole's prior presence would be visible only on the inside (fig. 10). The illustrated fragment shows a London example whose green glaze had sealed the outside but had run through the pinhole into the interior (fig. 11).

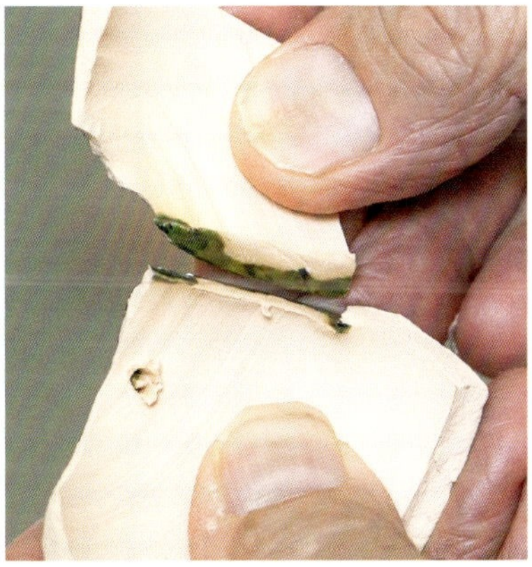

Figure 10 Fired fragments show the different impacts of piercing the wet clay and cutting the slot at the leather-hard stage.

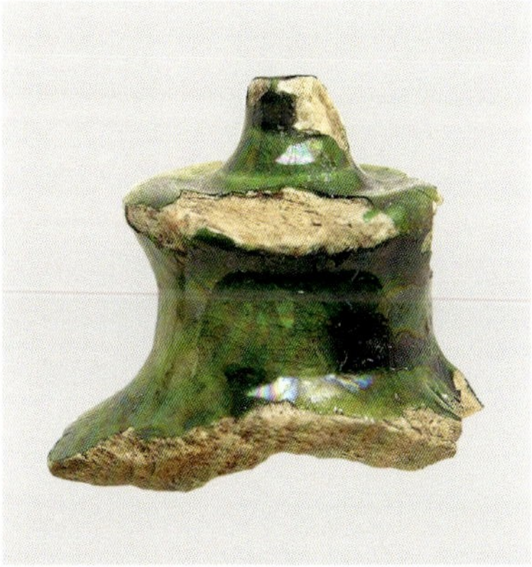

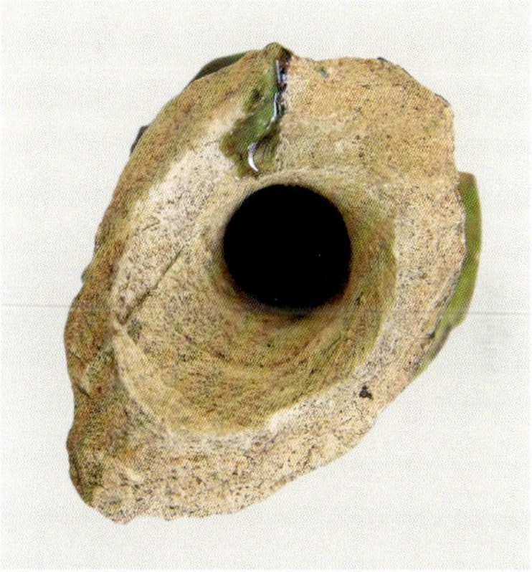

Figure 11 Lead-glazed knop fragment from The Theatre, Shoreditch (1576–1598), site reveals glaze bleeding through the concealed pinhole. (Museum of London; photo, Tim Braybrooke, © MOLA.)

To demonstrate that there was a time difference between the pinholing and the slitting, it was necessary to make a good copy of the box and then break it in order to study the inside. A tiny ridge around the hole showed that the clay was still wet when the pin had pushed it aside and thus had not punctured it cleanly. In contrast to the softness of the pinhole edges, with the box in its leather-hard stage the cutting of the slot neatly removed the displaced clay, leaving a sharp ridge alongside the interior of the cut. Although the powder disappeared, in some instances displaced clay fragments remained loose inside the boxes.

Pursuit of the enigmatic pinhole led Michelle to study other details, such as the way in which the coin slot was cut, and with what. It was slightly wider at the bottom than at the top, and the left side had a rough exterior edge while the right was smooth and slightly depressed. She found that it took a single slash with a knife from a right-handed potter to reproduce it. She also found that the button knop, which has sometimes been described as molded, was, in fact, reproducible by a skilled finger and thumb (fig. 12). We may not have discovered what the Elizabethan potter was thinking when he threw the little box, but we had learned how he did it!

Although a prodding pin may have been the genesis of this study, like the opening of any door, surprises can be waiting on the other side. We had not realized that globular Tudor money boxes preferred to bounce rather than break. After one of Michelle's reproductions was thrice thrown down onto a stone slab without breaking, it took a hammer to do the job (fig. 13). Here was graphic evidence that when a customer's ordering instruction was to make a product that could be broken (or not), the thickness (or rather the thinness) of the wall had to be a determining factor. One learned, too, that when smashing money boxes with a hammer, protective goggles are a wise investment (fig. 14).

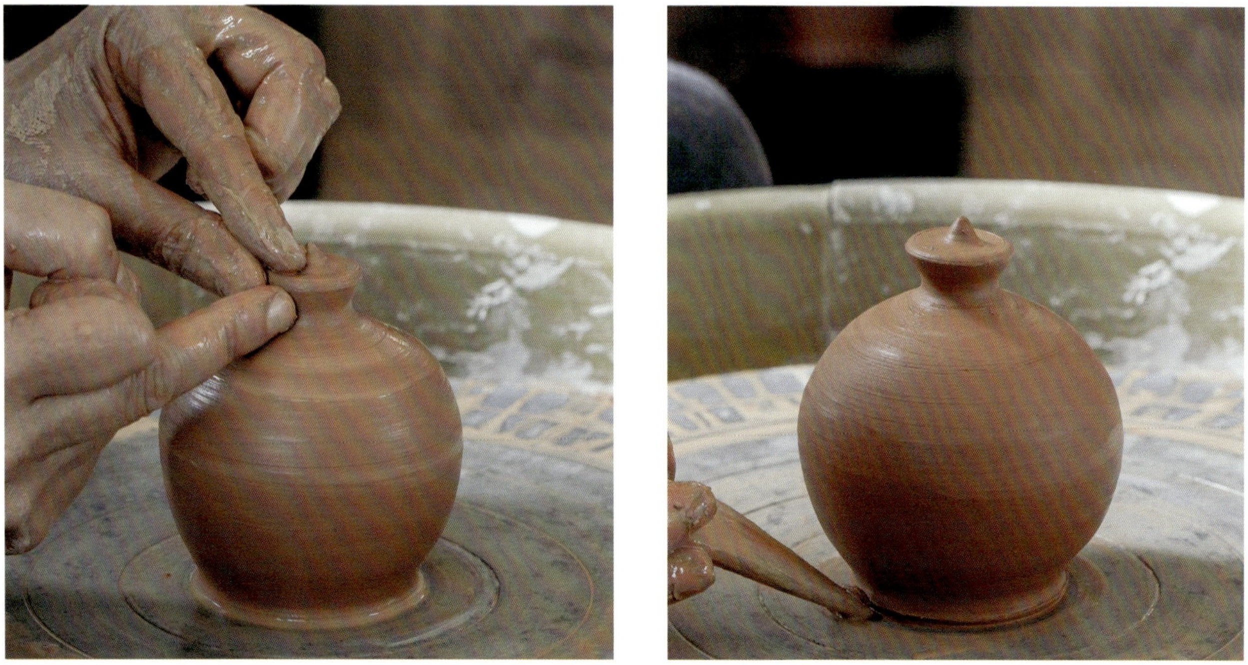

Figure 12 Turning the button knop and trimming the base before drying.

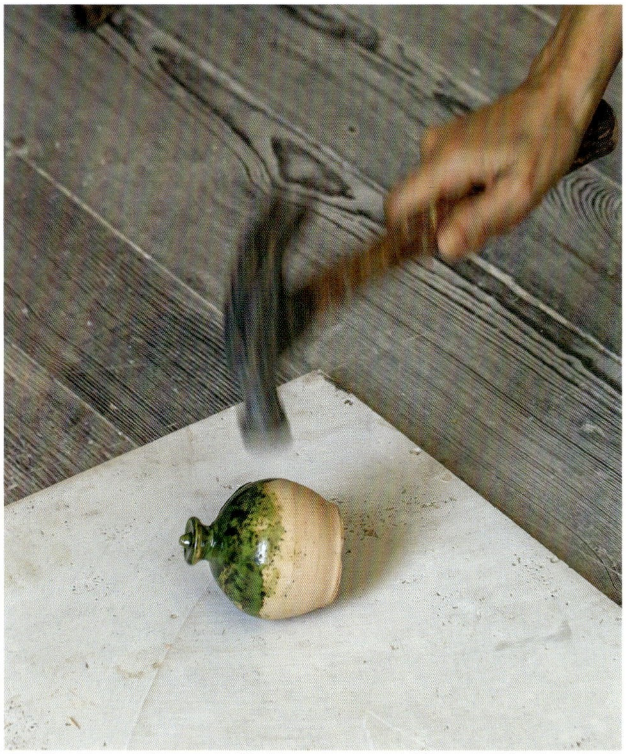

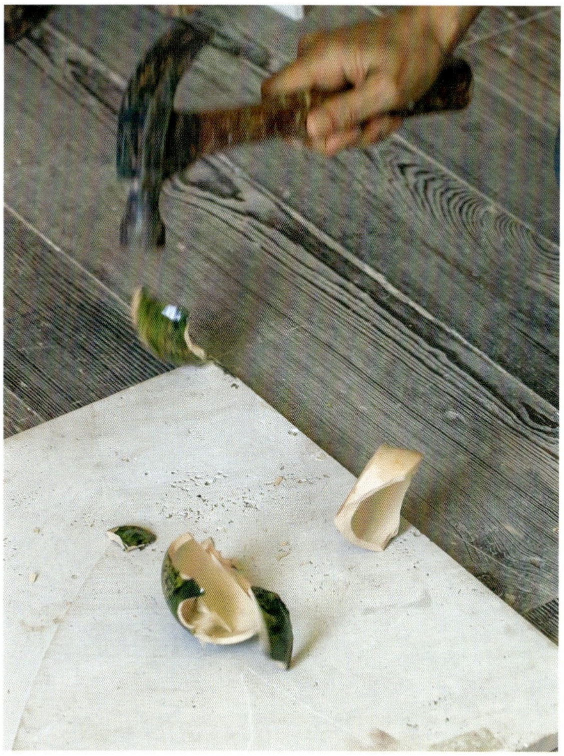

Figure 13 One method of opening the (reproduction) box.

Figure 14 The hammer crew.

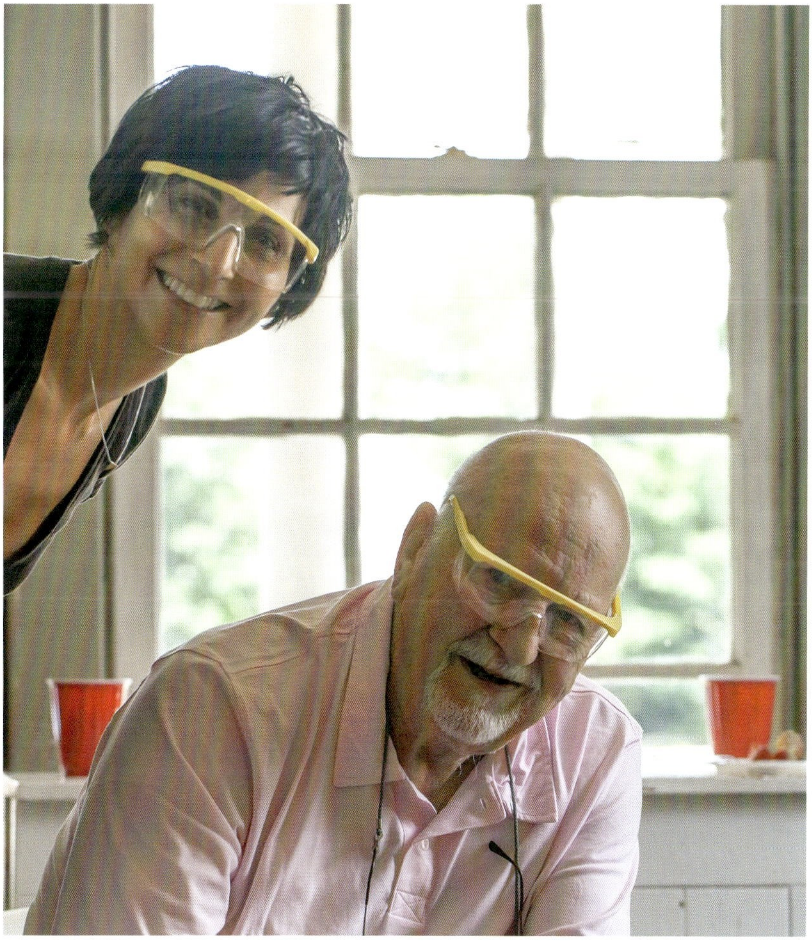

Figure 15 Unfired reproductions show concave (*left*) and flat (*right*) bases.

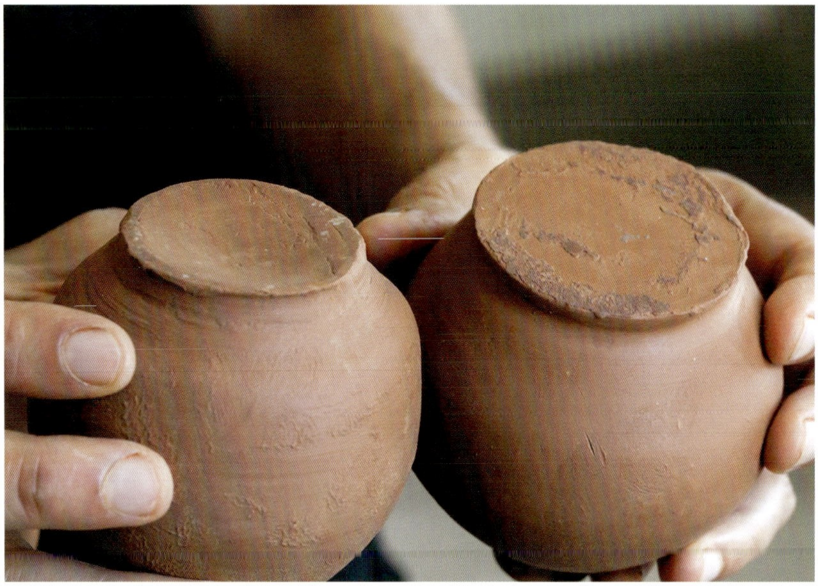

In her contribution to this study, Jacqui Pearce notes that some Tudor boxes exhibit "concentric or fan-shaped marks" on their flat bases caused by the wire employed to cut them from the wheel. However, our pin-pierced example has a gently domed bottom rather than a flat base.

Michelle's explanation goes like this: Open-mouthed vessels would not trap air inside them when being thrown, but money-boxes shaped without pinholes or coin slots would become clay balloons from which the air could not vent. As the clay shrank during the drying process, the volume of trapped air remained the same, and thus could thrust a flat base into becoming convex and unstable. To address that potential problem, some of Michelle's boxes were thrown atop a shallow dome and so emerged from the wheel with concave bottoms—or, as apprentices would have said, "Bottoms up!" (fig. 15).

Michelle notes that her first attempts to make air-sealed boxes did cause the bases to bulge, and that the tendency was countered by the mound. However, she is clear that the reasoning associated with the reproduction boxes does not imply any connection with open-mouthed vessels that have similarly concave bases.

Although we have established that the pinhole was an intentional step in our box's manufacture, another visible feature was not. A small vertical scar atop the glaze on one shoulder said that the box had abutted against something else while in the kiln. As the scar did not extend to the box's girth, its neighbor could not have been a jar or other straight-sided vessel. I suggested, therefore, that the companion pot had to have had a projecting spout or handle that touched the wall of the money box. Again I was wrong.

In her experiments Michelle had made several trial boxes and was able to show convincingly that when a second tier was set in the kiln, an inverted box sat precisely where the scar had been (fig. 16). However, the scar appears not to have been caused by the abutting box but by a small clay spacer placed between them.

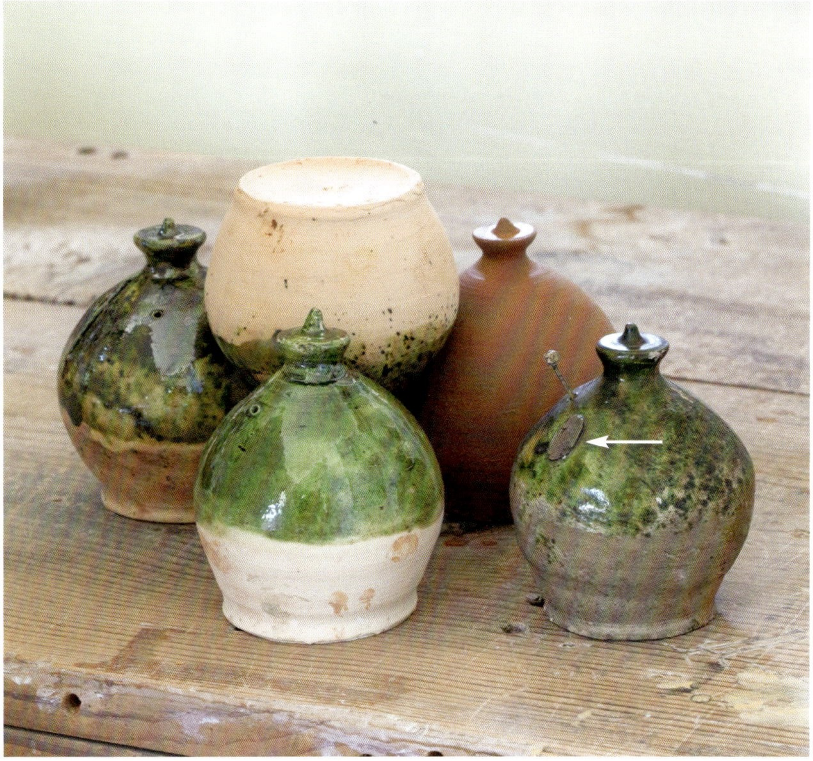

Figure 16 The white arrow points to the scar on the period money box, which provided a clue to the method of kiln stacking.

Further support for the kiln-stacking interpretation comes from the base of our tell-tale box. On it are scars left by its contact with a triangular clay spacer used to separate it from the inverted base of the box in the stack's previous layer (fig. 17). The evidence is saying, therefore, that our original box was not a single item popped into the kiln to fill up space. Instead, it was showing that the potter had fired several at a time, and might even have thrown enough to fill a small kiln.

That raises an entirely different question: How many potters were involved? The wide variations among both the knop finials and base widths suggest that there had been almost as many potters involved as there are

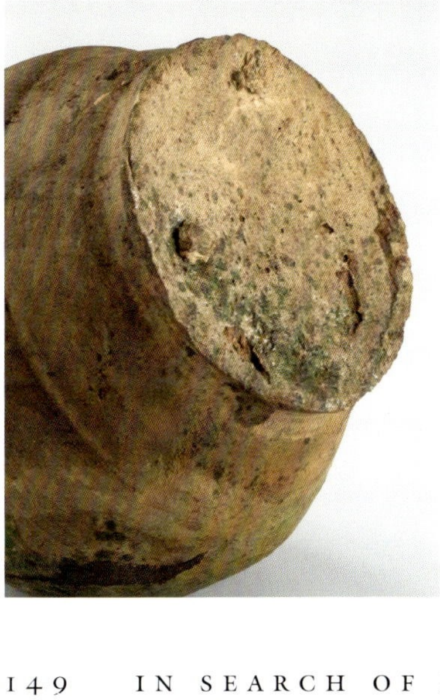

Figure 17 Tiny lumps on the base of the period money box showed that it had stood on top of another, separated by a trivet-style stilt.

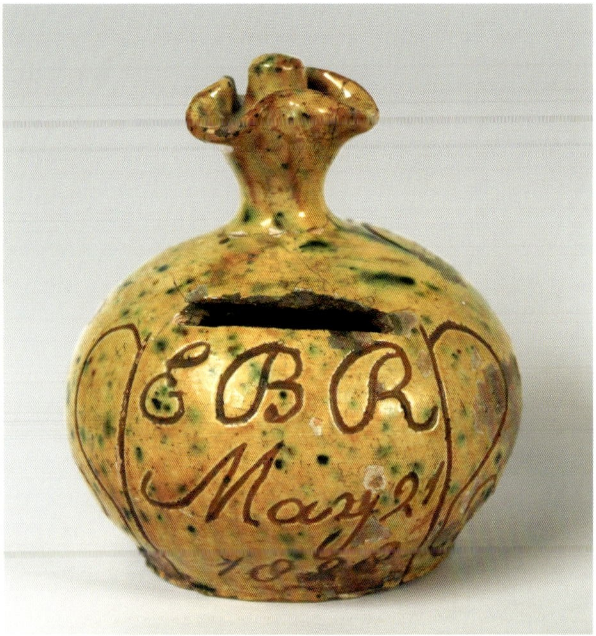

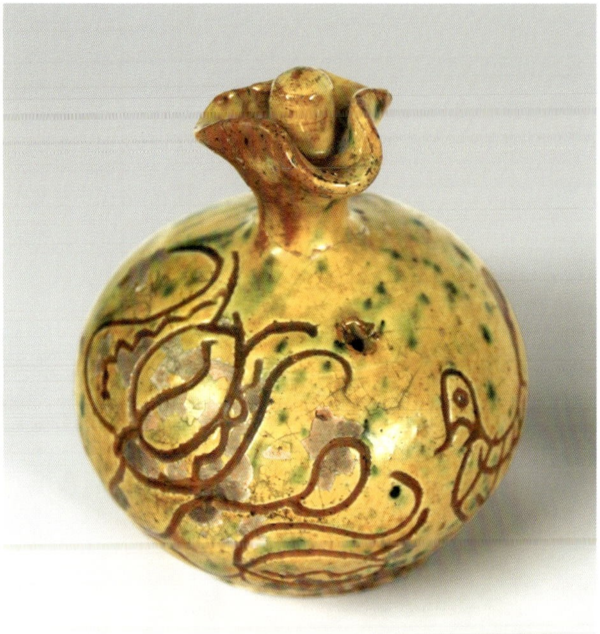

Figure 18 Money box, Donyatt, Somersetshire, England, 1828. Slipware. H. 4", slot length 1½", slot width 3⁄16". Inscription: E B R / May 21 / 1828. (Courtesy, Jacqui Pearce.) This pin-pierced Donyatt sgraffito slipware money box is the earliest dated example yet recorded from that Somersetshire factory.

surviving examples. With that said, one has to add that there is no proof of the numbers made at any one establishment or their dates of manufacture.

One thing is certain, however. The pinhole technique was not an anomaly limited to one potter, to one factory, or to one century. Jacqui Pearce owns a money box with the telltale hole that she had never previously asked to explain itself (fig. 18). Her box is dated May 21, 1828, and carried the pinholing forward across two centuries. By that time, money boxes were no longer for smashing. In consequence, the majority of survivors had their contents carefully extracted and so provided no opportunity to study their interiors. The exception illustrated in figure 19 had been subjected to the smash-and-grab technique that revealed the pinhole inside yet remained invisible on the exterior. Its companion (see fig. 27), which dates to about 1800, may have been made at the same Derbyshire stoneware factory, its pinhole showing externally as a hitherto unrecognized dark dot. Thus, as Michelle Erickson has demonstrated, when air is trapped in a sealed space, no matter what the date, it has to be vented.

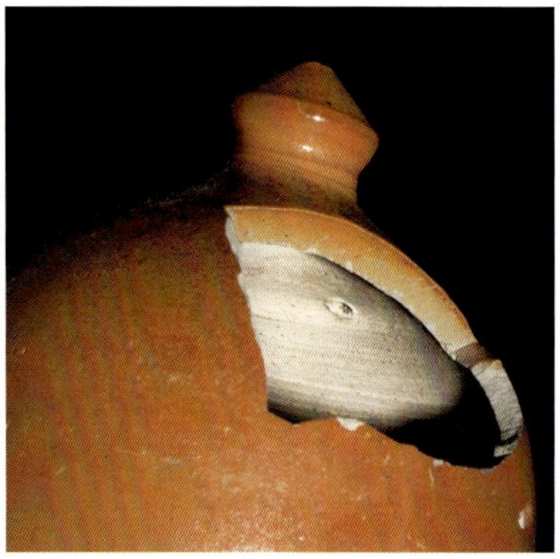

Figure 19 Pinhole revealed on the interior of the ca. 1800 brown stoneware money box illustrated in fig. 27.

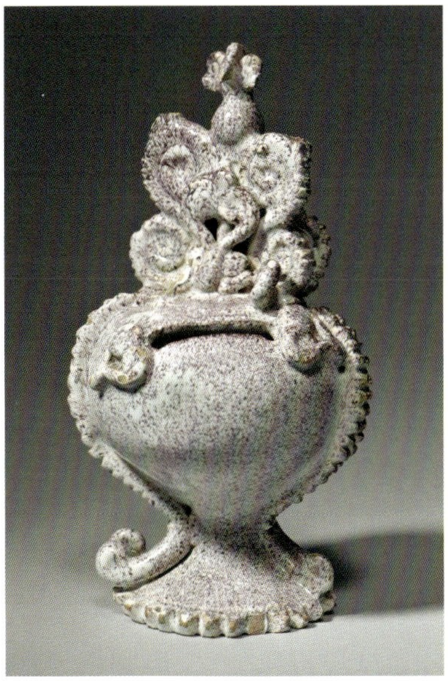

Figure 20 Money box, London, England, ca. 1650. Tin-glazed earthenware. H. 6½". (Private collection; photo, Alan Tabor, Robert Woolley Ceramics & Works of Art.)

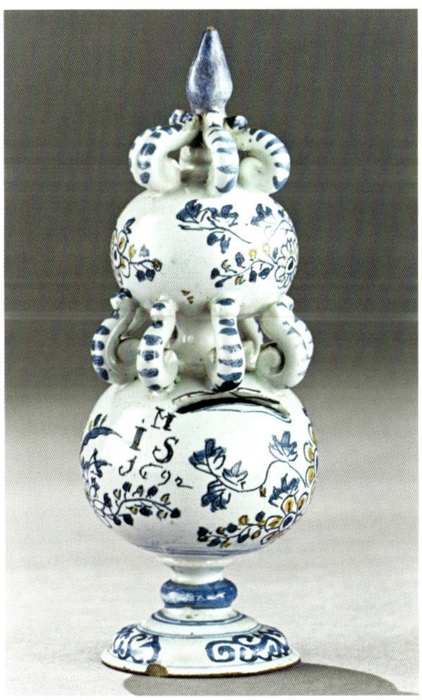

Figure 21 Money box, London, England, 1692. Tin-glazed earthenware. H. 7¾". (Courtesy, Colonial Williamsburg.)

Jacqui's box, though globular in shape, was personalized in sgraffito style but clearly was not to be smashed on Boxing Day.[15] So when did that practice cease? I suggested earlier that the rise of the Puritan ethic and the virtues of frugality may have been a factor, as might the upheavals of the English Civil War (1642–1651). Jacqui's archeological timeline (see p. 161, fig. 2, in this volume) would seem to support that conclusion, but so, too, does the scarcity of documentary references.

A Christmas carol published in 1653 carried the line "Our kitchen-boy hath broke his boxe," and a caption for an earlier woodcut (1634) reads, "Both with the Christmas boxe may well comply; It nothing yields till broke; they till they dye."[16] It is clear, therefore, that the practice of using breakable ceramic boxes continued until the mid-seventeenth century. However, it seems equally clear that boxes for breaking had been the virtual monopoly of the Border Ware potters. Boxes for saving, on the other hand, were intended for a better-quality market and seem to have been made to be sufficiently attractive to discourage owners with hammers. Two London delftware examples are known, the earlier from circa 1650, manganese stippled, and piled high with destruction-defying scrolls and curlicues (fig. 20). An even more elaborate testimony to the Southwark delftware potters is in the Colonial Williamsburg collection, helpfully dated 1692 (fig. 21). The dearth of other such examples prompted curator John Austin to note, "The rarity of money boxes may be explained by the fact that it was necessary to break them to retrieve the coins."[17] It can be argued, however, that the "break me" and "keep me" boxes had no evolutionary relationship one to the other.

By the later years of the seventeenth century, the concept of bestowing minimal largesse concealed in pottery boxes had run its course. However, the Christmas gift itself, stripped of its pottery packaging, survived to be described as "formerly the bounty of well-disposed people." By the nineteenth century it had become entrenched much as, in our time, executives expect an end-of-year bonus regardless of whether they deserve it. By the 1850s one of the hitherto well-disposed people complained, "The gift is now almost demanded as a right, and our journeymen, apprentices, &c. are grown so polite, that instead of reserving their Christmas for its original use, their ready cash serves them only for pocket money; and instead of visiting their friends and relations, they commence the fine gentlemen of the week."[18]

It is evident, therefore, that the Christmas box was no longer a clay ball or pig needing to be cracked open, but a metaphor for a hand-out passed with minimal merriment from palm to palm. Christmas of 1773 at Nomini Hall in Virginia found tutor Philip Fithian carefully recording the gifts handed out to his master's servants.[19] In the morning Fithian had received his clean clothes "sent in with a message for a Christmas Box, as they call it; I sent the poor Slave a Bit, & my thanks." A bit, Fithian explained, was a "pisterene bisected, or an English sixpence, and passes here for seven pence Halfpenny."[20] He added that the total of his Christmas donations to the servants amounted to five bits, or three shillings, one penny, and a halfpenny.[21]

Accepting a bisected Spanish silver pistole as a bit, the poor slave would not have received enough to rattle a ceramic money box. It is clear, however, that Fithian's limited largesse was not presented in a box, either wooden or ceramic. He ascribed the Christmas box term to some formless "they," and evidently was content to pursue neither their identity nor their box any further. Not so author and publisher John Dunton, who in the early eighteenth century posed the question in his weekly periodical, the *Athenian Mercury*. "From whence comes the custom of gathering of Christmas-box money?" he asked. "And how long since?"

Dunton's answer was convoluted and quaint, as one might expect from a man whose mind was compared by contemporaries to "a table where the victuals were ill-sorted and worst dressed."[22] "It is as ancient as the word mass," he wrote, "which the Romish priests invented from the Latin word *mitto*, to send, by putting the people in mind to send gifts, offerings, oblations; to have masses said for everything almost, that no ship goes out to the Indies, but the priests have a box in that ship, under the protection of some saint." What that had to do with Christmas boxes is not entirely clear.

After further discourse Dunton came to the point, namely that the "box" referred to the money box given to servants to enable them to pay the priests for masses. Dunton explained to his readers that "a priest will not say a mass or anything to the poor for nothing; so charitable they generally are."[23] It is scarcely necessary to add that Dunton's family were Presbyterians.

Scottish folklorist and publisher Robert Chambers took up the quest in 1862, dismissing Dunton's explanation as "an ingenious conjecture." The Boxing Day gift giving was not to be ascribed to the Roman Catholic clergy, he said, but dated back to the Roman feast of Saturnalia, which included a ritual of exchanging New Year's gifts. Contradicting Dunton, Chambers declared that the early Christian church denounced it on the grounds of its pagan origin. He added that regardless of clerical admonitions Christmas giving continued and grew into a "universally recognized institution." Being a frugal Scotsman, Chambers complained that "This custom of Christmas-boxes, or the bestowing of certain expected gratuities at the Christmas season, was formerly, and even yet to a certain extent continues to be, a great nuisance."[24]

Here again, as with Fithian in Virginia, the deed is emphasized, not its packaging. In spite of the fact that Boxing Day continues to be marked on the calendars of many English-speaking people, the day of gifting can be at any time from Christmas Day to New Year's Day. In the Victorian era, however, when social distinctions separated master from man, the family exchanged its large gifts on the 25th and gave its little ones to the servants on Boxing Day. That is not to say that gifts of ceramic money boxes had been abandoned. Some were made for specific people and inscribed as being intended for the festive Yuletide occasion. This continued into the twentieth century, as a Donyatt hen-and-chickens example marked "Ralph Coles / Xmas / 1905" testifies (fig. 22).

The Coles "box" is not shaped like a pig but rather as a pedestaled ball surrounded by chickens and topped by a sitting hen. It was made in the pre-

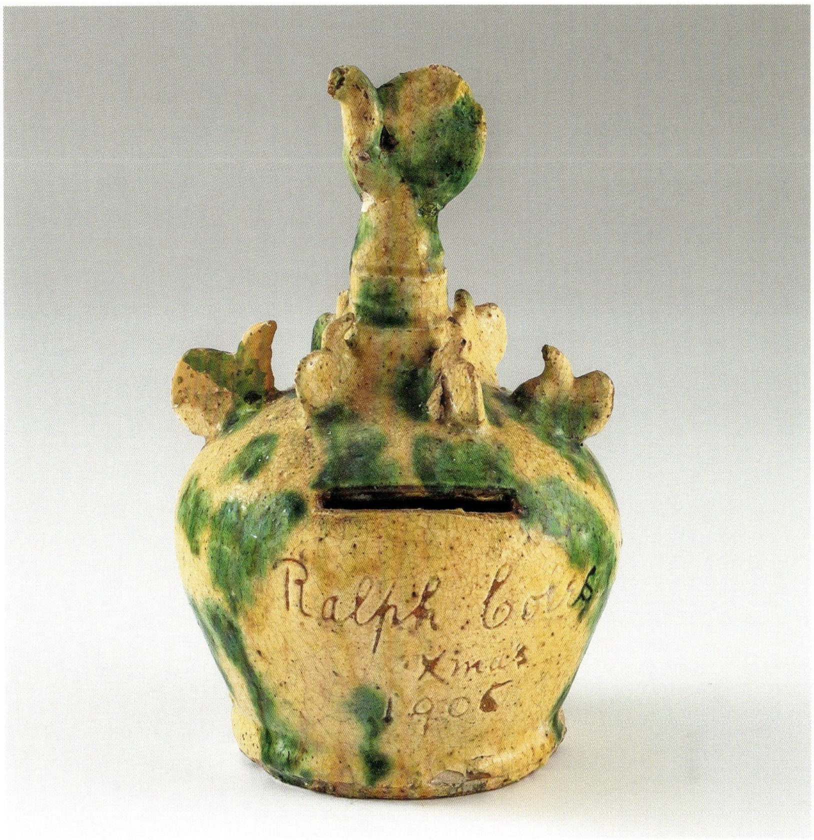

Figure 22 Money box, Donyatt, England, 1905. Slipware. H. 6"; slot 1¾ x ¼". Sgrafitto inscription: "Ralph Coles / Xmas / 1905." (Courtesy, Noël Hume Collection.)

viously cited Somerset village of Donyatt, whose potting industry had continued through six hundred years. The applied techniques had changed through the generations, but the known Donyatt money boxes are all of a pink-bodied ware coated with white slip spattered with copper oxide and incised in sgraffito style with names, initials, and dates, all under a yellowing lead glaze.

Jacqui Pearce's globular example of 1828 is the earliest known, and the incised chicken on one side might be symbolic of a "nest egg," a description in use by 1700 as "a sum of money put by as a reserve or serving as the nucleus of a fund."[25] Although there is no assurance that the money-box decorator had that in mind, the three-dimensional hen-and-chickens design was established at Donyatt by 1836. Another, like the much later Coles box, is dated 1893 and incised "Wishing Edith / A merry Xmas."[26]

Money-box collecting, like any other antique-hunting pursuit, has pitfalls larger than pinholes. Twentieth-century copies of nineteenth-century originals are common, but so, too, are genuine early examples imported from Europe masquerading as English. Large-scale construction projects in Amsterdam, Rotterdam, and Delft as well as others in Belgium have yielded quantities of late-medieval and sixteenth-century earthenwares found in wells and cesspits that have survived more or less intact.

An example, a cross between the miter style and the button-knopped English boxes, was excavated at Utrecht and attributed to 1450–1500.[27] Like the Cheam-style box, this has a narrow vertical slot, a detail not found on later Dutch money boxes. It would seem, therefore, that European and English money-box potters shared the same evolutionary trends.[28]

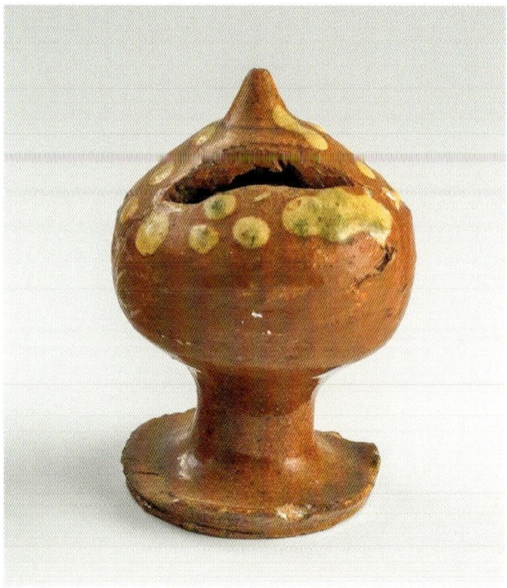

Figure 23 Money box, northern Netherlands, 1600s–1700s. Lead-glazed and white-slipped redware. H. 3½"; slot 1½", width damaged. (Courtesy, Noël Hume Collection.) This pedestal-based money box has been restored.

The Dutch export trade in restored ceramics recovered from construction sites is extensive. Once cleaned and repaired, they beguile novice collectors from the shelves of British antique shops. Worse, they regularly turn up on eBay, where one must take the seller's (no returns please!) word for what one is buying. I should add that the Dutch are extremely versatile with plaster and paint, as was found with the box in figure 23, bought on eBay. The small, pedestal-based, thick redware box, coated with lead glaze dappled with yellowish clay slip, rises to a point. Unlike earlier examples, the slot is lateral and has evidently been widened to extract the contents. Taken at face value, such a pot might pass for Metropolitan (Essex) slipware of the late seventeenth century, though careful (if belated) examination shows that it had been broken apart at its girth and skillfully restored. A shape parallel for it has been found in Amsterdam and attributed to circa 1675–1700.[29]

Another money box in the Noël Hume Collection bears no resemblance to the Dutch slipware and is thought to be French (fig. 24). With olive-green glaze over white-slipped buffware, its form resembles the English Tudor types but, instead of having a sharply defined disc-and-nipple knop, the top of this one lacks sharp contours, resulting in a more clumsy profile that does not safely lend itself to suspension. Nevertheless, the coin slot is vertical

Figure 24 Money box, Normandy, France, 1600s. Lead-glazed earthenware. H. 4⅜"; slot 1¼ x ⅛". (Courtesy, Noël Hume Collection.) This button-knopped, green-glazed, white-bodied money box has a pedestaled foot and vertical coin slot. It is believed to be Norman French in the English style.

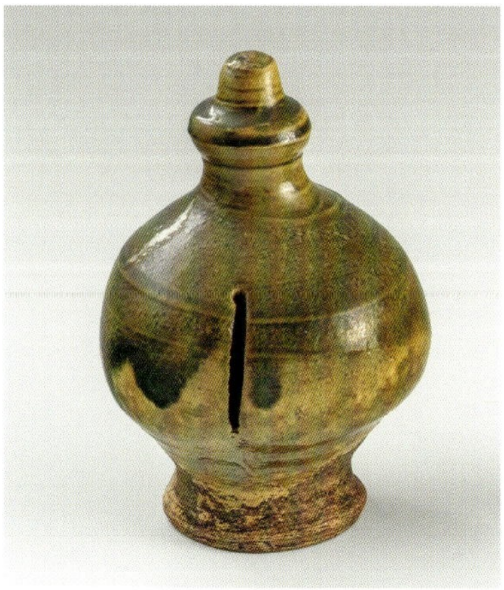

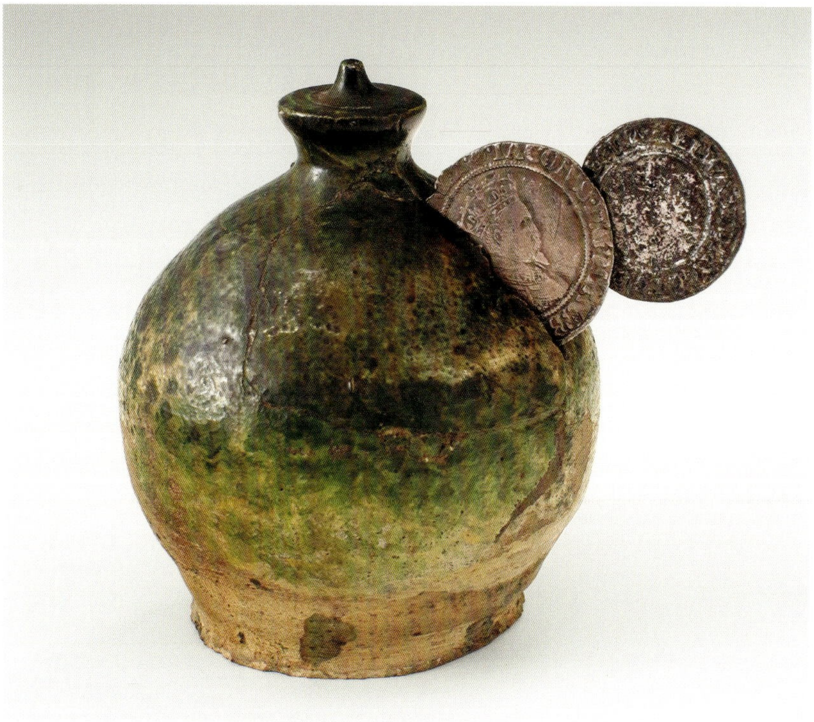

Figure 25 Testing coin receptivity of the money box illustrated in fig. 2 shows that its slot will accept neither a 1607 James I shilling nor a 1566 Elizabethan sixpence.

and, though shorter than the English example, could readily receive the French copper ***double tournois*** of the first half of the seventeenth century.[30]

The slot length on Tudor button-knopped boxes might be misleading. On the box illustrated in figure 2, it is too short to take a silver shilling

Figure 26 An undated Elizabethan silver penny is of the size that the money box illustrated in fig. 2 will accept, which was the price of admission to The Rose, where most of London's boxes were found.

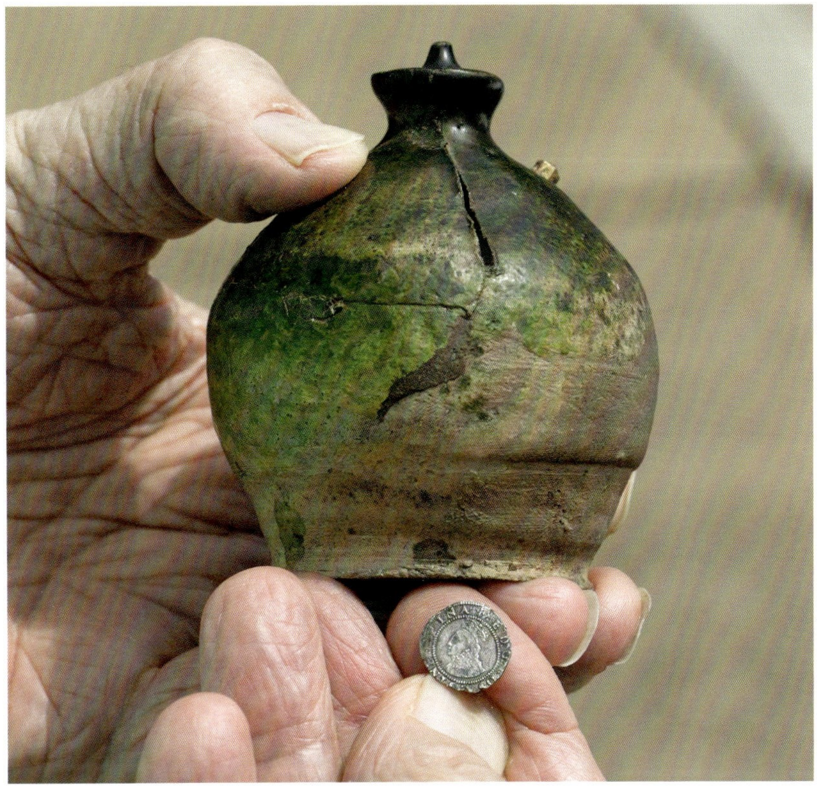

Figure 27 Money boxes, Derbyshire, England, 1800s. Brown stoneware. *Left*: H. 6¼"; slot 1¾ x ¼"; *right*: H. 4½". (Courtesy, Noël Hume Collection.) The slots of these examples were made to receive George III twopences (left) and pennies (right), which were issued only at the Soho mint in 1797.

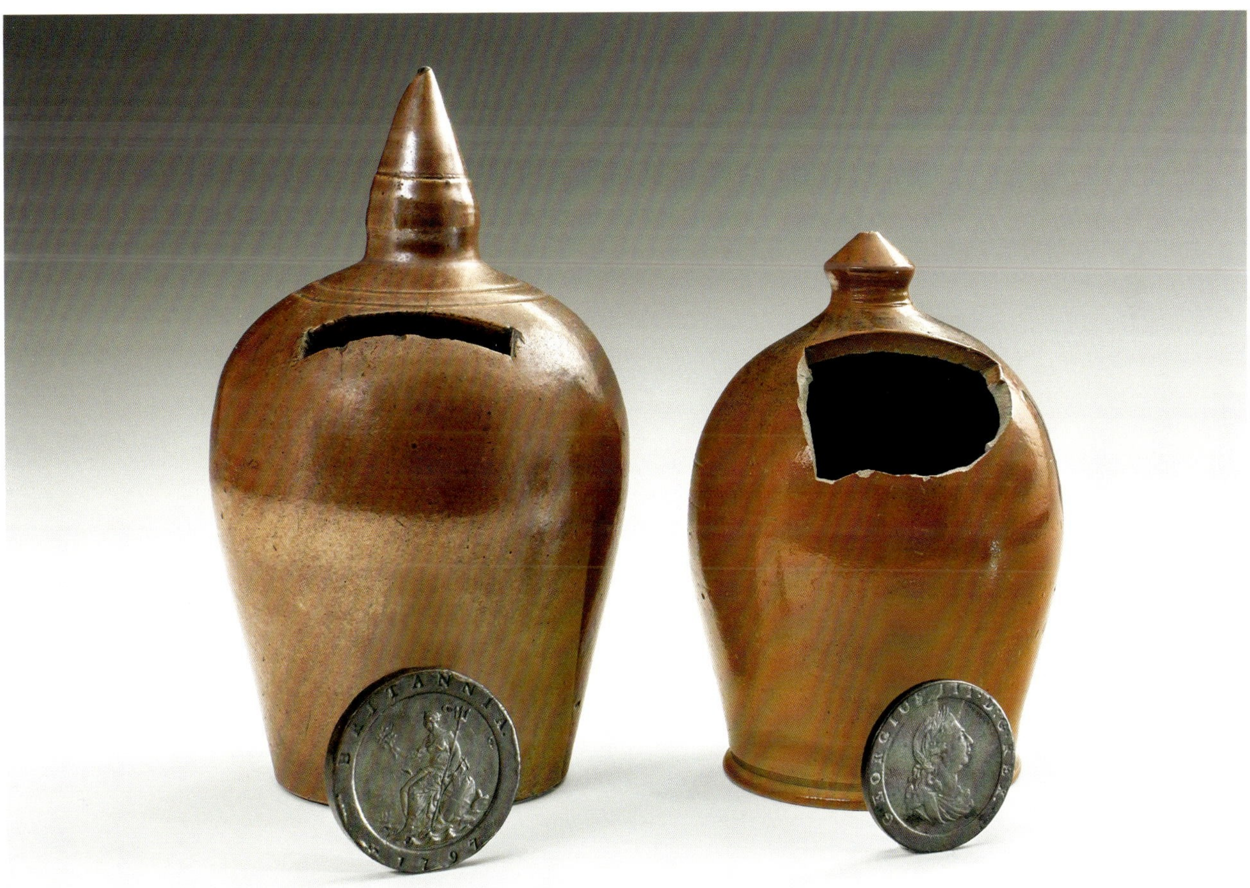

(unless heavily clipped) and too narrow to take a sixpence. The inference, therefore, is that in spite of the slot's length it could only accommodate coins of very low denomination, such as an Elizabethan penny (fig. 25). However, as Jacqui Pearce reminds us, most of the archeologically discovered boxes came from The Rose theatre (1587–1606), where the standing-room-only crowd could get in for a penny (fig. 26).

The diameter and thickness of coins in relation to the length and breadth of slots can be clues to dating boxes. Thus the thinness of Elizabethan and Stuart silver dramatically contrasts with the heavy copper pence briefly issued by Birmingham's Soho mint in 1797 (fig. 27). Made with penny and two-penny slots, it is likely that these stoneware boxes came into use because the unwieldly coins were better stored than spent. Not until 1806, when George III's new copper coinage was issued, did the sizes again become manageable. That seems to have coincided with a new age of money-box manufacturing that ranged from Staffordshire dogs' heads to old men, funny faces, and Victorian villas that cried "Keep me!" rather than "Break me!"

But that is another story (fig. 28).

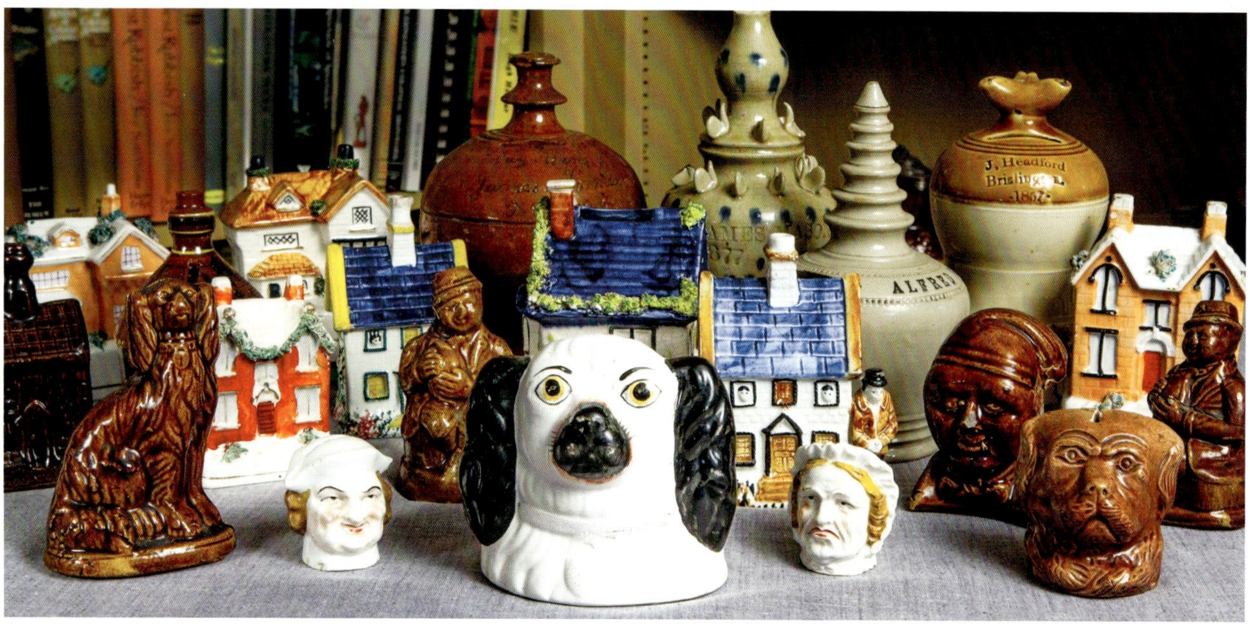

Figure 28 The 1806 return to numismatic normalcy inspired a century and more of turned and molded but nonpinholed money boxes. H. of Staffordshire dog 4¼". (Courtesy, Noël Hume Collection.)

1. *Catalogue of the Collection of London Antiquities in the Guildhall Museum*, 2nd ed. (London: Corporation of the City of London, 1908), p. 208, no. 557.

2. Wolf Mankowitz and Reginald G. Haggar, *The Concise Encyclopedia of English Pottery and Porcelain* (1957; repr. New York: Frederick A. Praeger, 1968), p. 214.

3. Contrary evidence is suggested by a pig-like bank recorded from the northern Netherlands, but as it closely resembles others attributed to ca. 1875–1900, I am prompted to dismiss it. E. M. Ch. F. Klijn, *Loodglazuuraardewerk in Nederland* (Lead-glazed Earthenware in The Netherlands) (Arnhem: Nederlands Openluchtmuseum, 1995), p. 281, top.

4. Jerzy Gawronski, *Amsterdam Ceramics: A City's History and an Archaeological Ceramics Catalogue, 1175–2011* (Amsterdam: Lubberhuizen, 2012), p. 173, no. 339 (thought to be German), and p. 235, no. 701 (attributed to the western Netherlands).

5. Charles Wriothesley, *A Chronicle of England during the Reigns of the Tudors, from A.D. 1485 to 1559*, edited by William Douglas Hamilton, 2 vols. ([Westminster]: Printed for the Camden Society, 1875–1877), 1: 75.

6. *Catalogue of the . . . Guildhall Museum*, p. 189, no. 201.

7. *Oxford English Dictionary*, s.v. "Christmas-box," 1612–1615, Joseph Hall, *Contemplations on the New Testament*, pp. iv, xi.

8. John Brand, *Observations on the Popular Antiquities of Great Britain*, 3 vols. (London: Henry G. Bohn, 1849), 1: 495.

9. *Catalogue of the . . . Guildhall Museum*.

10. Ibid., p. 198, no. 400.

11. Noël Hume, *Discoveries on Walbrook, 1949–50* (London: Guildhall Museum, 1950), n.p.

12. Erwin Baumgartner and Ingeborg Krueger, *Phönix aus Sand und Asche: Glas des Mittelalters*, exh. cat. (Munich: Klinkhardt and Biermann, 1988), p. 249, no. 257, where it is attributed to the fourteenth century.

13. Peter C. D. Brears, *The Collector's Book of English Country Pottery* (Newton Abbot, North Pomfret, Vt.: David & Charles, 1974), p. 56.

14. Jacqui Pearce cites a waster from a pottery site at Kingston upon Thames that did just that. Pearce, *Pots and Potters in Tudor Hampshire: Excavations at Farnborough Hill Convent 1968–72* ([Guildford, Eng.]: MoLAS/Guildford Borough Council, 2007), pp. 64–72.

15. The inscription suggests a christening rather than a birthday, and was made before the exact date was set.

16. Brand, *Observations on the Popular Antiquities of Great Britain*, pp. 489, 494. Information about his source can be found online at www.gutenberg.org/files/19979/19979-h/19979-h.htm.

17. John C. Austin, *British Delft at Williamsburg* (Williamsburg, Va.: Colonial Williamsburg Foundation in association with Jonathan Horne Publications, 1994), p. 288, no. 719.

18. Brand, *Observations on the Popular Antiquities of Great Britain*, p. 494.

19. Philip Fithian, *Journal and Letters of Philip Vickers Fithian, 1773–1774*, edited by Hunter Dickinson Farish, new ed. (Williamsburg, Va.: Colonial Williamsburg, 1957), p. 40 (entry for December 25, 1773).

20. A pisterene is described in the *Oxford English Dictionary* as a small coin used in America and the West Indies; 1774 is cited as the earliest recorded date of its use. Fithian's references were at least a week earlier.

21. In 1777 there was no English penny, requiring it to be created from halfpennies.

22. Robert Chambers, *The Book of Days, A Miscellany of Popular Antiquities*, 2 vols. (1862–64; repr. Detroit: Omnigraphics, 1990), 1: 626.

23. John Dunton (1659–1733) published the *Athenian Gazette*, renamed the *Athenian Mercury*, from 1691 to 1697.

24. Ibid.

25. *Oxford English Dictionary*, the earliest edition describing false eggs to promote laying, dates from 1606.

26. Richard Coleman-Smith and T. Pearson, *Excavations in the Donyatt Potteries* (Chichester: Phillimore, 1988), pp. 273–75. A similar hen-and-chicken box in my collection is in blue-dotted gray stoneware and inscribed "WALTER JAMES MASON 1877".

27. Klijn, *Loodglazuuraardewerk in Nederland*, p. 282.

28. Slip-decorated money boxes in elongated pig forms were being made in the northern Netherlands in the last quarter of the nineteenth century. Ibid., p. 281.

29. Gawronski, *Amsterdam Ceramics*, p. 234, no. 700 (buff body, yellow lead-glazed).

30. Joe Cribb, Barrie Cook, and Ian Carradice, *The Coin Atlas: The World of Coins from Its Origins to the Present Day* (New York: Facts on File, 1990), p. 38, nos. 26, 27.

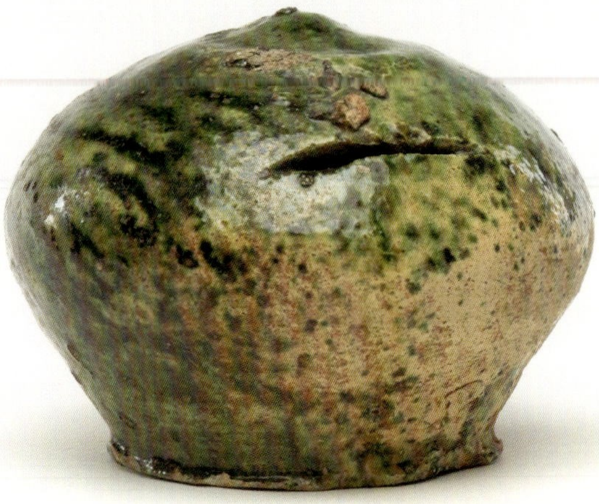

Figure 1 Money box, Surrey, England, fourteenth century. Lead-glazed earthenware. H. 2½". (Courtesy, Museum of London, acc. no. A3855; photo, © Museum of London.) This example of Kingston-type ware was found in Paternoster Row, City of London.

Jacqui Pearce

Money Boxes: The London Evidence

▼ WHEN NOËL HUME first emailed me about pinholes in money boxes from London, I had to confess I had never noticed any, but the next day I went to see what I could find in the Museum of London's Ceramic and Glass Collection. A quick examination brought to light six money boxes in Surrey-Hampshire Border Ware with the tiny telltale hole, still perfectly visible, just as Noël had said.

The finds in the Museum's reserve collection—the most complete and displayable examples from sites across London—are just the tip of the iceberg. The collection includes 83 ceramic money boxes in a variety of wares, spanning the medieval period through to the nineteenth century. A much larger body of evidence is to be found in the archive amassed by the Museum of London Archaeology (MOLA) and its predecessors over more than forty years. The computerized database records sherds from at least 556 money boxes excavated on 142 sites in London, mostly from within the square mile of the City of London and from the borough of Southwark, across the Thames. A number of questions quickly came to the fore: When did ceramic money boxes first appear, and how did the form evolve over time? What were the main centers of production? Where were they being used? And, of course, how common are examples with pinholes?

The earliest money boxes recorded in London excavations are in green-glazed white earthenware made at Kingston upon Thames, in the medieval Surrey whiteware tradition.[1] These small, enclosed containers have a low-domed or pointed top, like a bishop's miter, and usually have a narrow, vertical coin slit, although a few with horizontal slits are known (fig. 1).[2] A few examples have a more elaborately shaped knob or finial, resembling a money bag tied at the neck.[3] Several kilns have been excavated in Kingston, but so far have yielded only one money box.[4] Clearly that example was thrown away because the potter had not made adequate allowance for shrinkage and the coin slit had become too narrow during firing (see p. 142, fig. 6, in this volume for a similar example from the Farnborough kilns in Hampshire). Dating evidence from these kilns coincides closely with the earliest finds from London, recovered from Thames waterfront sites and dated by dendrochronology to the beginning of the fourteenth century.[5]

Money boxes were also made in at least two other ceramic industries within the medieval Surrey whiteware tradition. One of these was centered on Cheam in Surrey, its products used in London circa 1350–1500, and although no money boxes were found at any of the excavated kilns, a few have been recovered from consumer sites in London, including examples cited by Ivor Noël Hume (see pp. 139–41, figs. 3–4 and accompanying text in this volume).[6] One of these was excavated in 1949–1950 at St. Swithin's House, 30–37 Walbrook.[7] Work by MOLA on the same site in 1996 brought to light a second complete Cheam money box in a late-fifteenth-century context.[8] Another nearly complete miter-shaped money box was found during excavations at Gateway House, Watling Street, directed by Noël while he was the Guildhall Museum's City Archaeologist.[9]

Surrey-Hampshire Coarse Border Ware was perhaps the most significant of the whiteware industries in the developing story of the money box. Between circa 1350 and 1500 this heavy-duty, practical pottery dominated London's ceramic supply.[10] Money boxes formed a minor part of the potters' repertoire, and all were made in the familiar domed or miter shape.[11] At the end of the fifteenth century, however, the potteries of this region began to focus on producing a narrow range of high-quality, thin-walled drinking vessels in a much finer, green-glazed white earthenware, at the expense of coarseware production.[12] These finewares became the mainstay of the industry, which thrived on its market connections with London, supplying a number of major institutions, such as the Inns of Court.[13] During the second half of the sixteenth century, the industry was transformed under the influence of Harmon Raignold (Herman Reynolds), an immigrant German potter, developing new forms made in a more durable whiteware fabric.[14] By the early 1600s the Surrey-Hampshire potteries had become one of the most important sources of London's everyday pottery.

Among forms inherited from the medieval tradition, the money box evolved into its characteristic sixteenth–seventeenth-century shape, with a knob or finial formed above a narrow constriction at the top.[15] It invariably has a flat or slightly dished base, often with concentric or fan-shaped marks left by the wire used to remove it from the wheel. Money boxes also frequently have small kiln scars in two or three places around the shoulder or maximum girth, resulting from contact with other vessels in the firing stack (see p. 149, fig. 16, in this volume).

Money boxes made in Surrey-Hampshire Border Ware, including the early-sixteenth-century finewares, account for 52 percent of the Museum of London's reserve collection, and 84 percent of all examples from MOLA excavations; 13 percent of the remaining boxes were excavated finds made in medieval Surrey whitewares. Most Border Ware money boxes have a copper-stained green glaze covering the top half of the vessel, although a few have a clear glaze that appears yellow over the buff fabric.[16] Although the green-glazed whiteware was the most widely used, a small number of money boxes are also known in contemporaneous redware fabrics.[17]

The main concentration of finds from London dates between 1550 and 1630 (fig. 2), peaking at 1580 and falling off sharply after the 1640s, which Noël suggests could be the result of a Puritan clamp-down on apprentices' Christmas boxes during the Commonwealth. By far the greatest number of money boxes excavated in London is focused on the sites of two Tudor playhouses: The Theatre in Shoreditch and the Rose in Southwark, with a further concentration around the animal-baiting arenas of Southwark's

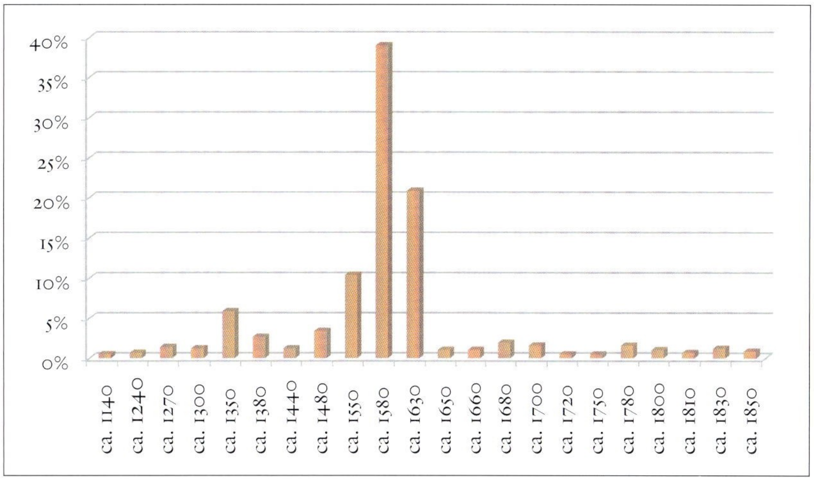

Figure 2 The chronological distribution of ceramic money boxes from London excavations, in minimum number of vessels from dated contexts.

Bankside.[18] Recent MOLA excavations in Shoreditch uncovered evidence for the playhouse long associated with early productions of Shakespeare's plays. Opened in 1576, The Theatre was dismantled in 1598 and its timbers reused in the construction of the Globe, in Southwark.[19] Sherds from at least twenty-four money boxes in Surrey-Hampshire Border Ware were recovered from the site, but these pale into insignificance compared with the massive collection from The Rose: 378 sherds from at least 162 money boxes (fig. 3).[20] All are made in green-glazed whiteware and share a common

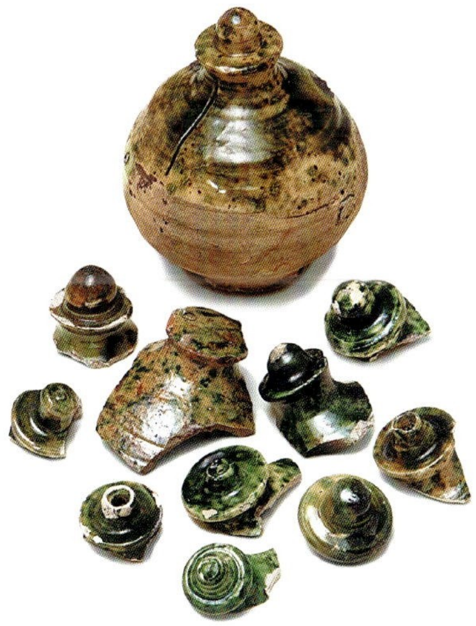

Figure 3 Money-box fragments, from Surrey-Hampshire border, England, 1580–1630. Lead-glazed earthenware. (Photo, A. Chopping, © Museum of London Archaeology.) These fragments were found on the site of The Rose theatre, Southwark, London (site code SBH88).

form, although there are minor variations in the shape of the finial.[21] Remarkably, money boxes formed the third largest group of ceramic vessels found on the site, after those used for food and drink.[22] They all came from contexts within the playhouse, dating to its lifetime and shortly after, from the 1580s to the 1630s. Although no examples were recovered from the site of the Globe, the excavation area was much smaller and access limited.

At least one money box from The Theatre and a number of money boxes from The Rose have a small pinhole pierced through the body (fig. 4). It is

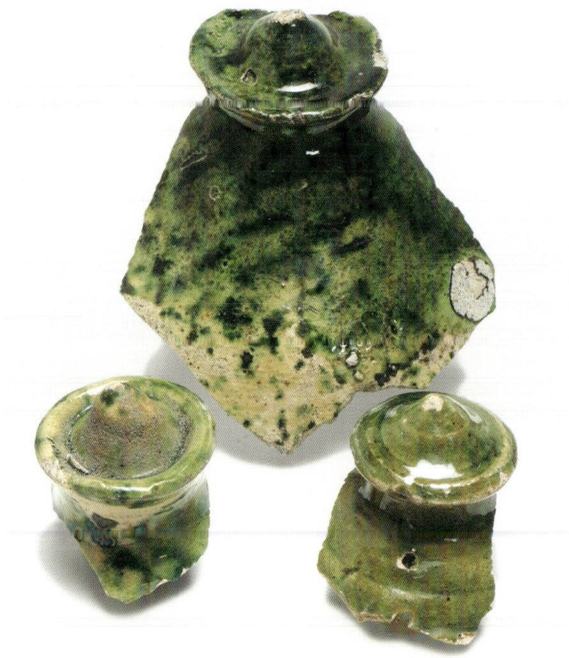

Figure 4 Money-box finials, from Surrey-Hampshire border, England, 1580–1630. (Photo, A. Chopping, © Museum of London Archaeology.) One of the finials, all of which were taken from the site of The Rose theatre, Southwark, London (site code SBH88), shows a pinhole made before firing.

impossible to tell how widespread the practice was because so many London finds are identified solely by their bases and finials. Only one of the medieval money boxes that were examined, a miter-shaped form in Kingston-type ware (Museum of London, acc. no. 5767), showed any evidence of a pinhole, although this does appear on several of the more complete sixteenth- and seventeenth-century forms. Closer examination of the Border Ware money boxes in the Museum of London reserve collection revealed an additional eleven examples with a pinhole, making seventeen in all (or 40 percent of all examples in this fabric). In most cases the hole is very small and not easily spotted, stabbed just below the finial on the side opposite the coin slit. A few examples have a slightly larger, neatly made hole, or have been stabbed closer to the coin slit. Some of the broken money boxes have a small spot of glaze on the inside, spreading from a tiny pinprick that is otherwise invisible from outside because it was filled in with glaze. Without breaking them it is hard to know how many more of the museum's complete money boxes without an obvious pinhole had been stabbed during manufacture.

Why were so many money boxes found at The Theatre and The Rose? One suggestion is that they were used at various access points within the playhouses, controlled by "gatherers" or "collectors" to receive monies from the patrons; the proceeds were then kept under lock and key in the "Commone boxe."[23] Ceramic money boxes were cheap to make and easily replaced following their inevitable breakage to remove the contents. However, as Noël has pointed out, this kind of container hardly seems suitable for collecting coins from large, bustling crowds of people and it positively invites deception. Another suggestion is that they were used by local traders who provided refreshments for the playgoers, although the same objections would apply to their suitability as money containers in a crowded venue.

Could they even have been a speculative bulk purchase, perhaps of surplus stock brought in by one of the livery companies, that proved impractical, leading to their ultimate replacement with more suitable containers? The proprietor of The Rose, Philip Henslowe, was also a member of the Dyers Company and this could provide a possible link.

We know that some of the potteries on the Surrey-Hampshire border had established business relationships with various London institutions, such as the Inns of Court, and supplied them with regular consignments of specific forms.[24] Perhaps some of the playhouses, or traders who held concessions within them, also had contracts with the potteries to supply them in bulk with pots designed to order. Similar arrangements were in place with some of the City's livery companies, who ordered easily replaceable, good-quality ceramic drinking vessels for use at their feasts and messes, as well as other forms designed to meet a variety of needs.[25] This brings us back to the apprentices and their earthenware boxes, which would need to be purchased each Christmas, presumably in bulk and from the same sources that supplied fine tablewares. This link is reinforced by a reference in *A Calendar of the Inner Temple Records* to an account for Christmas 1620, which lists ". . . in money boxes, 11*s*. 6*d*."[26] Were these given annually by the "Benchers" to the young gentlemen of the inn, their pupils?

The concentration of finds in Southwark, especially around the entertainment venues along Bankside, and on sites in the City of London, where Livery Halls were to be found at every turn, is telling. It may well be presenting us with the key to the development, use, and popularity of the simple money box, beginning in the late medieval period and reaching a peak in Tudor and Stuart London. Perhaps in these finds we can see yet another example of the entrepreneurial connections of the astonishingly enterprising potters of the Surrey-Hampshire border and their medieval antecedents. And what form could be more fitting than the money box to symbolize these flourishing connections?

1. There are eleven substantially complete Kingston-type ware examples in the Museum of London Ceramic and Glass Collection, and twenty-four in the MOLA database. These figures are based on data available up to July 2013; the MOLA ceramic database may be consulted by application to Museum of London Archaeology, Mortimer Wheeler House, 46 Eagle Wharf Road, London N1 7ED, UK.

2. Jacqueline Pearce and Alan Vince, *A Dated Type-Series of London Medieval Pottery, Part 4: Surrey Whitewares*. Special Paper (London and Middlesex Archaeological Society), 10 (London: London and Middlesex Archaeological Society, 1988), fig. 99, nos. 373–79.

3. Ibid., fig. 99, nos 381–83; three near-complete examples were found at Bull Wharf in 1990 (site code BUF90, context [3091], dated to ca. 1350–1500).

4. Excavated in 1995 at 70–76 Eden Street (site code EDN95). Pat Miller and Roy Stephenson, *A 14th-Century Pottery Site in Kingston upon Thames, Surrey: Excavations at 70–76 Eden Street*. MoLAS Archaeology Studies Series, 1 ([London]: Museum of London Archaeology Service, 1999), p. 27, fig. 29.

5. Pearce and Vince, *A Dated Type-Series*, fig. 41. Archaeomagnetic dates for Kiln 1 are 1300–1325 at 68% level of confidence, or 1290–1330 at 95% level of confidence, and for Kiln 2, 1310–1330 (68%) and 1300–1340 (95%). Ibid., p. 11.

6. Ibid., pp. 84–88. See Clive Orton, "The Excavation of a Late Medieval/Transitional Pottery Kiln at Cheam, Surrey," *Surrey Archaeological Collections* 73 (1982): 49–92; see also Clive Orton, personal communication, July 15, 2013. Only thirteen money boxes in Cheam white-

ware are listed in the MOLA database; one is in the Museum of London Ceramic and Glass Collection.

7. The site has subsequently been given the code GM158.

8. Then Museum of London Archaeology Service (MoLAS), site CAO96: Nicholas Elsden, *Excavations at 25 Cannon Street, City of London: From the Middle Bronze Age to the Great Fire*, MoLAS Archaeology Studies Series 5 ([London]: Museum of London Archaeology Service, 2002), fig. 39.

9. Site GM160.

10. Pearce and Vince, *A Dated Type-Series*, pp. 84–85.

11. Nineteen examples are recorded in the MOLA database; see ibid., fig. 119, no. 523.

12. Jacqueline Pearce, *Pots and Potters in Tudor Hampshire: Excavations at Farnborough Hill Convent 1968–72* ([Guildford, Eng.]: MoLAS/Guildford Borough Council, 2007), pp. 64–72.

13. Jacqui Pearce and Peter Tipton, "How Technology Transfer from the Continent Transformed English Ceramic Manufacture in the 16th Century," *Transactions of the English Ceramic Circle* 22 (2011): 179–214.

14. Pearce, *Pots and Potters in Tudor Hampshire*, pp. 13–16.

15. Jacqueline Pearce, *Post-Medieval Pottery in London, 1500–1700. Part 1: Border Wares* (London: HMSO, 1992), fig. 43, nos. 368–89; Pearce, *Pots and Potters in Tudor Hampshire*, p. 139, fig. 83, nos. 639–44, and fig. 84.

16. The MOLA database includes at least 221 money boxes in green-glazed Border Ware but only 33 examples with clear glaze (including examples in which the color appears olive through reduction in the kiln). An additional eight money boxes are made in the redware fabric.

17. There are four examples in the MOLA database in London-area post-medieval redware and four in the Museum of London Ceramics and Glass Collection, as well as nine money boxes in Essex-type post-medieval fine redware.

18. See Anthony Mackinder and Simon Blatherwick, *Bankside: Excavations at Benbow House, Southwark, London SE1*, MoLAS Archaeology Studies Series, 3 ([London]: Museum of London Archaeology Service, 2000), p. 16, fig. 10.

19. Site NIN08; J. Pearce, "A Face from the Past: A Unique Surrey-Hampshire Border Ware Find from Shoreditch, London," in Amanda Dunsmore, ed., *This Blessed Plot, This Earth: English Pottery Studies in Honour of Jonathan Horne* (London: Paul Holberton, 2011), pp. 211–15.

20. Julian Bowsher and Pat Miller, *The Rose and The Globe—Playhouses of Shakespeare's Bankside, Southwark: Excavations 1988–90*, MOLA Monograph, 48 ([London]: Museum of London Archaeology, 2009), pp. 72–73, fig. 63.

21. Ibid., p. 77, fig. 66.

22. Ibid., p. 133.

23. Ibid.

24. Leslie G. Matthews and H. J. M. Green, "Post-Medieval Pottery of the Inns of Court," *Post-Medieval Archaeology* 3 (1969): 1–17.

25. Ongoing research is being conducted by Mandy Hunt based on research for a B.A. dissertation at the University of Leicester.

26. *A Calendar of the Inner Temple Records*, edited by F. A. Inderwick, Q.C., 9 vols. (London: Chiswick Press), vol. 2 (1603–1660), p. 122. I am grateful to Mandy Hunt for this reference.

Diana Edwards

English and Continental Delft at Temple Newsam House, Leeds

▼ TEMPLE NEWSAM is a Tudor-Jacobean manor house in Leeds, West Yorkshire, England. The house was built in 1500 by Thomas, Lord Darcy. Sir Arthur Ingram added two wings in 1628 to make it a four-sided courtyard structure retaining the central block of the Tudor building.

First mentioned in the *Domesday Book* of 1086 as "Neuhusam," meaning New House, the property's original structure, which once belonged to the Knights Templar, was situated approximately half a mile south of the current one. In 1909 a portion of the land was taken over by the Leeds City Council for the construction of a sewage plant. The rest of the park and Temple Newsam House itself, however, remained in private ownership until 1922, when Edward Wood, 1st Earl of Halifax, deeded it to the city, with covenants to protect it for future generations; many of the contents were dispersed at that time. Well-known to collectors of English creamware and furniture, the house is maintained and open to the public by the Leeds City Council.

The vast collections at Temple Newsam contain furniture, silver, and paintings of high quality. In addition, the collection has many fine ceramics, given in the twentieth century by collectors such as Thomas E. Hollings (d. 1947) and Mrs. Arthur Smith, who for the most part collected locally. The ceramics gifted by Hollings were particularly strong in English lead-glazed earthenware and salt-glazed stoneware.[1] Some surprising and rare English and Continental pieces were given by other Yorkshire collectors, among them John Holmes (in 1892) and Frank Savery, Esq. (1883–1965).

The collections of delft were not extensive, however. Whereas Yorkshire had its own pottery and porcelain manufacturers, including the renowned Leeds and Swinton Potteries and the lesser known Castleford and Don Potteries, none of them manufactured delftware. Temple Newsam thus began to augment its collections, through gift or purchase, to fill the gaps in early English pottery.

Delft, or tin-glazed earthenware, is a low-fired clay body with a lead glaze that includes about ten percent tin oxide, which creates an opaque surface to which enamel decoration could be added if desired. The primary delftware manufacturing centers in England were initially in London and Southwark, in Brislington from the 1640s, and, later, in Bristol and Liverpool. Delft was also manufactured in Scotland and Ireland.

The tin-glazed earthenware technology was taken from the Middle East and translated to Europe following the conquest of the Muslims in the eighth century. Wherever the wares were made, decorated examples survive in vastly greater numbers than undecorated ones, which were less expensive to produce and less valued by the consumer. In Europe the decorative apex for tin-glazed earthenware was reached in Italian maiolica in the sixteenth century, but Spain was also a serious player, albeit in a much smaller context. The Netherlands and England both played major roles in the seventeenth and eighteenth centuries. Although it was relatively cheap, delft was friable and chipped and cracked easily. The demise of delft in England was due to the advent of the more durable and equally inexpensive salt-glazed stonewares beginning in the 1720s, although the final blows to the industry were not delivered until the later part of the century, by which time creamware had literally swept the market in the 1770s.

Most of the delft in the Temple Newsam collection is of English origin. A few pieces of European manufacture are illustrated because of their possible history of use in Yorkshire and their potential interest to readers.

From the seventeenth century is a dated 1647 bottle on which is written "SACK" (fig. 1). Sack, an inscription seen often on these bottles, was a dry white wine produced in Spain.[2]

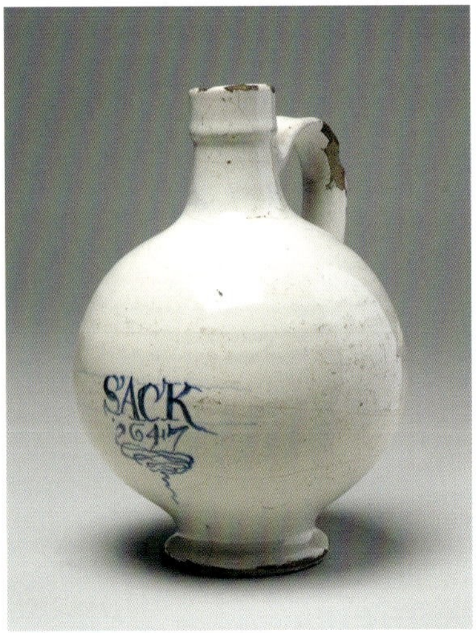

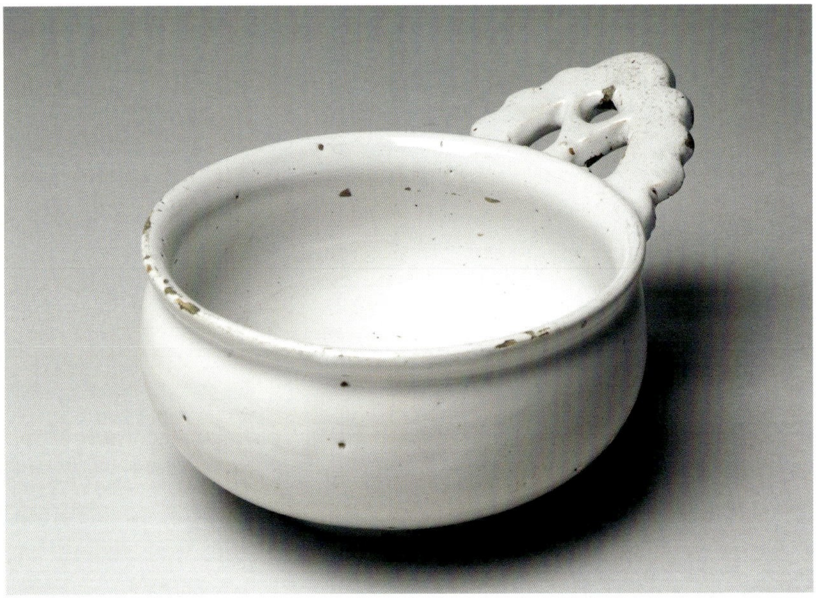

Figure 1 Bottle, probably Southwark, London, England, 1647. Tin-glazed earthenware. H. 8". (Temple Newsam House, Holmes collection, acc. no. 2011.0329; unless otherwise noted, all objects and photography courtesy of Temple Newsam House.)

Figure 2 Porringer, probably Southwark, London, England, 1675–1700. Tin-glazed earthenware. L. including handle 7¾". (Holmes collection, 1980.0051.)

Undecorated delftware porringers (fig. 2) were a common household item in the seventeenth century, but relatively few have survived the passage of time. Porringer fragments have been excavated from sites on Maryland's Western Shore (fig. 3) and from unrecorded sites in Williamsburg.[3]

A few examples of "embossed" ware—delft with bobs pushed out of the clay by hand from the inside—are in the collection. They appear to have been made in Southwark and, thanks to two dated examples, can be firmly placed within 1650–1660.[4] The mug in Temple Newsam House was an early purchase from the John Holmes collection by the Municipal Art Gallery Committee in 1892 (fig. 4).

Another purchase from the Holmes collection was a posset pot (fig. 5) with Chinese-style painted figures and landscapes in blue made in London circa 1680. Posset was a concoction of hot curdled milk and wine or ale and

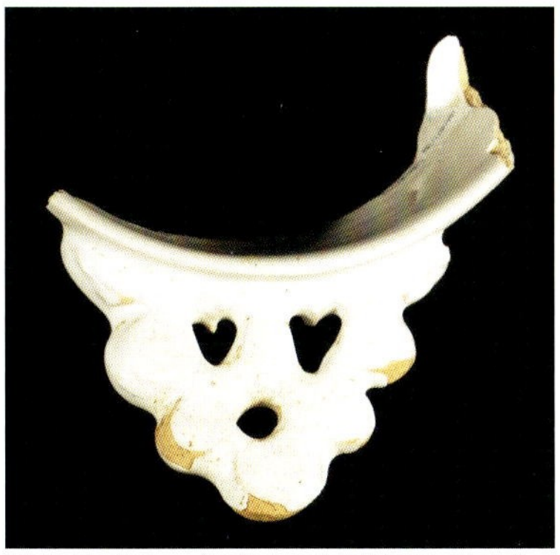

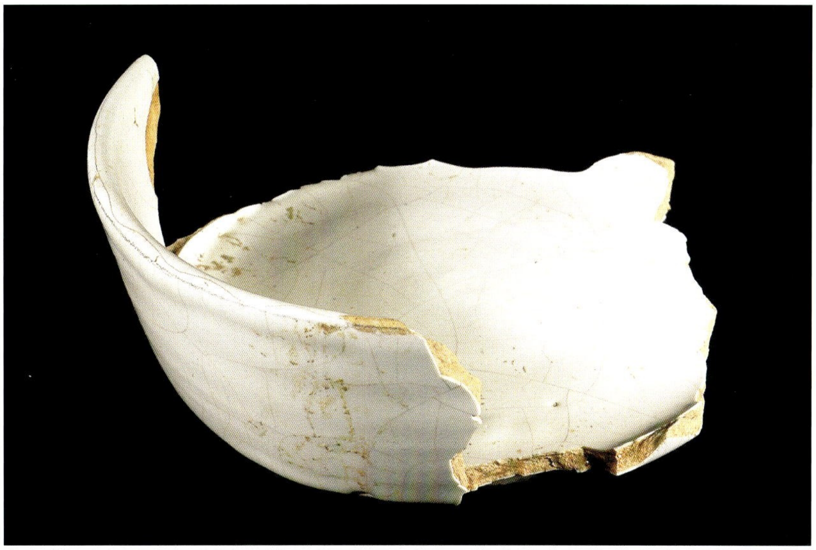

Figure 3 Porringer fragments, probably Southwark, London, England, 1670–1710. Tin-glazed earthenware. (Courtesy, Lost Towns Project, Annapolis, Maryland.) Porringers with the double-heart and round-hole handle have been uncovered at several archeological sites on the Mid-Atlantic seaboard, including Williamsburg, Virginia, and the Western Shore of Maryland. This one was excavated from Homewood's Lot.

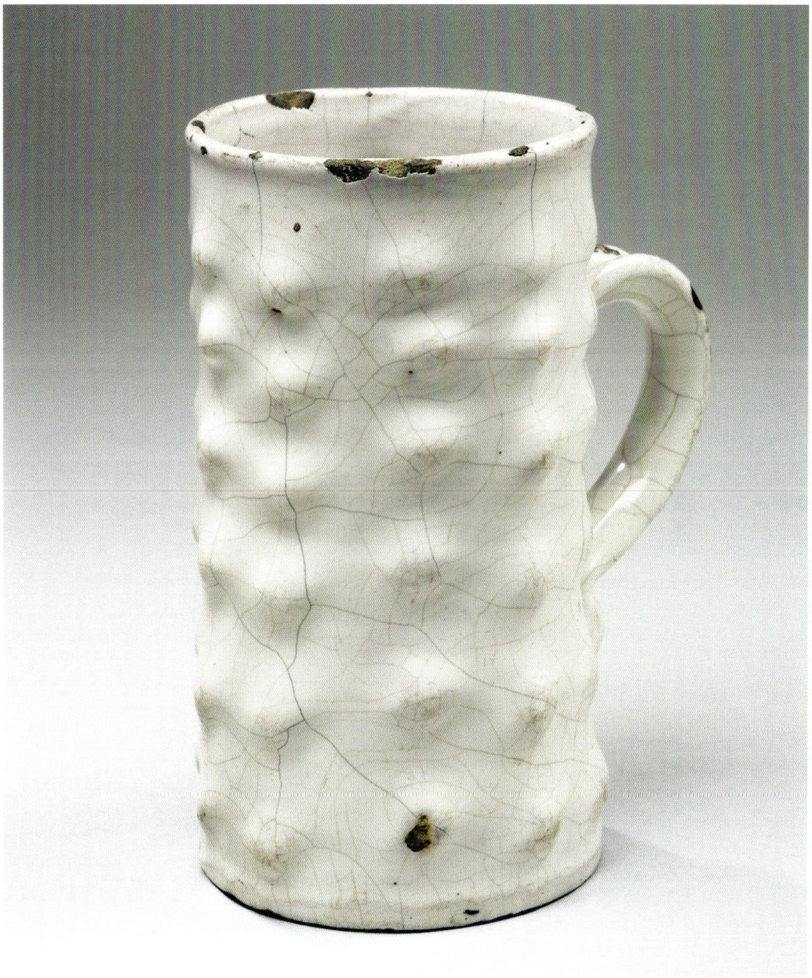

Figure 4 Mug, probably Southwark, London, England, 1650–1660. Tin-glazed earthenware. H. 6½". (Holmes collection, 1890.0006.)

other spices; its popularity seems to have waned after the seventeenth century with the introduction of punch. Although there are mentions of posset in the writings of Sir Walter Scott (1771–1832) and Fanny Trollope (1779–1863), the posset pot as we know it was not produced after the seventeenth century

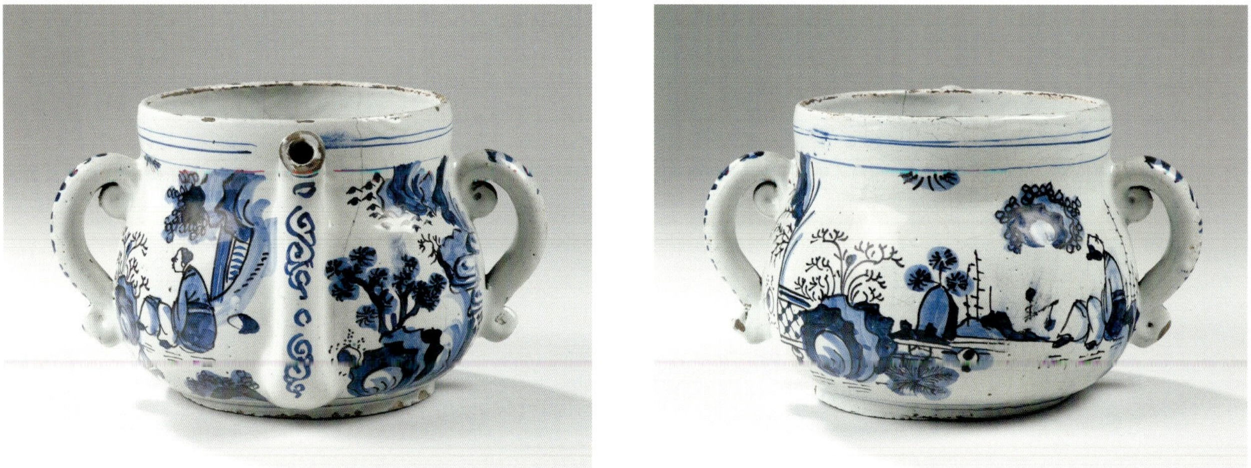

Figure 5 Posset pot, London, England, ca. 1680. Tin-glazed earthenware. H. 5". (Holmes collection, 1890.0004.)

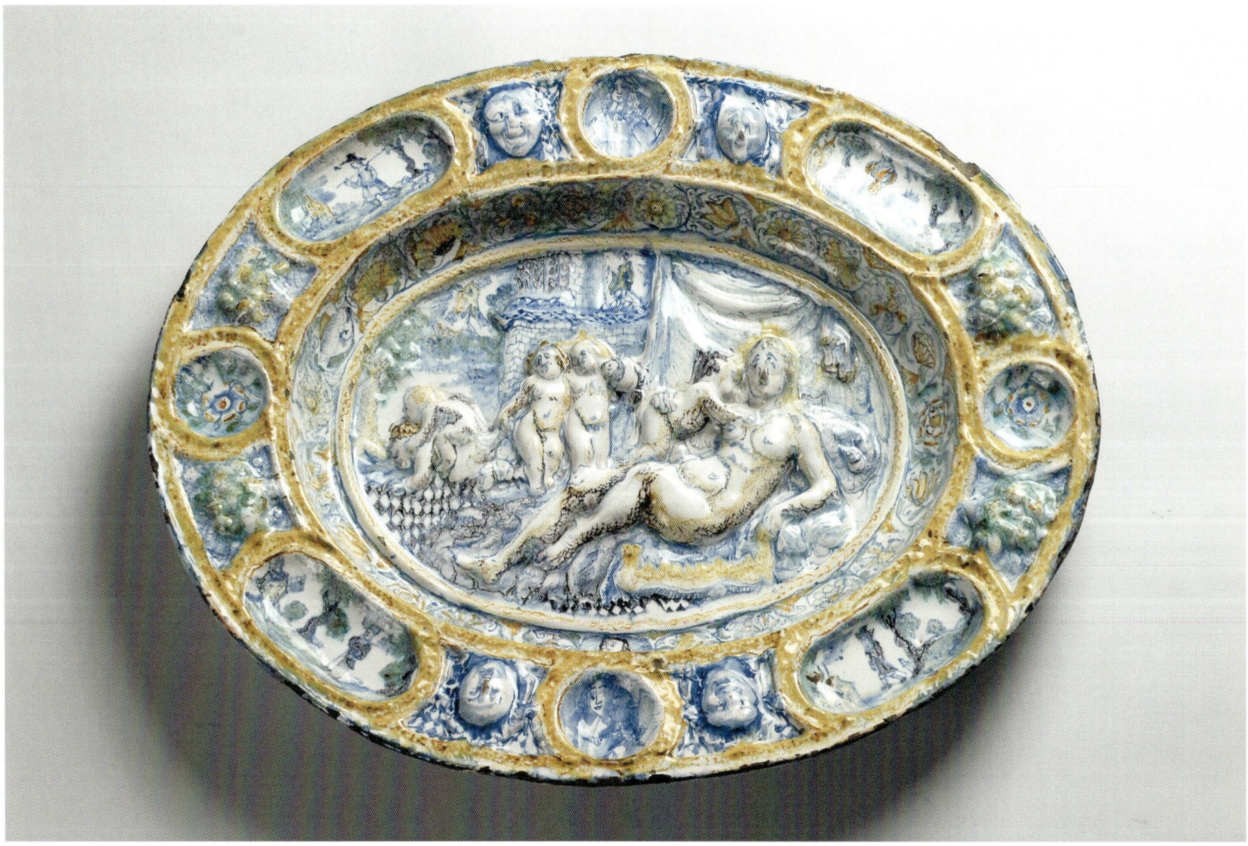

Figure 6 Dish, possibly "WP" potter, Pickleherring or Montague Close Potteries, Southwark, London, England, 1661–1688. Tin-glazed earthenware. D. 18½". (Holmes collection, 1985.0035.) The dish depicts La Fécondité and includes portraits of Charles II, James II, and Oliver Cromwell. Dishes like this were modeled after ceramics made by Bernard Palissy in the sixteenth century.

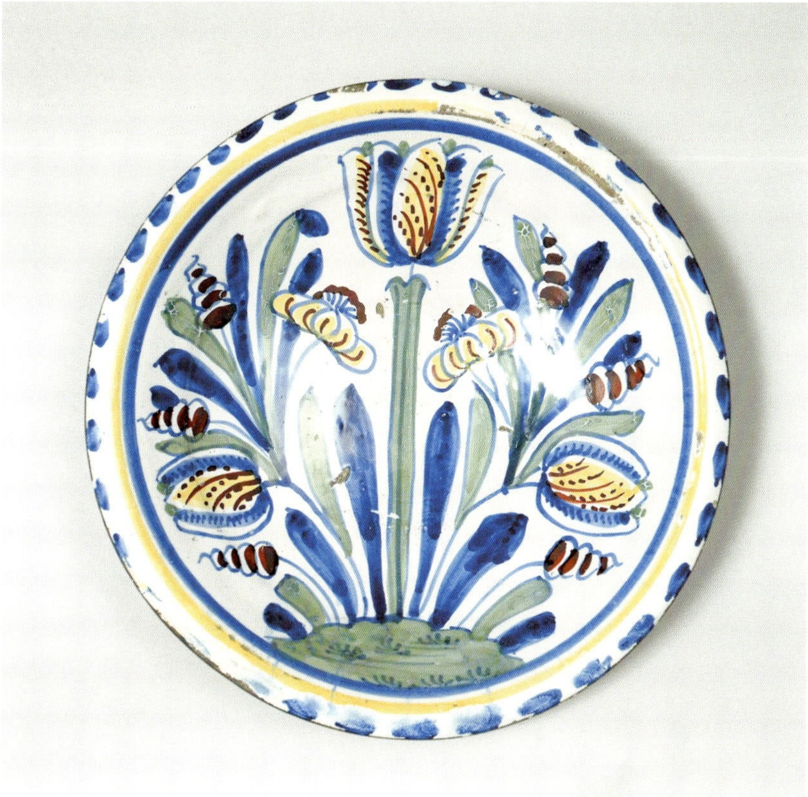

Figure 7 Dish, tin-glazed earthenware, Bristol or Brislington, England, 1675–1725. D. 13¼". (Hollings bequest, 1947.0016.0486.) Painted with tulips and lesser botanicals in varying stages of bloom. The rim is decorated with blue dashes.

and one does not find examples of the form in porcelain or the later earthenwares.[5]

A group of important dishes after the French style of Bernard Palissy were manufactured in London in the seventeenth century, with dates occurring on examples from 1633 to 1697.[6] The so-called fecundity dish in the Temple Newsam collection (fig. 6) is a particularly important example as it portrays three prominent political figures of the time: Charles II, on the upper rim; James II, on the bottom; and Oliver Cromwell as the nemesis within the broad marley border.[7] The potter might be William Price or William Pocock, each of whom lived in the Parish of St. Saviour in Southwark and worked nearby, at either the Pickleherring or Montague Close Potteries.[8]

The tulip charger with a blue-dash border illustrated in figure 7, made in Bristol or Brislington, was one of several stylized versions of the tulip motif that was popular on English delft circa 1700. The tulip became popular in England in the late sixteenth century. The first English author to do justice to the tulip was John Parkinson (*Paradisus* 1629), who enumerated 140 varieties. The first known reference in Western history to the tulip was by Ogier de Busbecq, ambassador from Emperor Ferdinand I to Sulieman the Magnificent.[9] On his way from Adrianople to Constantinople in late 1554, de Busbecq observed: "We came across quantities of flowers—narcissi, hyacinths, and *tulipans*, as the Turks call them. . . . The tulip has little or no scent, but it is admired for its beauty and the variety of its colours."[10] Western paintings, textiles, and pottery do not depict tulips until the end of the sixteenth century, placing de Busbecq on the cutting edge of what was to become tulip mania.

Another blue dash charger (fig. 8), also likely made in Bristol, depicts sponged trees flanking a full-length portrait of Queen Anne holding a scepter in her right hand and standing on a verdant surface of green fronds. Portraits of reigning English monarchs are iconic in English delft production.

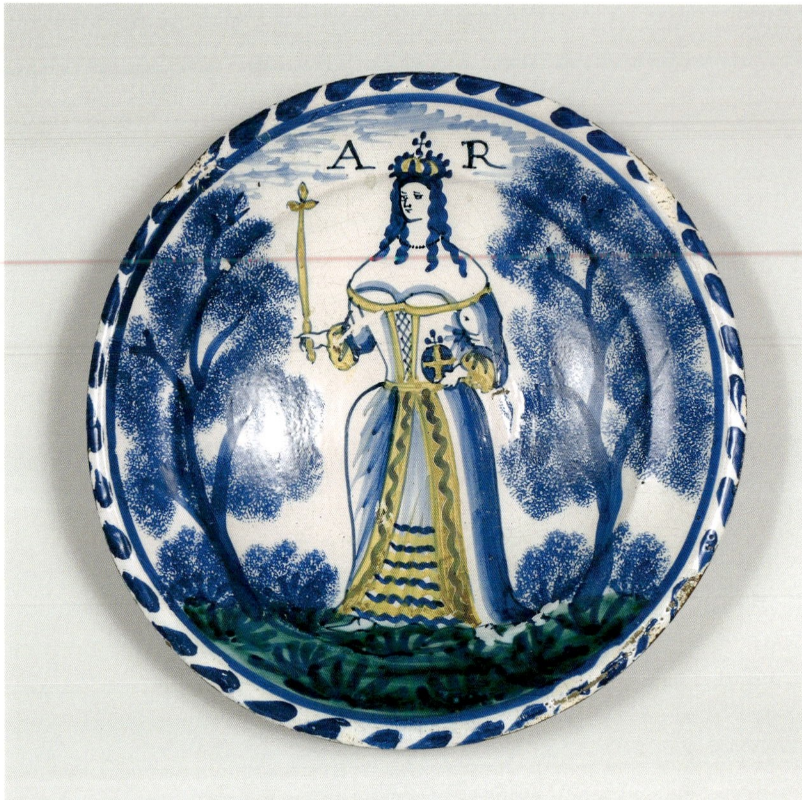

Figure 8 Dish, probably Bristol, England, 1705–1712. Tin-glazed earthenware. D. 13¾". (Hollings bequest, 1947.0016.0485.) Decorated with a blue-dashed rim and a full-length polychrome portrait of Queen Anne.

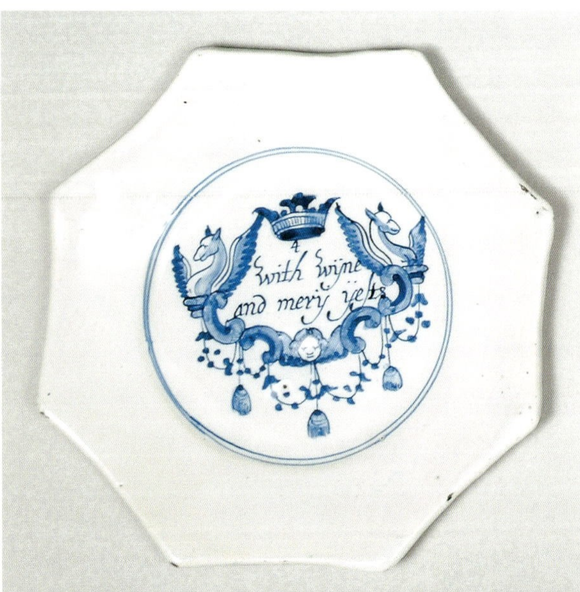

Figure 9 Plate, London, England, ca. 1680. Tin-glazed earthenware. D. 7½". (Hollings bequest, 1947.0016.0483.4.) This plate is one of a set of six tin-glazed earthenware plates decorated with the so-called Merryman rhymes.

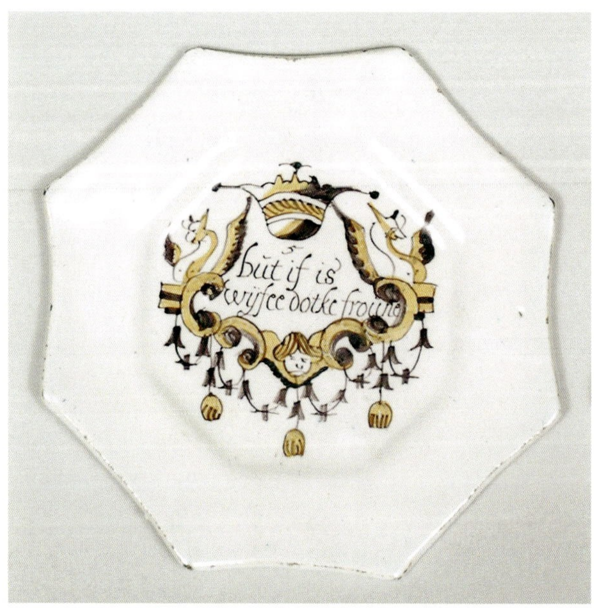

Figure 10 Plate, London, England, or the Netherlands, ca. 1680. Tin-glazed earthenware. D. 7½". (Hollings bequest, 1947.0016.0483.5.) The set from which this plate comes is unusually colored in ochre and manganese, possibly by Dutch artists working in London.

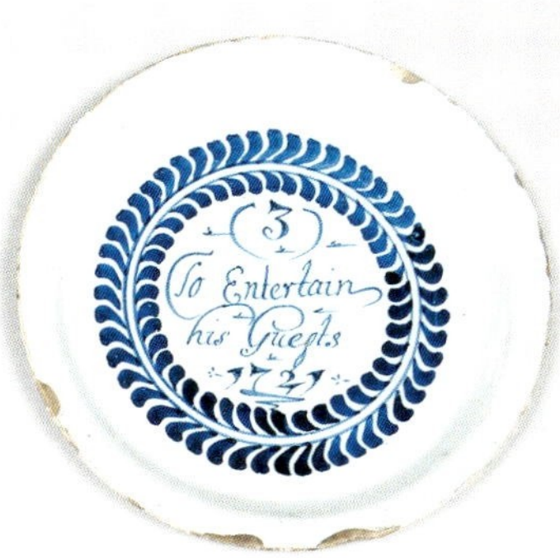

Figure 11 Plate, Bristol or London, England, 1721. Tin-glazed earthenware. D. 8⅛". (Hollings bequest, 1947.0016.0484.3.) One of six plates from another Merryman set.

There are many large dishes depicting Queen Anne, despite her relatively brief twelve-year reign, and most of them show a standing figure. A few, for example one in the Victoria and Albert Museum and two in the Glaisher collection at the Fitzwilliam Museum, show the queen seated. A dish identical to the Temple Newsam example was formerly in the Longridge collection, and another, formerly in the Freeth collection, is currently in the Bristol Museum.[11]

The collections at Temple Newsam House have three sets of the popular delft "Merryman" plates. Two sets are from the seventeenth century, possibly of Dutch origin or made by Dutch potters in London (figs. 9, 10); a third set is dated 1727 (fig. 11). Merryman plates satirize the perennial tussle between husband and wife. They were produced in sets of six, each inscribed with its own line from the six-line rhyme:

> *What is a / merrey man*
> *let him doe / what he cane*
> *to Entertain / his Guests*
> *with wyne / and mery jests,*
> *but if is / wyfee dothe froune*
> *all merriment / goes downe.*

The spelling varies on different sets; on some, it is closer to Dutch than English, possibly indicating that those versions are of Dutch manufacture.

A punch bowl painted in blue bears a portrait of a man on the exterior flanked by "K" "I" and an interior oval with the same portrait labeled "KING JAMES" and, underneath the portrait, "T. LILLY 1732" (figs. 12, 13). As Louis Lipski and Michael Archer point out, it is not a portrait of James II of England, who died in 1701, but a portrait of his son, James Edward Stuart, Prince of Wales, known as "The Old Pretender."[12] The suggestion is that the political and economic interests of the time may have favored the house of Hanoverian and Jacobite supporters, if they still existed, and that life might have been better had the crown remained on British heads.[13]

Figure 12 Punch bowl, possibly Bristol or Brislington, England, ca. 1732. Tin-glazed earthenware. H. 13½". (Holmes collection, 1973.0012.) The bowl is painted with blue enamel decoration and an oval cartouche of James Edward Stuart, Prince of Wales (see fig. 13).

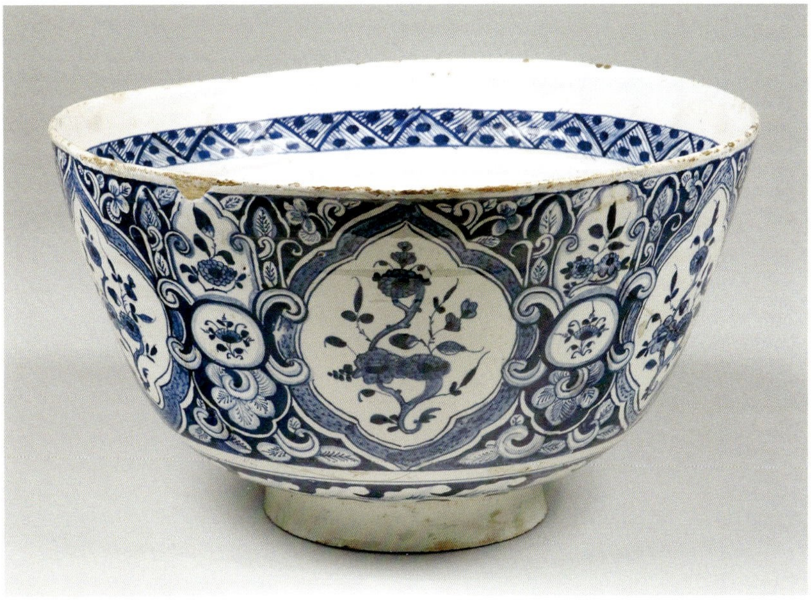

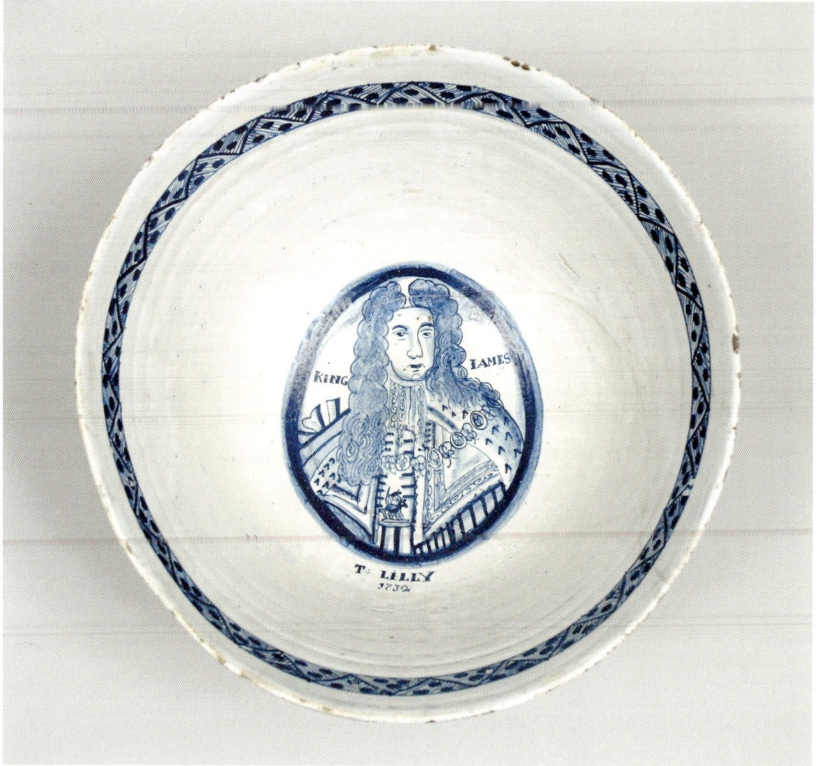

Figure 13 Oval cartouche inside the punch bowl illustrated in fig. 12. It depicts "The Old Pretender" James Edward Stuart, after a 1732 painting by T. Lilly.

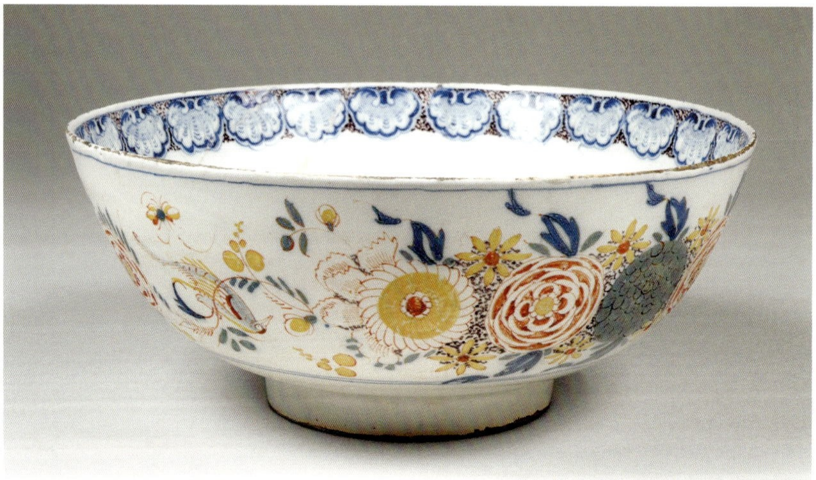

Figure 14 Punch bowl, Liverpool, England, ca. 1760. Tin-glazed earthenware. D. 12". (Holmes collection, 1979.0037.) The exterior is decorated with Fazakerley colored birds and flowers.

Another punch bowl (figs. 14, 15) purchased by the museum has flowers painted in the so-called Fazakerley palette of red, blue, yellow, and orange on the outside. The interior decoration, taken from an engraving, has the inscription "DUM VIVE THRIVO" in the banner linking the two lower sides of a mock coat of arms and beneath it the words "TO THE GLORIOUS INCERTAINTY OF THE LAW." In place of the shield is a parchment with five seals hanging from it. The crest is a fox wearing legal robes and the supports are two male figures—one holds a money pouch and stands on a plough, the other stands on a harrow. In the upper part of the shield is a lawyer in his office accepting bribes, and in the lower half is a book entitled *Coke upon Littleton*, a reference to Sir Edward Coke's commentary on the

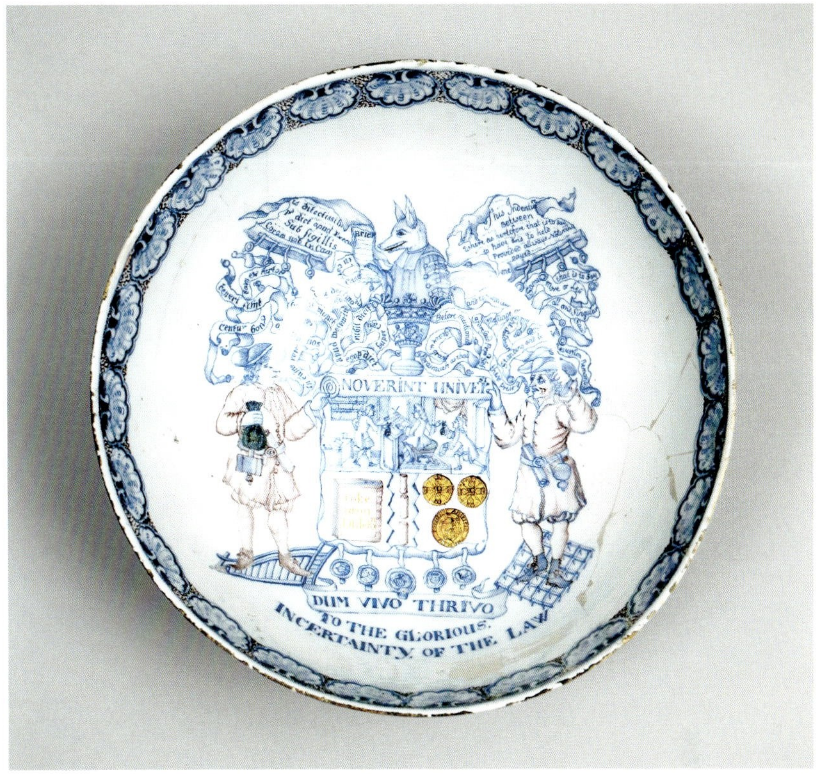

Figure 15 Interior of the punch bowl illustrated in fig. 14.

common law from Bracton and Littleton of 1628, which formed the basis of the British constitution.[14] A similar bowl with a variation of the print but quite different exterior decoration is in the Schreiber collection at the Victoria and Albert Museum.[15]

Teapots made in the delft body were particularly susceptible to chipping and crazing with the hot tea, and as a result were never particularly functional or popular. A blue-and-white painted example from the Hollings collection (fig. 16) has a faceted notched handle and crabstock spout and is painted with a chinoiserie landscape.

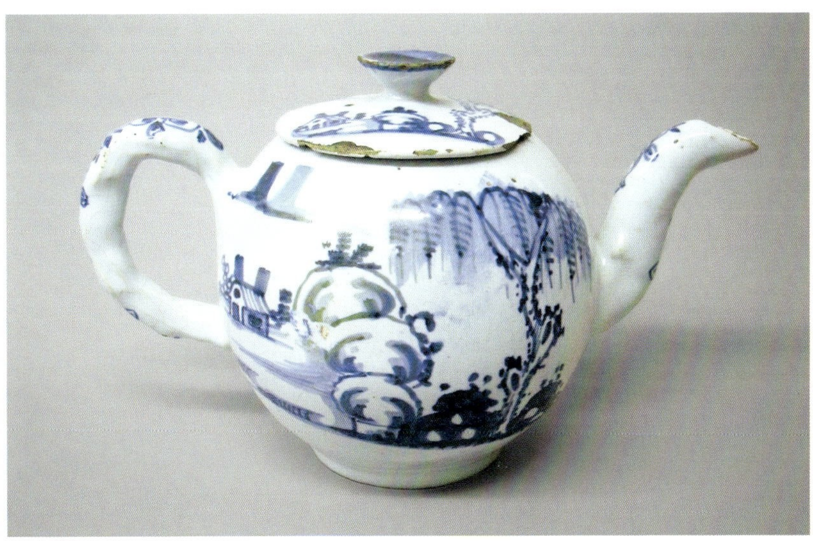

Figure 16 Teapot, probably London, England, 1750–1760. Tin-glazed earthenware. H. 4½". (Hollings bequest, 1947.0016.0478.) The teapot with molded handle and spout is painted in blue with chinoiserie scenes.

Figure 17 Pair of wall pockets, Liverpool, England, 1760–1770. Tin-glazed earthenware. H. 7¾". (Hollings bequest, 1947.0016.0480.) Each wall pocket is molded and painted in enamels with birds and foliage.

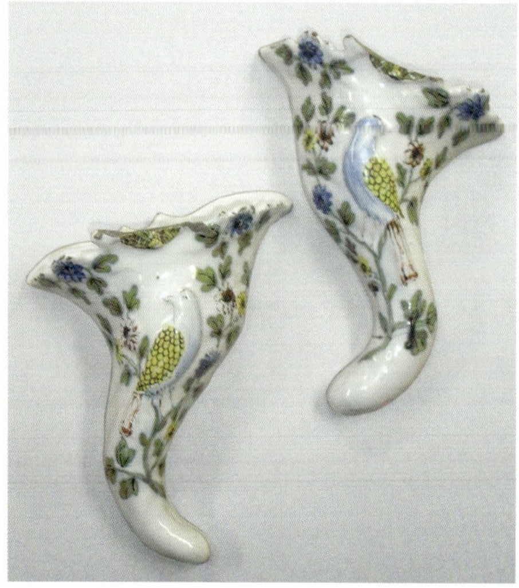

Figure 18 Mug, Liverpool, England, ca. 1760. Tin-glazed stoneware. H. 5⅛". (W. Ambrose Harding collection, 1984.0040.) The painted enamels depict exotic birds and foliage.

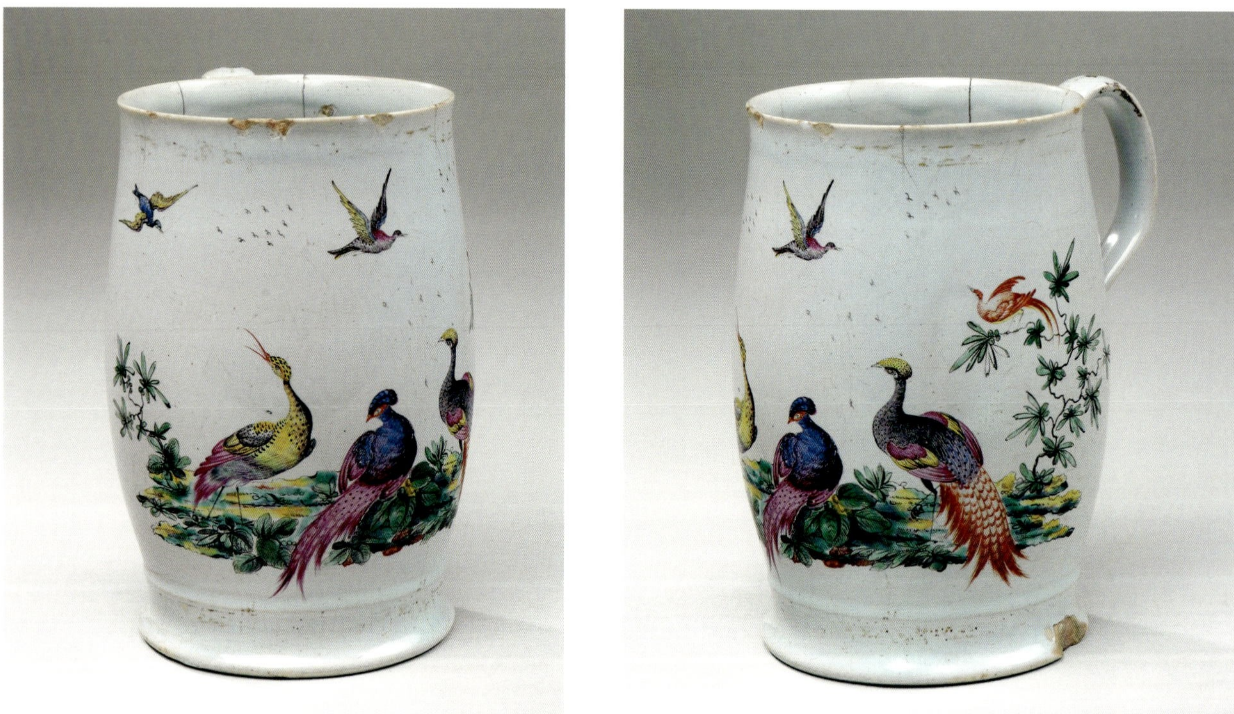

A pair of wall pockets in the form of cornucopias (fig. 17), another Hollings bequest, was molded with birds and painted in teal, greens, yellow, and blue. These are similar to a pair in the collection of Historic Deerfield in Massachusetts, both made in Liverpool 1760–1770.[16] In 1984 the museum purchased a rare enameled mug of tin-glazed stoneware (fig. 18). The mug is part of a small group of tin-glazed stonewares—fewer than two dozen—about which scholars have yet to determine how they were made or for what purpose.[17] The distinguishing feature of this mug is the polychromed enamel decoration of birds in the style of Worcester or Chelsea porcelain (most of the other identified pieces are decorated principally in blue and white). The rendering of the birds, foliage, and insects is sophisticated and could compete successfully with any porcelain example were it not for the fragility of the glaze, which plagued tin-glazed earthenwares from the onset.

A few important delft examples exist in the Temple Newsam collection that were not made in England. The hand warmer in the form of a shoe (fig. 19) is extremely rare and possibly from Faenza, Italy. Hand warmers, or *scalamani*, were filled with hot water and placed in a lady's carriage under a cloak or shawl to warm hands or feet. They are found in the form of books, shoes, and boots, but this particular shape is uncommon. There is a similar shoe in the Stein collection of the Philadelphia Museum of Art, but the form here differs and the fabric type is more ornate. In a watercolor by Jan II van Grevenbroeck (fig. 20) a courtesan in Renaissance costume lifts her skirt and displays a similarly shaped shoe, which is open at the sides to reveal her

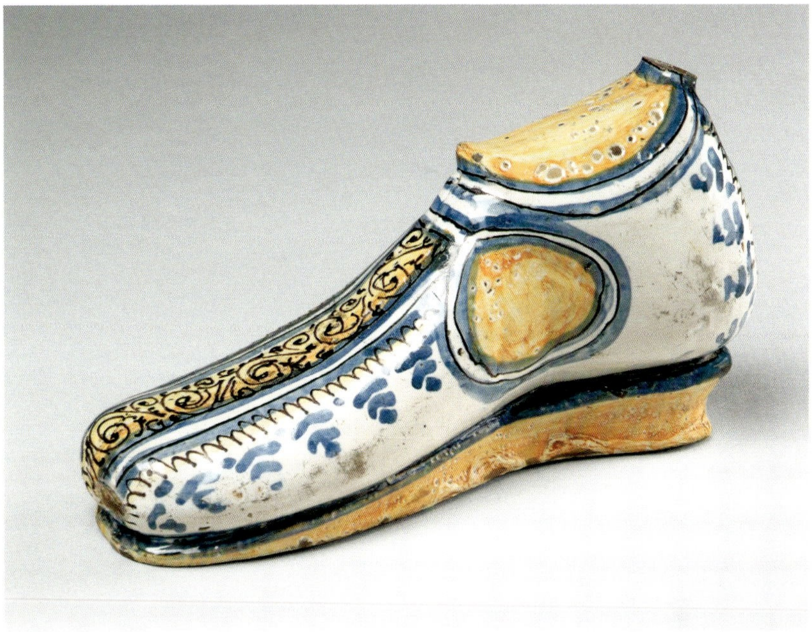

Figure 19 Hand warmer, possibly Faenza, Italy, 17th century. Tin-glazed earthenware. L. 8½". (Holmes collection, 2011.0327.)

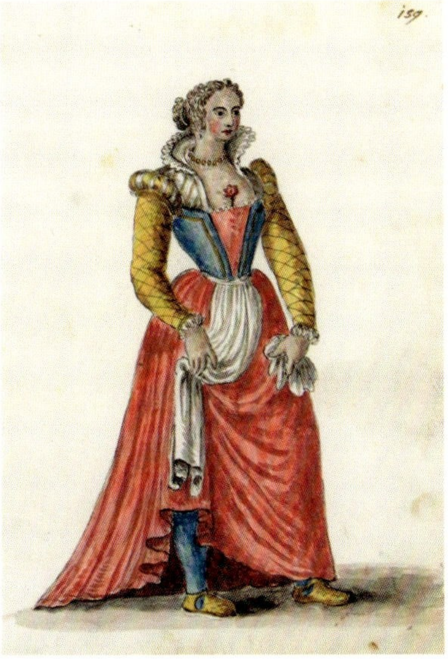

Figure 20 Jan II van Grevenbroeck (Italian, 1731–1807), "Courtesan" from *Gli Abiti De Veneziani*, vol. 3, pl. 159, 1750–1807. Watercolor on laid(?) paper, 11 x 11". (Courtesy, Museo Correr, Venice.) The Italian Renaissance courtesan depicted in the painting wears a shoe similar in style and decoration to the hand warmer illustrated in fig. 19.

hose.[18] Running down the front of the vamp of the shoe in Temple Newsam is a decorative band depicting a metal lace trim popular from the mid-seventeenth century into the 1680s. Shoes with square toes, latchets, and applied ribbon were fashionable in the late seventeenth century.[19] The palette suggests a northern Italy manufacture, and the Philadelphia shoe is attributed to Emilia-Romagna, probably Faenza.[20]

The last offering of tin-glazed wares in the collection is a group of three plaques made at the Alcora Ceramic Factory in Spain and given to Temple Newsam by Frank Savery in 1954. With their elaborate rococo borders and pictorial panels painted in vibrant colors, these plaques must have been produced in some numbers, as there are three in the Victoria and Albert Museum and others are known. Two of the plaques are horizontally oriented and depict European pastoral scenes.[21] One is adapted from a François Boucher (1703–1770) print, *Les amours pastorales* (fig. 21), the other horizontal plaque depicts the Allegory of the Earth, the reverse of a paint-

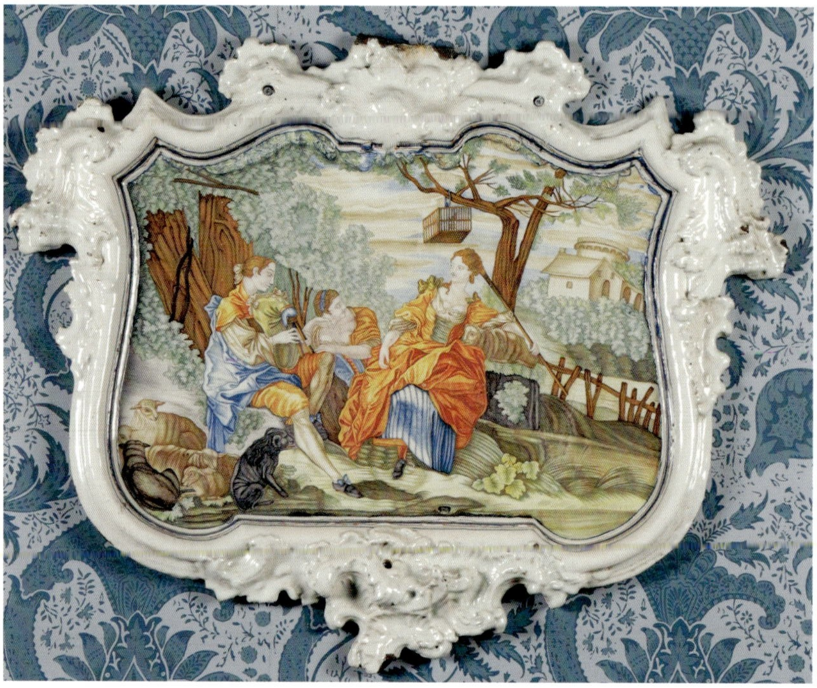

Figure 21 Wall plaque, Alcora Ceramic Factory, Alcora, Spain, ca. 1750. Tin-glazed earthenware. 21½ x 34". (Frank Savery, Esq. bequest, 2011.0324.0002–003.) This plaque depicts *Les amours pastorales*, after a scene by French artist François Boucher (1703–1770).

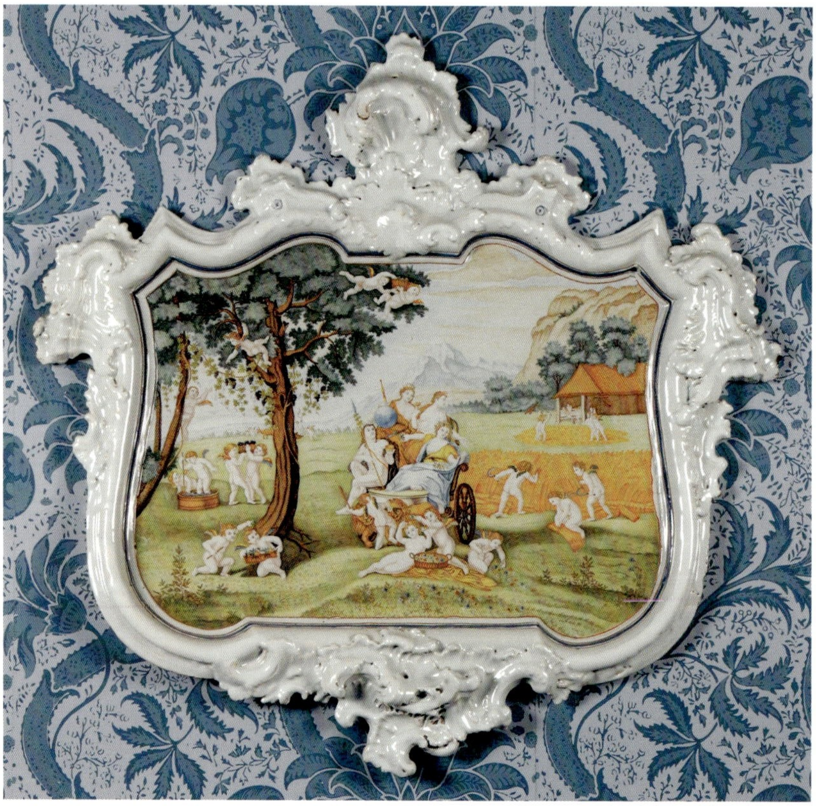

Figure 22 Wall plaque, Alcora Ceramic Factory, Alcora, Spain, ca. 1750. Tin-glazed earthenware. 21½ x 34". (Frank Savery, Esq. bequest, 2011.0324.0002–003.)

ing by Francesco Albano in the Turin Museum (fig. 22). The vertical plaque presents the biblical scene of Christ Disputing with the Doctors (fig. 23).

The collections of English and Continental tin-glazed earthenwares at Temple Newsam are explored here only in a cursory way. This brief overview publishes for the first time illustrations of many of the wonderful

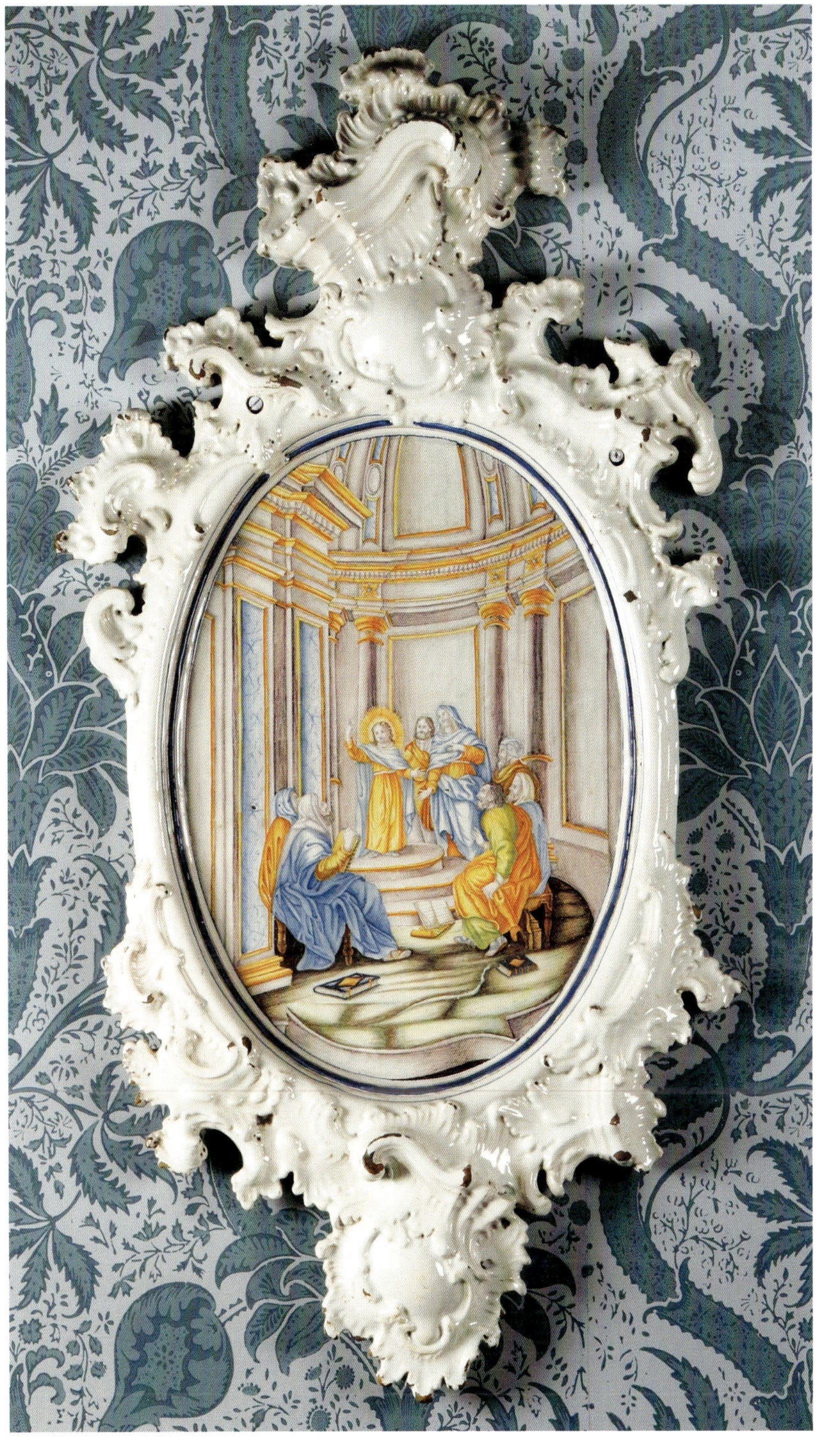

Figure 23 Wall plaque, Alcora Ceramic Factory, Alcora, Spain, ca. 1750. Tin-glazed earthenware. 33 x 19". (Frank Savery, Esq. bequest, 2011.0324.0001.)

examples of delft housed in a collection rich in furniture, paintings, porcelains, and other earthenwares as well as textiles.

1. See Diana Edwards, "Salt-glazed Stoneware in Temple Newsam House, Leeds," *Antiques* (July 2006): 70–77.

2. The term *sack* may have derived from the German *seck* but seems to have moved rapidly into general terminology for those wines produced in Jerez de la Frontera, Málaga, and the Canary Islands. In *Henry IV*, Shakespeare mentions sack in several passages: "If sack and sugar be a fault,

God help the wicked!" (II, iv); "Unless hours were cups of sack and minutes capons and clocks the tongues of bawds and dials the signs of leaping-houses and the blessed sun himself a fair hot wench in flame-colored taffeta, I see no reason why thou shouldst be so superfluous to demand the time of the day" (I, ii); and "I'll purge and leave sack, and live cleanly . . ." (I, v).

3. Al Luckenbach and Diana Edwards, "Ceramics of the Lost Towns Project: 17th- and Early 18th-Century Wares from Maryland's Western Shore," *English Ceramic Circle Transactions* 18, pt. 3 (2004): 567–69; Ivor Noël Hume, *Early English Delftware from London and Virginia* (Williamsburg, Va.: Colonial Williamsburg Foundation, 1977), pp. 90–91.

4. In his *London Delftware* (London: Jonathan Horne, 1987), p. 119, Frank Britton illustrates a similar mug given to the Museum of London by J. C. Joicey, and indicates that it was made in Southwark. There is a dated example of a posset pot with embossed decoration in the Wellcome Museum of the History of Medicine at the Science Museum, London, but according to Louis L. Lipski and Michael Archer it was originally undecorated and the date was added later. Lipski and Archer, *Dated English Delftware: Tin-glazed Earthenware, 1600–1800* (London: Philip Wilson; New York: Harper & Row, 1984), no. 958.

5. "A gold posset-dish to contain the night-draught," in Sir Walter Scott, *Kenilworth; a Romance* (Edinburgh: A. Constable, 1821), chap. 6; "I do wish he would try a hot posset of a night, just before going to bed," in Fanny Trollope, *Charming Fellow II* (London: Chapman & Hall, 1876), p. 205.

6. Graham Slater, "English Delftware 'Palissy' Dishes Based on 'La Fécondité' Design," *English Ceramic Circle Transactions* 17, pt. 1 (1999): 47.

7. This crudely painted portrait might have been taken from Sir Peter Lely's ca. 1675 portrait of Charles II (1630–1685) in the National Portrait Gallery, London (NPG 153). See Kai Kin Yung, comp., *National Portrait Gallery: Complete Illustrated Catalogue, 1856–1979*, edited by Mary Pettman (New York: St. Martin's Press, 1981), no. 153. Charles II became the focus for Royalist sympathizers after his father's execution and numerous images of him were circulated. James II, brother of Charles II and his successor, immediately sought sittings for portraits directly after the Restoration. The portrait of James II (1633–1701) may have been taken from a painting by Sir Peter Lely that is in the National Portrait Gallery (NPG 5211) or possibly from a miniature in ivory after Sir Godfrey Kneller, L152(15), currently on loan to the NPG. A full-length portrait by Henri Gascus, ca. 1672–73, in the National Maritime Museum, Greenwich, resembles the clothing and the stance from the shoulders upward in the crudely painted miniature portrait on the dish. The portrait of Oliver Cromwell resembles a miniature portrait in the Victoria and Albert Museum painted by Andrew Benjamin Lens (b. 1708).

8. Slater, "English Delftware 'Palissy' Dishes," pp. 58–62.

9. Wilfrid Blunt, *Tulipomania* (Harmondsworth, Eng.: Penguin Books, 1950), p. 7.

10. Ogier Ghislen de Busbecq, *Turkish Letters* (1927; repr., London: Eland, 2005), p. 16.

11. Frank Britton, *English Delftware in the Bristol Collection* (London: Sotheby Publications, 1982), p. 65, no. 3.53; Leslie B. Grigsby, *The Longridge Collection of English Slipware and Delftware*, 2 vols. (London: Jonathan Horne, 2000), vol. 2, *Delftware*, p. 66, no. D33.

12. Lipski and Archer, *Dated English Delftware*, no. 1087.

13. Ibid.

14. Peter Walton, "Acquisitions of English Pottery made for the Temple Newsam Collection since 1976," *Leeds Arts Calendar* 90 (1982): 17–19.

15. The probable print source for the interiors of both of these bowls as published by Roger Massey is in the British Museum. *English Ceramic Circle Transactions* 16, pt. 2 (1997): 235–36.

16. Amanda E. Lange, *Delftware at Historic Deerfield, 1600–1800*, exh. cat. (Deerfield, Mass.: Historic Deerfield, 2001), no. 73.

17. Alan Smith, "An Enamelled, Tin-glazed Mug at Temple Newsam House," *Leeds Arts Calendar* 82 (1978): 14–16.

18. Wendy M. Watson, *Italian Renaissance Ceramics: The Howard I. and Janet H. Stein Collection and the Philadelphia Museum of Art*, exh. cat. (Philadelphia: Philadelphia Museum of Art, 2001), pp. 141, 144.

19. I am grateful to Oriole Cullen and Susan North of the Victoria and Albert Museum's textile department for advice pertaining to the dating of this shoe.

20. Watson, *Italian Renaissance Ceramics*, p. 144.

21. Hugh Honour, "The Ceramics of Alcora," *Leeds Arts Calendar* 17, no. 24 (Winter 1954): 17.

Diana Stradling

Bonnin & Morris: Reading between the Lines

▼ DESPITE ALL THE words written by and about Gousse Bonnin and George Anthony Morris and their pre-Revolutionary porcelain works in Philadelphia, there remains an unwritten text that begs to be explored.[1]

When Gousse Bonnin brought his family from Britain to Philadelphia in 1768, the colonies were in the throes of resistance to taxes imposed by the Townshend Acts. British imports were boycotted and nonimportation fever was in the air. It was clear that manufacturing in America would be the wave of the future and native materials would be essential. As raw minerals were refined in crucibles, Bonnin was so certain that these melting pots would be in constant demand that he returned to England to purchase a Royal Patent granting him a monopoly on their manufacture. On his return, he warned off any potential competition in two bold announcements of his intent to make "Black Lead [i.e., graphite] Crucibles," placing one advertisement in Philadelphia on September 18, 1769, and another in New York City, which ran well into November of the same year.[2]

Barely two months later, however, he was in a different business, having entered into partnership with George Anthony Morris, a Philadelphia neighbor. With the ink barely dry on their agreement, they announced on December 29, 1769, that their manufactory of "New China Ware" was already under construction. This first notice hinted that it would be a very sophisticated factory (fig. 1): "[T]he proprietors . . . have proved to a certainty, that the clays of America are productive of as good PORCELAIN as any heretofore manufactured at the famous factory in Bow . . . ." But why Bow? If this was not an idle boast, then the arduous testing and tweaking of formulas had already been done, which suggests that there was on hand a knowledgeable "master of pastes," as he was known in the porcelain works of Europe, someone who knew which clays to use and how to test them. The workers sought were specialists—throwers, turners, modelers, molders, pressers, painters—indicating that this plant would be organized according to what later became known in America as "The Staffordshire Plan"—assembly-line production, with each man trained to perform a single function, a system first tentatively implemented in the 1730s but still being perfected in England.[3]

There is no way that Bonnin or Morris, two sons of wealth who had led lives of leisure, could possibly have had the knowledge to build or superintend such a venture. So who was running the place? And what had happened to make Bonnin change course so drastically?

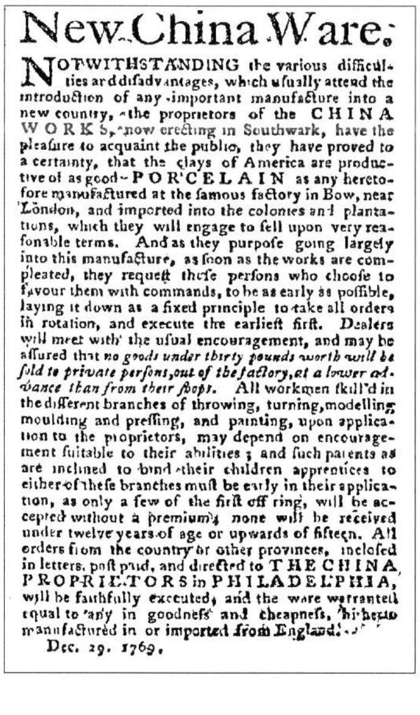
New China Ware.

NOTWITHSTANDING the various difficulties and disadvantages, which usually attend the introduction of any important manufacture into a new country, the proprietors of the CHINA WORKS, now erecting in Southwark, have the pleasure to acquaint the public, they have proved to a certainty, that the clays of America are productive of as good PORCELAIN as any heretofore manufactured at the famous factory in Bow, near London, and imported into the colonies and plantations, which they will engage to sell upon very reasonable terms. And as they purpose going largely into this manufacture, as soon as the works are compleated, they request those persons who choose to favour them with commands, to be as early as possible, laying it down as a fixed principle to take all orders in rotation, and execute the earliest first. Dealers will meet with the usual encouragement, and may be assured that *no goods under thirty pounds worth will be sold to private persons, out of the factory, at a lower advance than from their shops.* All workmen skill'd in the different branches of throwing, turning, modelling, moulding and pressing, and painting, upon application to the proprietors, may depend on encouragement suitable to their abilities; and such parents as are inclined to bind their children apprentices to either of these branches must be early in their application, as only a few of the first offering, will be accepted without a premium; none will be received under twelve years of age or upwards of fifteen. All orders from the country or other provinces, inclosed in letters, post paid, and directed to THE CHINA PROPRIETORS in PHILADELPHIA, will be faithfully executed, and the ware warranted equal to any in goodness and cheapness, hitherto manufactured in or imported from England.

Dec. 29. 1769.

Figure 1 Notice in the *Pennsylvania Chronicle and Universal Advertiser*, January 1–15, 1770. (© NewsBank and American Antiquarian Society.) The December 29, 1769, announcement of a new china manufactory.

In the eighteenth century, most craftsmen, artisans, and mechanics had the skills to set themselves up in business but no money, so there evolved an economic system wherein a man of means would team up with a skilled craftsman, often building and equipping a factory to the craftsman's specifications. The craftsman, or "artist," as he was known, remained anonymous. If lucky enough to become a silent partner, he became the "& Co." at the end of the company name. He might be recognized as "the Manufacturer," but the money man was always "the Proprietor," and it was his name on the sign over the door. If the craftsman failed to fulfill his promise, he was usually branded a charlatan and sent packing with full blame. If successful, he might be lured away by a competitor with promises of higher pay. In either case, the proprietor still owned a brand-new, fully equipped plant and could hire another artisan to run the works, if he could find one.[4]

The correspondence of Philadelphians left many clues to the inner workings of the Bonnin & Morris porcelain manufactory. In a letter detailing what history he knew, Dr. James Mease, who gave his father's china basket to the Franklin Institute, wrote, "[T]he Clay . . . was procured from white-

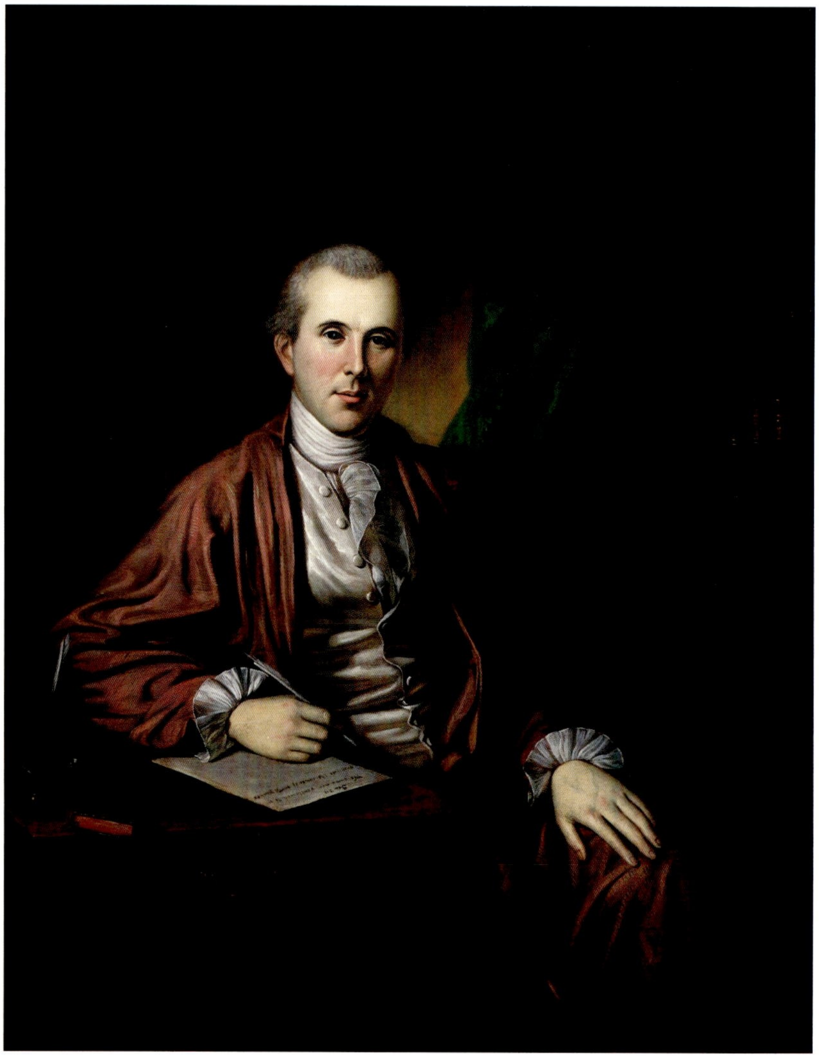

Figure 2 Charles Willson Peale (American, 1741–1827), *Portrait of the Young Benjamin Rush*, 1783–1786. Oil on canvas, 49⅞ x 39⅞". (Courtesy, The Henry Francis DuPont Winterthur Museum. Gift of Mrs. T. Charlton Henry.) In this portrait, Rush is featured beside his cabinet of books.

Figure 3 Advertisement in the *Pennsylvania Packet*, September 4, 1771. (© Accessible Archives.) Benjamin Rush advertised his "Course of Lectures and Experiments on Chemistry," including "Salts" and "Earths." Rush was elected professor of chemistry at the City Medical College soon after his return from Edinburgh in 1769, a position he retained even while practicing medicine.

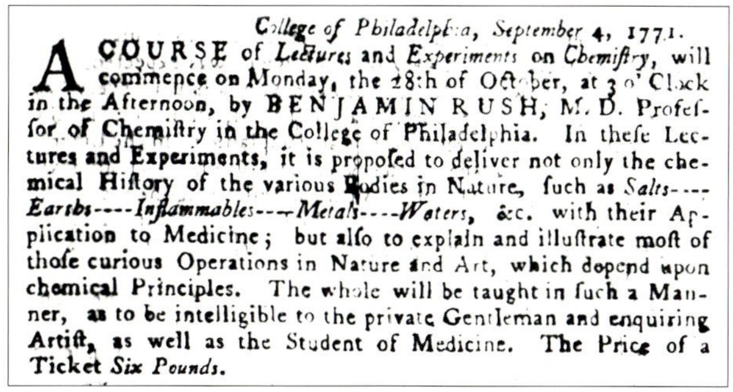

College of Philadelphia, September 4, 1771.
A COURSE of *Lectures* and *Experiments* on *Chemiſtry*, will commence on Monday, the 28th of October, at 3 o'Clock in the Afternoon, by BENJAMIN RUSH, M.D. Profeſſor of Chemiſtry in the College of Philadelphia. In theſe Lectures and Experiments, it is propoſed to deliver not only the chemical Hiſtory of the various Bodies in Nature, ſuch as *Salts*----*Earths*----*Inflammables*----*Metals*----*Waters*, &c. with their Application to Medicine; but alſo to explain and illuſtrate moſt of thoſe curious Operations in Nature and Art, which depend upon chemical Principles. The whole will be taught in ſuch a Manner, as to be intelligible to the private Gentleman and enquiring Artiſt, as well as the Student of Medicine. The Price of a Ticket *Six Pounds*.

clay Creek in the State of Delaware, a few miles from the City of Wilmington."[5] And Thomas Gilpin Jr. wrote, "[T]he clay . . . was obtained from the banks of the Delaware between Newcastle and Wilmington and it was mixed with calcined bones; the whole was under the direction of a Swiss or French Artist and a number of beautiful articles particularly of tea ware well shaped & painted were in use among the best families in America. . . ."[6]

That the source of the clay was local echoes information published earlier by Benjamin Rush (figs. 2, 3). Ardent patriot and scientist, Rush had long been promoting the idea of American manufacturing and had encouraged experimentation even while studying medicine in Edinburgh. He wrote the Bradfords, publishers of the *Pennsylvania Journal and The Weekly Advertiser* in June 1768: "From late intelligence I have had from America, I am now fully convinced of the possibility of setting up a china manufactory in Philadelphia." Within a year, in a letter of January 1769 published anonymously on April 6, 1769, he wrote: "We have all the materials for making china or porcelain ware in our province and in New-Castle county . . . ."[7]

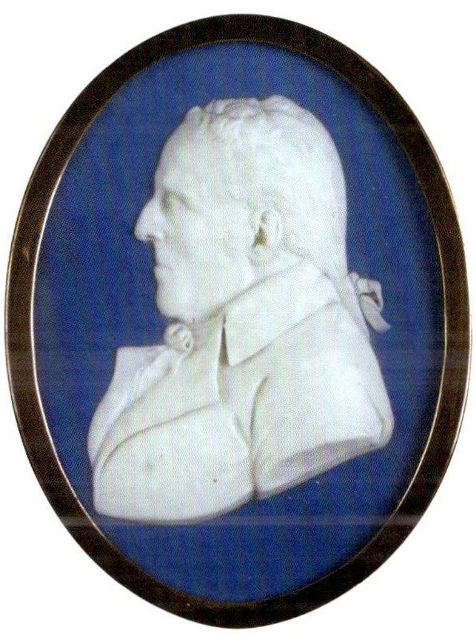

Figure 4 William Theed, R.A. (1764–1817), Wedgwood portrait medallion of Thomas Byerley, ca. 1800. (Courtesy, Wedgwood Museum.) When Byerley was named a partner in 1790, the firm became Wedgwood, Sons & Byerley.

If Dr. Rush was aware of experiments already conducted with the very clays used by Bonnin & Morris, he also knew who had conducted them. He returned to Philadelphia in the fall of 1769, and his burning interest in manufactures must have brought him into contact with Bonnin. It may well have been Rush's personal influence and commitment to porcelain that persuaded Bonnin to abandon crucibles. Rush would have introduced his own, invisible artisan (possibly the "Swiss or French Artist" of Gilpin's recollection) to the new partnership, who became the indispensable "Master of Pastes."[8]

Thomas Gilpin Jr. provided another important bit of information: "[I]t is a curious fact that the acting partner of Mr. Wedgewood of Etreria many years afterwards told the writer of this article that he was first employed in this infant establishment at Philadelphia which was more of French than English origin."[9] Thomas Byerley, Josiah Wedgwood's nephew, is easily identified from Gilpin's description (fig. 4). In later life Byerley took charge of the London salesrooms after the death of Thomas Bentley in 1780, and in 1790 became a partner in the firm Wedgwood, Sons & Byerley—and he was in Philadelphia in those early years. Byerley had worked for his uncle when younger, but was afflicted with wanderlust. He had wanted to be an actor and tried the Dublin stage, then an Irish company of strolling players, but gave it up and went to London. Finally, he was shipped off to Philadelphia

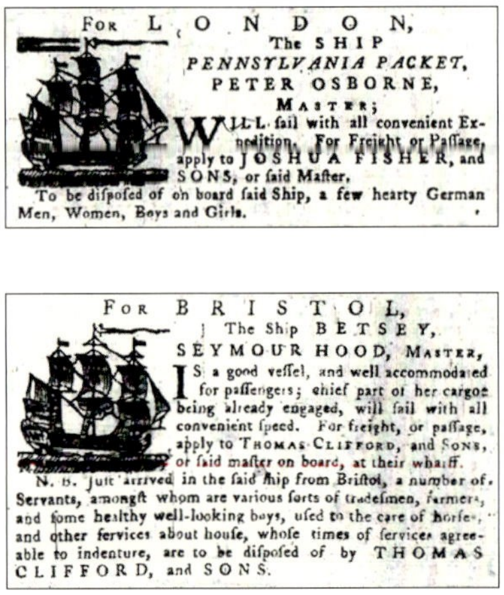

For LONDON,
The SHIP
PENNSYLVANIA PACKET,
PETER OSBORNE,
MASTER;
WILL sail with all convenient Expedition. For Freight or Passage, apply to JOSHUA FISHER, and SONS, or said Master.
To be disposed of on board said Ship, a few hearty German Men, Women, Boys and Girls.

For BRISTOL,
The Ship BETSEY,
SEYMOUR HOOD, MASTER,
IS a good vessel, and well accommodated for passengers; chief part of her cargo being already engaged, will sail with all convenient speed. For freight, or passage, apply to THOMAS CLIFFORD, and SONS, or said master on board, at their wharff.
N. B. Just arrived in the said ship from Bristol, a number of Servants, amongst whom are various sorts of tradesmen, farmers, and some healthy well-looking boys, used to the care of horses, and other services about house, whose times of services agreeable to indenture, are to be disposed of by THOMAS CLIFFORD, and SONS.

Figure 5 Two advertisements from the *Pennsylvania Gazette*, May 12, 1773. (© NewsBank and American Antiquarian Society.) Indentured servants arrived on every ship. Immigrants who had mortgaged their future labors to ship captains or brokers for the price of their passage to America were a potential source of labor at the factory for the mere price of their contracts. Captain Osborne's notice reads, "To be disposed of on board said Ship, a few hearty German Men, Women, Boys and Girls." The ship *Betsey* offered "various sorts of tradesmen, farmers, and some healthy well-looking boys . . . whose times of services agreeable to indenture, are to be disposed of . . . ."

during the summer of 1768 to sow his wild oats far from home.[10] Whatever the length of his employment by Bonnin & Morris, this episode of his stay in America has been missed by Wedgwood scholars.

What role might Byerley have played in the infant establishment? He was no potter but had, after all, grown up at the knee of the man universally believed to be the architect of the factory system that brought vast improvements to The Potteries.[11] Bonnin and Morris would have needed little persuasion to believe that Byerley knew how to build the plant, which, properly designed, would facilitate the flow of work from one specialist to the next in the production line. They might also have asked Byerley to take charge of hiring the workers, but where to find them? There was no lack of indentured immigrants, available for the price of their contracts (fig. 5), although they needed training in the trade, as Bonnin would discover; and when they ran away, he would have to advertise far and wide, as he did on October 18, 1770, and offer a reward for their capture or forfeit his investment.[12]

Another pottery had been established in Cain Hoy, about nine miles from Charleston, South Carolina, by John Bartlam, an insolvent master potter who had left Staffordshire in debt. He had been in America since at least 1763 and had sent agents to England to seduce workers to America with promises of higher pay.[13] Wedgwood found this intolerable. Not only was Bartlam's English-looking earthenware a threat to all of Staffordshire if he succeeded, but as has been proven recently, he was also producing porcelain and was thus a threat to the Philadelphia china factory as well (fig. 6).[14] Bartlam planned a move to Charleston and early in 1770 returned to Staffordshire looking for workers.

Byerley was keenly aware of Bartlam's presence and of his uncle's antipathy toward him. If hiring workers was Byerley's job, one can only imagine his glee when he realized he might kill two birds with one advertisement: he could decimate Bartlam's work force while staffing the new Philadelphia company with skilled workers. Accordingly, a request for workers was run in the *South-Carolina Gazette* on February 20, 1770. An agent in Charleston would arrange their travel, but "[n]one will be employed who have not served their apprenticeship in England, France, or Germany."[15]

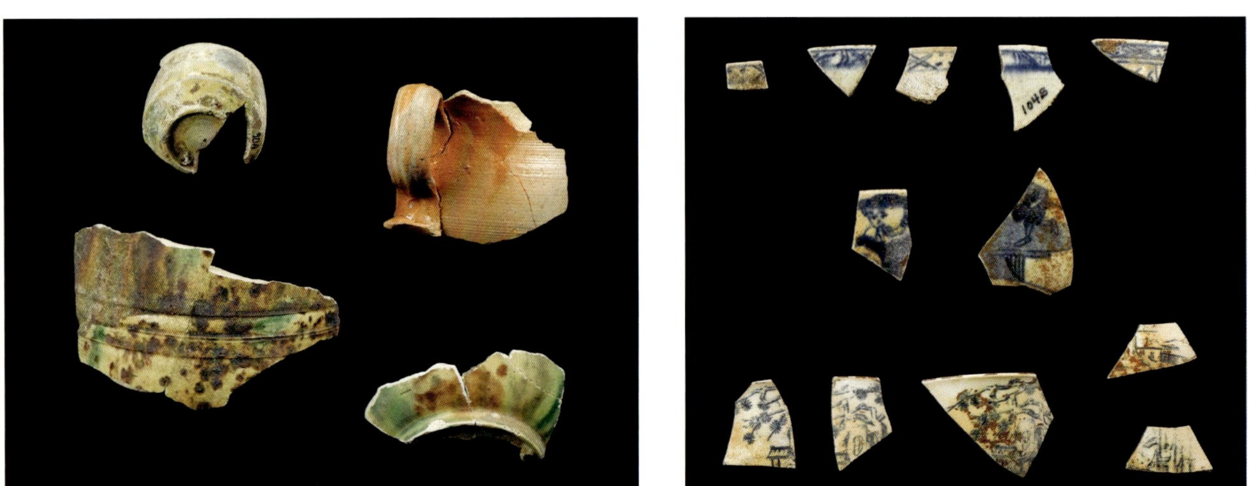

Figure 6 Waster fragments, John Bartlam, Cain Hoy, South Carolina, 1765–1770. *Left*: Bisque earthenware; *right*: soft-paste porcelain. (Courtesy, South Carolina Institute of Archaeology and Anthropology; photo, Gavin Ashworth.)

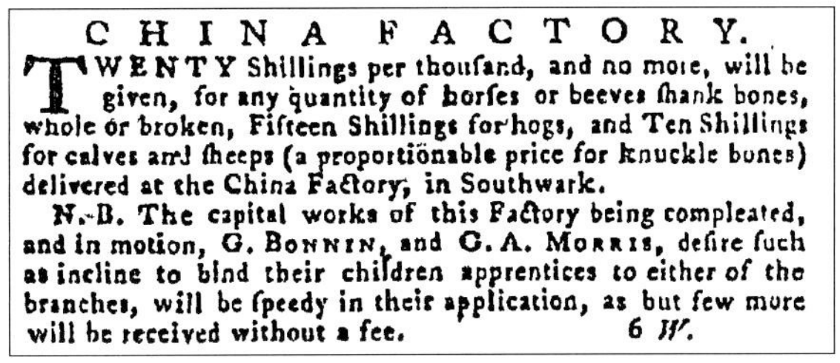

CHINA FACTORY.

TWENTY Shillings per thousand, and no more, will be given, for any quantity of horses or beeves shank bones, whole or broken, Fifteen Shillings for hogs, and Ten Shillings for calves and sheeps (a proportionable price for knuckle bones) delivered at the China Factory, in Southwark.

N. B. The capital works of this Factory being compleated, and in motion, G. BONNIN, and G. A. MORRIS, desire such as incline to bind their children apprentices to either of the branches, will be speedy in their application, as but few more will be received without a fee. 6 W.

Figure 7 Advertisement in the *Pennsylvania Gazette*, July 26, 1770. (© NewsBank and American Antiquarian Society.) Note that bones and apprentices were wanted: "any quantity of horses or beeves shank bones . . . calves and sheeps . . . knuckle bones . . . ."

How successful this advertisement was in luring workers to Philadelphia is not known, but by the end of July the Philadelphia factory was completed and "in motion" with at least some workers in place (fig. 7). Someone on staff was capable of teaching the apprentices, and someone was able to evaluate the bones piling up in the yard.[16] Bone ash was an important component of English porcelain, comprising 40 percent or more of the ingredients. The prices offered in this particular ad were scaled according to usefulness, as some animals had too much iron in their bones. This arcane knowledge is further evidence that a "paste man" was at work.

Apprentices were a source of cheap labor, and certainly America needed to train its own pool of skilled artisans, but who was there to teach them? If already in operation, as claimed, the factory must have had workers in place by the end of July. Seemingly undeterred, Bartlam & Co. made its move to Charleston and announced on October 4, 1770, that the "proper hands" had arrived from England and they were about to manufacture earthenware and china. One week later, they were open and ready to do business.[17]

Meanwhile, the head painter—"whose Name is Fry"—and "about a dozen prime hands" from the Bow China Factory were already on the high seas, on their way to work in the Philadelphia factory. This fact was unknown until the recent discovery of a tantalizing bit in a news column from London dated September 24. The six-line paragraph announcing their departure from Bow went unnoticed by the indefatigable compiler Alfred Coxe Prime, by the Philadelphia press, and by more than a century of ceramic history (fig. 8). The column, which was found among digitized images of historic newspapers, had been republished that December and in early January 1771 in New York City and New England but not in Philadelphia. The workers must have been contracted at the very outset of the enterprise, hence the proprietors' audacious comparison of their as-yet-unmade porcelain to Bow. Their arrival in October would have been planned for when the factory was completed and ready for them to go to work.[18]

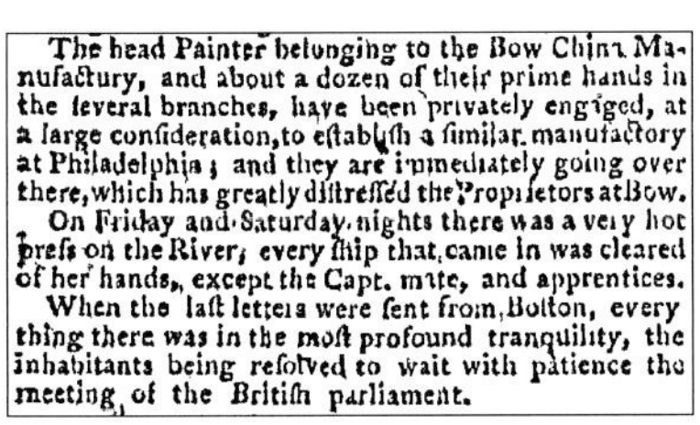

The head Painter belonging to the Bow China Manufactury, and about a dozen of their prime hands in the several branches, have been privately engaged, at a large consideration, to establish a similar manufactory at Philadelphia; and they are immediately going over there, which has greatly distressed the Proprietors at Bow.

On Friday and Saturday nights there was a very hot press on the River; every ship that came in was cleared of her hands, except the Capt. mate, and apprentices.

When the last letters were sent from Boston, every thing there was in the most profound tranquility, the inhabitants being resolved to wait with patience the meeting of the British parliament.

Figure 8 Extract from column originally published in London, September 24, 1770, and republished in the following: *New-York Journal or General Advertiser*, December 6, 1770; *Essex Gazette* of Salem, Massachusetts, December 11–18, 1770; *Providence Gazette*, January 12–19, 1771; and *Boston Evening Post*, January 15, 1771. (© NewsBank and American Antiquarian Society.) The extract informs readers that Bow Porcelain Manufactory workers have departed for Philadelphia, Pennsylvania.

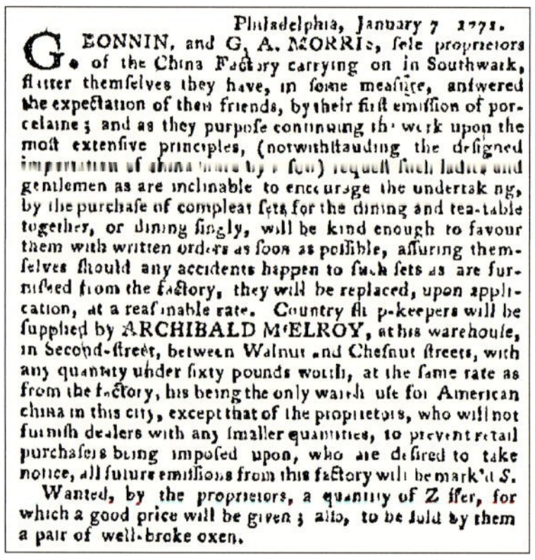

Philadelphia, January 7 1771.

G. BONNIN, and G. A. MORRIS, ſole proprietors of the China Factory carrying on in Southwark, flatter themſelves they have, in ſome meaſure, anſwered the expectation of their friends, by their firſt emiſſion of porcelaine; and as they purpoſe continuing the work upon the moſt extenſive principles, (notwithſtanding the deſigned im[illegible] by a few) requeſt ſuch ladies and gentlemen as are inclinable to encourage the undertaking, by the purchaſe of compleat ſets for the dining and tea-table together, or dining ſingly, will be kind enough to favour them with written orders as ſoon as poſſible, aſſuring themſelves ſhould any accidents happen to ſuch ſets as are furniſhed from the factory, they will be replaced, upon application, at a reaſonable rate. Country ſhop-keepers will be ſupplied by ARCHIBALD M'ELROY, at his warehouſe, in Second-ſtreet, between Walnut and Cheſnut ſtreets, with any quantity under ſixty pounds worth, at the ſame rate as from the factory, his being the only warehouſe for American china in this city, except that of the proprietors, who will not furniſh dealers with any ſmaller quantities, to prevent retail purchaſers being impoſed upon, who are deſired to take notice, all future emiſſions from this factory will be mark'd *S*.

Wanted, by the proprietors, a quantity of Zaffer, for which a good price will be given; alſo, to be ſold by them a pair of well-broke oxen.

Figure 9 Notice in the *Pennsylvania Chronicle*, December 31, 1770–January 7, 1771. (© NewsBank and American Antiquarian Society.) "Zaffer" was the label given one phase of refining cobalt ore, which, after roasting, was fused with sand into a glass. The resulting coloring agent could be used for underglaze painting. An *S* was the intended mark, as indicated in the advertisement, but underglaze *P* and *Z* have also been found.

The arrival in America of the "prime hands" or "capital workmen," now numbering nine, was not ignored by the Philadelphia press, but the name of the Bow factory from whence they came was omitted.[19] Sadly, the names of seven of the men might remain forever unknown because the movements of British citizens traveling within the colonies on a British ship were not recorded. Apprentice Thomas Frye, however, about sixteen years of age, is known because he ran away about October 10, 1771.[20] Frye was the nephew and namesake of one of the founders of the factory at Bow, the noted Irish portraitist and mezzotint engraver of the same name. And Thomas Gale is known because he was jailed. Graham Hood sent fragments and photographs from his excavations of the Bonnin & Morris site to Bernard Watney, the preeminent expert on English blue-and-white porcelain, for comparison. Watney recognized the hand of an artist and wrote, "[T]he painted insects and floral sprays on a pair of documentary baskets were almost certainly by the Bow painter who sometimes marked his work of about 1765 with a crescent. . . ."[21] Nearly every writer about Bonnin & Morris porcelain has noted its similarity to Bow; here, at last, is proof of the connection.

On April 12, 1770, in response to the boycott of English goods in major ports along the Eastern seaboard, the British Parliament had "graciously" repealed the duties on glass, paper, and painters' colors imposed by the Acts of 1767, which had so rankled the colonies, yet it retained one tax on tea. Thomas Gilpin Sr., in a series of letters to Benjamin Franklin, predicted that this mean-spirited action would drive Americans into frugality and manufacturing their own goods. On June 8, 1770, he wrote that repeal was not to take effect until December, and on July 19 that "New York has relaxed from the non-importation agreement. . . ." On September 28 he wrote, "[A] relaxation has also taken place here commencing on the 20th instant . . ." and "orders for goods have been sent . . . with the exception of tea and the vessels are not to sail from London before the 15th of January [1771]. . . ."[22] Thus, imported goods had not yet arrived. Finally, on December 31, 1770, announcement of the first emission from the kilns of Bonnin and Morris was published (fig. 9), exactly one year after their first declaration of intent to make porcelain, and a mere two months after the arrival of the workers from Bow—fast work, or perhaps the porcelain had already been biscuit-fired and was waiting to be decorated and glazed. The merchant Archibald McElroy reserved a large portion of the output and took a warehouse in which to sell it. For a time there was a certain honeymoon effect, and it was impossible to keep up with the demand.

One notable customer who wanted some Philadelphia porcelain was Sir William Johnson (1715–1774), a hero of the French and Indian Wars (fig. 10). Born in Ireland, he was twenty-three years old when an uncle (later to become Admiral Sir Peter Warren) invited him to manage and settle tenant-farming families on thousands of wilderness acres in western New York State that he had acquired through marriage to an heiress of the New York City De Lancey family. Six feet tall and charming, Johnson learned to maneuver among the worlds of German settlers, Albany Dutch and Irish farmers, British soldiers, and suspicious Indian tribes. He learned the native languages and dealt with the tribes so fairly that they eventually traded only with him; he was even inducted as a blood brother into the Wolf Clan of the Mohawks, an Iroquois tribe.[23]

When the French and Indian Wars reached New York State, Johnson enlisted the Iroquois nations in the British cause and raised his own army, discovering in himself a latent talent for battle and for building. He named Lake George, reinforced Fort Edward, built Fort William Henry, and made it possible for the British to rout the French from Canada. King George rewarded Johnson with knighthood, £5,000, and a commission as "Sole Superintendant of Indians and their Affairs in North America." Johnson

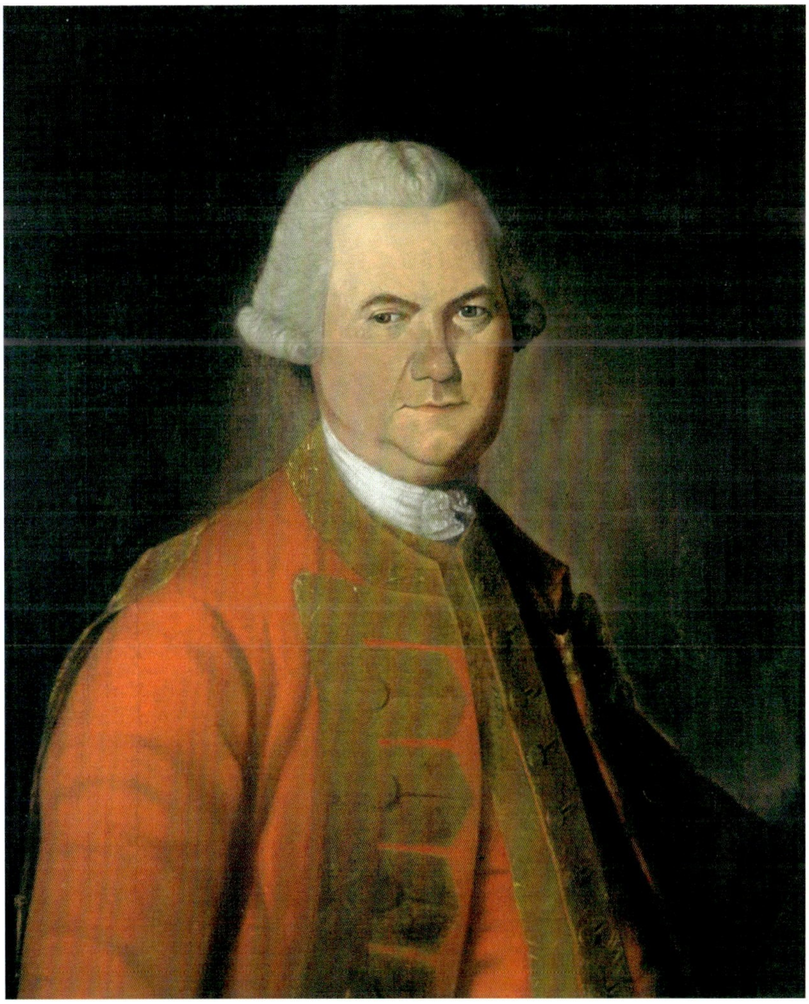

Figure 10 Attributed to Matthew Pratt (American; 1734–1805), *Portrait of Sir William Johnson* (1715–1774), ca. 1772. Oil on canvas. 36½ x 31½". (Courtesy, Johnson Hall State Historic Site, Johnstown, New York, New York State Office of Parks, Recreation & Historic Preservation.)

acquired vast lands north of the Mohawk River and founded Johnstown at the center, where he built for himself a fine Georgian mansion that he named Johnson Hall (figs. 11, 12). His furnishings included all the amenities of an Irish country gentleman, as well as samplings of American manufacture and crafts.[24] Johnson's voluminous correspondence, accounts, and papers fill fourteen volumes, much of which has been digitized. Philadelphia merchant Carpenter Wharton wrote to Johnson on October 14, 1771, apologizing for his delay in obtaining a complete breakfast set. He should have written long since but had been

> frequently disappointed by the Managers of the China Factory in the receiving a breakfast Set . . . (Which I beg leave to Present You for the use of the Hall,) as I flatter my Self it is the very best of the Kind they have yet exhibited. I have applied at Several Stores for a Set of Decanters and Glasses [Stiegel glass] . . . as I am Senceable it would afford you a Peculiar Satisfaction in observing the progress made in the China and Glass Manufactories, the demand for them is So Great, that the proprietors of the Manufactories are not able to supply the orders from the different Colonies. . . .[25]

Indians were always welcome in Johnson's homes, but by 1774 it was obvious to them that war injuries and lingering illnesses were taking a toll on their friend. Some six hundred Indians of many nations met on his lawn, fearing a future without Johnson as mediator. In the act of delivering a speech to them, he collapsed. Carried to his bed, he pulled up his beaver skin blankets and died. Two years later America was at war with the British. Johnson's sons remained loyal to the Crown, and on the night of May 21, 1780, they burned, massacred, and laid waste their own tenants and neighbors, and fled to Canada.

Johnson Hall was stripped of its furnishings in confiscation sales that took place about 1787; surprisingly, however, many remained in the vicinity, handed down in families with reverence for their history, and some have

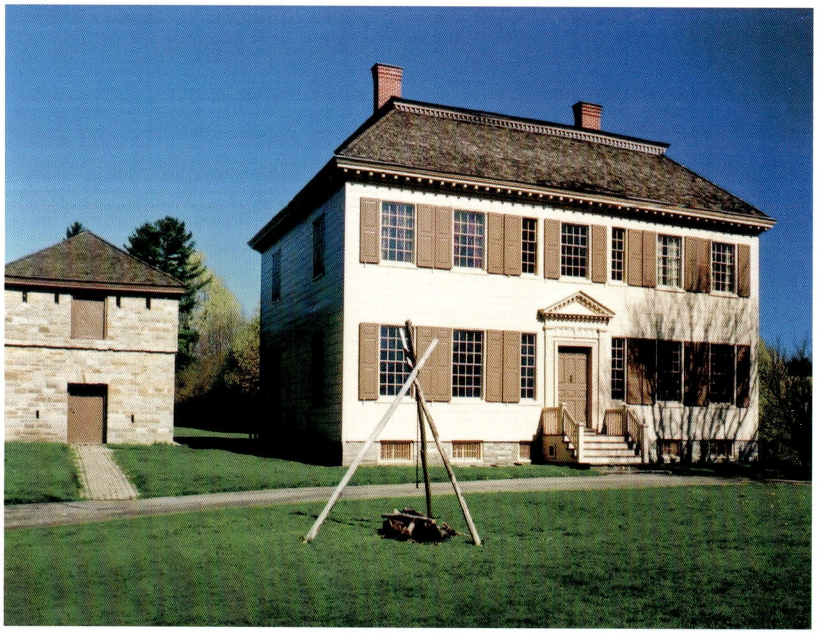

Figure 11 Johnson Hall, Johnstown, New York. (Photo courtesy Anne Cassidy, Bureau of Historic Sites, Peebles Island, New York.) A Philadelphia-china breakfast set once graced Johnson Hall, the final residence of Sir William Johnson. The Georgian mansion was conceived by Johnson and builder Samuel Fuller. Constructed of wood boards made to look like stone, it was finished in 1763. It is now designated a New York State Historic site and open to the public.

been restored to Johnson Hall. Nearly thirty miles south of Johnson Hall, a porcelain fruit basket marked with a "P" on the bottom was found by a local antiques dealer (figs. 13–15).

Once luxury goods arrived in quantity from abroad, Bonnin and Morris were forced to lower their prices, and subsequent ads used language such as "will engage to deliver China at as low a rate as any imported in the last ships from England." Inevitably, quality suffered and complaints about the ware began to be heard. Tinkering with the formula becomes evident in their advertisements, along with announcements of great strides, as on April 18, 1771, for example: "The Proprietors of the China Factory have great pleasure in acquainting the Public, that notwithstanding their declining the reception of any more bones, they are determined to carry on the works upon the most extensive principles . . . ."[26]

Had Bonnin and Morris quit making bone china? It is not likely they had too large an inventory. As bone ash comprised 40 percent of the mix of bone china, any stockpile would have dwindled quickly.[27]

There were changes to the factory's porcelain body noted on January 2, 1772, and January 13, 1772. "FLINT GLASS" was wanted, "all Sorts of broken Flint Glass [and] clean and clear WHITE FLINT STONE . . ." (fig. 16).[28] On January 20, 1772, they asked again for broken flint glass and solicited orders for sets of dressing boxes decorated in blue and enamel.

Figure 12 John Mitchell (1711–1768), detail from *A Map of the British and French Dominions in North America, with the Roads, Distances, Limits, and Extent of the Settlements, Humbly Inscribed to the Right Honourable The Earl of Halifax*, London, 1755. Line engraving with period hand color. (Courtesy, Colonial Williamsburg Foundation, Museum Purchase, 1955-407.)

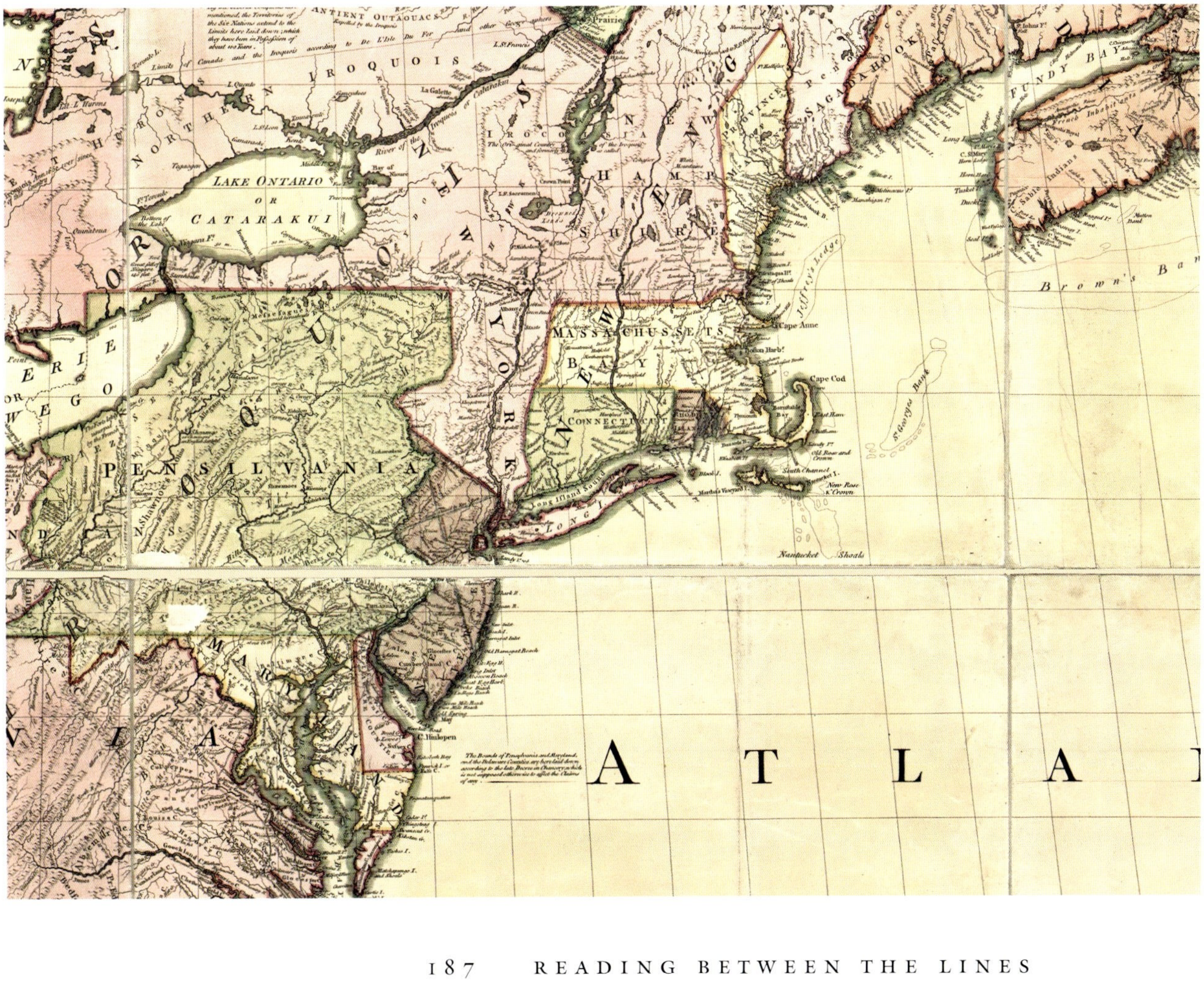

Figure 13 Openwork fruit basket, American China Manufactory (Bonnin and Morris), Philadelphia, Pennsylvania, 1770–1772. Soft-paste porcelain. D. 5⅞". (Collection of Eric Noah; on loan to The Metropolitan Museum of Art; photo, Gavin Ashworth.)

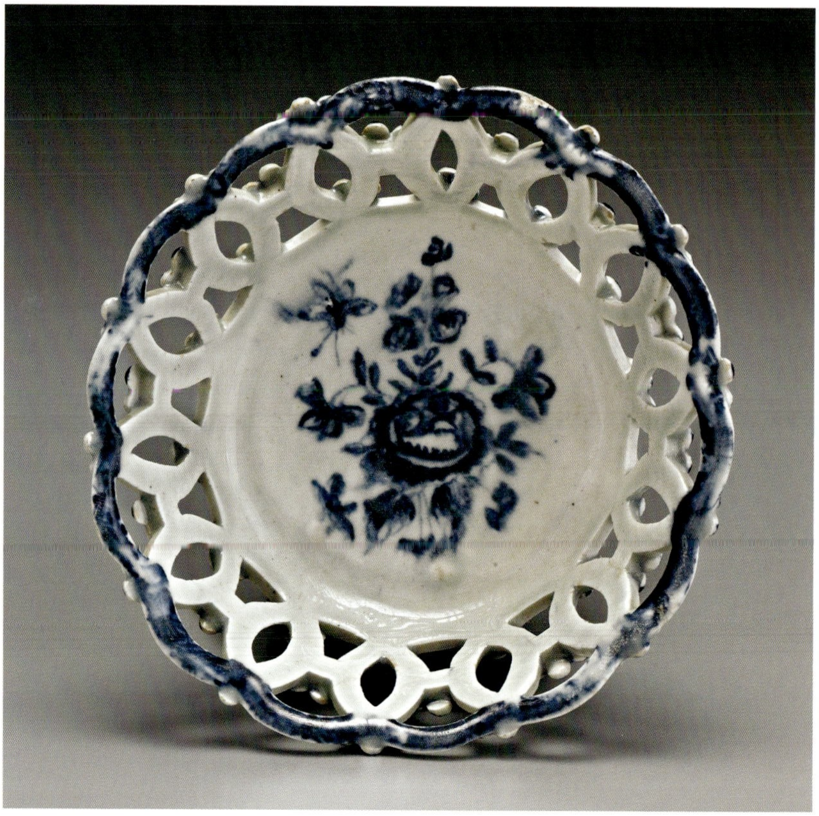

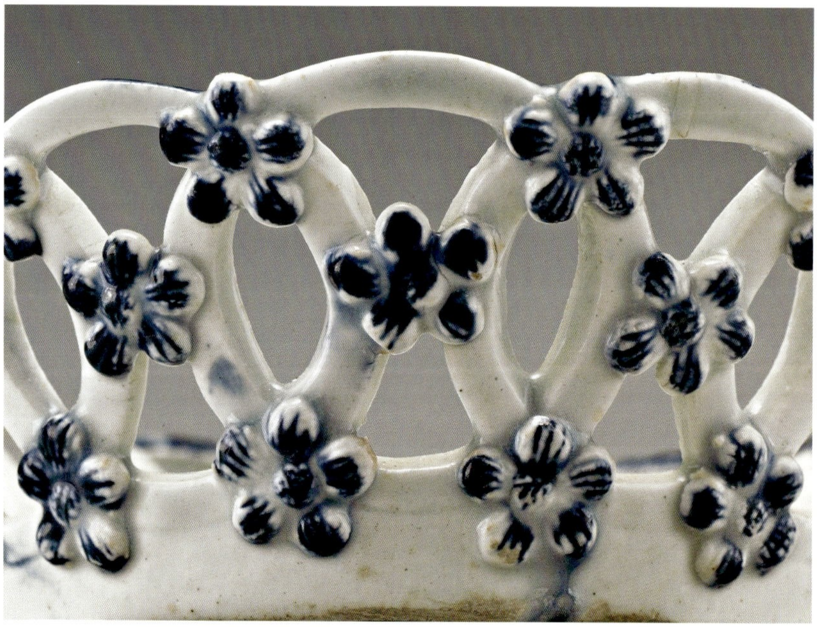

Figure 14 Detail of the mark on the underside of the basket illustrated in fig. 13.

Figure 15 Detail of the open fretwork on the basket illustrated in fig. 13.

"Flint glass" was a misnomer that everyone understood to mean "lead glass," as it was made with a high proportion of lead oxide in its mix. "Flint stone" was quartz, a natural mineral and an almost pure form of silica that when calcined (roasted), ground to a fine powder, and added to the body formula, would impart strength and solidity to it—making it less fragile and less susceptible to thermal shock, though also more opaque.

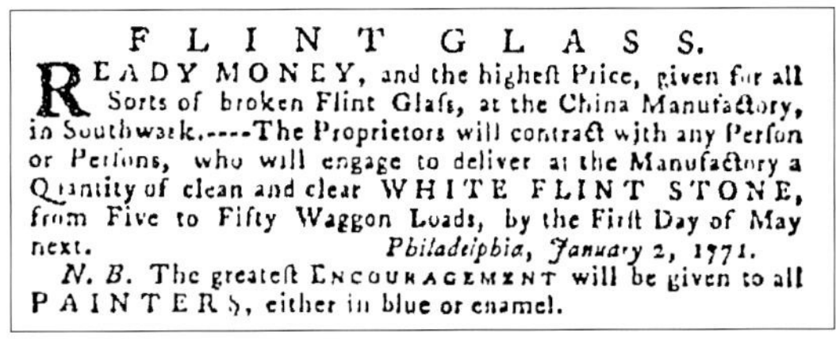

FLINT GLASS.

READY MONEY, and the higheſt Price, given for all Sorts of broken Flint Glaſs, at the China Manufactory, in Southwark.----The Proprietors will contract with any Perſon or Perſons, who will engage to deliver at the Manufactory a Quantity of clean and clear WHITE FLINT STONE, from Five to Fifty Waggon Loads, by the Firſt Day of May next. *Philadelphia, January 2, 1771.*

N. B. The greateſt ENCOURAGEMENT will be given to all PAINTERS, either in blue or enamel.

Figure 16 Notice in the *Pennsylvania Gazette*, January 2, 1772. (© NewsBank and American Antiquarian Society.)

The American China Manufactory, driven by the need to cut costs, seems to have followed the same path as Bow, ultimately leading to the deterioration of its porcelain. Ceramics historians Adams and Redstone observed a loss of quality to Bow porcelain beginning in the 1750s, when body translucency diminished and fragility increased, and suggested that underfiring might have been the cause, either as the result of a shorter firing time or firing at lower temperatures. Chemical analysis of a plate made 1759–1762 determined that the proportion of calcium was increased in relation to phosphorous, due probably to an increase of silica (from sand or flint) and a reduction of clay and bone ash in the body mix.[29] Watney dated the last stage of deterioration to 1764–1776, when Bow porcelain became so opaque, gray, lacking in brilliance, and underfired as to resemble earthenware: "In this debased state the production of Bow porcelain came to an end after over thirty years of continuous endeavor to make useful and ornamental china which would be less expensive and more acceptable than foreign wares."[30]

Those words might apply to Bonnin & Morris porcelain as well. Bonnin and Morris introduced glass to the mix, probably hoping that the transparency of glass would counter the opacity of flint.[31] On April 16, 1772, a notice in the *Pennsylvania Journal* informed the public that a china decorator was on hand to teach apprentices, and for the week of April 13–20 the *Pennsylvania Chronicle* carried the notice as well.[32]

Within weeks Bonnin and Morris appeared ready to give up their New Castle clays in favor of clay presented to them by "a Gentleman of Charles-Town, South-Carolina." They said experiments proved it could withstand great heat, and claimed the product was superior to the china imported by the East India Company (fig. 17).

Bonnin and Morris had, in fact, received that clay at least six months earlier from that "Gentleman of Charles-Town," one Henry Laurens, a man who had a continuing interest in porcelain production and contributed £50 to the fund allotted to Bartlam's pottery.[33] On November 28, 1771, Laurens wrote to his friend William Williamson, belatedly informing him that he had delivered a batch of his own clay (its source unidentified), along with a keg of Cherokee clay belonging to Williamson,

> into the hands of Mr. George Morris, the Manager or one of the principal Managers of the China Manufactory Philadelphia under these different directions[:] to make an Essay on each Material, and to inform me of the Quality respectively. My own Clay you are welcome to, tho' you may if you please send me a dozen Cups and Saucers raised from it. But that Keg WW is the Property of my Friend William Williamson, Esquire in Charles Town, So. Carolina . . . . To which Mr. Morris Promised to pay due Attention, and before I came from Philadelphia, he acquainted me of a Trial

which he had made of each Clay, by which he had proved that they were both very fine, that of mine somewhat finest. . . . Mr. Morris is a genteel Young Man and will give Satisfaction in your Commands.

PHILADELPHIA, *July 29.*

From Charleſtown, in Maryland, we are informed, that on Saturday, the 18th inſt. a moſt inhuman murder was committed within a few miles of that place, on the body of Mr. William Simpſon, a gentleman from Georgia, on his travels through North-America for his health. He was ſuppoſed to have had a conſiderable quantity of money about him, which induced the villain or villains to perpetrate this horrid crime. The body is not yet found, or his bags and papers; but the horſe on which he rode was taken, with the ſaddle all bedawbed with blood.—Two men have been ſince apprehended and committed to Charleſtown goal, on ſuſpicion of the above murder.

A few weeks ago, about twenty Indians, in paſſing through Ann Arundel to Annapolis, in Maryland, in order to complain to the Governor of the incroachments of the back ſettlers on their territories, took up their lodgings in a gentleman's barn, when the Chriſtian inhabitants aroſe in the night time, and maſſacred every one of them.—*Tell it not in Gath, neither publiſh it in the ſtreets of Aſkelon!*

Auguſt 3. We have the pleaſure to acquaint our readers, that the proprietors of the china manufactory in this city, have lately made experiments with ſome clay preſented to them by a gentleman of Charles-Town, South-Carolina, which produces china ſuperior to any brought from the Eaſt-Indies, and will ſtand the heat beyond any kind of crucibles ever yet made.

Figure 17 Advertisement in the *Providence Gazette*, August 15, 1772 (first published in July 29). (© NewsBank and American Antiquarian Society.)

Laurens does not say whether he ever received his cups, but Bonnin and Morris must have thought long and hard before informing the public about this clay, because it was virtually an admission that their present china could not withstand high temperatures. Every housewife understood that china seldom survived the sudden shock of boiling water poured into cups in mid-winter. Indeed, of the nineteen extant objects produced by the Bonnin & Morris manufactory, there are no tea or coffee pots and no cups. Only the two surviving sauceboats might have been subjected to any kind of heat. Of the dinner set owned by Dr. Mease's father (all that he got for his £500 loan to Bonnin), Mease wrote that the quality of it was about equal to the Delftware of Holland, and, by 1841, only the one broken basket remained, which he donated to the Franklin Institute.

Whether the partners ever produced tableware made of the Charleston clay is unknown, but on August 15, 1772, they advertised their "BLUE and WHITE WARE, both useful and ornamental," and asked again for broken glass, although flint glass was no longer specified. It seems that now any kind of glass would do. "Pot and Pearl Ashes" were also wanted (fig. 18).

The night of November 2, 1772, was one of merriment in Southwark. An advertisement in the *Pennsylvania Gazette* that was published two days later, on November 4, revealed that Tom Byerley had returned to Philadelphia and appeared in two opera performances at the theater presented by The American Company; among the cast members were a Mr. Morris and a Mrs. Morris. In the adjacent column, however, was an ominous paragraph that spelled the end of the Bonnin & Morris china manufactory. It would seem the workers had taken to the streets, and were claiming to have received no pay for months. The proprietors took to the press to protest: "WHEREAS sundry persons are soliciting the charitable contributions of the inhabitants of the city . . . and whereas the Proprietors, think it as well their duty to forewarn the Public of such impositions, as to vindicate their own characters, they request the general opinion may be suspended . . . ." (fig. 19).

Despite hinting that the proprietors would issue a statement giving their side of the story, none had appeared by November 11, when a scathing letter addressed anonymously to the *Pennsylvania Journal* appeared. An excoriating exposé, it effectively damned the proprietors. All the details of salary and terms of employment were laid bare, including a reference to "the failure of

Figure 18 Notice in the *Pennsylvania Gazette*, September 26, 1771. (© NewsBank and American Antiquarian Society.) Several other Pennsylvania papers and the *Providence Gazette* also advertised the American China Manufactory and its "new assortment of blue and white, useful and ornamental ware . . . ."

Philadelphia, September 26, 1771.

A FRESH aſſortment of blue and white AMERICAN CHINA is this morning opening, by BONNIN and MORRIS, proprietors of the manufactory; at their warehouſe, the eaſt ſide of Second-ſtreet, a few doors below Market-ſtreet; who will furniſh ware at a much cheaper rate, both wholeſale and retail, than any imported in the laſt ſhips from England. As they are determined to uſe their utmoſt endeavours to give ſatisfaction to all kinds of cuſtomers, they requeſt no perſons will be influenced by the malevolent attempts of thoſe who think it their intereſt to depreciate the quality of American China.

N. B. There will ſhortly be an emiſſion from this manufactory, an aſſortment of both uſeful and ornamental ENAMELLED CHINA.

2 W.

Figure 19 Notice in the *Pennsylvania Gazette*, November 4, 1772. (© Accessible Archives.) The announcement, which is dated November 2, 1772, indicates that the American China Manufactory is coming to an end.

To the PUBLIC.

WHEREAS ſundry perſons are ſoliciting the charitable contributions of the inhabitants of the city of Philadelphia, under pretence of labouring under great diſtreſſes, from the conduct and behaviour of the Proprietors of the China Houſe towards them; and whereas the Proprietors, think it as well their duty to forewarn the Public of ſuch impoſitions, as to vindicate their own characters, they requeſt the general opinion may be ſuſpended, until a plain and impartial ſtate of the caſe, as far as relates to the Workmen coming over from England, and their treatment here, is made out, and publiſhed, which ſhall be done with all poſſible diſpatch. *Southwark, Nov. 2, 1772.*

the late China Manufactory of this city." Seemingly premature, this was information that only a confidante—or someone who was there from the beginning—could have known.[34] Years later, Josiah Wedgwood (1730–1795) identified the anonymous writer when, in 1783, the war with America finally over and afraid of losing his well-trained workers to overseas enticements, he addressed his company on the dangers of emigration. Deliberately scaring the daylights out of his employees, he told them the sad tale of the Philadelphia workers, "all of whom died." He said, "Mr. Byerley, a nephew of mine, being then upon the spot, published in the newspapers, a letter in behalf of the poor survivors . . . ." Wedgwood enlarged upon what he had heard from Byerley, adding "the proprietors, soon finding that they had no chance of succeeding, not only gave up the undertaking, but silenced the just complaints of the poor injured workmen, by clapping one of them (Thomas Gale) into a prison: the rest who had never received half the wages agreed for were left entirely to shift for themselves. Thus abandoned . . . without a penny . . . [they] were reduced to begging in the public streets for a morsel of bread."[35] Of course, Byerley had a flair for the dramatic and Wedgwood's speech was written some years after the fact. However, if this was a true account of events, the workers had not *walked* out on Bonnin but had been *locked* out by him.

Bonnin responded three days later, on November 16, 1772 (fig. 20). He was quitting and intended to leave with his family on one of the first ships in the spring of 1773. He had placed the entirety of the potworks for a sale scheduled for mid-December, and was making no attempt to pay his men. The property did not sell, however, and Bonnin was forced to stay longer. Another auction was scheduled for the evening of May 31, 1773. But a notice published in May 1773 mentions someone new on the scene—someone who, under a new owner, might help rescue the factory (fig. 21):

> Any Gentleman inclining to engage in the China Business, may now enter on very advantageous terms, as these works are completely fitted, and a young man of sobriety and integrity in town, from Germany, who is completely skilled in the whole process of compounding the materials, upon a plan fully equal to the best in England, and who would readily undertake the management, upon reasonable terms, either in partnership or otherwise.

A basket inscribed "PHILADELFIA / the 23 April 1773" on the bottom might have been made by the young German just days before the above notice was placed.[36]

The pottery failed to sell on May 31, however, and Bonnin finally set sail for England late in the year, abandoning all to the sheriff. The November

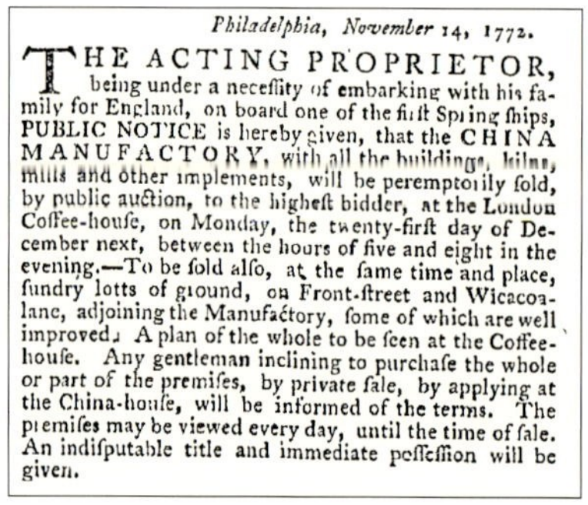

Philadelphia, November 14, 1772.

THE ACTING PROPRIETOR, being under a neceſſity of embarking with his family for England, on board one of the firſt Spring ſhips, PUBLIC NOTICE is hereby given, that the CHINA MANUFACTORY, with all the buildings, kilns, mills and other implements, will be peremptorily ſold, by public auction, to the higheſt bidder, at the London Coffee-houſe, on Monday, the twenty-firſt day of December next, between the hours of five and eight in the evening.—To be ſold alſo, at the ſame time and place, ſundry lotts of ground, on Front-ſtreet and Wicacoa-lane, adjoining the Manufactory, ſome of which are well improved. A plan of the whole to be ſeen at the Coffee-houſe. Any gentleman inclining to purchaſe the whole or part of the premiſes, by private ſale, by applying at the China-houſe, will be informed of the terms. The premiſes may be viewed every day, until the time of ſale. An indiſputable title and immediate poſſeſſion will be given.

Figure 20 Notice in the *Pennsylvania Packet and General Advertiser*, November 16, 1772. (© NewsBank and American Antiquarian Society.) The announcement is dated November 14, 1772.

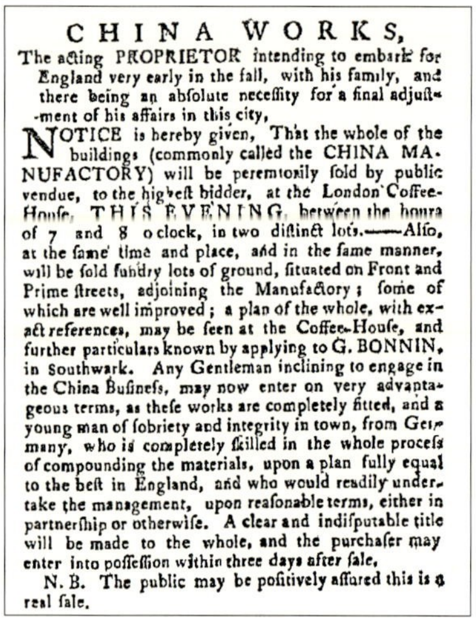

CHINA WORKS,

The acting PROPRIETOR intending to embark for England very early in the fall, with his family, and there being an abſolute neceſſity for a final adjuſtment of his affairs in this city,

NOTICE is hereby given, That the whole of the buildings (commonly called the CHINA MANUFACTORY) will be peremtorily ſold by public vendue, to the higheſt bidder, at the London Coffee-Houſe, THIS EVENING, between the hours of 7 and 8 o clock, in two diſtinct lots.——Alſo, at the ſame time and place, and in the ſame manner, will be ſold ſundry lots of ground, ſituated on Front and Prime ſtreets, adjoining the Manufactory; ſome of which are well improved; a plan of the whole, with exact references, may be ſeen at the Coffee-Houſe, and further particulars known by applying to G. BONNIN, in Southwark. Any Gentleman inclining to engage in the China Buſineſs, may now enter on very advantageous terms, as theſe works are completely fitted, and a young man of ſobriety and integrity in town, from Germany, who is completely ſkilled in the whole proceſs of compounding the materials, upon a plan fully equal to the beſt in England, and who would readily undertake the management, upon reaſonable terms, either in partnerſhip or otherwiſe. A clear and indiſputable title will be made to the whole, and the purchaſer may enter into poſſeſſion within three days after ſale,

N. B. The public may be poſitively aſſured this is a real ſale.

Figure 21 Advertisement from the *Pennsylvania Chronicle*, May 24–May 31, 1773. (© NewsBank and American Antiquarian Society.)

17, 1773, issue of the *Pennsylvania Gazette* carried news from North Carolina of the sudden death from bilious fever of George Anthony Morris on October 5, 1773, in the prime of life, having resided there for eighteen months and having made many friends.[37]

This revelation raises many questions. Had Morris moved away to avoid the inevitable embarrassment of his company's failure? Or had he been negotiating for the Cherokee clay, located in North Carolina? Eighteen months before Morris's death would date back to a time after he had received Laurens's clay but three months before it was advertised. It would indeed have been sensible to procure the clay before offering its product to the public, but the reason for his move remains a mystery.

When even the sheriff failed to sell the manufactory property, Joseph Morris purchased it in July 1774, divided it, and sold it piecemeal.[38] New owners continued to advertise it, each succeeding notice revealing more detail about the facatory as it was dissembled. From excerpts of advertisements from November 1772 to December 1774, a picture of the factory and its fittings emerges:

> a three-story frame building, the whole width of the lot, about 80 feet by about 16 feet in depth— . . . compleatly sashed, having about 700 panes of glass . . . . [T]here is a cart-way under the middle of the building, which communicates with the rear of the lot . . . . [T]here is a rolling press, for copper-plate printing . . . two iron stoves, with their pipes . . . . [T]here are 3 kilns, 2 furnaces, 2 mills, 2 clay vaults, a pair of large mill-stones, each stone weighing about two tons and an half . . . all the moulds, models, laths, and wheels . . . together with a great quantity of painted and plain unglazed China Ware . . . .

By these descriptions, the china factory bore close resemblance to Josiah Wedgwood's brand new Staffordshire pottery called "Etruria" (figs. 22, 23). Certainly, there was nothing like it in America, nor would its like be seen again for another fifty years. Byerley (surely it *must* have been Byerley) had transplanted to Philadelphia the very model of Etruria.

Were Byerley and Gousse Bonnin two-of-a-kind in dreaming grand schemes? Had Bonnin and Morris squandered their resources on a plant far too grandiose for two novice managers with no experience in an industry they knew nothing about? It brings to mind Wedgwood's exasperated letter to his brother, when the young Byerley was working for him in 1765. He would have no objection to paying his nephew handsomely except that Tom had not "the least notion of the value of money, nor would keep a farthing

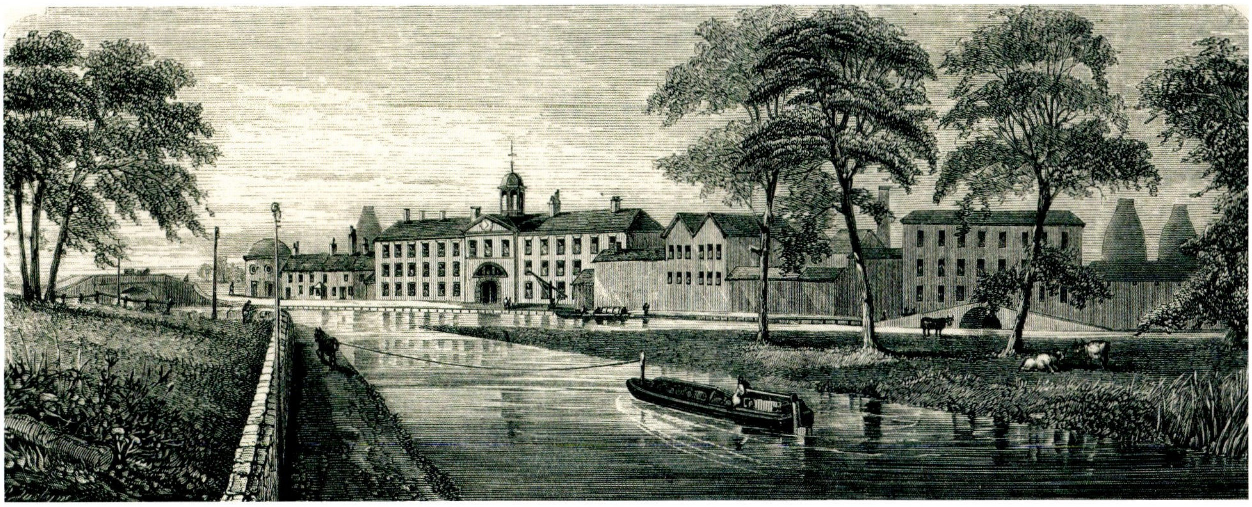

Figure 22 Wedgwood factory at Etruria, engraving from *The Life of Josiah Wedgwood*, 1866. (Courtesy, The Art Archive at Art Resource, N.Y.) Josiah Wedgwood's new factory, where the best of its ornamental wares were produced, was completed in 1769. From the descriptions (published in advertisements from November 1772 to December 1774) of the Philadelphia china factory as it stood, fully equipped, waiting to be sold, it bore an uncanny resemblance to Etruria. The Etruria factory was designed to enable clay brought from the canal to be fully processed, from each specialist potter to the next, step by step, until the finished product was packaged and ready to be loaded back onto the barges.

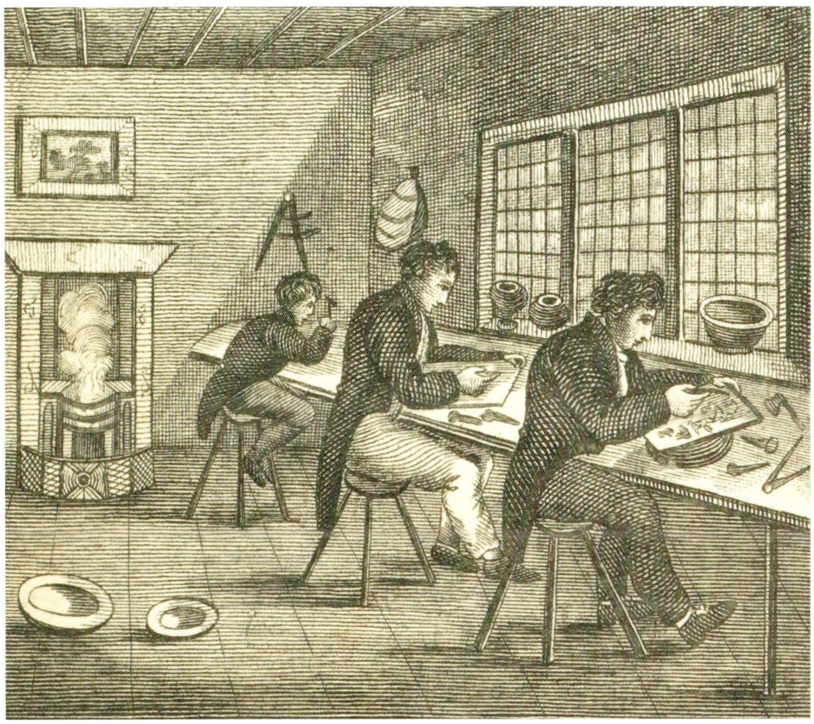

Figure 23 Plate 10 from *A Representation of the Manufacturing of Earthenware* by Ambrose Cuddon, London, 1827. (Courtesy, The Winterthur Library, Printed Book and Periodical Collection.)

any longer than an opportunity offer'd of parting with it."[39] Yet against all logic, this adventure into china manufacturing in Philadelphia very nearly succeeded.

The problems Bonnin and Morris encountered would be repeated in the future. Again and again, expert workmen would be brought over only to find, to their embarrassment, that minerals did not behave in America as

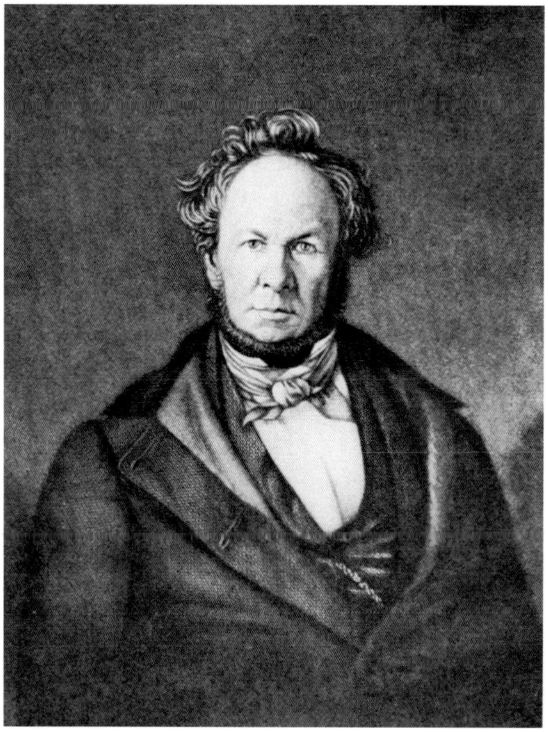

Figure 24 Portrait of Henry Rowe Schoolcraft (1793–1864), from the frontispiece of Schoolcraft, *Personal Memoirs of a Residence of Thirty Years with the Indian Tribes on the American Frontiers: With Brief Notices of Passing Events, Facts, and Opinions, A.D. 1812 to A.D. 1842* (1851), courtesy of and reproduced in *Antiques* (April 1939): 186–88.

they had in England. There was a growing conviction that native materials could never successfully yield English types of ware.

Henry Rowe Schoolcraft, himself a failed manufacturer of flint glass, proposed a mechanics' manual for Americans in 1817 (fig. 24):

> We are not deficient in the materials, but in the knowledge of them. . . . We must render ourselves better acquainted with the nature and properties of the substances that we employ, and habituate ourselves more to the conducting of experiments with scientific precision: We must become better chemists before we can become better manufacturers . . . .[40]

ACKNOWLEDGMENTS

This article began as a talk at the symposium "New Perspectives on Philadelphia's Role in the Production of Early American Porcelain," held at the Philadelphia Museum of Art on April 9, 2008. It was further developed as a paper delivered to the English Ceramic Circle at Bonham's, New Bond Street, London, on October 7, 2010. The published version first appeared in *Transactions of the English Ceramic Circle* 20, pt. 3 (2009 [printed April 2011]), and is updated here.

I am indebted to Joanna Pessa for locating the newspaper references and for restoring long-overdue credit to the workers from Bow; to Alexandra Kirtley and to British colleagues Pat Halfpenny, David Redstone, and Nick Panes, who goaded me into publishing what began as a talk; to Gale and Graham Hood for their encouragement; to Angela Snye and Jerrold Pepper of the Adirondack Museum, and to its former director Caroline Welsh for making its archives available to me; to my patient husband and mentor, J. Garrison Stradling, the best sleuth of all; most especially to the Kaufman Americana Foundation of Linda Kaufman and the late George Kaufman, whose constant support has made the publications and study of Bonnin & Morris possible; and to all the researchers and writers before me who have unearthed invaluable information—I have merely connected the dots.

1. Almost all press notices and advertisements placed by the proprietors of the Philadelphia china manufactory were captured first by Alfred Coxe Prime, comp., *The Arts & Crafts in Philadelphia, Maryland, and South Carolina, 1721–1785/Gleanings from Newspapers* (Topsfield, Mass.: Walpole Society, 1929), pp. 114–24. These were supplemented in Graham Hood, *Bonnin and Morris of Philadelphia: The First American Porcelain Factory, 1770–1772* (Chapel Hill: Published for the Institute of Early American History and Culture at Williamsburg, Va., by the University of North Carolina Press, 1972), a book reprinted in its entirety in *Ceramics in America*, edited by Robert Hunter (Hanover, N.H.: University Press of New England for the Chipstone Foundation, 2007), pp. 2–75. This issue, devoted to early porcelain in America, has added much valuable new information. The research of Joanna Pessa has yielded more contemporary information, much of it from Readex.com, a Division of NewsBank, and the American Antiquarian Society, an internet database service providing scans of America's historical newspapers.

2. These notices ran in the *Pennsylvania Chronicle and Universal Advertiser*, September 18–25 and October 5–12, 1769; the *New York Journal, or The General Advertiser*; and the *New-York Gazette and The Weekly Mercury*, October 5–November 20, 1769.

3. In "A New Insight into Staffordshire's 18th Century Ceramic Revolution," a lecture he delivered at the New York Ceramics Fair on January 19, 2002, David Barker indicated that the reorganization of the pottery industry had its beginnings as early as 1730. Further insights are included in his article "The Industrialisation of the Staffordshire Potteries." Barker was formerly Keeper of Archaeology at the City Museum and Art Gallery, Stoke-on-Trent. Barker, "The Industrialisation of the Staffordshire Potteries," in Barker and David Cranstone, eds., *The Archaeology of Industrialization*, The Society for Post-Medieval Archaeology Monograph 2 (Leeds, U.K.: Maney, 2004), pp. 203–21.

4. For a documented example of this relationship in America, see Arlene Palmer, "'To the Good of the Province and Country,' Henry William Stiegel and American Flint Glass," *American Craftsman and the European Tradition, 1620–1820*, edited by Francis J. Puig and Michael Conforti, exh. cat. (Minneapolis: Minneapolis Institute of Arts, 1989), p. 204. British examples are cited in Barker, "The Industrialisation of the Staffordshire Potteries." A German named Martin Greiner contracted in 1752 to "instruct and inform . . . his backers in . . . the art & Mistery of erecting & Building a Glass House and also in blowing and makeing of glass." He was to help construct the glasshouse and fabricate the ovens and pots needed to melt the glass batch. His voyage from Saxe-Weimar to New York City was paid, he was granted housing and land, and he was paid a wage based on his production. Yet, after six years of apparent success, production at the glasshouse in New York City came to a standstill early in 1763. Their so-called desertion was said to be the cause, as Greiner and the other "servants" were at work the next year at Henry William Stiegel's Manheim works.

5. Hood, *Bonnin and Morris of Philadelphia*, p. 27, in which the handwritten letter of 1841 from Dr. James Mease to the Franklin Institute is quoted in full.

6. "Memoir of Thomas Gilpin (Found among the papers of Thomas Gilpin, Jr.)," *Pennsylvania Magazine of History and Biography* 49, no. 4 (1925; repr., 1965, Johnson Reprint Corporation, N.Y.), pp. 309–10. Thomas Gilpin Jr. (1776–1853) had prepared a memoir of his father, Thomas Gilpin Sr. (1728–1778), from letters and papers, sometimes extracting and updating quotations, sometimes paraphrasing in his own voice. The mention of the Swiss/French superintendent occurs after a letter of September 28, 1770, but as it was not a quotation from Gilpin Sr., it is impossible to date the worker's tenure accurately. I thank Jack Gumbrecht, director of research services for the Historical Society of Pennsylvania, for this information.

7. Michael K. Brown, "Piecing Together the Past: Recent Research on the American China Manufactory, 1769–1772," *Proceedings of the American Philosophical Society* 133, no. 4 (1989): 557–59, reprinted in the 2007 issue of *Ceramics in America*. Brown cites as his source L. H. Butterfield, *Letters of Benjamin Rush*, 2 vols. (Princeton: Princeton University Press, 1951), 1 (1761–1792): 60–61, 74–75.

8. A diligent search by British scholar Nick Panes of the records of the Society of the Sons of St. George, a charitable organization for the relief of distressed Englishmen, has discovered

circumstantial evidence pointing to the peripatetic potter William Ball. It was recorded in the minutes of July 23, 1773, that a certain "Mary Ball, the Widow of a China Manufacturer lately deceased," had petitioned for a weekly stipend until passage to England for her daughter, granddaughter, and herself could be arranged. The wife of potter William Ball was indeed a Mary, but as her name also appears as Sarah Ball on petitions to the Overseer of the Poor, the evidence is ambiguous unless further research proves otherwise. Potter Ball, however, remains a strong possibility. He was the very embodiment of the phrase "wandering arcanist," having been associated with a variety of porcelain types and ventures in Bow, London, Limehouse, and Brownlow Hill, Liverpool (William Reid), and he was listed in Liverpool directories of 1766–1769 as "china maker" but was not listed there during the Bonnin & Morris years.

9. "Memoir of Thomas Gilpin," p. 310.

10. There are only scattered descriptions of Thomas Byerley's sojourn in America, summarized from the following sources: Eliza Meteyard, *The Life of Josiah Wedgwood: From His Private Correspondence and Family Papers*..., 2 vols. (London: Hurst and Blackett, 1865–66; facsimile ed., London: Cornmarket Press, 1970), 2: 424; Llewellynn Jewitt, F.S.A., *The Ceramic Art of Great Britain from Pre-Historic Times Down to the Present Day*, 2 vols. (London: Virtue and Co., 1878), 2: 373; *The Selected Letters of Josiah Wedgwood*, edited by Ann Finer and George Savage (New York: Born & Hawes, 1965), pp. 48, 66, 186; Robin Reilly, *Josiah Wedgwood 1730–1795* (London: Macmillan, 1992), p. 39. Information about the portrait medallion of Byerley illustrated in figure 9 is from "Josiah Wedgwood: 'The Arts and Sciences United,'" an exhibition held at the Science Museum, London, in 1978, and included in the accompanying catalog, *Josiah Wedgwood, 'the arts and sciences united'* (Barlaston, U.K.: Josiah Wedgwood & Sons, 1978), p. 75, no. 237.

Fond as he was of his sister's son, Josiah sent him off to America with a princely gift of £70 and a warning to make it last. He wrote Thomas Bentley (in London): "My nephew, T. Byerley, is now going to Philadelphia. . . . If he has a mind to take a few pieces, or £3 or £4 worth of my Ware with him, you may let him have so much on his own account, and pray advise him to ship himself in the Cheapest way for America as he may there come to want a Crown, or even a shilling, and know not where to have it." *Selected Letters of Josiah Wedgwood*, p. 66. Once in Philadelphia, he spent wildly, landing in jail, no doubt for debt. Reilly, *Josiah Wedgwood*, p. 39.

Thomas Byerley would have been just twenty-one years old that summer of 1768. (One wonders if he came into contact with that other new arrival, Gousse Bonnin, who had just brought his family to Philadelphia.) He seems to have lost no time in joining The American Company, theatrical producers who staged dramas, musicals, and poetry readings. A notice of a New York City performance which was for his benefit and that of another actor appeared in the *New-York Gazette or Weekly Post-Boy* of June 12, 1769. He probably traveled between New York and Philadelphia before his brief employment by Bonnin and Morris; then he seems to have wandered (or explored) New England while continuing to act. Many correspondents kept Wedgwood informed of his nephew's escapades. Finally, Byerley showed signs of settling down to the life of a schoolmaster in New York City, and Wedgwood, wanting to encourage this serious turn of mind, had Bentley send him a fine pair of globes in April 1774, expecting they would be a good teaching tool. Meteyard, *Life of Josiah Wedgwood*, p. 424. Instead, Byerley, who had opened a day school for young students with a Mr. Day, put them to better use in a course for adults on geography and astronomy, as their usual scholars were yet too young to learn about "Spheres." *New-York Gazette and Weekly Mercury*, July 4, 1774.

By then it was the eve of the Revolution, and the Battle of Concord forced a difficult choice on Byerley—to fight his own countrymen as an American, or to go home to Staffordshire—and in June 1775 the prodigal returned home and took up his duties as a clerk with his uncle's company. Reilly, *Josiah Wedgwood*, p. 40. "This time he stayed ... to become a dutiful and loyal partner in the firm, and, for some years after Josiah's death, he carried the entire burden of its management. . . . It fell to Byerley to steer the fortunes of the firm during the difficult years of transition from the first to the second generation of Wedgwoods" (*Josiah Wedgwood, 'the arts and sciences united'*, p. 75)

Ironically, "the manufacture of [bone] china, which had never been attempted by the first Josiah, was commenced at Etruria about 1808 or 1809, in the time of Mr. Byerley . . . but was only carried on for nine or ten years." Jewitt, *Ceramic Art of Great Britain*, 2: 373.

11. Neil McKendrick, "Josiah Wedgwood and Factory Discipline," *Proceedings of the Wedgwood Society*, no. 5 (London: Wedgwood Society, 1963): 1–29. Wedgwood imposed order and

discipline on every action of his potters, from keeping regular hours and avoiding waste to learning the skills of each specialty and accepting a ban on drinking.

12. Prime, *Arts & Crafts of Philadelphia, Maryland and South Carolina*, p. 116. He placed just such a "notice of a runaway indented servant" in the *Pennsylvania Journal*, October 18, 1770, a reminder that not only were slaves being bought and sold, but also that immigrants frequently sold their own future labor in exchange for their passage to America.

13. Bradford L. Rauschenberg, "John Bartlam, Who Established 'new Pottworks in South Carolina' and Became the First Successful Creamware Potter in America," *Journal of Early Southern Decorative Arts* 17, no. 2 (November 1991): 6; available online at www.mesdajournal.org/online-issues/.

14. See articles by Robert Hunter, Stanley South, Lisa R. Hudgins, and J. Victor Owen in the 2007 issue of *Ceramics in America*.

15. Prime, *Arts & Crafts of Philadelphia, Maryland and South Carolina*, pp. 115–16. The notice dated Philadelphia, February 20, 1770, ran in the *South-Carolina Gazette*, March 15, 1770.

16. For discussions of bones, see Andrew Ure, *A Dictionary of Arts, Manufactures, and Mines* (Philadelphia: George S. Appleton, 1844), p. 1029; also J. Victor Owen, "Antique Porcelain 101: A Primer on the Chemical Analysis and Interpretation of Eighteenth-Century British Wares," *Ceramics in America*, edited by Robert Hunter (Hanover, N.H.: University Press of New England for the Chipstone Foundation, 2002), pp. 39–61, esp. 42.

17. See Prime, *Arts & Crafts of Philadelphia, Maryland and South Carolina*, p. 112, for Bartlam's ads from the *South-Carolina Gazette*. Rauschenberg, "John Bartlam," p. 17, adds to those: on November 8 1770, Bartlam urgently requested apprentices; on January 31, 1771, he asked for "samples of any Kinds of fine Clay . . . in order to make them Trials of. He already makes what is called Queen's Ware, equal to any imported . . . ." Inexplicably, the threat of competition from Bartlam never materialized. He had left Charleston by May 1771, and reappeared in Camden, South Carolina, more than a year later, attempting once again to make earthenware.

18. *Public Register or Freemans Journal*, Dublin, December 4, 1770. Fry presumably was related to the founder, Thomas Frye. Nicholas Panes, "Mr Ball the English Potter and the American China Manufactory," *English Ceramic Circle Transactions* 20, pt. 3 (2009), pp. 633ff. William Bradford Reed, *The Life of Esther De Berdt, Afterwards Esther Reed, of Pennsylvania* (Philadelphia: C. Sherman, 1853). Reed, a descendant and biographer of one of the celebrity passengers on board Captain Osborne's ship, dated its arrival as October 26.

19. Prime, *Arts & Crafts of Philadelphia, Maryland and South Carolina*, p. 116, cites Philadelphia's German-language newspaper, the *Staatsbote*, as the source of his quote: "In the same ship (Capt. Osborne's) nine master workers have arrived here for the porcelain manufactory of this city." As it appeared there in German, Prime may have wrongly translated *haupt* (defined in a German/English dictionary as "head, chief, principal, leader") and *arbeiter* ("workman, hand.") when calling them "master workers." In the London and New England press they were called "prime hands" or "capital workmen." He may thus have misled future scholars who referenced him, because in the hierarchy of labor, "Master Worker" has a special meaning: the master invents the plan; the journeyman executes the plan; and the apprentice learns the plan. In the pottery industry, master potters had charge of groups of workers. Bow would indeed have been "distressed" had it lost nine masters in as many branches.

20. Prime, *Arts & Crafts of Philadelphia, Maryland and South Carolina*, pp. 118–19; *Pennsylvania Journal*, October 10, 1771.

21. Bernard Watney, *English Blue and White Porcelain of the Eighteenth Century*, 2nd ed. (London: Faber and Faber, 1973), p. 25.

22. "Memoir of Thomas Gilpin," pp. 312–14. Many scholars have placed the end of nonimportation in Philadelphia as of September 1770. Gilpin's information gives four more months of breathing room to Bonnin & Morris, before competition closed in.

23. Johnson's exploits are the stuff of legends—perhaps the earliest literary reference being James Fennimore Cooper's Natty Bumpo, who fought under "Sir William"—to the latest, the 2005 history by Fintan O'Toole, *White Savage: William Johnson and the Invention of America* (New York: Farrar, Straus and Giroux, 2005). Still more recent, the rousing story by Elaine Cohen, "Chief Big Business," *Adirondack Life* (July–August 2007): 72–82, brought Johnson to my attention and with it the realization that Eric Noah's Bonnin & Morris basket must once have held fruit in Johnson Hall.

24. Wanda Burch, site manager, Johnson Hall, in conversation with the author, June 2007.

25. Sir William Johnson, *The Papers of Sir William Johnson* (Albany: University of the State of New York, 1933), pp. 293–94 (Carpenter Wharton to Johnson, October 14, 1771). Thanks

to George R. Hamell for this reference, which is also cited in Hood, *Bonnin and Morris of Philadelphia*, p. 17. Much of Johnson's *Papers* has been digitized and is available on a CD-ROM for purchase through the New York State Library and online through www.nysl.nysed.gov. Each volume is a separate PDF file.

26. Prime, *Arts & Crafts of Philadelphia, Maryland and South Carolina*, pp. 117–18; *Pennsylvania Journal,* April 18, 1771.

27. George Savage, *18th Century English Porcelain* (London: Spring Books, 1952), pp. 2–3.

28. Flint stone had been added to earthenware and stoneware since about 1720. Lead glass, made in Pennsylvania, was available in Philadelphia. Henry William Stiegel had been producing lead glass in Manheim since 1769, and it was being made at the Philadelphia Glass Works of Robert Towers and Joseph Leacock after 1771. See J. Garrison Stradling, "The John Troup Decanter: Saluting Freemasonry in 18th-Century America," *Journal of Glass Studies* 45 (2003): 151–65.

29. Elizabeth Adams and David Redstone, *Bow Porcelain* (London: Faber and Faber, 1981), pp. 98, 163, 241.

30. Watney, *English Blue and White Porcelain of the Eighteenth Century*, p. 26.

31. The question is always were these materials intended for the porcelain body, the glaze, the print colorant, or for all three? In his role as archeologist/analyst, Victor Owen works with the end product. Fragments, having passed through the fire, have become something other than the sum of their ingredients. Owen tells us what the composition is after fusion and vitrification have altered them. The written record of advertisements and notices placed by Bonnin and Morris provides information as to what went into the original batch and when these changes occurred. These notices are the link between historical reality and Owen's microscope. See Owen, "Antique Porcelain 101," pp. 39–61. For a discussion of bones, see ibid., p. 42. See also J. Victor Owen, "Geotechnical and Mineralogical Distinctions between Bonnin and Morris (Philadelphia, 1770–1772) Porcelain and Some Contemporary British Phosphatic Wares," *Geoarchaeology: An International Journal* 16, no. 7 (2001): 785–802.

32. Prime, *Arts & Crafts of Philadelphia, Maryland and South Carolina*, p. 119.

33. Rauschenberg, "John Bartlam," p. 76. Rauschenberg cites *The Papers of Henry Laurens*, vol. 8, *Oct. 10, 1771–April 19, 1773*, edited by George C. Rogers Jr. and David R. Chesnutt (Columbia: University of South Carolina Press, 1980), p. 55.

34. Hood, *Bonnin and Morris of Philadelphia*, pp. 60–61, appendix 9, cites in full the anonymous letter published in the *Pennsylvania Journal*, November 11, 1772. Extracts include: "To the Printers of the Pennsylvania Journal The feelings of a benevolent heart instruct us. . . . What then shall be said of a late advertiser, who has ungenerously and disingenuously attempted to check the rising benevolence of the public . . . . It is a subject so generally known, that perhaps it may be needless to say, the author of an advertisement in the Pennsylvania Gazette, dated Southwark, Nov. 2, 1772, is here meant." "The faults of these poor people . . . did, by no means, contribute to the failure of the late China Manufactory of this city, consequently, their application for relief, could not excite the resentment of the directors; surely therefore it is cruel, very cruel, to oppose the good intentions of the public." "I make no reflections to the disadvantage of anyone . . . nor can I deny that the Proprietors of the late China Work, tho' unsuccessful, are entitled to praise for their endeavours to introduce so valuable a Manufactory: But . . . altogether owing to want of judgment, or proper information previous to these endeavours, in the Managers; and if they cannot fulfil the first engagement they entered into, which, from the pleasing prospect it afforded, drew these people from an easy situation in life; they ought, at least, to do what they can, to restore them to that situation . . . the workmen, who were lately engaged in the China Manufactory of this city, when first applied to in England, were in easy and comfortable circumstances; esteemed as artificers, respected as men: but allured by the very tempting offers of the conductors of that work, they quitted their native country; separated themselves from agreeable connexions, and left a certain competence of fortune for their condition . . . . Induced . . . by the following agreement . . . [t]he said Proprietors agreed to pay their passage to America, and for five years after a weekly salary of One Guinea and an Half; in consideration of their doing such work as the said Proprietors should have occasion for, twelve hours in the day, allowing two hours for the time of meals; and for any part of the remaining ten hours, which the workmen should waste, a due proportion of said salary was to be deducted. The Proprietors, on their part, have exacted a full obedience from the men to this agreement, even so far as frequently to deduct for a quarter of an hour, in which a workman has attended the call of an occasional acquaintance; but they have not so scrupulously fulfilled their own engagement . . . but modestly

wish and desire only . . . to be restored to their own, where they can still live in plenty by their honest endeavours." "This . . . the Proprietors owe to justice and to the community, for though they too have been sufferers in the common misfortune, they have not, as is the case of these poor people, lost their All; but are still very able to make some amends . . . ."

35. Josiah Wedgwood's 1783 "Address to the Workmen in the Pottery, on the Subject of Entering into the Service of Foreign Manufacturers," was printed and circulated to other potteries as well. It is cited in full in Rauschenberg, "John Bartlam," pp. 50–57, appendix 1. It is worth noting that the last British troops set sail from New York Harbor on November 25, 1783, and Wedgwood was all too aware that his carefully trained men might find many enticements on the American side of the Atlantic.

36. This basket was illustrated and discussed in Diana Stradling and J. Garrison Stradling, "Bonnin and Morris Revisited," in *Ceramics in America*, edited by Robert Hunter (Hanover, N.H.: University Press of New England for the Chipstone Foundation, 2007), pp. 188–92. Subsequent analysis of it has shown its composition to be a lead-bearing, silicious-aluminous-calcic (S-A-C) that contrasts with the earlier phosphatic (bone-ash) porcelains (the lead derived from the addition of flint glass). See J. Victor Owen and Robert Hunter, "Too Little, Too Late: The Geochemistry of a 1773 Philadelphia Porcelain Openwork Basket," *Journal of Archaeological Sciences* 36 (2009): 333–42.

37. Brown, "Piecing Together the Past," p. 573. Apparently the *Gazette* did not mention the town where George Morris had been living for the previous eighteen months.

38. Ibid., p. 573.

39. Meteyard, *Life of Josiah Wedgwood*, 2: 394.

40. Harry Hall White, "Henry Rowe Schoolcraft, Glassmaker, Part III," *Antiques* (April 1939): 187 (courtesy of Ian Simmonds).

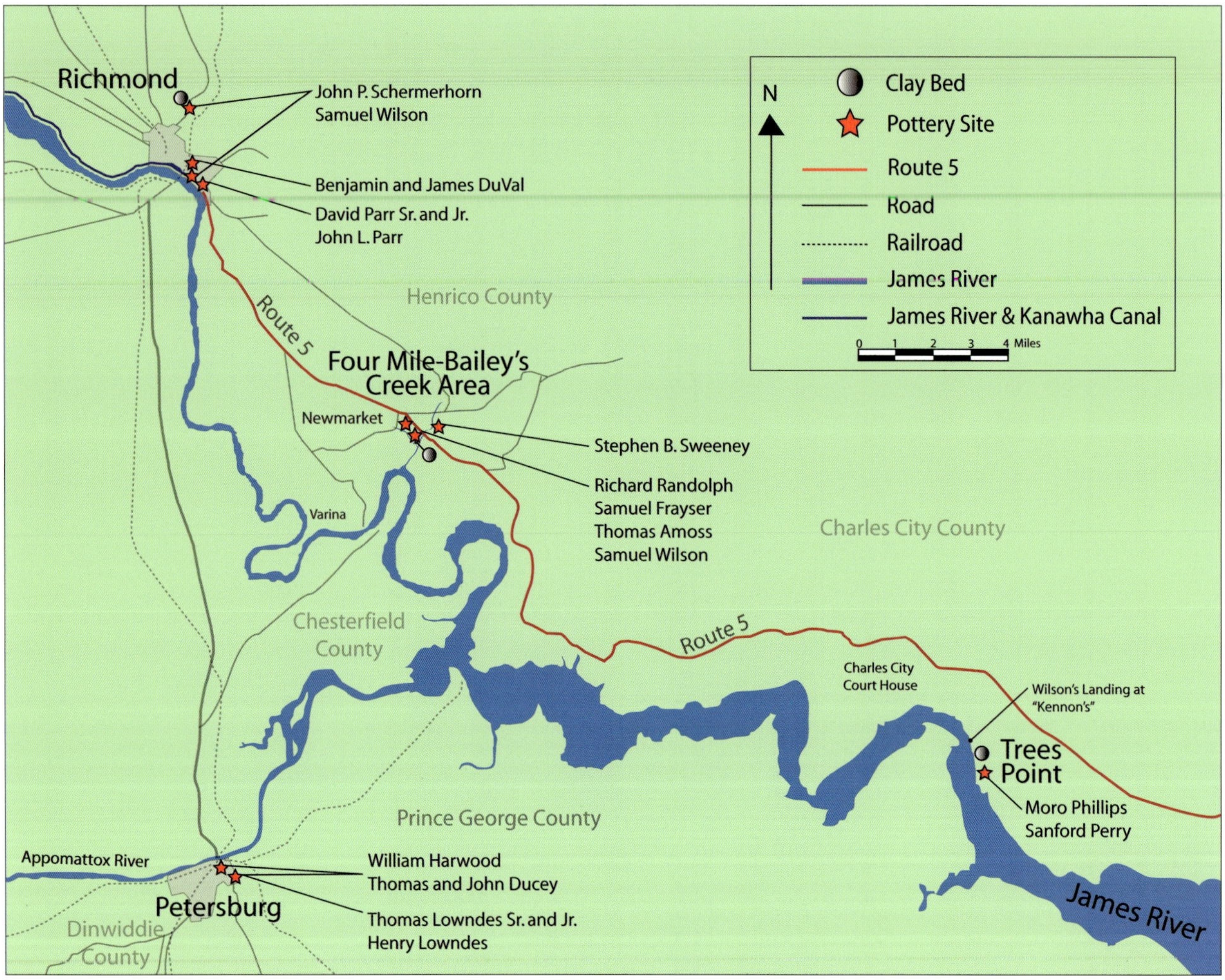

Figure 1 Map of Richmond–Petersburg area of Virginia, showing the locations of known and suspected stoneware potteries and clay beds. (Drawn by Oliver Mueller-Heubach.)

The Remarkable 19th-Century Stoneware of Virginia's Lower James River Valley

Kurt C. Russ, Robert Hunter, Oliver Mueller-Heubach, and Marshall Goodman

▼ NINETEENTH-CENTURY salt-glazed stoneware of the lower James River Valley has been actively collected for more than half a century, though little information about the potters and their wares has been documented in the literature. This article provides much-needed attributions, combining archaeological fragments from the production sites with surviving pieces in public and private collections. Most of the stoneware is embellished in blue cobalt, in styles from neoclassical to naturalistic. A surprisingly number of examples contains well-developed pictorial statements in themes from comic to political. Techniques include cobalt slip trailing, brushwork, and incising in the Germanic tradition with occasional nods to English brown stoneware and its forms. Many of the pieces are published for the first time. Viewed collectively, they reflect a bold decorative aesthetic rivaling long-recognized and celebrated schools of stoneware production seen elsewhere in eastern North America.

Although Yorktown, Virginia, was the home of America's first successful stoneware pottery (1720–1745), it was not until the turn of the nineteenth century that expanding populations and recently discovered clay sources heralded widespread production of Virginia stoneware.[1] Potters hailing from England, Ireland, Maryland, New Jersey, New York, and Pennsylvania created a rich hybrid tradition with forms and decorations adapted from their regional styles. Their training of the native sons—white, enslaved, and free black—carried the industry through nine decades. Trade networks, such as roads, canals, and railroads, grew with the potteries, ensuring the spread of James River stoneware into southern, western, and coastal markets. The clay that brought these artisans to Virginia features proudly in geographic locales aptly named Kaolin and Claymount.

Whereas the stoneware industries of Alexandria and the Shenandoah Valley have been the subjects of extensive research and publication and have been nationally recognized for their contribution to American ceramic history, ceramics researchers have largely overlooked the stoneware pottery industry of eastern Virginia.[2] In the last decade, however, this began to change, and research on the stoneware potters of the Richmond and Petersburg areas has advanced rapidly. The topic was the focus of a 2004–2005 exhibit at the Virginia Historical Society and a related article in *Antiques* magazine in 2005 highlighting the diversity of stoneware produced in eastern Virginia from 1720 until 1890.[3] In 2012–2013 the Museum of Early Southern Decorative Arts hosted the symposium and exhibit "From Kaolin to Claymount: Demystifying James River Valley Stoneware."[4] Most important, perhaps, is the research of Oliver Mueller-Heubach, contained in his 2013 dissertation "From Kaolin to Claymount: Landscapes of the 19th-Century James River Stoneware Industry."[5]

The culmination of this research is a more complete picture of the historical depth and complexity of the stoneware manufactories once operating in Richmond, Petersburg, and the adjacent counties of Henrico and Charles City along the James River (fig. 1). As a result, a broader context can be established for understanding the significance of eastern Virginia stoneware, its potters, and production sites.

Earthenware manufactories centered in the Richmond area in the late eighteenth century probably involved only two operations and perhaps four potters. In contrast, stoneware manufacture in Richmond, southern Henrico County, Charles City County, and the Petersburg/Dinwiddie County areas involved as many as sixteen different operations and some forty identified potters. Those potters, who labored at various places at various times, alternating between cooperation and competition, were responsible for the production of a unique body of wares, tangible evidence of connections across space and time.

Richmond Pottery Production

Stoneware manufacture began in Richmond, Virginia, about 1807, having been preceded by a pair of little-understood earthenware endeavors (fig. 2). An intriguing early reference to Richmond white clay is found in a 1784 letter by James Brindley (1745–1820)—a canal builder who immigrated that year from England and settled in Wilmington, Delaware.[6] Brindley, who was involved in several canal projects, among them Virginia's James River and Kanawha Canal, wrote to his uncle in England:

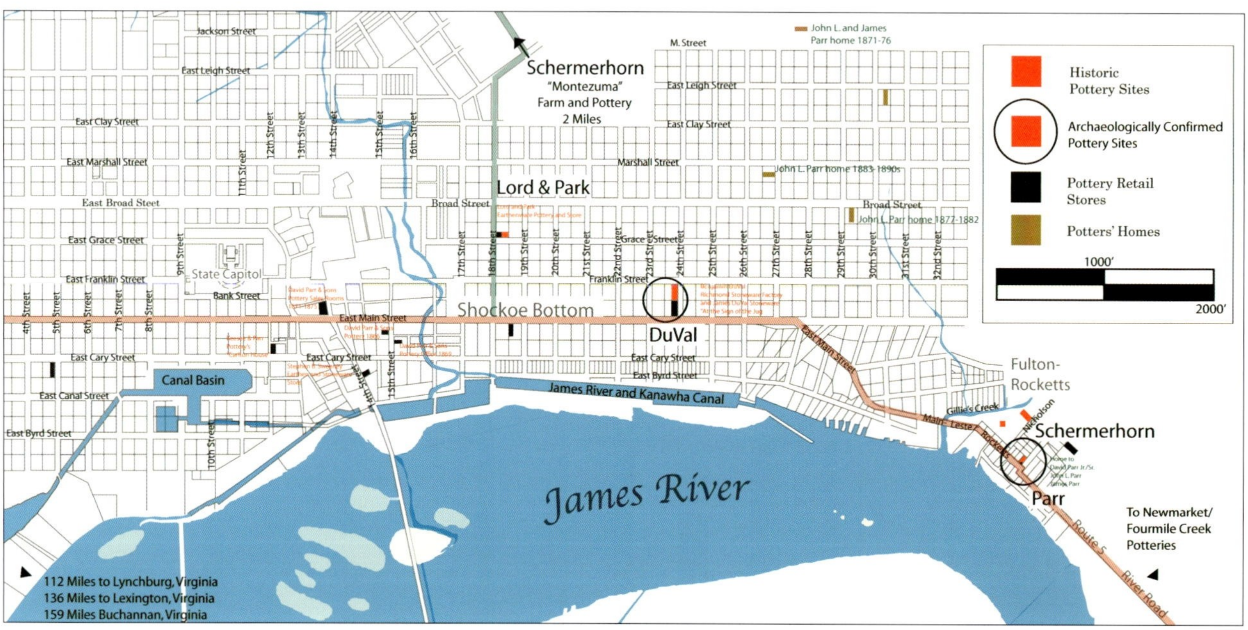

Figure 2 Map of Richmond, Virginia, showing nineteenth-century pottery sites and stoneware retail stores. (Drawn by Oliver Mueller-Heubach.)

> Wilmington, Aug.[t] 10th 1784
> Dear Unkle—I rec'd your kind letter dated Aprill 5th 1784 with Cousin Taylors and 12 crates of Ware, come safe to hand. Cousen mentions sending a farther assortment which I hope will also arive safe. when request, You'l not send any more for the present, untill I can dispose of these and make good my remitance, when I will write you for the most saleable sorts and in the mean time make enquiring and try to fix you in connexion with some stanch able men. [Illegible] in this country which is probable will correspond with you to good advantage on all sides. Exclusive of what I can dispose of. As my time will be for the most part taken up in the canal business and presume you would wish to dispose of Large quantities of your manufacture of which I will inform you by first opportunity. You write me to send you some white clay that I mentioned in my last is at Richmond 300 miles from here nearly, but will endeavor to get some with some other that I have lately here of that's any white.[7]

Just as English manufacturers were eager to market their wares to the region, so too were they seeking to extract raw materials with an eye to supplying new manufactories. These new works required advanced scouting for both "stanch able men" and identification of the "most saleable sorts" of wares. This late-eighteenth-century reference, though probably specific to kaolin clays for use in refined wares, is a prelude to later historical documentation of the significant clay resources for brick, coarse earthenware, and, especially, stoneware within the Richmond-Henrico County area.

Local production of wares, "equal or superior," as they put it, to those previously imported from England, Germany, or the New York–New Jersey region, allowed Virginia potters to successfully expand their enterprises by offering competitive pricing. This enabled them to eventually dominate the market, with Richmond becoming a hotbed of stoneware production during the second and third decades of the nineteenth century. The influx of potters was no doubt influenced by a variety of factors—the restrictions of the Embargo Act of 1807 and the Non-Intercourse Act of 1809, the War of 1812, and the potters' desire to seek out new markets. Growing populations on navigable southern waterways would allow for market expansion and greater distribution of their wares.

Expanding Virginia businesses—pharmacies, distilleries, dairies, and mercantile operations—also created a demand that encouraged local pottery manufacture. The late-eighteenth-century advertisements placed by Richmond pharmacist Benjamin DuVal, for example, offered employment to potters able to supply wares for his business. Samuel Allinson was the first to answer this call, building and operating a kiln on a lot owned by Isaac Younghusband and managed by DuVal and partner Thomas Warren. William Harwood, the holder of a roof-tile patent, partnered with DuVal in 1809, furthering the enterprise begun by Allinson. Harwood may have been instrumental in expanding the operation into salt-glazed stoneware production two years later.[8]

EIGHTEENTH-CENTURY RICHMOND POTTERS

Richmond was still a relatively young town in the late 1700s with imports largely satisfying local demand. As the city grew into its new role as the capital of Virginia after 1780, craftsmen arrived to feed the growing market. Earthenware potter Gresham (Gersham) Lord had arrived in Richmond about June 1781 in the company of John Robinson, another trained potter who resided with him.[9] The 1782 city census identifies one "Dresham Lord," a twenty-nine-year-old potter, as having been "in residence for 9 months" in the area of Wardship No. 3 with John Robinson, potter, age twenty-eight.[10] Lord is not taxed for ownership of either real estate or other property.

By March 1782 Lord had partnered with a tanner's son, Jonathan Park. Jonathan had continued the family tannery business, but after his father's death offered it for lease in April 1783 and eventually dissolved it in October of the same year, perhaps in favor of devoting his full attention to the pottery.[11] It appears that Park used the proceeds (£20) from the sale of a thirty-acre lot "on the waters of Allen's Branch, binding on Rocket's line" in Henrico, to help finance his newly formed pottery partnership with Gersham.[12] Theirs was the first documented attempt to meet the growing local demand for utilitarian wares.

An advertisement in Richmond's *The Virginia Gazette, and the Weekly Advertiser*, March 16, 1782, announces the establishment by Lord and Park of an "Earthen-ware Manufactory":[13]

> 1782, March 16
> Earthen-ware Manufactory.
> The subscribers make and sell on the most reasonable terms, all kinds of course earthen ware, wholesale and retail. Orders will be carefully attended to, and furnish- on the shortest notice.
> JOHNATHAN PARK.
> GERSHAM LORD.
> N. B. A likely boy or two will be taken as apprentices.[14]

Despite the fairly prominent location of their pottery on lot 85 at the corner of 18th and Grace streets, scant historical evidence of their operation is available today. Jonathan Park advertises as a tanner during this period and is likewise listed in the 1790 city census, suggesting his involvement with the pottery was primarily as a business partner.[15]

In February 1791 Benjamin DuVal and Thomas Warren placed the first advertisement for a potter "who understands working at the wheel," setting the stage for an influx of northern potters to Richmond.[16] The next reference is from June 20, 1791:

> "On the application of the proprietors of the pottery in this city, Resolved, that leave be granted them under the direction of the Committee of streets to make use of the dirt in the cross street between Scherer and Richardson's houses."
> X Warren, Thomas[17]

The "pottery in this city" almost certainly is that of Benjamin DuVal and Thomas Warren (the latter undersigned the announcement). DuVal had advertised for potters only four months earlier, and this reference perhaps describes the first lot of clay procured for the pottery. Soon New Jersey Quaker potter Samuel Allinson and his journeymen John Carty and Richard Esdall of New York were producing ware, including earthenware cups, bowls, milk pans, and pitchers. Three years fraught with disputes between Allinson and DuVal may have left Allinson yearning for independence. In 1793 Allinson married Francis Johnson, and the next year he is noted "for building a potter's kiln on a lot of Isaac Younghusband and burning earthen ware therein."[18] A surviving 1784 deed to Isaac Younghusband shows his purchase of a 400-acre tract of land in the New Market vicinity, which could potentially have provided a clay source for his new tenant Allinson. As little as we know about these earthenware operations, less is known of their wares. No firmly attributed specimens of eighteenth-century Richmond-made earthenware have come to light.

BENJAMIN DUVAL & CO.

Benjamin DuVal produced clay roofing tile as early as 1808 on property north of Main between 23rd and 24th streets. By 1811 he expanded this operation to include the manufacture of salt-glazed stoneware. A number of newspaper advertisements detail a range and variety of products rivaling those of the best stoneware potteries in Manhattan and New Jersey. DuVal himself was not a potter, but hired potters from the north.

More is known about the DuVal pottery and its products than other Richmond-area stoneware manufactories. The site was first identified in 1978 through historical research conducted by Brad Rauschenberg, who initiated his research when the Museum of Early Southern Decorative Arts (MESDA), with which he was associated, acquired a large salt-glazed stoneware jar marked "B. DuVal & Co / Richmond." Preliminary archaeological investigation directed by Norman Barka of the College of William and Mary subsequently confirmed the survival of this important historical site's physical remains. In 2002 the site was destroyed by commercial development, but during the initial demolition of the property Robert Hunter and Marshall Goodman led an emergency archaeological salvage effort. An illustrated analysis of the vessel forms represented in the several thousand fragments they recovered is recorded in the 2005 volume of *Ceramics in America*.

The shapes of many of the DuVal pots are similar to those produced in the New York area. Based on the number listed in his advertisements and the quantity of recovered fragments, the most common vessel form produced at the DuVal pottery was the storage jar, in graduated capacities from one to five gallons. Jugs were also produced in graduated capacities. Pitchers of at least five sizes were recorded—a previously unrecorded one-gallon example is illustrated in figure 3. Other forms included bottles, flasks, chamber pots, and inkstands. As of this writing, however, fewer than ten marked examples of DuVal stoneware are known to survive.

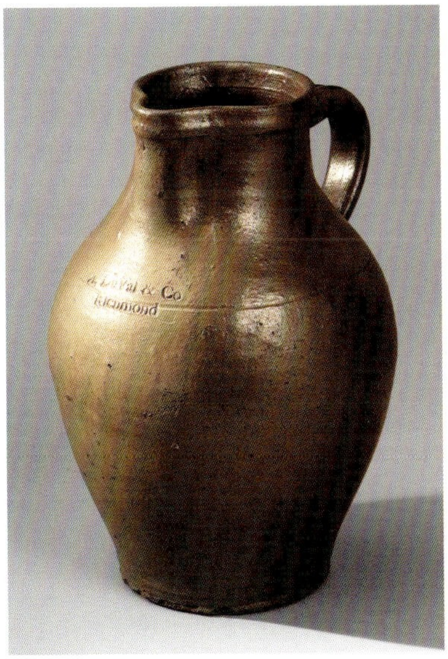

Figure 3 Pitcher, Benjamin Duval & Co., Richmond, Virginia, 1811–1817. Salt-glazed stoneware. H. 12". Capacity: 1 gallon. Stamped mark: "B. DuVal & Co. / Richmond" (Courtesy, Colonial Williamsburg Foundation.)

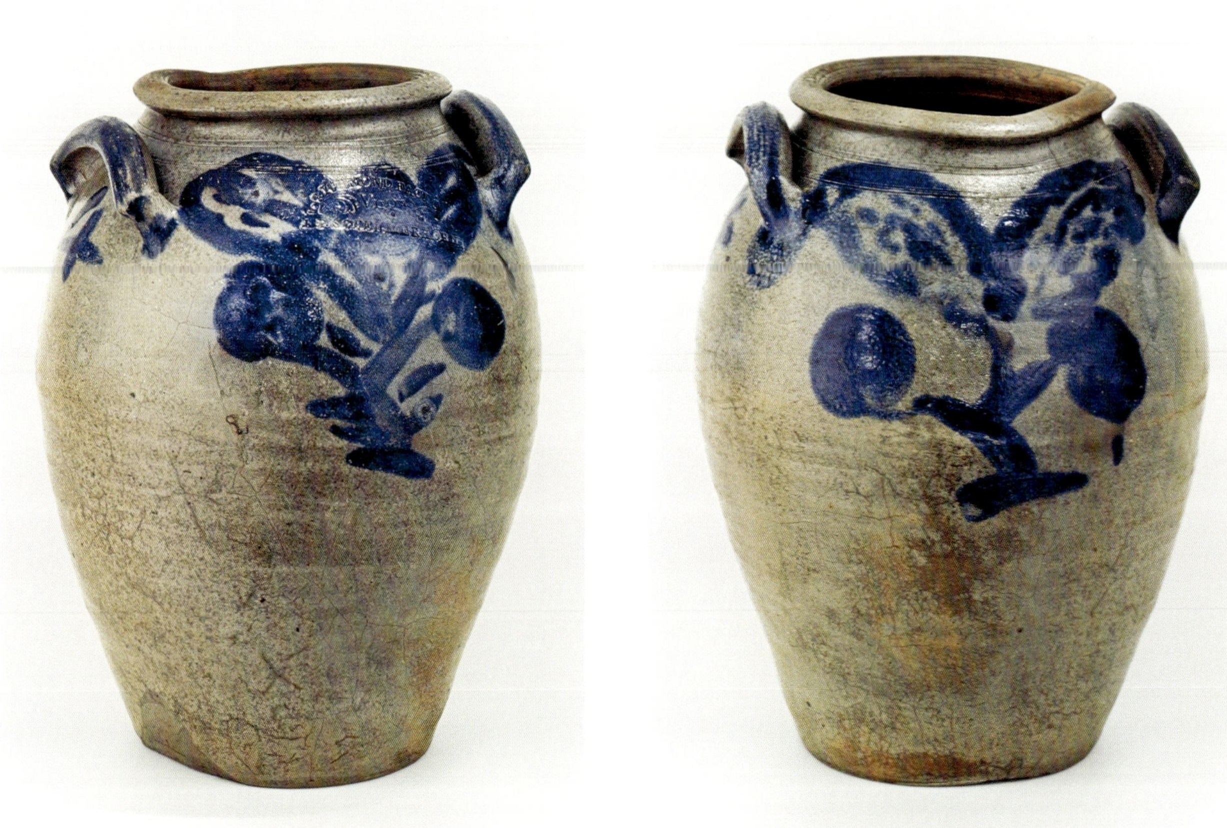

Figure 4 Storage jar, Schermerhorn Pottery, Richmond, Virginia, 1817–1837. Salt-glazed stoneware. H. 15⅜". Capacity: 4 gallons. Stamped mark: "RICHMOND FACTORY / Virg[a] / J. P. SCHERMERHORN" (Courtesy, Colonial Williamsburg Foundation.)

JOHN POOLE SCHERMERHORN (1788–1850)

John Poole Schermerhorn's association with DuVal was apparently shortlived. By 1817 he had left DuVal's employ to pursue pottery manufacturing under his own name in Richmond (fig. 4). Schermerhorn's well-situated manufactory at Rocketts, Richmond's primary port, facilitated his production, marketing, and distribution, and he became the city's most successful potter from 1817 until 1837 (fig. 5).[19] In addition to serving Richmond clients, Schermerhorn took advantage of water transport to sell up and down the James River and into the Carolinas.[20]

In the 2005 volume of *Ceramics in America*, Kurt C. Russ and W. Sterling Schermerhorn presented a well-illustrated summary of John P. Schermerhorn's work in Richmond. Since that time, much new information has come to light about his endeavors and the attributions of some of his previously illustrated pots. Of particular note is the archaeological confirmation of a potting business at Schermerhorn's home, Montezuma (fig. 6). Advertisements in the *Richmond Enquirer* between November 1828 and January

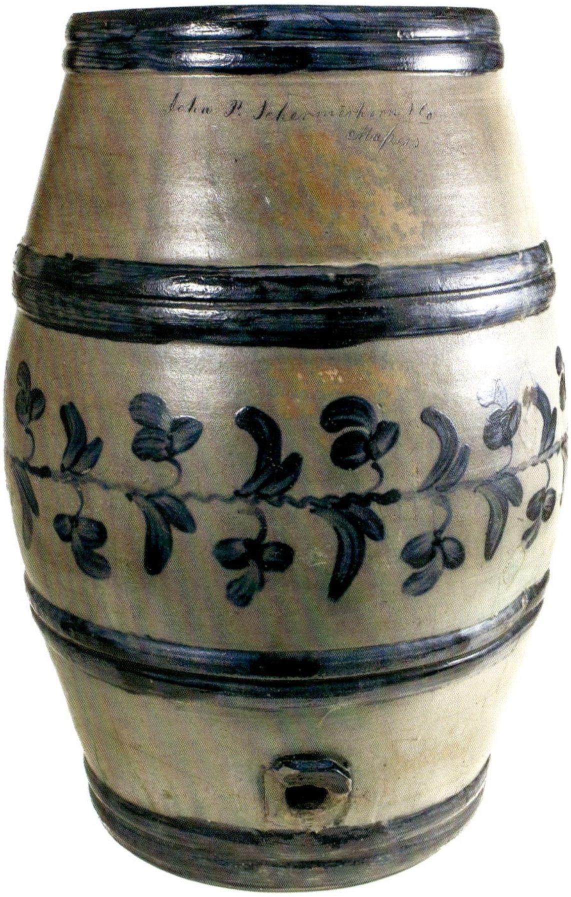

Figure 5 Water or beer cooler, Schermerhorn Pottery, Richmond, Virginia, 1817–1837. Salt-glazed stoneware. H. 27½". Capacity: 30 gallons. Incised marks: [on front] "John P. Schermerhorn & Co./, Makers" (Private collection; photo, Robert Hunter.) This massive cooler was recently discovered in an eastern Virginia estate and may have seen use in a local tavern or business. The form clearly relates to Albany stoneware, whereas the cobalt floral vines recall Baltimore decorative motifs. This is the only known vessel with Schermerhorn's hand-incised signature.

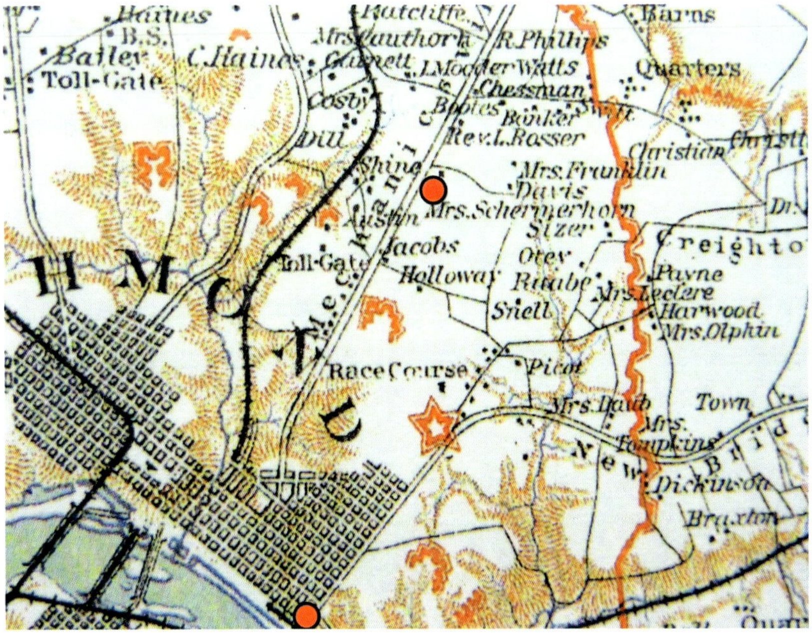

Figure 6 Map showing location of two J. P. Schermerhorn pottery manufactories, one at Rocketts (1817–1837), the other adjacent to his residence at Montezuma (1828–1850), two miles north of Richmond on the Mechanicsville Turnpike. Reproduced from George B. Davis et al., *The Official Military Atlas of the Civil War* (1891–95; repr., New York: Gramercy Books, 1983), pl. XIX, no. 1.

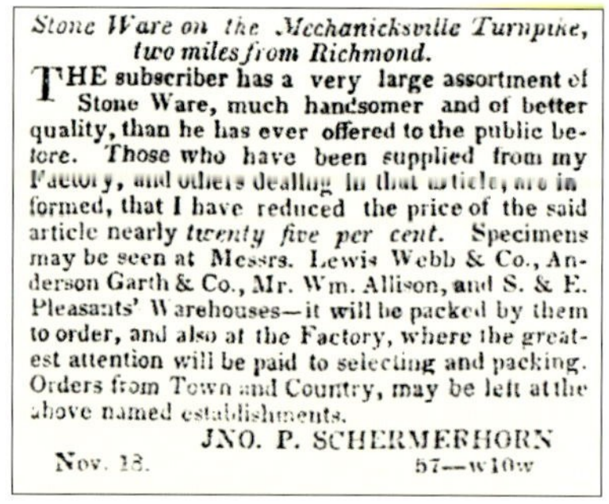

Stone Ware on the Mechanicksville Turnpike, two miles from Richmond.

THE subscriber has a very large assortment of Stone Ware, much handsomer and of better quality, than he has ever offered to the public before. Those who have been supplied from my Factory, and others dealing in that article, are informed, that I have reduced the price of the said article nearly *twenty five per cent.* Specimens may be seen at Messrs. Lewis Webb & Co., Anderson Garth & Co., Mr. Wm. Allison, and S. & E. Pleasants' Warehouses—it will be packed by them to order, and also at the Factory, where the greatest attention will be paid to selecting and packing. Orders from Town and Country, may be left at the above named establishments.

JNO. P. SCHERMERHORN

Nov. 18. 57—w10w

Figure 7 Advertisement identifying J. P. Schermerhorn's "Mechanicksville Turnpike" stoneware factory. *Richmond Enquirer*, January 1, 1829, p. 4.

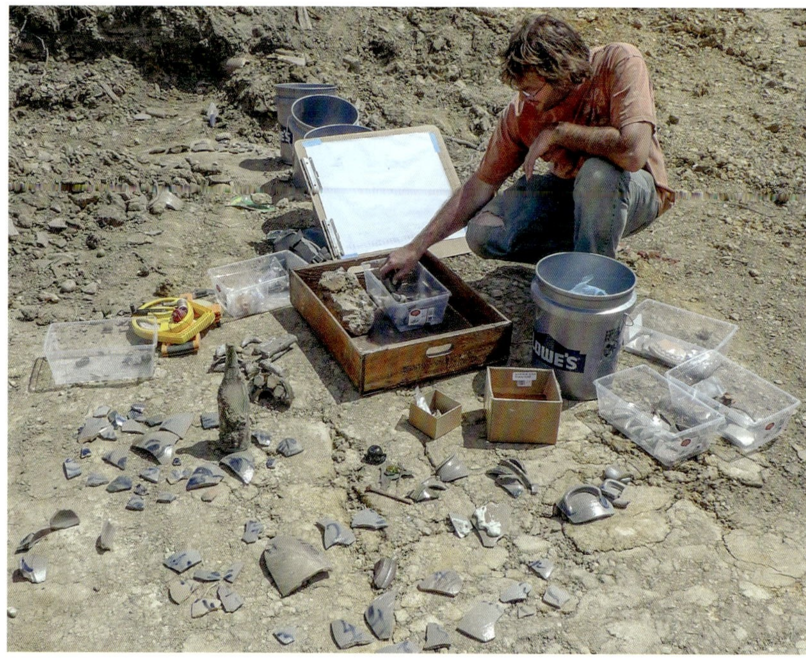

Figure 8 Archeological salvage work at Montezuma, the home and workplace of John P. Schermerhorn, 2010. (Photo, Robert Hunter.) Oliver Mueller-Heubach records evidence and gathers artifacts from Schermerhorn's pottery, established sometime after his purchase of the land in 1823. According to advertisements, the operation was up and running by 1828.

Figure 9 Stoneware fragments, attributed to John P. Schermerhorn, Henrico County, Virginia, 1828–1850. (Private collection; photo, Robert Hunter.) These fragments were found at Montezuma.

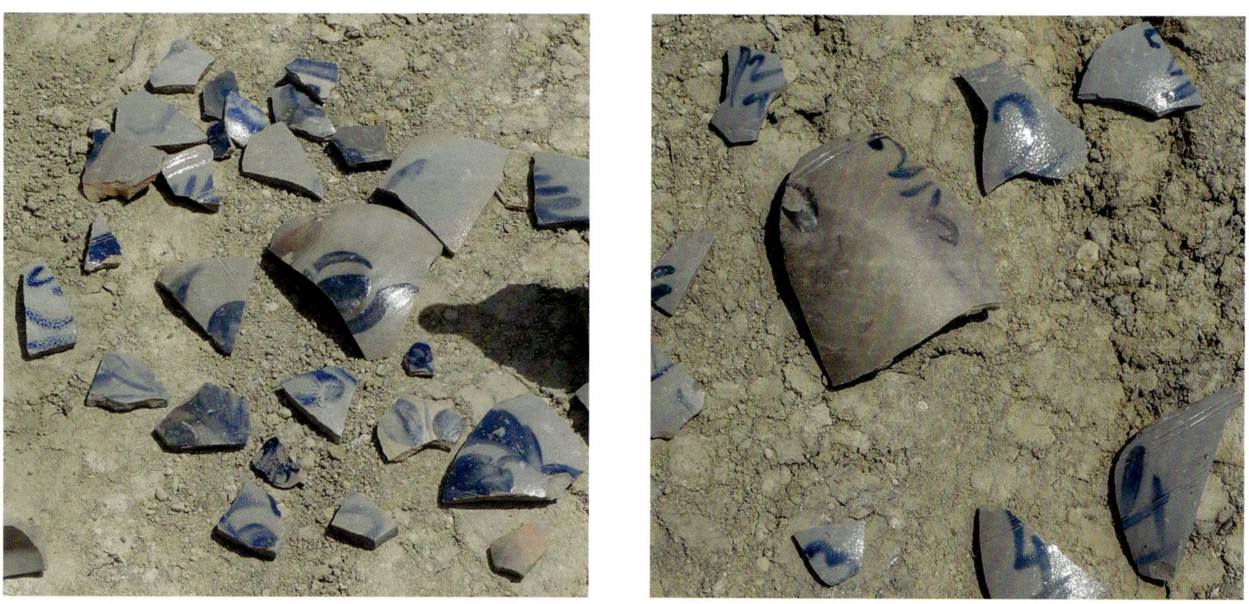

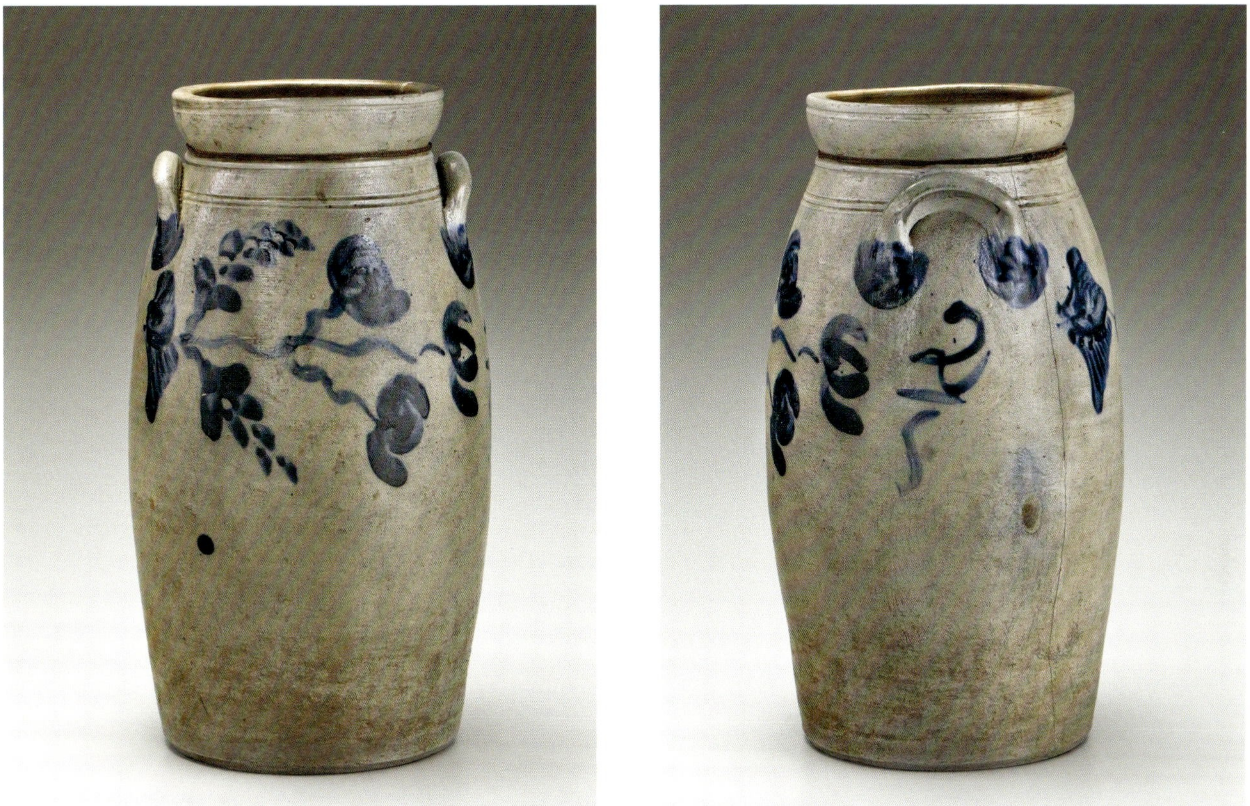

Figure 10 Churn, attributed to John P. Schermerhorn, Henrico County, Virginia, 1828–1850. Salt-glazed stoneware. H. 14¾". Capacity: 2 gallons. (Private collection; photo, Gavin Ashworth)

1829 first identify "Stone Ware on the Mechanicksville Turnpike, two miles from Richmond" (fig. 7). In them, Schermerhorn touts "a very large assortment of Stone Ware, much handsomer and of better quality" than offered previously. The wares were listed as available at both his factory and several retail outlets where orders were accepted from "Town and Country."[21]

The discovery of the Schermerhorn pottery at Montezuma was reported as early as 2008. Oliver Mueller-Heubach conducted salvage archaeological work in 2010 after remains indicating the presence of a pottery kiln (heavily glazed bricks and kiln furniture) were found during development of the property (fig. 8). Although no maker's marks were noted, the distinctive brushed-cobalt capacity designations known from extant Schermerhorn wares were well represented (fig. 9). Recovered fragments exhibit identical flaring rims, rather distinctive floral motifs, and brushed capacity numbers similar to those as found on a two-gallon tabletop churn attributed to Schermerhorn (fig. 10).

Previously unrecognized Schermerhorn vessels have come to light since the 2005 *Ceramics in America* articles. Of particular note is a group of three associated with Andrew Jackson (fig. 11). While already held as a military hero for his defeat of the British in the Battle of New Orleans in 1815, Jackson's

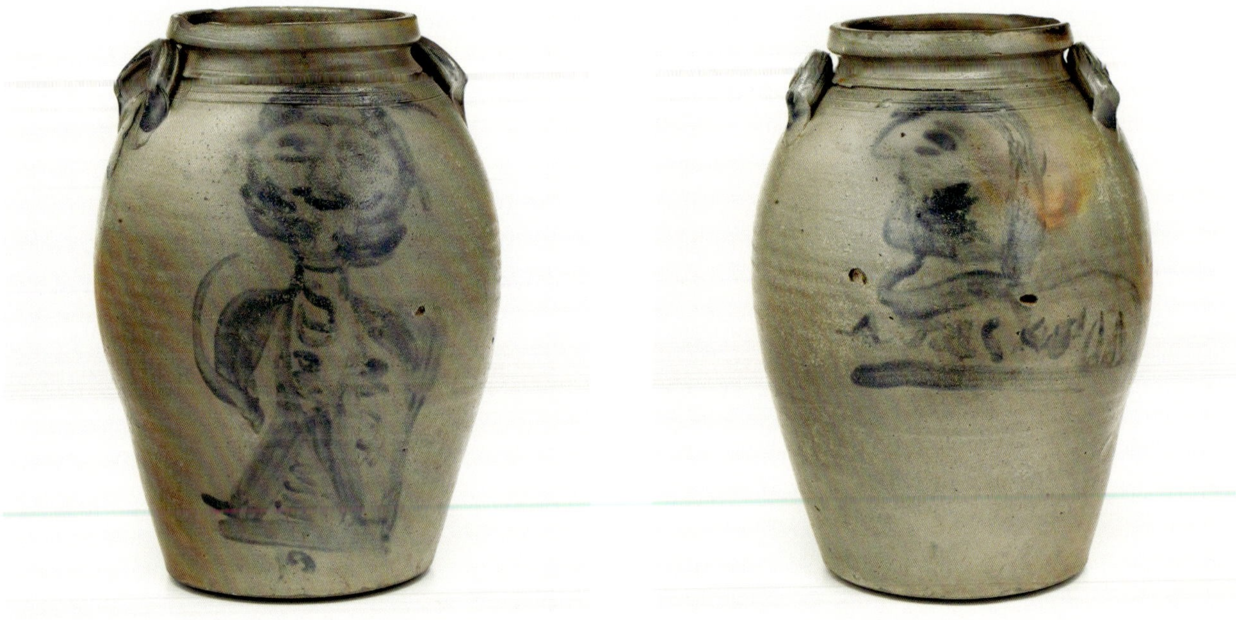

Figure 11 Storage jar, attributed to John P. Schermerhorn, Henrico County, Virginia, 1828–1850. Salt-glazed stoneware. H. 14¾". Capacity: 3 gallons. Inscribed, in brushed cobalt: "A. JACKSON" (Private collection; photo, Gavin Ashworth.) The brushed-cobalt portrait of a bearded Andrew Jackson in military uniform wielding his sword in his right hand dominates the front of the vessel; on the reverse, above the inscription, is Jackson's profile.

perceived opposition to elitism led the yeomen, farmers, and laborers—potters among them—to think of him as the "greatest man of his age." His wielding of the presidential veto, bordering on dictatorship in the eyes of opponents, further endeared him to loyal supporters. The Rev. John Freeman Schermerhorn (d. 1851), a close friend of President Jackson and one of two agents to sign the treaty of the Cherokee Removal, or "Trail of Tears," was a distant cousin of J. P. Schermerhorn and is buried in the family cemetery at Montezuma.[22]

New research into Schermerhorn's workforce has revealed the names of enslaved and free blacks who may have worked in the pottery. In 1828 he inherited Hall and Watt, two male slaves from the estate of Francis Wilson. At his death, Schermerhorn owned seventeen slaves, including six "adult" males—Archey, Ben, Charles, Henry, Ned, and William—as well as "one old man," Davy.

Knowledge of which of Schermerhorn's children were involved in this business is sketchy at best. One candidate is William Francis ("Frank") Schermerhorn, born in 1817 to Schermerhorn's first wife, Sarah Christian Wilson. Frank was in his teens when the pottery and house at Montezuma were built, and he would inherit the estate in 1850. John Poole Jr. was born to his father's second wife, as was James Cornelius (b. 1832), who was eighteen at the time of his father's death and the last of the sons conceivably trained in potting by John Poole Sr.

DAVID PARR POTTERY AND KEESEE & PARR POTTERY

A later Richmond operation was established circa 1850 by native Marylander David Parr Jr. (1819–1882). The *Richmond City Directory* first listed this "pottery at Rocketts" in 1852, then again in 1855 and 1860.

Parr had been associated with pottery manufacture in Baltimore from 1834 to perhaps as late as 1849. In 1860 he is recorded as living on Front Street (aka Rocketts/Lester/Main) with his sons John L. (1837–1895), David Preston (1838–1880), and James (b. 1843 or 1844), all of whom were listed as stoneware potters in Richmond's Marshall Ward in 1860 and 1870; David appears as a stoneware potter in the 1860 Eastern Division Henrico census.

Potter George N. Fulton worked at Parr's Virginia factory as early as 1855, and perhaps as late as 1862. He later went on to a long and well-documented career in the western part of the state, Alleghany County specifically. His brother Robert, also a potter, is listed as being in Richmond's first ward and briefly worked with Parr. Another potter associated with Parr is Joseph B. Ramey (1830–1909).

In 1858 Parr entered into retail partnership with auctioneer Thomas W. Keesee (1819–1882), who provided an influx of capital into Parr's operation. Keesee operated an auction house across from the Parr pottery as early as 1854, which he continued through the years of partnership.[23] Vessels marked Keesee & Parr (figs. 12, 13) were showcased at Keesee's store at 12th and Cary streets, known as Carlton House, while business continued at the Parr factory at "Front Lower End Rocketts." The *Richmond City Business Directory* of 1856–1857 does not mention either Parr or Rocketts; in the 1858–1859 directory, Keesee & Parr is the only pottery listed. The city business directory of 1860 describes the Keesee & Parr factory, located at "Front lower end Rocketts," as an earthenware manufactory.[24] The 1860 industrial census describes it as a significant stoneware pottery, with $5,000 capital invested, sizable quantities of raw materials on hand (clay, sand, blue, and wood), and ten male workers, who were paid $200 in wages and produced stoneware worth $12,000.[25] David Parr Jr. was named as the manager of the business.

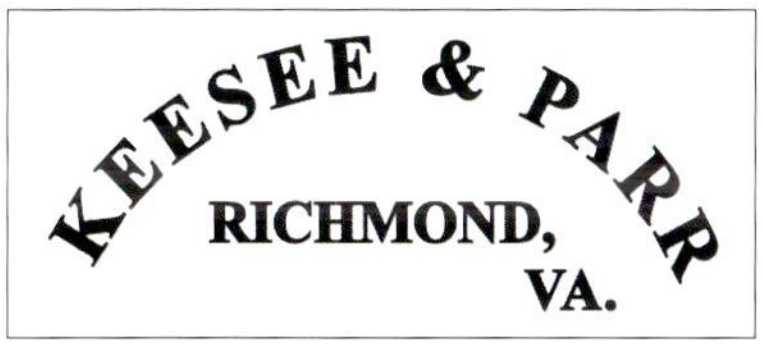

Figure 12 Reconstruction of the stamp used by Keesee & Parr. (Artwork by Oliver Mueller-Heubach.)

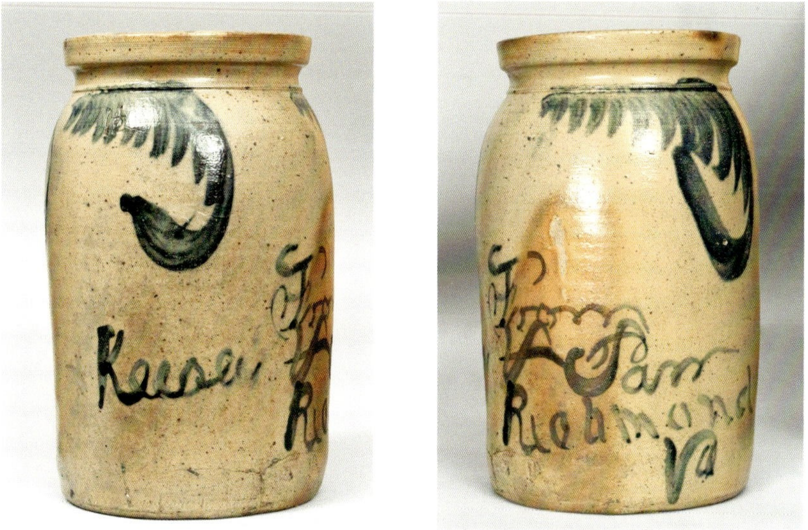

Figure 13 Jar, Keesee & Parr, Richmond, Virginia, 1860–1865. Salt-glazed stoneware, H. 14". Capacity: 3 gallons. Inscribed, in cobalt: "Keesee & Parr / Richmond / Va" (Courtesy, Valentine Museum; photo, Crocker Farm Auctions.)

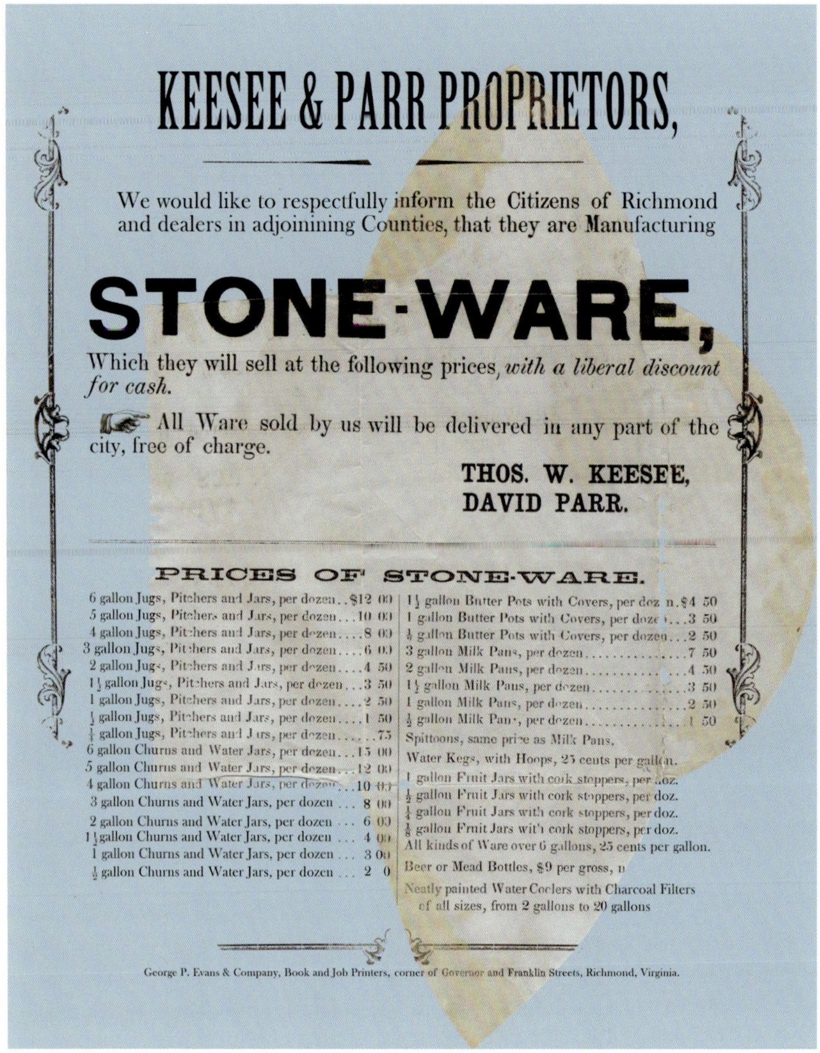

KEESEE & PARR PROPRIETORS,

We would like to respectfully inform the Citizens of Richmond and dealers in adjoining Counties, that they are Manufacturing

STONE-WARE,

Which they will sell at the following prices, *with a liberal discount for cash.*

☞ All Ware sold by us will be delivered in any part of the city, free of charge.

THOS. W. KEESEE,
DAVID PARR.

PRICES OF STONE-WARE.

6 gallon Jugs, Pitchers and Jars, per dozen..$12 00
5 gallon Jugs, Pitchers and Jars, per dozen...10 00
4 gallon Jugs, Pitchers and Jars, per dozen....8 00
3 gallon Jugs, Pitchers and Jars, per dozen....6 00
2 gallon Jugs, Pitchers and Jars, per dozen....4 50
1½ gallon Jugs, Pitchers and Jars, per dozen...3 50
1 gallon Jugs, Pitchers and Jars, per dozen....2 50
½ gallon Jugs, Pitchers and Jars, per dozen....1 50
¼ gallon Jugs, Pitchers and Jars, per dozen......75
6 gallon Churns and Water Jars, per dozen...15 00
5 gallon Churns and Water Jars, per dozen...12 00
4 gallon Churns and Water Jars, per dozen...10 00
3 gallon Churns and Water Jars, per dozen ... 8 00
2 gallon Churns and Water Jars, per dozen ... 6 00
1½ gallon Churns and Water Jars, per dozen ... 4 00
1 gallon Churns and Water Jars, per dozen ... 3 00
½ gallon Churns and Water Jars, per dozen ... 2 0

1½ gallon Butter Pots with Covers, per dozen.$4 50
1 gallon Butter Pots with Covers, per dozen...3 50
½ gallon Butter Pots with Covers, per dozen...2 50
3 gallon Milk Pans, per dozen...............7 50
2 gallon Milk Pans, per dozen...............4 50
1½ gallon Milk Pans, per dozen..............3 50
1 gallon Milk Pans, per dozen...............2 50
½ gallon Milk Pans, per dozen...............1 50
Spittoons, same price as Milk Pans.
Water Kegs, with Hoops, 25 cents per gallon.
1 gallon Fruit Jars with cork stoppers, per doz.
½ gallon Fruit Jars with cork stoppers, per doz.
¼ gallon Fruit Jars with cork stoppers, per doz.
⅛ gallon Fruit Jars with cork stoppers, per doz.
All kinds of Ware over 6 gallons, 25 cents per gallon.
Beer or Mead Bottles, $9 per gross, n
Neatly painted Water Coolers with Charcoal Filters of all sizes, from 2 gallons to 20 gallons

George P. Evans & Company, Book and Job Printers, corner of Governor and Franklin Streets, Richmond, Virginia.

Figure 14 Digital restoration of Keesee & Parr handbill, 1858–1865. (Private collection.) This handbill, printed by George P. Evans and Co. in Richmond, Virginia, details the various forms offered by the business and the price of each. When found, the handbill had been cut and folded to create an envelope, a relic of Richmond's wartime paper shortage. Two Jefferson Davis stamps were affixed to the outside. It is addressed to Franklin Davis, Loudon County, Virginia, sent by "D. Parr," whose signature appears in the upper left edge of the envelope.

Like a number of their southern Henrico potter contemporaries, the Parrs were members of the Society of Friends. The Quaker Parrs had a profound impact on stoneware production and marketing in the Richmond area during the second half of the nineteenth century. Tilman [or Tilighman] Vestal, born to a Quaker potting family in North Carolina, was another craftsman who worked with Parr, in 1864.

Having kept his business going during the lean years of the Civil War, Parr was unable to protect his factory from the conflagration initiated during the evacuation of Richmond in April 1865. A large portion of the city, along with numerous other warehouses and businesses along Cary Street, suffered partial or wholesale destruction.[26] Not only was the pottery temporarily out of commission, but three months later the Carlton House store was consumed—and with it the Keesee & Parr partnership.

Evidence suggests that Parr's operation was revived circa 1866 and continued until circa 1890. Parr newspaper advertisements grew sparse in the late 1860s, but in the 1870s they expanded their business with a flowerpot-molding machine, which they exhibited at state fairs and—as revealed by flowerpot wasters in a broad range of sizes—put to steady use. The *Richmond City Directory* of 1889–1890 lists John L. Parr as general manager of the Richmond Pottery Company at 3734 and 3736 Lester Street.

The business had wide-ranging and prolific production from which many examples survive, though relatively few are marked. They range from plain vessels to those bearing elaborate cobalt-brushed florals. Forms include straight-sided jars, jugs, pitchers, milk pans, cream and butter pots, fruit jars, bread risers, churns, bottles, water kegs and coolers, chamber pots, and spittoons, along with firebrick crucibles, furnaces, pipes, chimney sleeves, and so forth (fig. 14). Parr is also known to have produced special commission pieces.[27]

Parr may have taken over or occupied John P. Schermerhorn's earlier Port Mayo, Rocketts, location.[28] The site of the Parr factory was investigated by archaeologists from the College of William and Mary's Center for Archaeological Research in 2009. The extensive waster dump contained a wide range of vessel forms and decorative treatments as well as capacity and maker's marks (fig. 15). An initial collection of this material was made and catalogued. No stoneware fragments were identified that suggested that the Parrs continued operations at Schermerhorn's earlier Rocketts kiln. Nor were any *in situ* kiln remains identified, although a probable location across the road atop the bluff was suggested by the research.[29] The site continues to be threatened by urban development, and it is hoped that additional research can be undertaken before it is lost forever.

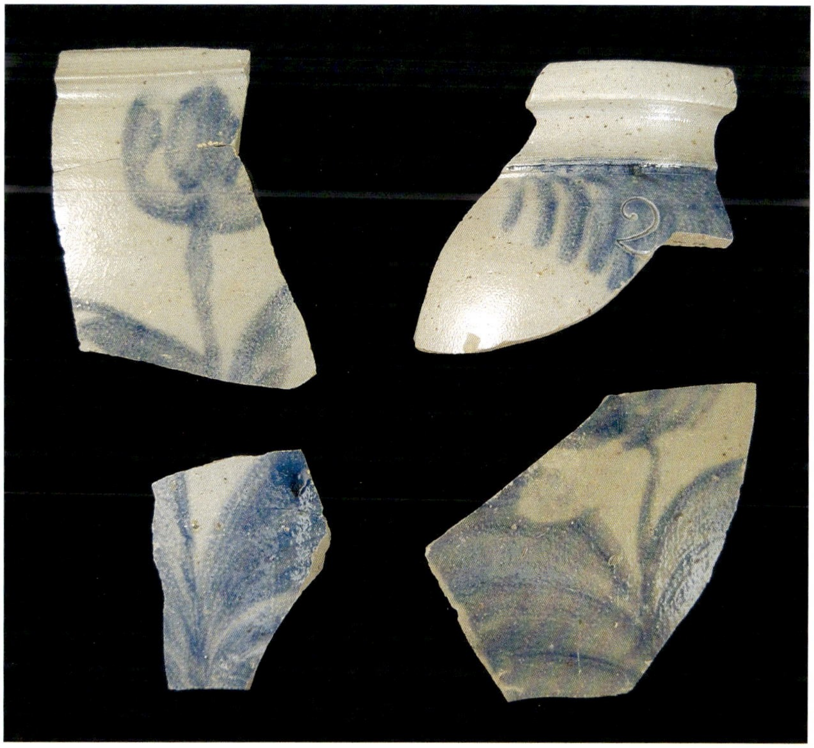

Figure 15 Stoneware fragments from the David Parr Pottery site with flower decoration typically sold under the Keesee & Parr name. (Courtesy, William and Mary Center for Archaeological Research.)

HISTORICAL OVERVIEW OF FOUR MILE CREEK/BAILEY'S CREEK AREA

Above a sharp bend in the James River ten miles southeast of Richmond, in southern Henrico County, Virginia, lies one of the unheralded heartlands of Tidewater history (fig. 16). Early on, John Rolfe's plantation, Varina, arose nearby, becoming one of the eight original Virginia shires and serving as the Henrico County seat from 1634 to 1752.

As Richmond became the new seat of government, Varina's fortunes waned, and much of the land became part of the Cox family plantation. In 1698 Curles, or Curles Neck, the next peninsula to the east, became a seat of Virginia's Randolph family.

The area's low-lying topography was improved with canals and levees, while natural elevations such as New Market Heights and Claymount became centers first of development and, later, of conflict. The area of Four Mile and Bailey's creeks hosted movements of troops during the Revolution and particularly in the Civil War, when the battles of Deep Bottom, Malvern Hill, and New Market Heights encircled and scarred the landscape.

The establishment of numerous individual potteries drove the area's growth and transformation. In the case of the Four Mile Creek/Bailey's Creek area, a prominent and substantial mill seat was transformed with the creation of a pottery manufactory that could exploit nearby resources including fuel, water, rich clay, and mineral deposits. The channelization of streams and construction of canals facilitated the efficient transport of products via the James River. Soon, other potters were drawn to the area and what developed was one of the most significant nineteenth-century ceramic manufacturing complexes in the state, with four potteries and a half-dozen kilns in operation. Despite the absence of any signed wares from several of these potteries, comparisons of regionally collected extant wares with archaeological sherds from local sites provided tentative attributions for these identified operations.

RICHARD RANDOLPH'S POTTERY

Norwich Mills was an important eighteenth-century milling operation on Four Mile Creek. By the 1800s, however, it had fallen into disrepair. Speculator David Ross restored the mill and in 1809 sold the works, together with a potential contract for the miller, Samuel Ellyson, to Richard Randolph and his partner, Michael W. Hancock.

A 1781 map of the James compiled by Major Capitaine to aid American and French troops against Cornwallis includes the lands later owned by Ross (fig. 17). This map had been interpreted as showing the seat of Norwich Mills on Four Mile Creek just west of the split with Bailey's Creek.[30]

At the time of Ross's purchase, the tract was referred to as the "plantation called Norwich Mills" situated "between Bailey's run on the east & fore mile creek south and the Lands of Thos. Matthews North in the county of Henrico."[31] It contained eight buildings including a four-story merchant mill, a storehouse, a bake-house, a barn, a miller's house, and three dwellings, one located "on the road" (figs. 18, 19). All of the buildings were of

Figure 16 Map showing the Four Mile Creek/Bailey's Creek area of southeastern Henrico County, Virginia. (Artwork by Oliver Mueller-Heubach.)

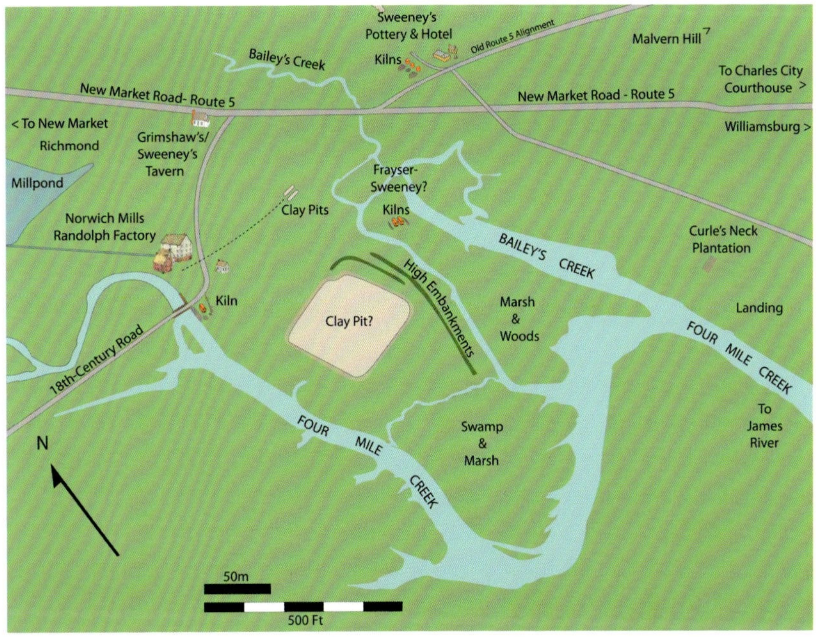

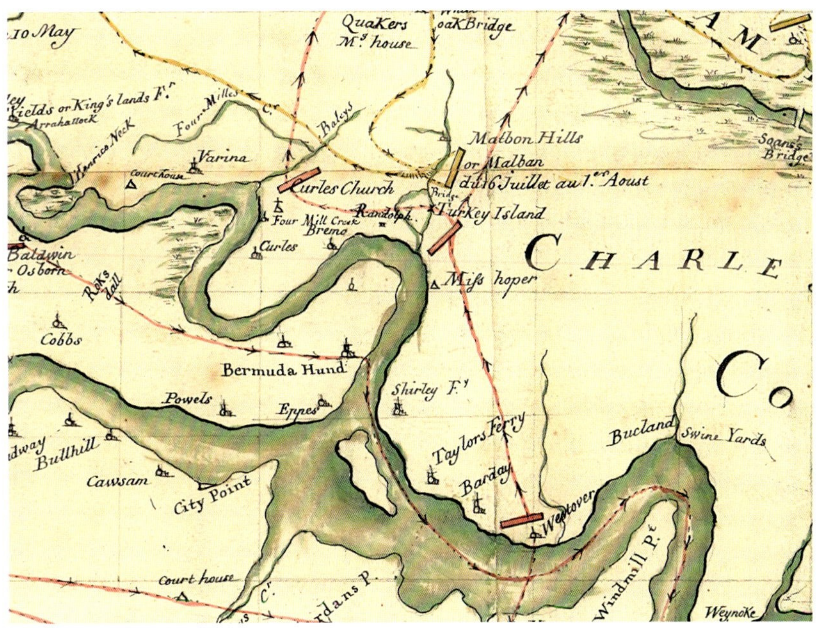

Figure 17 Detail, *Campagne en Virginie du Major Général M'is de LaFayette: ou se trouvent les camps et marches, ainsy que ceux du Lieutenant Général Lord Cornwallis en 1781*, by Major Michel Capitaine du Chesnoy, aide-de-camp of General LaFayette. (Courtesy, Library of Congress.)

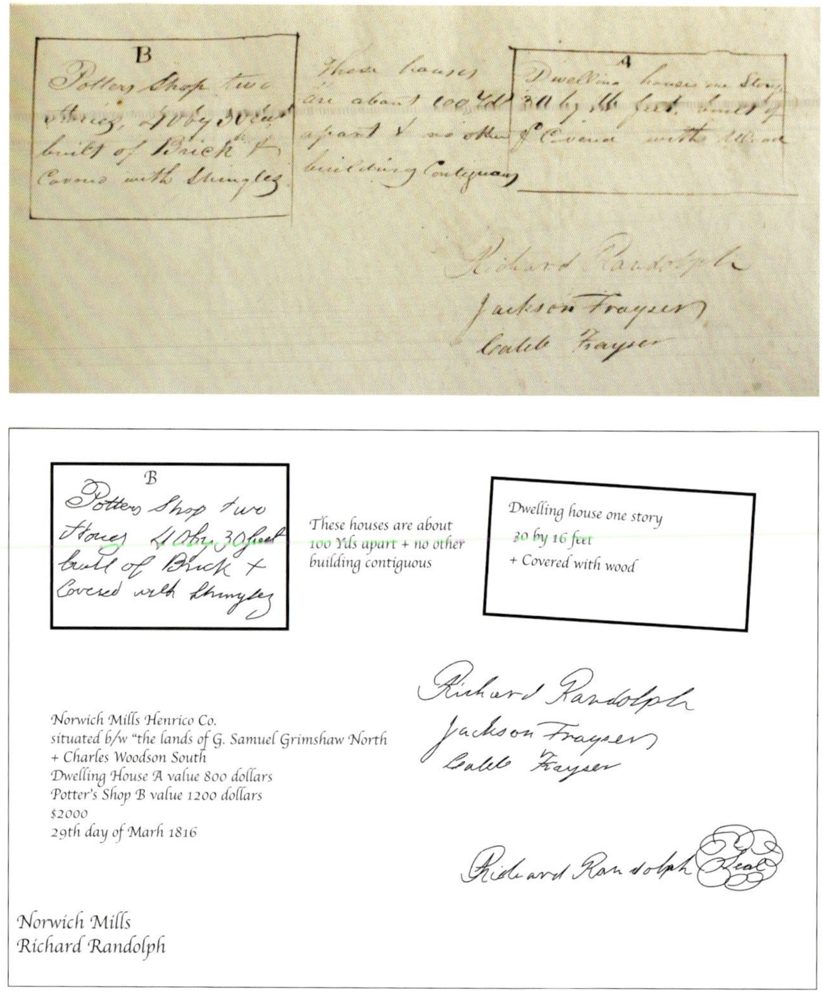

Figure 18 *Top*: Mutual Assurance Society sketches of insured buildings comprising David Ross's Norwich Mills complex in 1816. (Courtesy, Library of Virginia.) *Bottom*: Black-and-white schematic sketch of the insured buildings. (Artwork by Oliver Mueller-Heubach.)

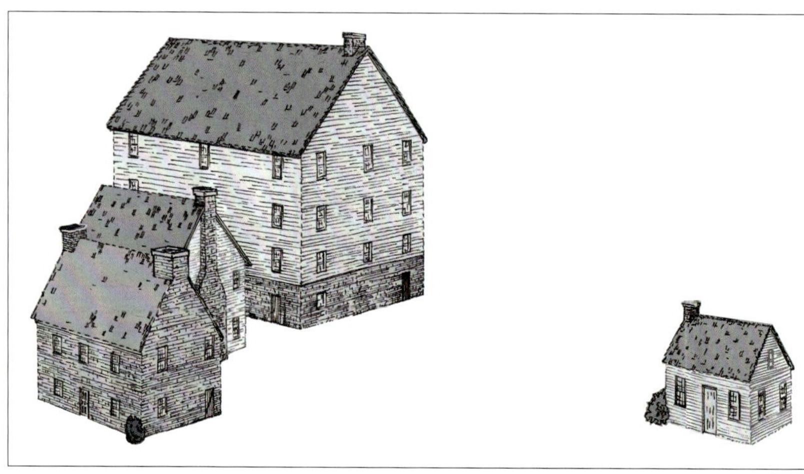

Figure 19 Hypothetical illustration of eighteenth-century Norwich Mills complex showing the buildings on site from the time of Ross's 1809 transfer (reflecting those insured in 1806) to the evolution or conversion in 1816 of the site to a pottery manufactory (ca. 1812–1813) under Randolph's ownership. (Artwork by Oliver Mueller-Heubach.)

frame construction with the exception of the mill, which was brick and frame, and the bake-house and storehouse, which were of solid brick.[32]

By June 1809 Richard Randolph and partner Michael W. Hancock were the proprietors of Norwich Mills. In addition, and perhaps in an effort to ensure their continued successful operation, between the years of 1809 and 1815 Richard Randolph, in theory, oversaw the upkeep of some of the roads

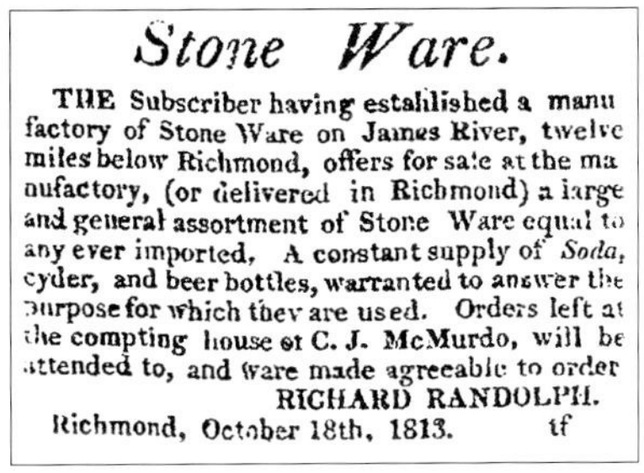

Stone Ware.

THE Subscriber having established a manufactory of Stone Ware on James River, twelve miles below Richmond, offers for sale at the manufactory, (or delivered in Richmond) a large and general assortment of Stone Ware equal to any ever imported. A constant supply of *Soda*, cyder, and beer bottles, warranted to answer the purpose for which they are used. Orders left at the compting house of C. J. McMurdo, will be attended to, and ware made agreeable to order.

RICHARD RANDOLPH.

Richmond, October 18th, 1813. tf

Figure 20 Advertisement for Randolph's "Soda, cyder, and beer bottles" available for sale at C. J. McMurdo's "compting house" in Richmond, October 18, 1813. *Enquirer*, Richmond, October 22, 1813, p. 3-4.

and bridges in the immediate vicinity of his land holdings. The upkeep of the local roads was essential to travel and the continuing economic success of the area.

After Randolph's purchase the property was described as a full 500 acres, perhaps because of an updated survey, and located ten miles southeast of the courthouse and near New Market.[33] Randolph disposed of 136.5 acres in 1812, but held the remaining 363.5 acres (containing $500 worth of improvements) until 1820. The improvements included a pottery. On October 22, 1813, Randolph advertised as follows (fig. 20):

> The Subscriber having established a manufactory of STONE WARE
> On James River, twelve miles below Richmond, offers for sale at the manufactory (or delivered in Richmond,) a large and general assortment of stone ware, in quality equal to any ever imported. A constant supply of Soda, cyder and beer bottles, warranted to answer the purpose for which they are used. Orders left at the coumpting house of C. J. McMurdo
> Will be attended to, and ware made agreeable to order.
> October 23. Richard Randolph[34]

By 1816 the portion of the Norwich Mills tract insured by Richard Randolph consisted of a frame dwelling measuring 30 feet by 16 feet and a "two-storey potters shop" of brick, 40 feet by 30 feet. The improvements were described as lying some 100 yards apart and removed from other structures.[35] An earlier insurance policy for the property of David Ross at his plantation, Norwich Mills, identified a building C, described as a "Bake house of Brick covered with wood, 40 by 30 feet two stories high."[36] On March 29, 1816, the building was reevaluated at $1,200 ($520 more than under Ross's ownership) and plotted as building B.[37]

In an advertisement from July 4, 1817, entitled "KAOLIN," Randolph offers for sale some 400 acres of his land and these pottery-related improvements (fig. 21). The Kaolin moniker promotes the property's greatest resource—clays—suitable for both brick and stoneware production. The existence of a stoneware manufactory "distinguished for its excellent wares" suggests that finished products as well as the clays were being shipped via Four Mile and Bailey's creeks. Randolph's mention of recently adding "one artist" to an operation whose production was equivalent to $400–$500 per month hints at the economic potential of the area. In addition to the touted agricultural value of the land for producing grain and grass crops, the approximately 200 acres of wood and timber promised a standing fuel supply for the pottery.

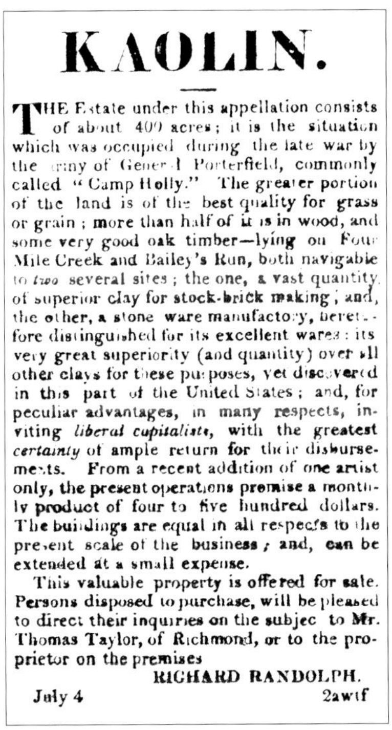

KAOLIN.

THE Estate under this appellation consists of about 400 acres; it is the situation which was occupied during the late war by the army of General Porterfield, commonly called "Camp Holly." The greater portion of the land is of the best quality for grass or grain; more than half of it is in wood, and some very good oak timber—lying on Four Mile Creek and Bailey's Run, both navigable to *two* several sites; the one, a vast quantity of superior clay for stock-brick making; and, the other, a stone ware manufactory, heretofore distinguished for its excellent wares: its very great superiority (and quantity) over all other clays for these purposes, yet discovered in this part of the United States; and, for peculiar advantages, in many respects, inviting *liberal capitalists*, with the greatest *certainty* of ample return for their disbursements. From a recent addition of one artist only, the present operations promise a monthly product of four to five hundred dollars. The buildings are equal in all respects to the present scale of the business; and, can be extended at a small expense.

This valuable property is offered for sale. Persons disposed to purchase, will be pleased to direct their inquiries on the subjec to Mr. Thomas Taylor, of Richmond, or to the proprietor on the premises

RICHARD RANDOLPH.

July 4 2awtf

Figure 21 Advertisement for Richard Randolph's Kaolin estate, July 4, 1817. *Enquirer*, Richmond, July 4, 1817.

The *Census of Manufactures* for 1820, however, does not list Randolph's name (a mere four years after he insured his "potters shop" and associated residence), which suggests his pottery was operated by someone else.[38] It seems the property was not sold in its entirety, for in 1821 Richard Randolph transferred 335 acres (which included the portion containing $500 worth of improvements) of his then 363.5-acre Norwich Mills tract to Francis Lewis, who quickly enhanced the improvements to a value of $1,174. Lewis retained the land until 1828, when it passed into the hands of William Depriest, who, in turn, conveyed the property to Francis Frayser in 1835. Although Frayser died within a year, the tract remained part of his estate through 1850; the portions sold off were devoid of improvements.[39]

Personal property tax rolls from Randolph's years as owner of the Norwich Mills tract indicate that he owned slaves.[40] Additionally, in June 1815 the overseers of the poor for Henrico County bound out to Randolph two orphaned brothers and free persons of color, Tom and Joe Givin (fourteen and twelve years old, respectively).[41] They are thought to have served as apprentices in his pottery business.[42] Coincident with this indenture, the overseers of the poor also bound out to Randolph's New Market neighbor Samuel P. Parsons two brothers from the same family, Jack (age eighteen) and Frederick Givin (age fourteen).[43]

While it is unknown who was the first potter on Richard Randolph's premises, it seems likely that Samuel Frayser leased from Randolph prior to establishing his own facilities (discussed below). Randolph may well have been an absentee landlord after 1817, particularly in light of the fact that just three days after Randolph advertised "Kaolin" for sale, Frayser was named overseer of the road from "four Mile Creek bridge to Little Cornelius's." This clearly indicates Frayser's residence or employment in the immediate vicinity.[44] His appearance in the 1820 manufacturer's census bears witness to his new, more significant role in the pottery business, while Randolph's lack of inclusion suggests a diminished role, possibly exclusively that of landlord/lessor pending completion of the sale. Also of note is the fact that both Jackson and Caleb Frayser witnessed Randolph's 1816 insurance policy for the pottery shop and house.[45] The presence of his family members living in the southern Henrico area, the fact that they were present as witnesses for the insuring of the pottery, and the subsequent purchase of the property by a relative, makes a strong case for Frayser being Randolph's master potter.

In addition to Frayser and the Givins, it appears likely that Thomas Amoss was involved at Norwich Mills under Randolph. Randolph's mention of "a recent addition of one artist" in his 1817 Kaolin property ad may refer to either Frayser's or Amoss's slightly later affiliation with his operation. Others who may have worked there during the Randolph years include Samuel Wilson, who worked for the DuVals and with John P. Schermerhorn. Isaac Denson had been apprenticed to James DuVal in 1818, but he may have left with Wilson when the DuVal Pottery closed the next year.

Surface finds from the property provide tantalizing clues as to the stone-

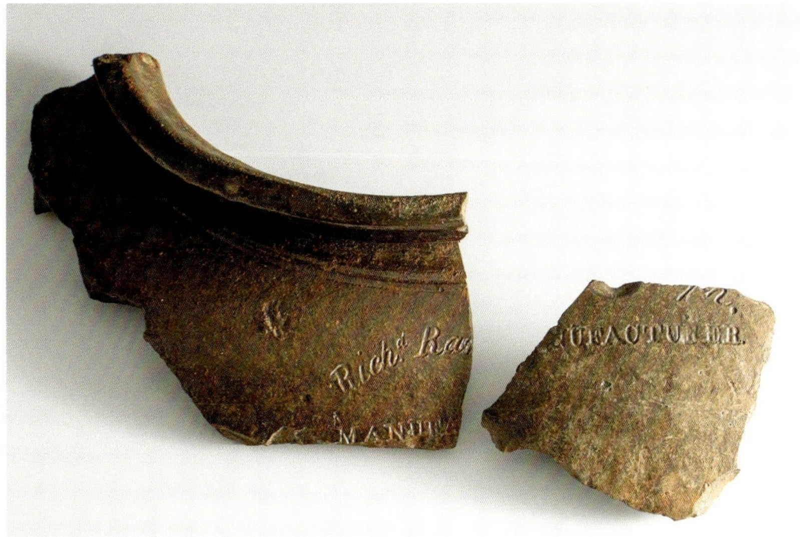

Figure 22 Storage jar fragments, Richard Randolph Manufactory, Four Mile Creek, Henrico County, Virginia, 1813–ca. 1821. Salt-glazed stoneware. Impressed mark: "*Richd Randolph* / MANUFACTURER" (Private collection; photo, Robert Hunter.) No intact examples of stoneware with this mark are known.

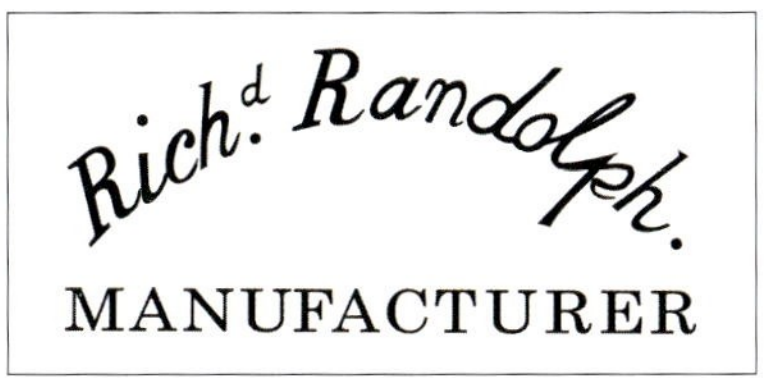

Figure 23 Reconstruction of the Richard Randolph mark. (Artwork by Oliver Mueller-Heubach.)

ware products of the Randolph-sponsored operation. Of particular interest are two fragments with a partial maker's stamp (figs. 22, 23). The design is very similar to that used by Schermerhorn in Richmond.

Randolph's initial 1813 advertisement announced the establishment of his "Stone Ware" factory and emphasized his "constant supply of Soda, cyder and beer bottles" (see fig. 20). Thomas Jefferson wrote to him in 1814 requesting stoneware jugs in two sizes—"quart jugs" and "pottles" (half gallons)—for his homemade beer at Monticello:

> Dear Sir [Randolph] Monticello Jan. 25. 14
> Will you be so good as to send me two gross of your beer jugs; the one gross to be quart jugs, and the other pottles d^{o}. they are to be delivered to a mr William Johnson a waterman of Milton, who will apply for them about a week hence. Mr Gibson will be so good as to pay for them on your presenting this letter. they should be packed in crates, or old hogsheads or such other cheap package as you use. Accept the assurance of my esteem & respect.
> Th: Jefferson[46]

Although he also wrote to England for beer bottles, Jefferson heard of or saw the ad for the new Four Mile operation and may have wished to favor native industry and familial ties. In fact, a second request was made in late September that year for "four gross more" jugs to accommodate Jefferson's new brew of "malt strong beer."[47]

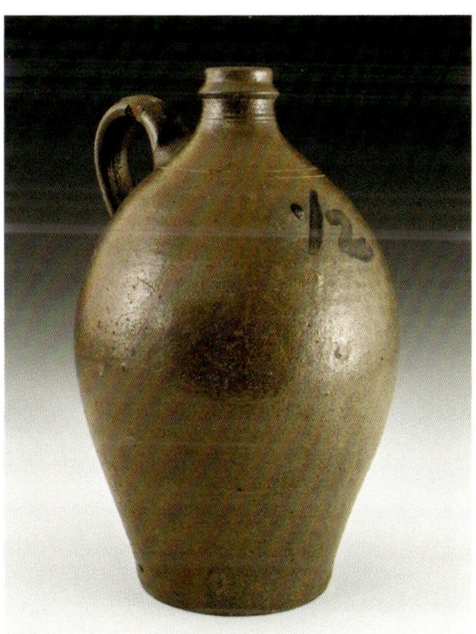

Figure 24 Pottle, attributed to the Richard Randolph pottery, Henrico County, 1813–ca. 1821. Salt-glazed stoneware. H. 9¾". (Private collection; photo, Robert Hunter.) The identical lip-and-neck configuration has been found among archaeological collections at Monticello. Those pottles were ordered by Thomas Jefferson in 1814 to hold beer brewed by Peter Hemmings.

A recent comparison of bottle fragments recovered at Randolph's pottery with those found at Monticello reveal matching forms with distinctive neck treatments and iron-oxide washed surfaces, confirming Randolph's operation as their source. As this article was going to press, an intact example of a pottle was identified (fig. 24).

The most important information from the Jefferson letters is gleaned from Randolph's response to the first request for beer jugs, in which he stated that "my best workman was in New York when I received your letter, he returned yesterday and will make your jugs next week, when they shall be forwarded agreeable to your directions."[48] Viewed in the context of

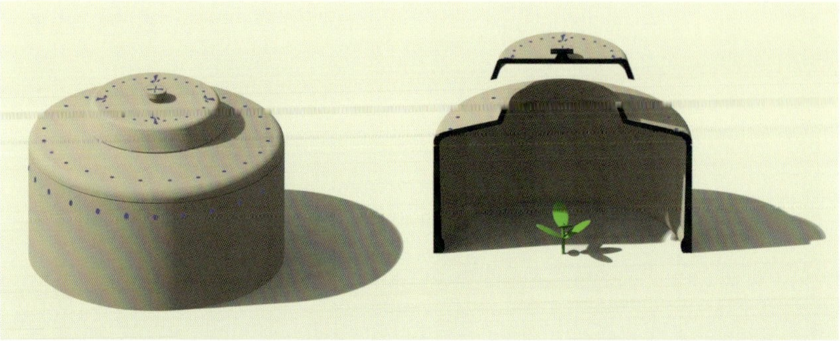

Figure 25 Illustration and cross-section of a sea kale pot based on fragments recovered from Monticello and now attributed to Thomas Amoss's production at the former Randolph pottery on Four Mile Creek in southeastern Henrico County, Virginia. (Artwork by Tabitha Lavis.)

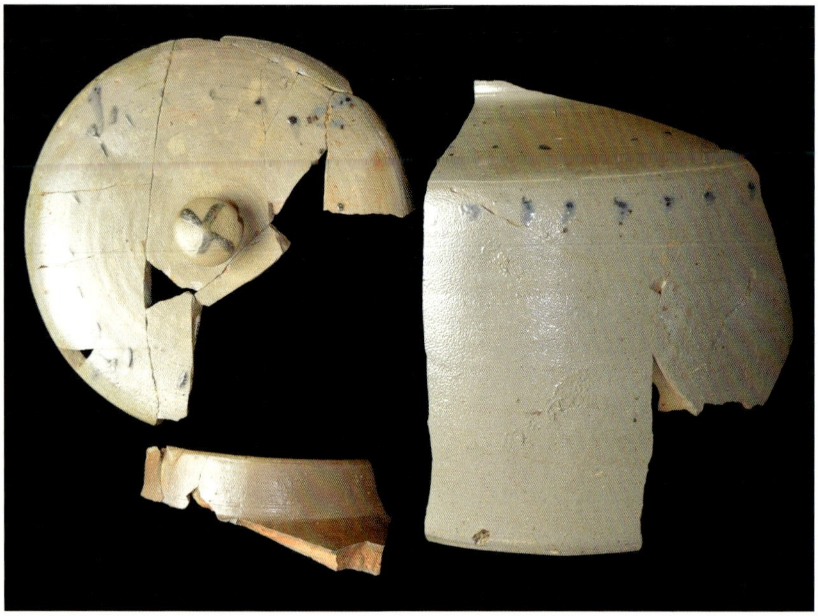

Figure 26 Kale-pot fragments, attributed to Thomas Amoss, Henrico County, Virginia, 1818–1822. Salt-glazed stoneware. (Courtesy, Department of Archaeology, Monticello; photo, Kurt Russ.) Note the cobalt slip decoration of dots and sea kale sprouts around the circumference and the top of the lid. The clay and the style of the cobalt slip application match Amoss's production at Four Mile Creek.

his 1817 addition of "one artist" to his Kaolin operation, it appears Randolph had respect for these craftsmen.

Randolph also provided kale pots for Monticello's gardens. A sort of lidded bell jar, they were used to blanch or "bleach"— and sometimes force— young sea kale plants by sheltering them from cold and direct sun.[49] Packed around with heat-generating manure, the lidded pots allowed gardeners to inspect and tend the young plants inside (fig. 25). On February 21, 1821, seven years after his first request for bottles, Jefferson wrote to Edmund Peyton inquiring about "some earthen pots" for covering sea kale in his garden, which he had been told were made "at a pottery in or near Richmond.[50] Having requested at that time to be supplied with "½ hundred," it was not until September 16, 1821, that he recorded "two pennies" being paid for bringing the pots to Monticello.[51] In his monograph *Archaeology at Monticello*, William M. Kelso notes a March 29, 1821, letter from Peyton to Jefferson referencing the original order "from Wickham from Richard Randolph's pottery."[52]

In May 1822 Jefferson again inquired of Randolph: "I do not know whether you continue your pottery. If you do, I will request 50. pots for the sea kale such as you saw here, which indeed are made on the exact model of mr Wickham's. If delivered packed in hogsheads to the order of Colo. Peyton, he will, on sight of this letter, pay for them."[53] Randolph did not elaborate on the status of his pottery—he was at this date no longer its owner—but responded to Jefferson as follows: "I have given directions for your bleaching pots to be made, and as soon as they are done, shall be delivered to Colo. Peyton."[54]

Randolph forwarded Jefferson's order to Thomas Amoss, who was then potting at the factory. Examination of fragments in archaeological collections at Monticello reveal both the plain Wickham examples and some that are attributable to Thomas Amoss's production at Four Mile Creek (fig. 26). Decorated in Amoss's characteristic slip-trailing, the pots were ringed in fine dots interspersed with what appear to be stylized kale-plant sprouts.

Randolph also had an association with the DuVal family. His daughter, Beverly, married druggist Philip DuVal (1789–1847), Benjamin DuVal's son. Preliminary analysis of fragments found at Randolph's pottery shows that many of the vessels share attributes with those produced at DuVal's Richmond pottery, suggesting they may have made by the same potter, a past apprentice, or someone with similar training, perhaps in New York.

SAMUEL FRAYSER'S POTTERY

Samuel Frayser (ca. 1787–1849), identified in the 1820 census of manufacturers as the proprietor of a pottery factory, purchased 63.5 acres from William Hobson in 1818.[55] The acreage lay twelve miles southeast of the Henrico courthouse, probably on Four Mile Creek or Bailey's Creek, near Newmarket. His appointment as overseer of the road from Four Mile Creek to Little Cornelius's on July 7, 1817, suggests he was living in the vicinity prior to the establishment of his own operation.[56] On December 12, 1821, he married Elizabeth Francis Weymouth and they had two children, Samuel Jr. and Ann E.

Despite Samuel Frayser's listing as the proprietor of a pottery factory in 1820, improvements on his own land were not erected until 1824, and his official residence in Henrico County was not established until 1826, suggesting his probable employment working with either Schermerhorn in Richmond and/or for Randolph in Henrico prior to this time. Like Schermerhorn, Frayser was a private in the Virginia Militia during the War of 1812; they both served for over a year in 1812–1813 in Captain Robert Gamble's troop of Calvary, 19th regiment, Virginia Militia, commanded by Lt. Col. John Ambler.

One possible scenario is that when Schermerhorn first visited the region in 1810–1811, he encountered Randolph and perhaps Frayser and learned of Randolph's plan to establish a pottery on his recently acquired Camp Holly acreage. Schermerhorn, or one or more potters traveling with him, might have agreed to stay and assist Randolph. Although Schermerhorn was soon employed by DuVal, a fellow northern potter might have remained to become Randolph's first potter. On the other hand, it is equally plausible that one of the master potters involved in earthenware production in Rich-

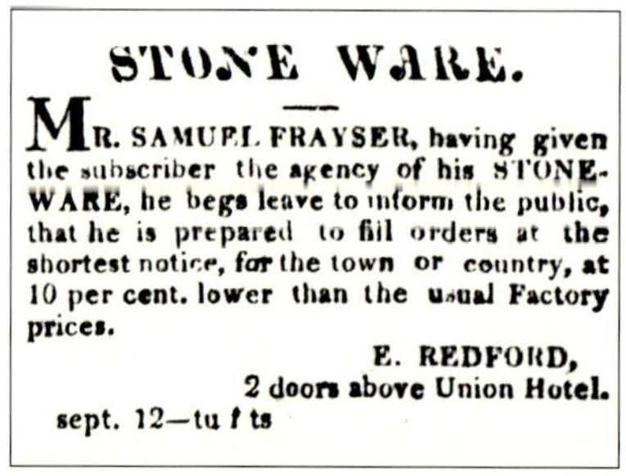

STONE WARE.

MR. SAMUEL FRAYSER, having given the subscriber the agency of his STONE-WARE, he begs leave to inform the public, that he is prepared to fill orders at the shortest notice, for the town or country, at 10 per cent. lower than the usual Factory prices.

E. REDFORD,
2 doors above Union Hotel.

sept. 12—tu f ts

Figure 27 E. Redford advertisement for Frayser stoneware appearing in the *Richmond Commercial Compiler*, September 12, 1820, p. 3-5

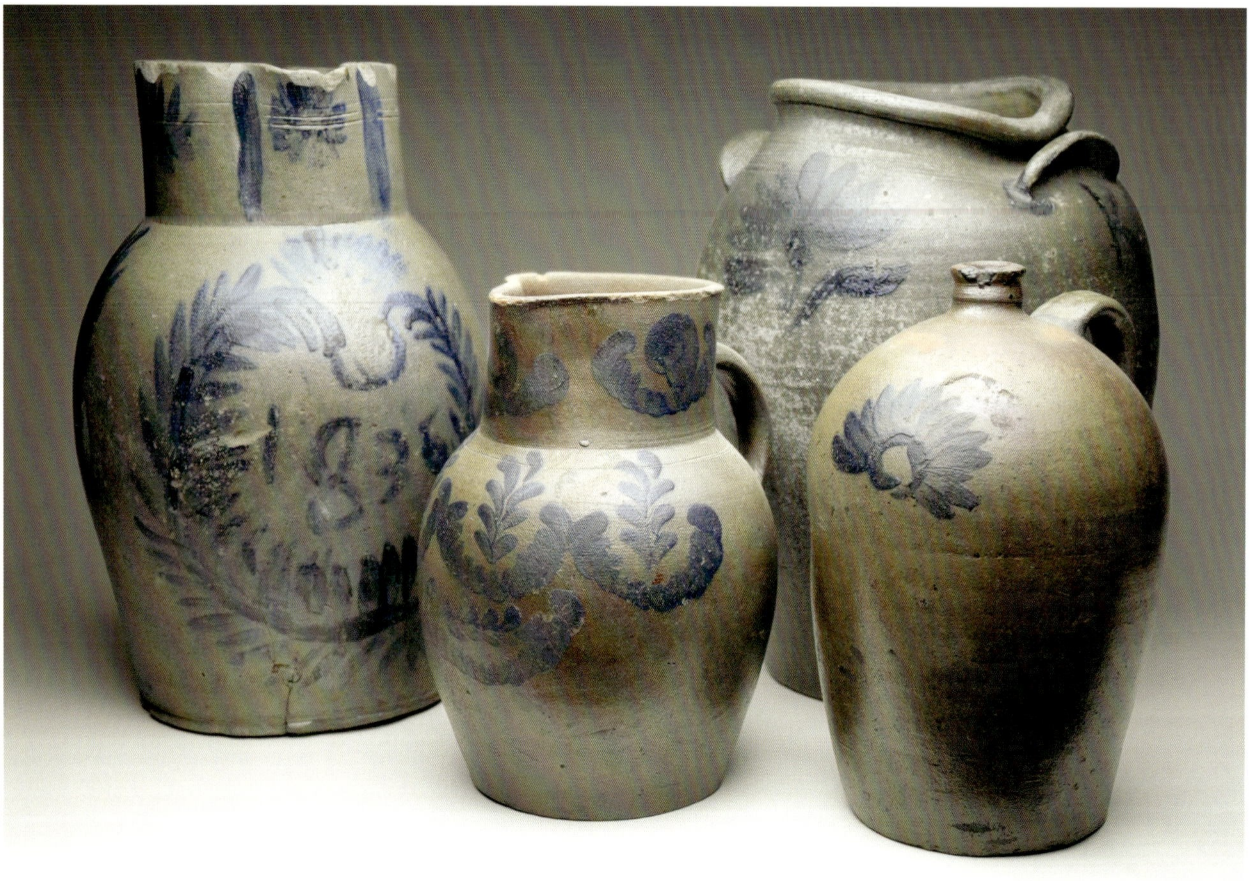

Figure 28 Vessels, attributed to Samuel Frayser, Henrico County, Virginia, 1817–1849. Salt-glazed stoneware. H. of tallest 15". The largest pitcher is dated 1836. (Private collections; photo, Kurt Russ.) The storage jar has a brushed-cobalt capacity mark, "4," under one handle and a "K" beneath the other.

mond circa 1782–1794 made his way downriver to establish and run Randolph's new pottery, perhaps instructing Frayser in the trade. A third, less likely possibility is that Frayser continued to work with Schermerhorn but, circa 1820, established his own operation on land purchased from Hobson in 1818, on his relation Francis Frayser's land in the vicinity of Deep Bottom and Varina Road, or at some other location. He might also have run the Amoss operation after the Quaker's death.

The Henrico County 1820 Census of Manufacturers recorded Samuel Frayser as using 50 tons of clay, 80 cords of wood, and 15 sacks of salt, and employing 3 men and no boys, with 1 kiln, 2 wheels, and an annual payroll of $200, to produce "Stone Ware of all Kinds."[57] His wares were advertised in Richmond by an agent, E. Redford, in 1820 (fig. 27).[58] Between 1822 and 1829 Frayser leased two adjacent buildings (valued between $3,000 and

$3,250) in Rocketts. The dwellings were located at the corner of Bloody Run and Elm Street and were owned by widow Mary Weymouth, who may have been Frayser's mother-in-law. Whether these served as a residence for him and associated workers or as a warehouse, or both, has not been determined.[59]

At Norwich Mills, Frayser may have worked alongside Randolph's New York–connected "best workman." Did he then become proprietor of the pottery as a lessee in 1820? If so, then the question becomes whether he established a separate facility on his own land circa 1824 or remained at Norwich Mills through its transition to his relative's ownership circa 1835, or even to potter Stephen Sweeney's purchase three years later.

From 1824 through 1849 Samuel Frayser was taxed on buildings that ranged in value from $125 to $225.75, and he continued to acquire additional land until his death, in 1849. Frayser's son Samuel Jr., who was seventeen when his father died, would have been entering apprentice age in the early 1840s and may have worked alongside his father during this period. Some of the land Frayser Sr. acquired was in the vicinity of Gravelly Hill, an area of rich clay and mineral resources exploited by several area potters.[60]

Two groups of extant vessels have been attributed to Samuel Frayser Sr. The first bears decorative elements strongly associated with Schermerhorn, in particular the festoons and wreaths (fig. 28). The second group exhibits incised names in combination with horizontal garlands and leafy flourishes.[61] Vessels associated with the first group include two pitchers with matching garlands, one bearing the date 1836 within a wreath (which appears on another three-gallon example), and a smaller example by the same hand. A three-gallon storage jar with the characteristic fan or headdress-like motif and a series of smaller jars from the same production have also been identified.[62] The storage jars have distinctive, well-defined rims and necks and are characterized by ovoid forms, applied U- or crescent-shaped handles, dark clay bodies (typically with rough, thinly salt-glazed exterior surfaces), and brushed-cobalt florals that sometimes are composed of elongated curvilinear leaves. It now appears that a cobalt-slip eagle jar previously attributed to Schermerhorn is attributable to Frayser's operation (fig. 29).[63]

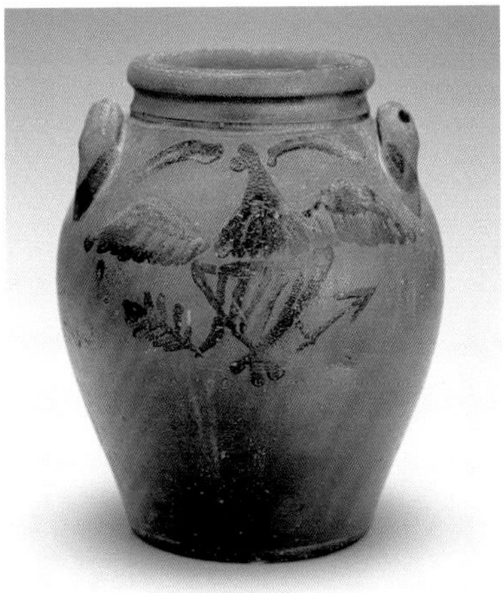

Figure 29 Storage jar, attributed to Samuel Frayser, Henrico County, Virginia, 1817–1849. Salt-glazed stoneware. H. 10". Capacity: 1 gallon. (Private collection; photo, Gavin Ashworth.) This ovoid storage jar displays a similar eagle on the reverse, but that version has olive branches in both talons.

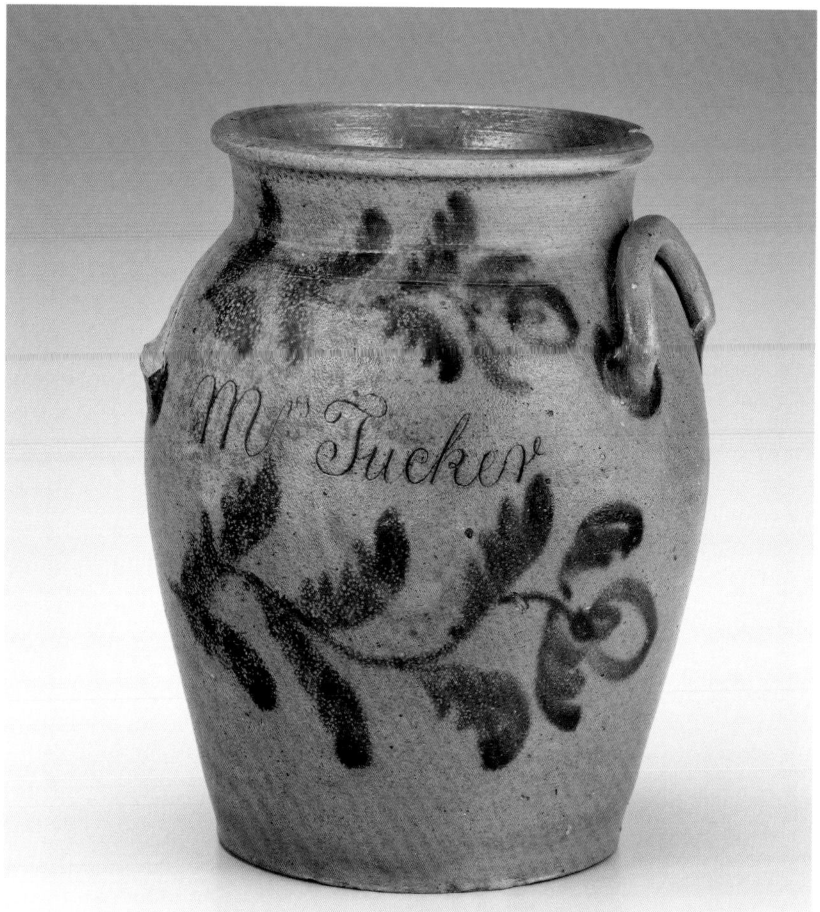

Figure 30 Storage jar, attributed to Samuel Frayser, Henrico County, Virginia, 1817–1849, Salt-glazed stoneware. H. 9⅝". Incised, in script: "M[rs] Tucker" (Private collection; photo, Gavin Ashworth.)

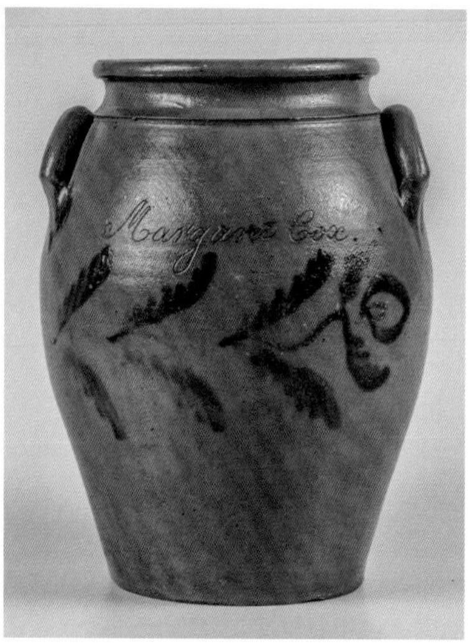

Figure 31 Storage jar, attributed to Samuel Frayser, Henrico County, Virginia, 1817–1849. Salt-glazed stoneware. H. 9½". Incised, in script: "Margaret Cox"; "Quinces" (Courtesy, Virginia Historical Society; photo, Kurt Russ.) The cobalt floral decoration matches that on the jars illustrated in figs. 30 and 33.

The second group of Frayser-attributed wares is made up of five similarly shaped stoneware jars with related decorations and incised names. Intended as presentation pieces, the incised names are clues to the area of their manufacture and the historical associations of potters with neighbors, adjacent landholders, and contemporary political and public figures. These wares have been attributed to Frayser's production at Randolph's Four Mile Creek pottery.

The most recently discovered vessel of this group is incised "Mrs. Tucker" (fig. 30). A "store of John Tucker" was located just to the northeast of Richmond, along Mechanicsville Turnpike. Although only a short distance to the southwest of the Schermerhorn estate and a residence associated with Schermerhorn's wife, the location was at least ten miles north of the Newmarket area potteries, including Frayser's operation. Tucker might have sold wares from local potteries, Frayser's among them. Presumably this jar was made as a presentation piece for John's wife, Mrs. L. H. Tucker, who according to the 1860 Federal Census was living in Richmond's third ward with her husband, John R., and five children.

A second jar bears the inscription "Margaret Cox" and relates to the Cox farm just a couple of miles from the pottery (fig. 31). A Civil War–era map shows Henry Cox's large tract of land at the southern end of Varina Road, west of a Mrs. Frazier's land and to the south of the Highland Road. West of the Cox holdings was a large tract along Turnpike Road near its intersection with Kingsland Road, identified as belonging to "E[dward] Cox." This substantial landholding included Cox's Ferry, located at the terminus of Turnpike Road on the banks of the James River. Another neighbor was William B. Randolph, whose one hundred acres adjoined land belonging to a Richard Cox. Despite its different form and clay color, the Cox jar exhibits horizontal garlands virtually identical to those on the Tucker jar (see fig. 30).

Similar in form and clay color is the third example, which is incised "Ann R. Randolph/August 23, 1836" (fig. 32). Several Randolph families lived in the vicinity of the potteries, the most significant being Richard Randolph, who was responsible for the establishment of a pottery on his property by 1813. Mid-century maps identify William B. Randolph with disparate land holdings, including land recorded as southwest of Richmond along the James River, westward of Turnpike Road, as well as a tract to the southeast of the Sweeney pottery several miles distant. William B. Randolph's wife was Anne R. (1783–1830). Although the form of this jar is comparable to the other examples, the clay is decidedly darker with a rougher surface, the glaze is thinner, and the distinct decoration suggests an associated group of wares that was thrown by a different potter and/or decorated by a different hand. The similarity of form and inscription, however, suggest manufacture at the same pottery.

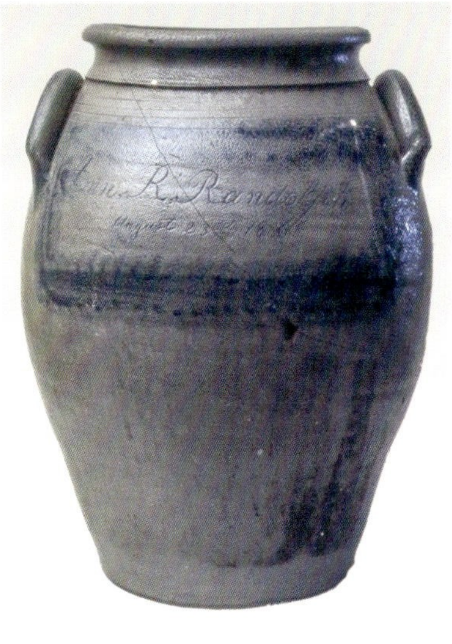

Figure 32 Storage jar, attributed to Samuel Frayser, Henrico County, Virginia, 1817–1849. Salt-glazed stoneware. H. not recorded. Capacity: 1 gallon. Incised, in script: "Ann R. Randolph / August 23, 1836" (Photo, courtesy MESDA Research Files.) Although the form of this jar matches those illustrated in figs. 31 and 33, including the sharp, slightly raised lip or ridge at the base of the neck, the decoration decidedly does not. The broadly amorphous brushed florals were executed by a different decorator.

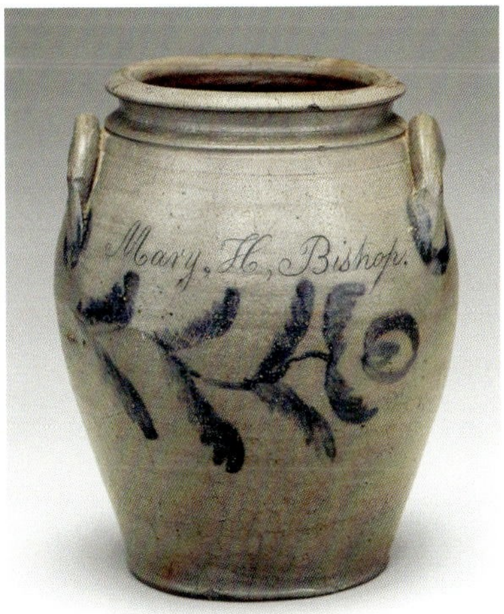

Figure 33 Storage jar, attributed to Samuel Frayser, Henrico County, Virginia, 1817–1849. Salt-glazed stoneware. H. 9 1/4". Incised and highlighted with cobalt: "Mary H. Bishop" (Private collection; photo, Gavin Ashworth.)

The fourth in this group is a semi-ovoid jar bearing the name "Mary H. Bishop" (fig. 33), who is identified in the 1870 U.S. Census as having been born circa 1811 and as a resident of the Varina area. At the time, she was the head of household and shared her household with Virginia (55), Andrew (24), Sarah (20), Nathaniel (21), Truman (16), Mary (2), and Susan (1). In terms of form and garland, the jar most closely parallels the Cox jar (see fig. 31). In fact, the Tucker, Cox, and Bishop jars undoubtedly were decorated by the same hand and most likely thrown by the same potter.

A squat, semi-ovoid, open-mouthed jar with applied crescent-shaped handles and a slightly flaring collar is the fifth and final vessel associated with this grouping (fig. 34). The name "Elizabeth Flournoy" is incised in handwriting remarkably similar to the other examples and appears on the front together with a series of five open circles across the upper shoulder and garlands or flourishes beneath and on the lower half of the vessel.[64] On the reverse is incised "Peaches" (accompanying similar cobalt florals), indicating

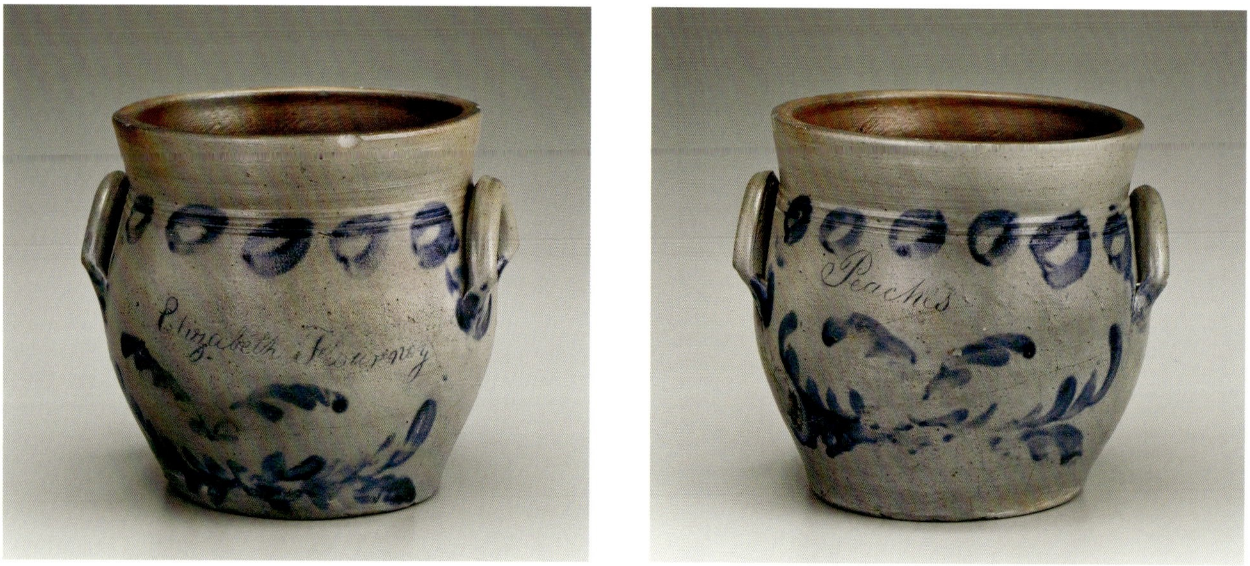

Figure 34 Storage jar, attributed to Samuel Frayser, Henrico County, Virginia, 1817–1849. Salt-glazed stoneware. H. 8". Incised: (*front*) "Elizabeth Flournoy"; (*reverse*) "Peaches" (Private collection; photo, Gavin Ashworth.)

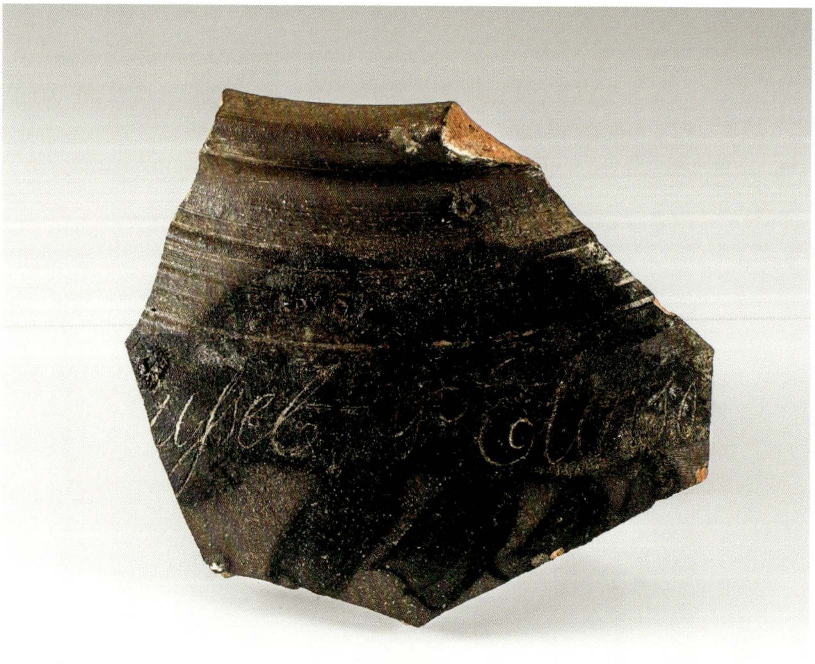

Figure 35 Storage-jar fragment, attributed to Samuel Frayser, Henrico County, Virginia, 1817–1849. Salt-glazed stoneware. Partial inscription: "Fussell & Elyson" (Private collection; photo, Robert Hunter.)

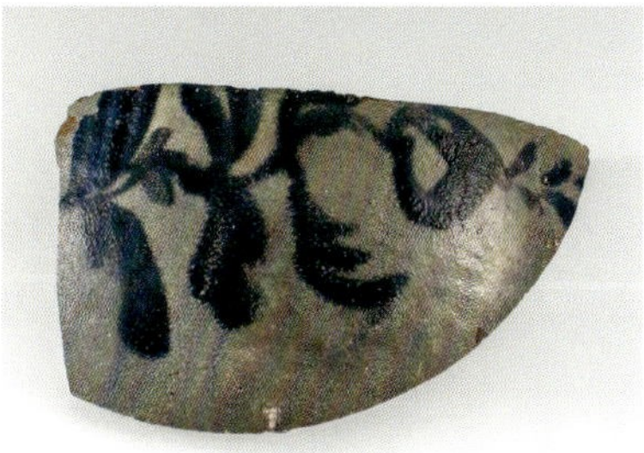

Figure 36 Jar fragment, attributed to Samuel Frayser, Henrico County, Virginia, 1817–1849. Salt-glazed stoneware. (Private collection; Robert Hunter.) Note the horizontal undulating-vine decoration.

the vessel functioned as a preserve jar, just as the Cox jar was intended for quinces.

Fragments recovered from the Four Mile Creek pottery help confirm the Frayser attribution of these inscribed presentation jars, while they also associate a decorative motif with a group of previously unattributed extant jars from the region. A recently recorded jar fragment bear a partial inscription interpreted as "Fussell & Elyson" (fig. 35). The names refer to two families with ties to both the Quaker community and the milling operations of southeastern(?) Henrico County. According to David Ross's advertisement for Norwich Mills, "Mr. Samuel Ellison [*sic*], who lives at the Mills . . . would undertake the management and engage a small interest in the business as a partner [to the new owner]." The "Fussell & Elyson" fragment may reference a trade partnership with Ben Fussell's mill a few miles to the northwest. Another recently found jar fragment bears the distinctive, horizontal vine garland associated with this group of pots (fig. 36).

THOMAS AMOSS POTTERY

Thomas Amoss, a Quaker, worked as a potter between 1818 and 1822 approximately ten miles southeast of the Henrico County courthouse on Four Mile Creek.[65] Yet another portion of Richard Randolph's tract of land—forty acres acquired by William Dandridge with no improvements—was by early 1820 sold to Amoss. This was probably another undeveloped portion of Randolph's original 500-acre Norwich Mills tract. The property is defined in the deed and associated indenture as

> all that Tract piece or parcel of land lying and being in the Said County of Henrico on four mile creek to the Junction With Baileys run thence up the run to the main road, thence up the road to Mrs. A. Kidds line thence down Kidds line to the creek.[66]

By 1821 Thomas Amoss had erected $550 worth of buildings on the property, although they were not insured. The buildings likely served as the pottery manufactory listed under his name in the 1820 census of manufacturers, given it was the only land that he owned in Henrico County.[67]

Personal property tax rolls for Henrico County in 1818 and 1819 confirm Amoss as a resident. Shortly thereafter others apparently joined him, for in 1820 the rolls record two free white males of tithable age (another man and Amoss), two slaves over the age of sixteen, and two horses or mules. This labor force built and operated the Amoss pottery, which consisted of one kiln and three wheels, and "used fifty tons of clay, eighty cords of wood, and

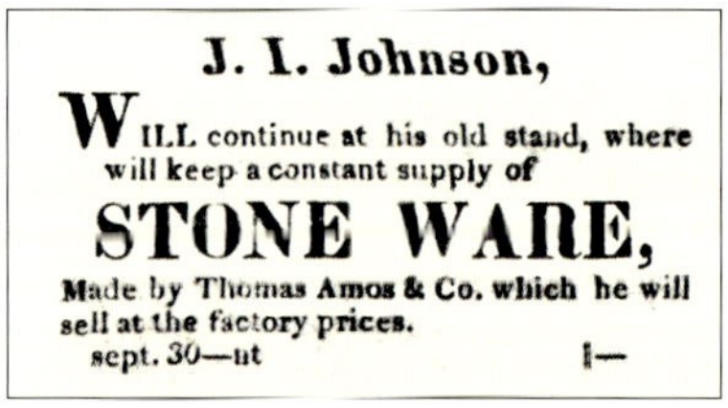

Figure 37 J. I. Johnson advertisement in the *Richmond Commercial Compiler*, September 30, 1819, p. 3-3.

fifteen sacks of salt."[68] J. I. Johnson, a Richmond merchant, advertised "a constant supply of STONE WARE Made by Thomas Amos & Co." available for sale at factory prices (fig. 37).[69]

Within a year, the Quaker Amoss was no longer incongruously credited with the two slaves. Perhaps these were workers from Norwich Mills who resided with Amoss during construction of his own establishment. In 1822, however, he paid taxes on himself and an adult white male, and, curiously, another slave over age sixteen, as well as the two equines. Although Amoss died in 1823 before the tax assessor's visit, his forty-acre parcel and its $550 worth of buildings were credited to his estate through 1832; his Henrico "stoneware factory" was bequeathed to his wife, Caroline, until her death, at which time it would go to their son, Edward N. Amoss. Edward may have been trained in potting by his father. His birth year is unknown, but Thomas's will indicates Edward was under the age of twenty-one in 1822. Amoss's estate paid taxes on two slaves in 1824, and two free white adult males and two slaves in 1825.[70]

Meanwhile, the value of the buildings on the late Thomas Amoss's forty acres declined in value from $550 in 1830 to $500 in 1831. His estate might have been encumbered by debt, for in 1833 his property was transferred to John Whitlock Jr. by a court-appointed commissioner. In 1834 Whitlock conveyed the forty acres and $500 worth of buildings to Richard Ellyson. Ten years later Ellyson's executor transferred widow Sarah Ellyson's interest in the property to potter Stephen B. Sweeney, who in 1845 acquired full ownership of Amoss's forty acres plus an additional 20.5 acres of Ellyson land.

Known as the Factory Tract, this 60-acre parcel, located just north of the juncture of Four Mile and Bailey's creeks, is identified on a recently discovered 1841 survey of Richard Ellyson's land completed at his death to identify his widow's dower (fig. 38). Illustrated on the survey are a dwelling house, a "Factory," a crib, and an old mill dam to the north; the plat depicts what are likely the remaining buildings of Randolph's former tract (a portion of which was subsequently owned by Amoss).[71] "Factory" relates to pottery production at Randolph's operation rather than to the earlier Norwich Mills. The old mill dam just to the north indicates that the former grist mill stood in the immediate vicinity.

Amoss's own pottery might be represented by the paired kiln remains and associated linear waster dumps some 450 feet to the east, near Bailey's Run, that seem to have formed the basis of Stephen Sweeney's later operations south of Route 5. These features were recorded by archaeologists from Virginia Commonwealth University in 1980. Still visible are the realigning of Bailey's Creek and the shoring up of its banks into sizable levees to prevent

Figure 38 1841 survey of Factory Tract, land owned by Richard Ellyson. (Courtesy, Library of Virginia.)

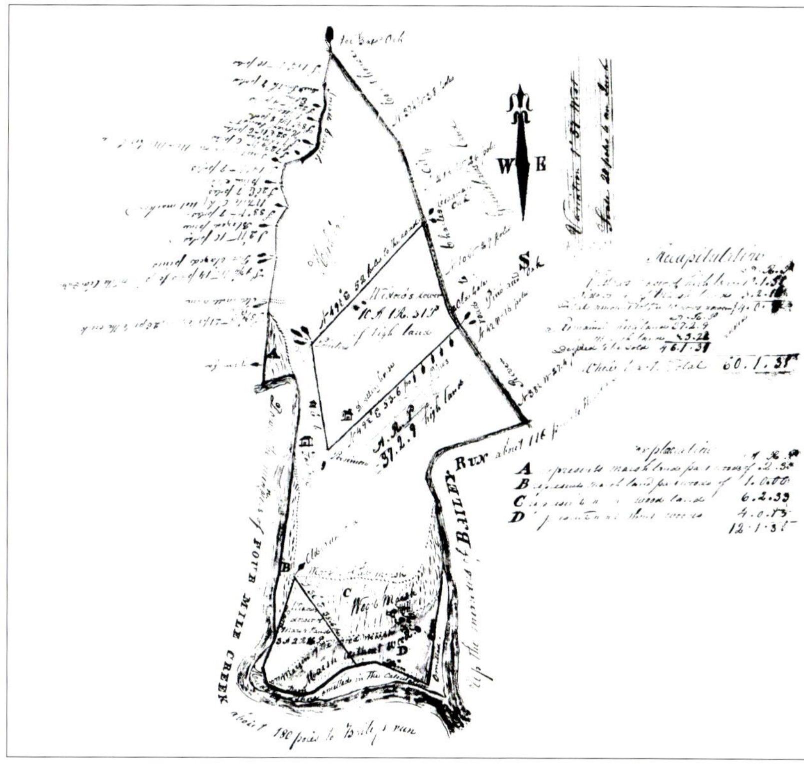

Figure 39 Fragments, attributed to the Thomas Amoss pottery, Henrico County, Virginia, 1818–1823. Salt-glazed stoneware. (Courtesy, Virginia Department of Historic Resources; photo, Robert Hunter.) These archaeologically recovered sherds from the Henrico County site show rim, handle, and decoration matching those on locally discovered vessels.

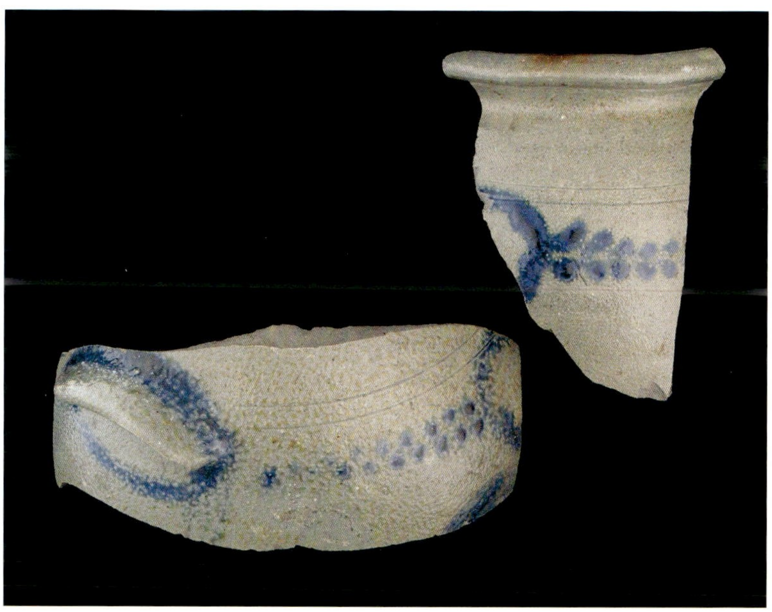

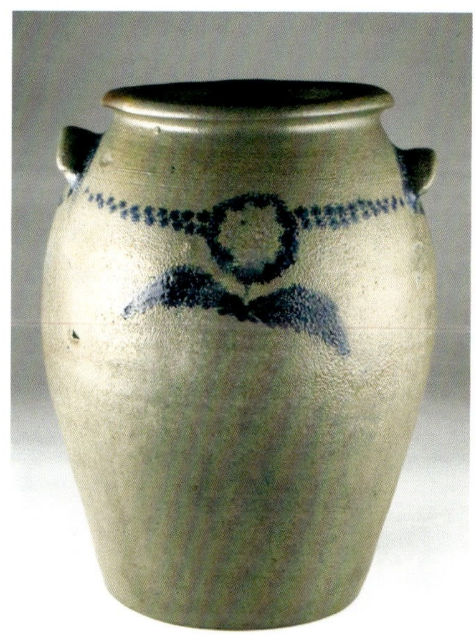

Figure 40 Storage jar, attributed to Thomas Amoss, Henrico County, Virginia, 1818–1822. Salt-glazed stoneware. H. 12½". (Courtesy, Colonial Williamsburg.)

flooding. Clay pits on the high ground are open to the west, toward the earlier Randolph operation.

Vessels by Thomas Amoss are distinguished by cobalt slip used in the style of the Morgan and Amoss potteries of Baltimore (figs. 39, 40). Indeed, historical evidence shows that several Baltimore potters influenced or were involved with the production of stoneware in Virginia from about 1818. In most cases, Henrico vessels exhibit darker clay and somewhat more robust, less elongated forms than their Baltimore counterparts. Amoss's forms, some of the scarcest of the extant wares from the region, include semi-ovoid storage jars, pitchers, and jugs (figs. 41, 42).

The influence of the Morgan family of potters working in Baltimore on Thomas Amoss's Virginia production was significant. The firm of Thomas Amoss and Co. owned by Thomas Morgan and Thomas Amoss operated

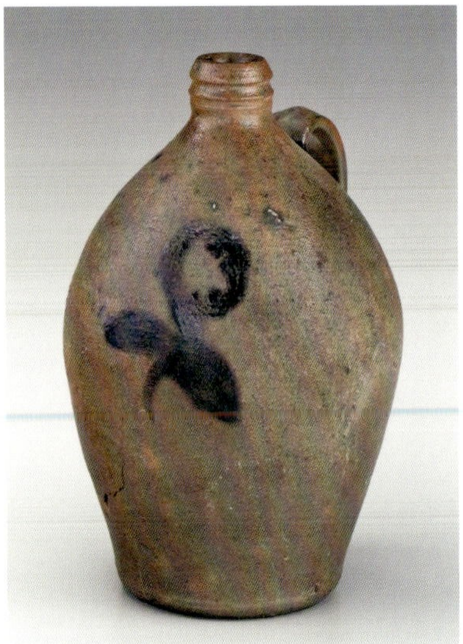

Figure 41 Jug, attributed to Thomas Amoss, Henrico County, Virginia, 1818–1823. Salt-glazed stoneware. H. 10". (Private collection; photo, Gavin Ashworth.)

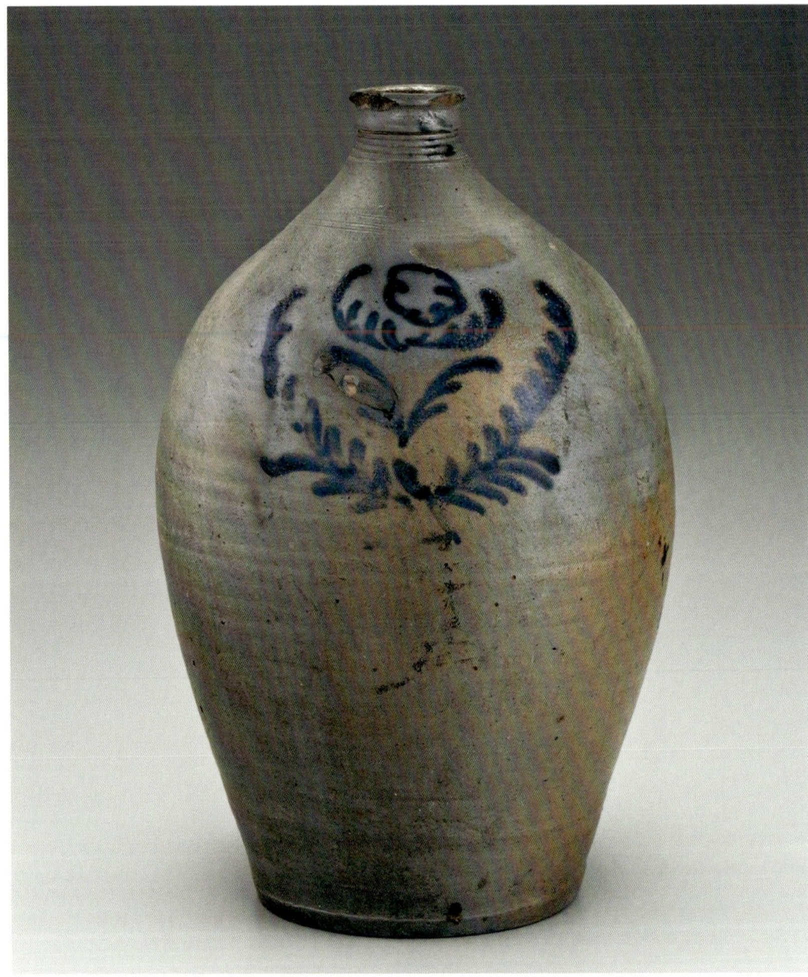

Figure 42 Jug, attributed to Thomas Amoss, Henrico County, Virginia, 1818–1823. Salt-glazed stoneware. H. 15⅝". (Private collection; photo, Gavin Ashworth.)

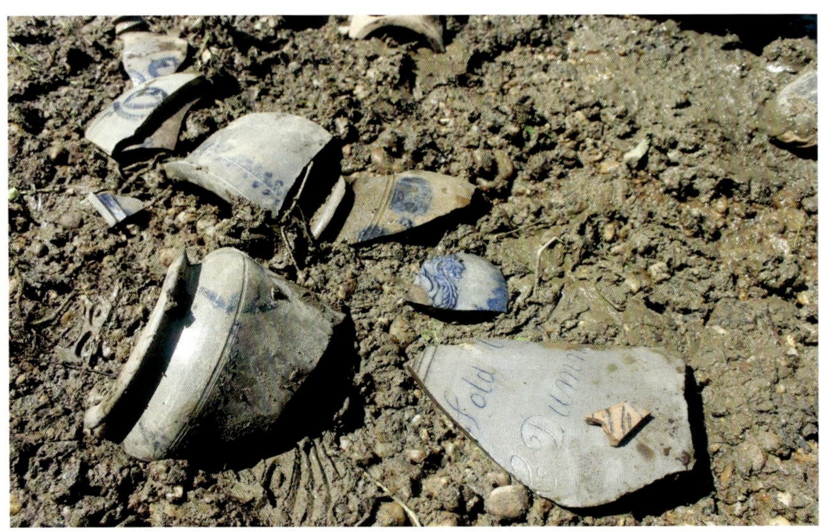

Figure 43 Collection of salt-glazed stoneware wasters, Henrico County, Virginia, 1818–1849. (Photo, Oliver Mueller-Heubach.) The variety of forms and decorative treatments help differentiate the makers. Amoss, Frayser, and other potters are represented in these examples.

in Baltimore from 1815 until 1819.[72] Amoss's Henrico products are semi-ovoid vessels with well-defined collars and somewhat diminutive, slightly rolled rims and a defined edge above curved tab handles with blunt ends. They often are brushed with cobalt at their shoulder joint. A characteristic design consists of swags of paired, cobalt-dotted lines hanging from a central flower head and the narrow ends terminating at the handles. No signed Amoss Virginia wares are known. Recovered fragments from the bed and bank of Four Mile Creek near the Randolph kiln display a diversity of new motifs (fig. 43).

STEPHEN B. SWEENEY'S POTTERY

During 1838, one Stephen B. Sweeney purchased four contiguous tracts of land consisting of 92.5 acres, 51 acres, 39.5 acres, and 8.5 acres, located near "Baileys Old Field," from Samuel Garthright Sr.'s executor. In 1839 the tax assessor indicated that $800 worth of buildings had been added to the 8.5-acre tract, which previously had only $150 worth of improvements. The following year, the assessor consolidated Sweeney's land into a 186.5-acre tract and recorded that Sweeney had sold off 5 acres to Jane Pleasants. By 1844 Sweeney had disposed of 2 more acres, retaining the 184.5 acres and the parcel's improvements, then worth $900.[73]

Working at or associated with the Sweeney pottery at midcentury were Stephen B. Sweeney Jr., and potters Edward J. Clark (b. ca. 1830 in Virginia) and Walt (Watt) Green (b. ca. 1829 in Virginia), a free black. Green might have worked for Sweeney for as long as sixteen years by that time, and there is a chance he was related to a slave named Watt inherited in 1828 by John Schermerhorn from Francis Wilson. Stephen Sweeney's will included nineteen slaves. The men listed therein include Anderson, Bill, Billy, Peter, Walker, Washington, and William, all valued at $800 each except Peter, who was assessed at $1,000 (perhaps he had some specialized training). It is unknown who among them worked in the pottery as opposed to Sweeney's contemporaneous agricultural pursuits, such as wheat, corn, and livestock.

By 1860 Stephen B. and Virginia Hughes Sweeney's youngest son, Charles Henry Sweeney (b. 1841), was potting, as was one Patrick Murphy (b. ca. 1835 in Ireland), identified as a stoneware potter. Sweeney's other sons could have worked in the shop as well. Martin (b. 1826) would have been entering his teens when his father established his first pottery. Stephen B. Jr. (b. 1839) and Charles might have begun work there in the late 1840s and 1850s, before they went on to their known careers.

Sweeney's 1844 purchase of the 40.5 acres with $500 worth of improvements from Ellyson's estate represents the addition of the former Randolph and Amoss pottery sites to his holdings. Sweeney held this acreage together with an additional 20.5 acres of Ellyson land well into the 1860s, by which time the tax assessor combined it with the large tract Sweeney had acquired in 1838.[74] In fact by 1854 the assessor had consolidated Sweeney's parcels into a 237-acre tract that Sweeney christened "Claymount" for its abundance of potters' clay. The property then had $2,000 worth of buildings and was described as situated thirteen miles southeast of the county courthouse. Among the buildings were "an excellent dwelling, kitchen, stables, barns, carriage house, sheds &c., all in good order."[75] In 1860 Sweeney's stoneware pottery was described as having $1,000 of capital invested with workers earning $150 per year resulting in stoneware production worth $3,000.[76] Period maps show the location of Sweeney's "Pottery & Hotel" north of New Market Road and east of Bailey's Run, as well as the later expansion of his holdings to the southwest to include the 1844 property just north of the fork of Four Mile Creek with Bailey's Run, labeled "Sweeney's Pottery" (fig. 44).

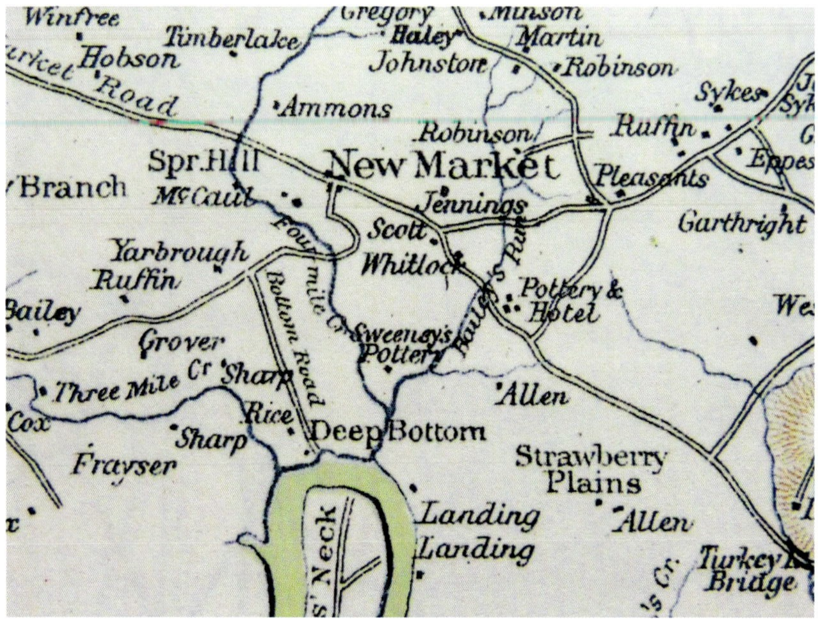

Figure 44 Detail, "Map of Southeast Virginia," 1862, U.S. Department of the Army, showing the location of both Sweeney's pottery of ca. 1844–1863, labeled "S. Sweeney's Pottery," between and just above the forks of Four Mile Creek and "Bailey's Run," and Sweeney's "Pottery and Hotel," shown just to the northeast of New Market Road. George B. Davis et al., *The Official Military Atlas of the Civil War* (1891–95; repr., New York: Gramercy Books, 1983).

Sweeney remained in possession of Claymount and its improvements until his death in 1862. As he prospered he purchased some small parcels near Gravelly Hill, among them a 13-acre and a 15-acre tract.[77] These purchases, together with Samuel Frayser's earlier purchase of land in the immediate vicinity of Gravelly Hill, suggest that there was a natural resource, presumably clay or temper, that was useful in manufacturing pottery.[78] By 1862 the Claymount lands included about 500 acres along the east side of Long Bridge Road and the north side of Route 5. In time, the northern branch of Bailey's Creek, which cut through the property, became known as Sweeney Creek.

Southeast Henrico potters' close ties to Richmond are evidenced by an 1860 listing showing that Sweeney was leasing a retail outlet on 14th between Cary and Dock streets. As early as November 1858, this building was insured under the title "Earthen and Stone Ware Store of Stephen B. Sweeney."[79]

Sweeney's will and the appraisal of his property provide some additional potting-related details. The administrator of his estate records several sales of stoneware: three separate lots sold in 1862 on October 3 (for $3.75), October

12 ($5.00), and November 12 ($2.50), with the largest lot going to David Parr on January 17, 1863, for $365.58. Among the other items listed were six "Turning Tables for Ware," which presumably also sold to Parr, for $30.[80] Sweeney is also credited with owning four lots of land in Charles City County with an aggregate value of $738.75. The fact that another Four Mile Creek/Bailey's Creek potter held land of significant value there supports the possibility of a pottery in that area, possibly involving Schermerhorn and/or Wilson and others.

Today, the area north of Route 5 (New Market Road) and east of Bailey's Creek across from the site of the Norwich Mills/Factory Tract is heavily wooded, yet it is certain that this is where Stephen B. Sweeney potted for some twenty-four years, beginning in 1838. A plethora of maps and documents confirm this location, which is often referred to as "Sweeney's Pottery and Hotel." Massive waster deposits and kiln remains north and south of the road are the only vestiges of his once-thriving pottery enterprise (fig. 45).

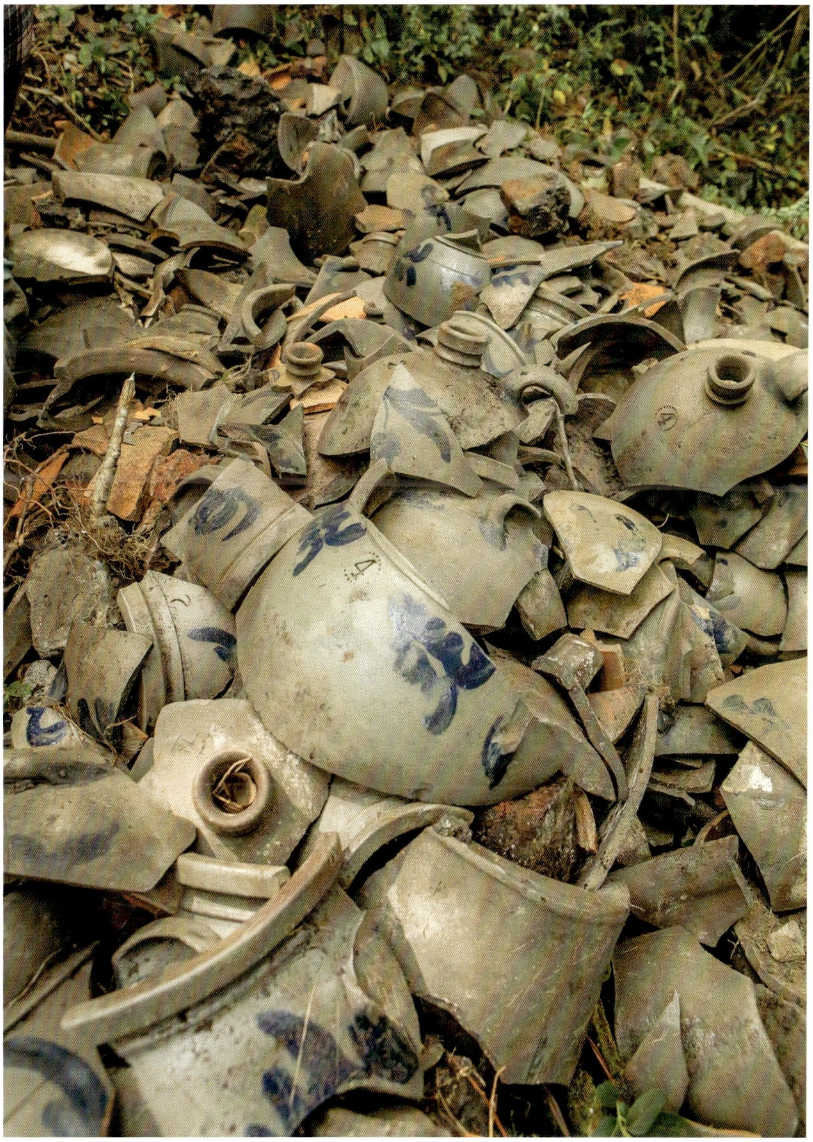

Figure 45 Salt-glazed stoneware fragments at the Stephen B. Sweeney pottery site. (Photo, Robert Hunter.)

Figure 46 Jug, attributed to Samuel Frayser, Henrico County, Virginia, 1817–1849. Salt-glazed stoneware. H. 15⅜". Capacity: 3 gallons. (Private collection; photo, Gavin Ashworth.) The concave neck with concentric reeding is common in the archaeological specimens from Four Mile Creek made by Frayser and Amoss, while the floral decoration is common to Frayser-attributed wares and likely precursors to later Sweeney motifs.

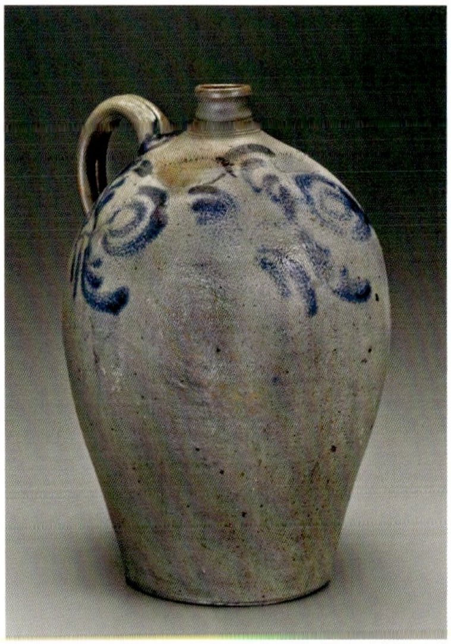

Figure 47 Churn, attributed to Stephen B. Sweeney, Henrico County, Virginia, 1838–1863. Salt-glazed stoneware. H. 21⅞". Capacity: 5 gallons. (Private collection; photo, Gavin Ashworth.)

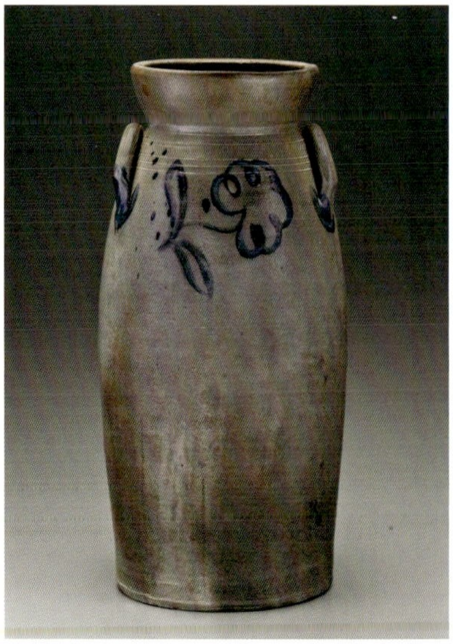

Figure 48 Pitcher, attributed to Stephen B. Sweeney, Henrico County, Virginia, 1838–1863. Salt-glazed stoneware. H. 17½". Capacity: 3 gallons. (Private collection; photo, Gavin Ashworth.)

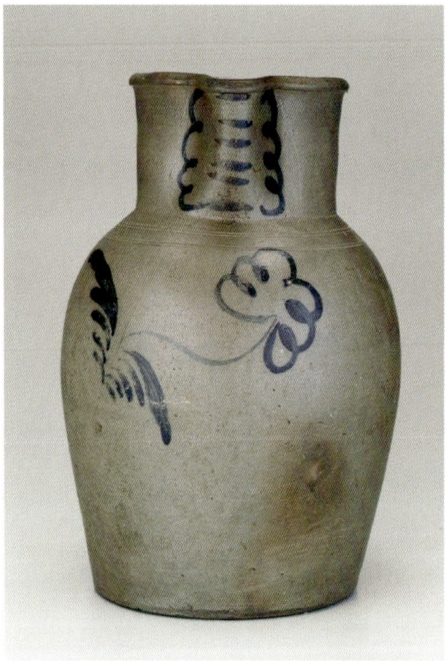

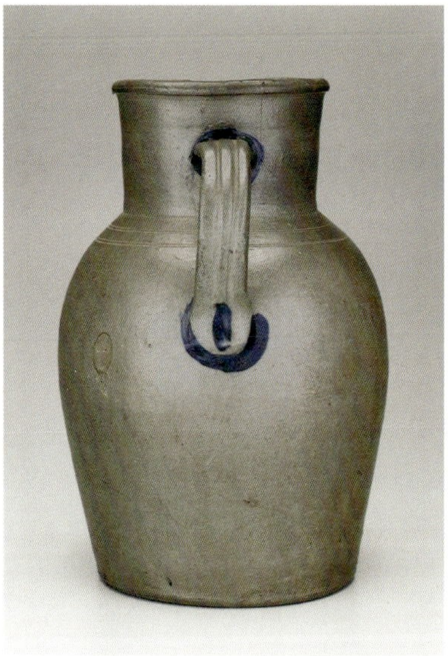

Figure 49 Jar, attributed to Stephen B. Sweeney, Henrico County, Virginia, 1838–1863. Salt-glazed stoneware. H. 9". Capacity: 1 gallon. (Private collection; photo, Robert Hunter.)

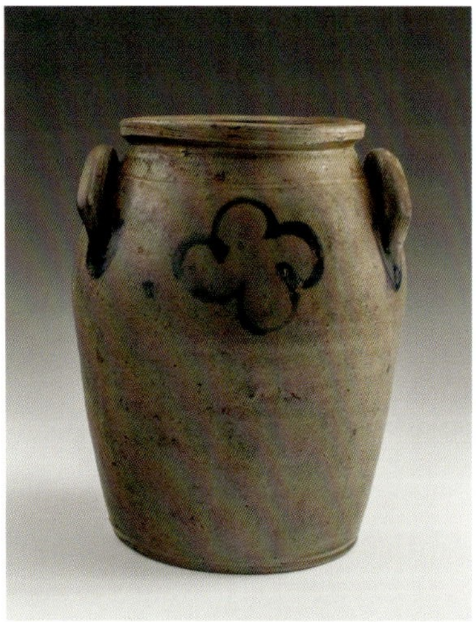

Figure 50 Jar, Stephen B. Sweeney, Henrico County, Virginia, 1838–1863. Salt-glazed stoneware. H. 14". Capacity: 3 gallons. (Private collection; photo, Robert Hunter.)

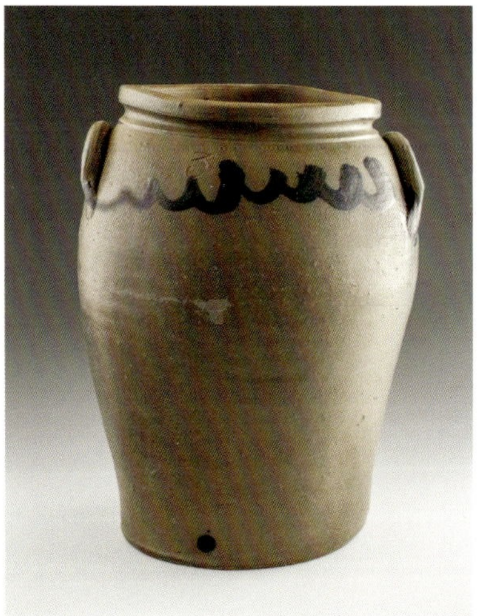

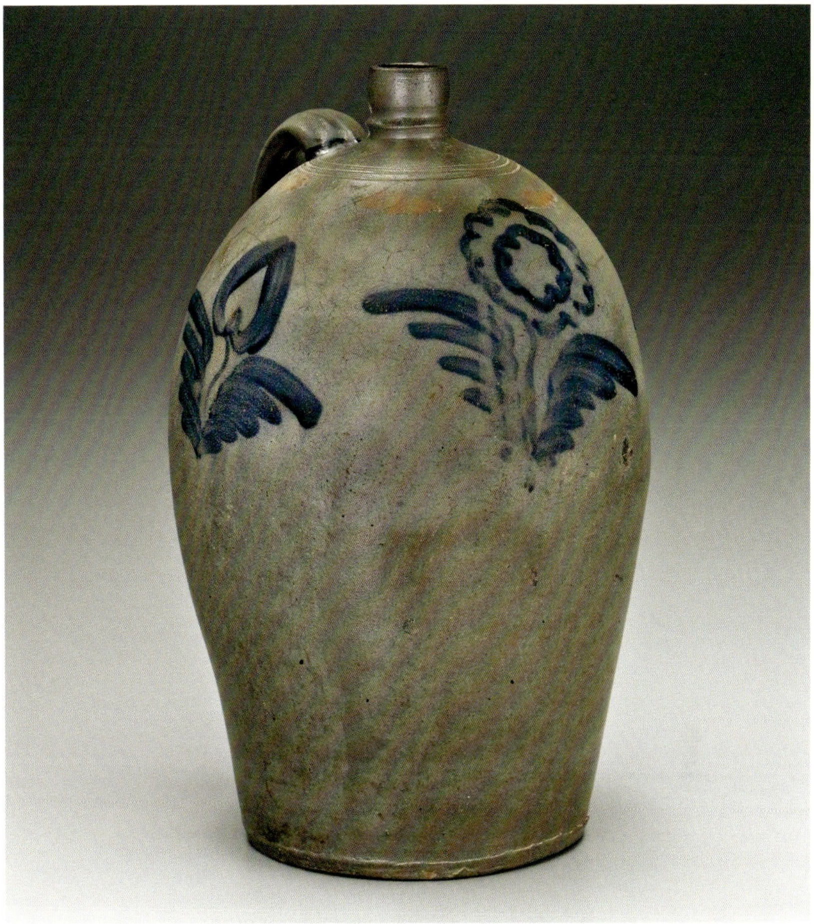

Figure 51 Jug, Stephen B. Sweeney, Henrico County, Virginia, 1838–1863. Salt-glazed stoneware. H. 16½". Capacity: 3 gallons. (Private collection; photo, Gavin Ashworth.) The blossom configuration of the brushed blue-cobalt floral motif on the front of this jug was undoubtedly influenced by Amoss. The blossoms of the lateral flanking motifs are more common in the Sweeney decorative repertoire.

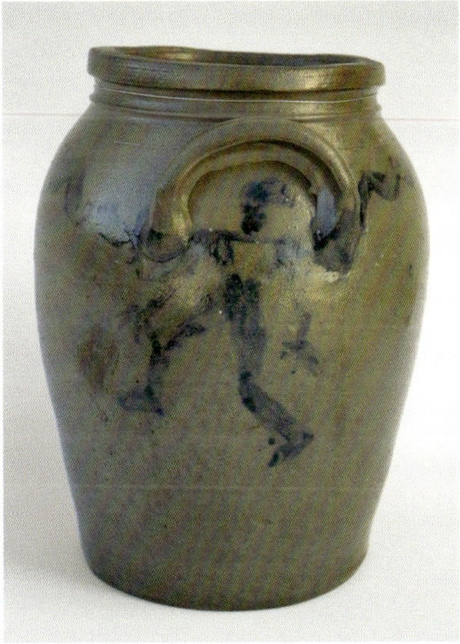

Figure 52 Storage jar, attributed to Stephen B. Sweeney, Henrico County, Virginia, 1838–1863. Salt-glazed stoneware. H. 9½". Capacity: 1 gallon. (Private collection; photo, Kurt Russ.) The decorative motif is a brushed-cobalt dancing man beneath one of the handles.

Although this site has been identified for decades, it has yet to benefit from any preservation protection or concerted archaeological research.

Sweeney stoneware vessels generally exhibit more variation in decorations than do Schermerhorn's wares. In addition to semi-ovoid storage jars ranging in capacity from one to five gallons, pitchers from one pint to three gallons, and jugs with forms closely resembling those produced in Richmond, churns have also been identified, all sharing similar decoration (figs. 46–48).

The most common decorative treatment consists of a single floral stem, extending from two opposing outlined leaves, that terminates in a circular blossom. In many cases the blossom is rendered without stem and leaves, appearing cloudlike (fig. 49). Ongoing research is documenting the many decorative variations used during Sweeney's tenure. A common decoration is a simple squiggle around the shoulders of storage jars that is brushed in blue cobalt and sometimes paired (fig. 50).

Sweeney's vessels exhibit increased standardization in form as well as fine-grained, well-processed clay reflective of a later period. Similarities in decorative motifs help link them to vessels attributed to Schermerhorn, Frayser, and, in some cases, Amoss (fig. 51). Incised animal decoration with cobalt is extremely rare; brushed-cobalt depictions of both animals and human characters have been noted, but they too are scarce (fig. 52).

One of the most significant pieces of stoneware known from Sweeney's production is a five-gallon jar with an incised and cobalt-painted rooster or game cock (fig. 53). A new, firm attribution to Sweeney is based on distinctive attributes of vessel form, a simple stamped five-gallon capacity mark, a characteristic dark gray clay, and a striking resemblance to decorated sherds recovered from Sweeney's earliest pottery site. The decoration on the reverse of the rooster/game cock vessel is a leafy-stemmed "spitting tulip" in brushed cobalt. Variations of this motif appear on Sweeney wares spanning the twenty-five years of his production (figs. 54, 55). On later pieces, the floral decoration is less elaborate and executed with broadly stroked leaves and simpler motifs both in terms of quality, size, and number of leaves.

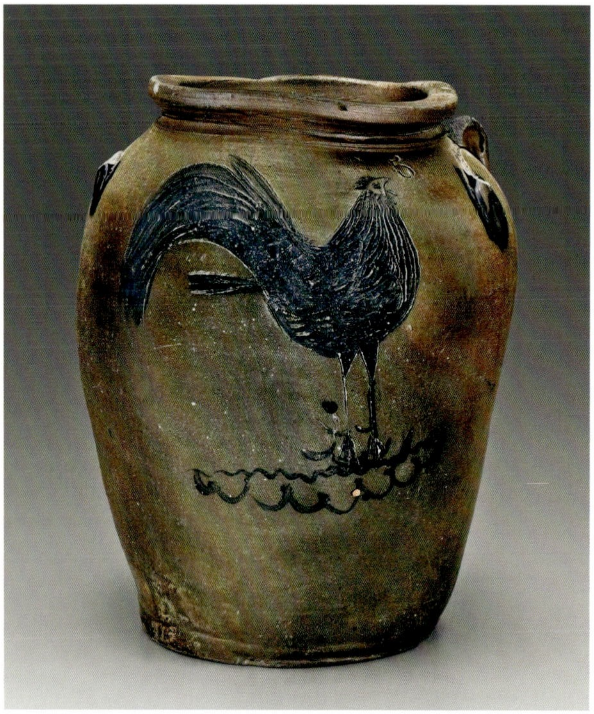

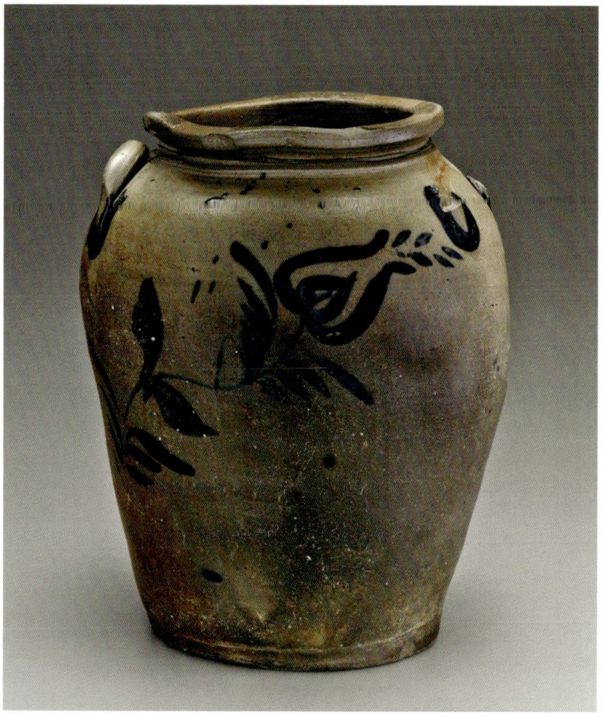

Figure 53 Storage jar, attributed to Stephen B. Sweeney, Henrico County, Virginia, 1838–1863. Salt-glazed stoneware. H. 15". Stamped capacity mark: "5" (Private collection; photo, Gavin Ashworth.) This ovoid vessel displays a well-defined incised and cobalt-filled rooster, or game cock, standing on cobalt festoons similar to those seen on other Sweeney and Schermerhorn vessels, such as the jar illustrated in Russ and Schermerhorn, "Rocketts' Red Glare," p. 72, fig. 14.

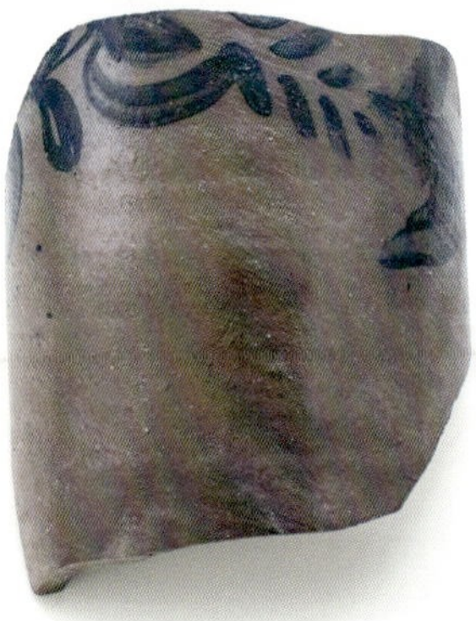

Figure 54 Storage-jar fragment, attributed to Stephen B. Sweeney, Henrico County, Virginia, 1838–1863. Salt-glazed stoneware. (Private collection; photo, Robert Hunter.)

A recently recovered butter-crock fragment appears to have been decorated by the same hand as the rooster/game cock jar. The incised markings probably represent the tail of a fish or a bird, in combination with the lettering "R. Cla . . ." and Sweeney's partial signature (fig. 56). Another, equally enlightening piece shows an incised and cobalt-painted bird that corroborates a level of artistic skill and range of decorative treatments employed during Sweeney's production that was unexpected given the dates of operation (fig. 57).

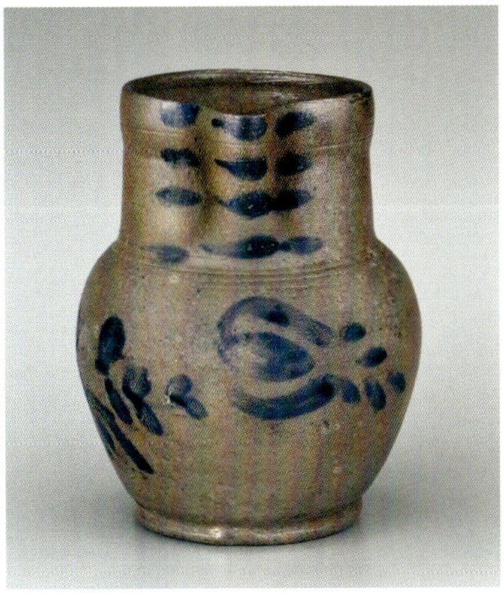

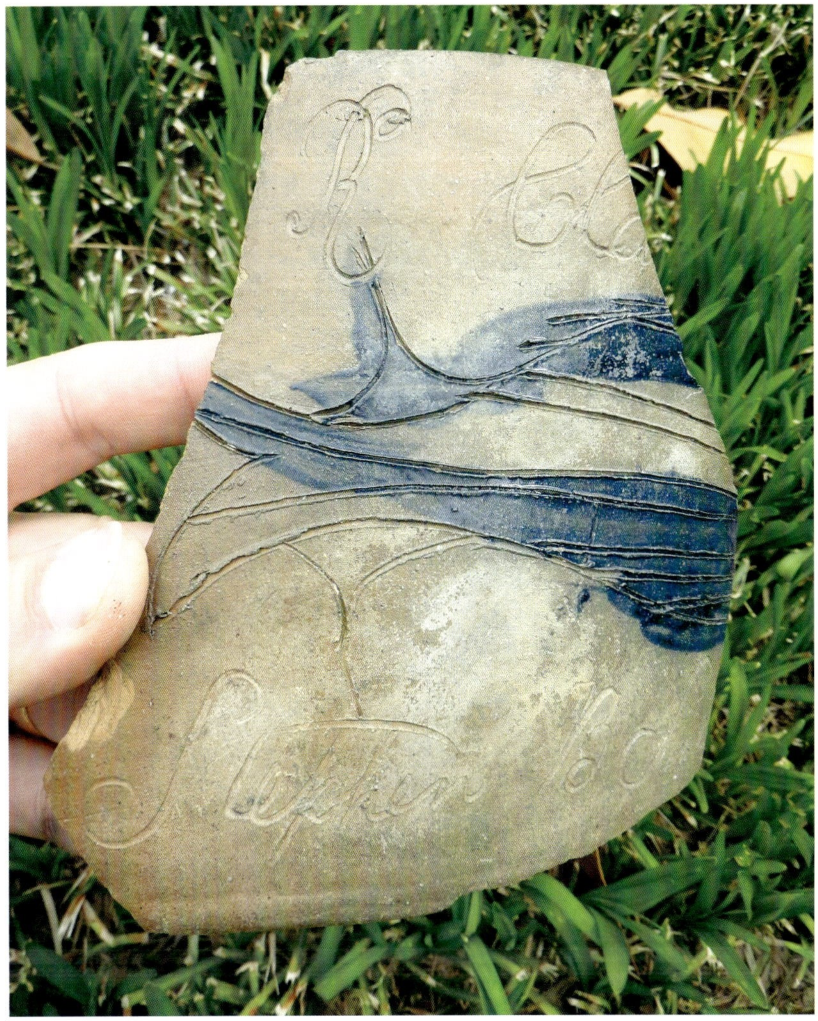

Figure 55 Pitcher, Stephen B. Sweeney, Henrico County, Virginia, 1838–1863. Salt-glazed stoneware. H. 6⅜". (Private collection; photo, Gavin Ashworth.)

Figure 56 Butter-crock fragment, Stephen B. Sweeney, Henrico County, Virginia, 1838–1863. Salt-glazed stoneware. Inscribed: "R. Cla . . ."; "Stephen B. Swe . . ." (Private collection; photo, Oliver Mueller-Heubach.) The decoration on this waster fragment is a fish or a bird between the incised inscription and Sweeney's partial signature. Speculation on what the inscription refers to includes names of local families, such as Claiborne or Clay, or of a nearby landowner, Clarke.

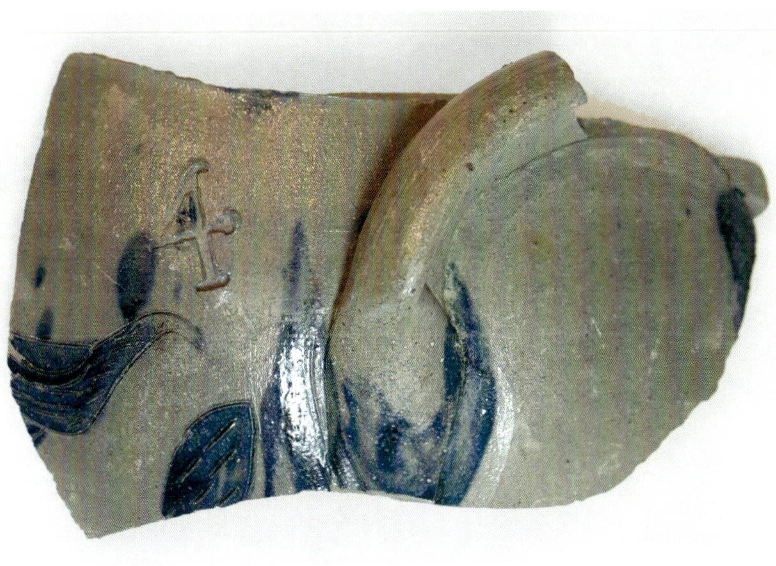

Figure 57 Storage-jar fragment, Stephen B. Sweeney, Henrico County, Virginia, 1838–1863. Salt-glazed stoneware. (Private collection; photo, Oliver Mueller-Heubach.) Incised and cobalt-filled bird and floral decoration embellishes this shoulder remnant from a four-gallon jar.

POTTER SAMUEL WILSON

Samuel Wilson was living and working as a potter in Albany, New York, between 1813 and 1816. It seems he was associated with both DuVal and Schermerhorn in the manufacture of stoneware in Richmond by 1817 at the latest. Given Schermerhorn's marriage to Samuel's sister, Sarah Christian Wilson, and their shared New York connections, it is also likely that he and Samuel knew one another prior to their removal to Virginia.

As suggested previously, it is possible that Wilson journeyed with Schermerhorn or others to Virginia. Personal property tax rolls reveal that Samuel Wilson moved to Henrico County sometime after the tax assessor's 1819 visit but before he returned in 1820. Wilson's household included two free white males of tithable age, and by 1821 he was equipped with a horse or mule. According to the 1820 Census of Manufacturers, Samuel Wilson was then producing pottery; the operation was probably owned by a third party and consisted of "one kiln and two wheels, making use of 100 loads of clay, 80 cords of wood, 3 loads sand, 15 sacks salt, and 40 pounds of cobalt at a total cost of $600 per year." With an investment (by either Wilson or the owner) of $1,000 capital, $500 in payroll, and $500 in other expenses, he employed three men and two boys in the production of "Stone Ware of all kinds."[81] He is also listed as having insured three buildings in "Shedtown" in Henrico in 1822; his name appeared in the personal property tax rolls for that year but disappeared thereafter.[82] Shedtown was a Richmond neighborhood northeast of Church Hill, relatively near the DuVal pottery and the Schermerhorn shop at Port Mayo-Rocketts. The name is said to trace back to the sheds of brickmakers who lived and worked there and, if true, this suggests a potential clay source and a possible pottery location.

It is interesting that Samuel Wilson did not own any real estate in Henrico County before or after this period, which strongly indicates he was plying his trade while occupying somebody else's property. It might be significant that three pieces of property of the late Thomas Wilson and of Daniel Wilson were adjacent to land owned by members of the DuVal family.[83] This raises the possibility that Samuel Wilson operated a pottery factory on land owned by one of the DuVals or another member of the Wilson family. Also, unlike some of his contemporaries (Amoss and Frayser, for example), no advertisements describing Wilson's production have been identified in local papers of the period. This seems to corroborate the idea that Wilson either worked for or in association with another pottery owner. Understanding precisely where and with whom Wilson was associated is problematic. The nature of his involvement at Four Mile Creek/Bailey's Creek is unresolved, as is his possible participation in an unrecorded operation in Charles City County, perhaps in connection with members of his family or his relation by marriage, John Schermerhorn.

The group of extant wares tentatively attributed to Wilson's production can be placed within four distinct subgroups, all related by decorative motifs, forms, and distinctive capacity marks.

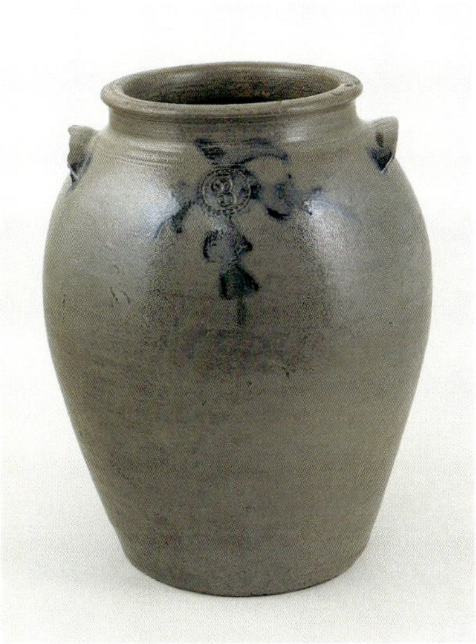

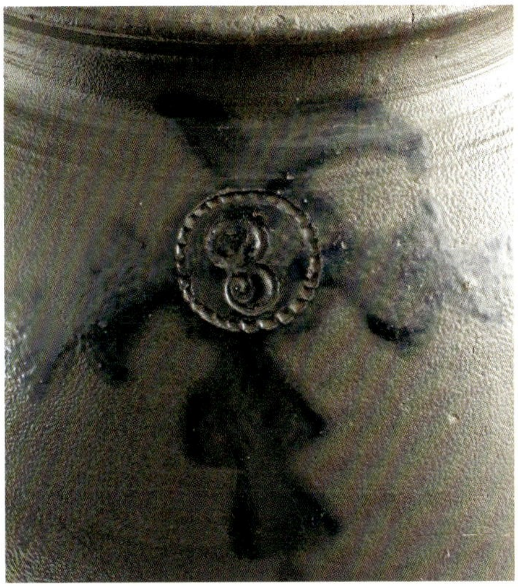

Figure 59 Detail showing the capacity mark on the jar illustrated in fig. 58. The distinctive capacity mark distinguishes these Group I wares attributed to Samuel Wilson working with John Schermerhorn.

Figure 58 Storage jar, attributed to Samuel Wilson and/or associates, Richmond, Virginia, 1817–1850. Salt-glazed stoneware. H. 14½". Capacity: 3 gallons. Incised on base: "Richmond" (Private collection; photo, Gavin Ashworth.)

GROUP I "RICHMOND TASSEL."—Serving as the Rosetta Stone for defining this group and its attribution to Samuel Wilson is the jar illustrated in figure 58. A number of these jars have surfaced in the Richmond-area antique market over the years but this is the only known example that is inscribed "Richmond" on the base. This group is composed of robust ovoid jars, jugs, or pitchers of a medium to dark gray clay with cobalt decorative motifs ranging from swag-and-tassel or graduating tassels to floral motifs composed of tripartite tassel-like elements. These motifs surround a central circular blossom or simple leafy-stemmed flowers with circular blossoms and sometimes are accompanyied by a distinctive capacity stamp (fig. 59).

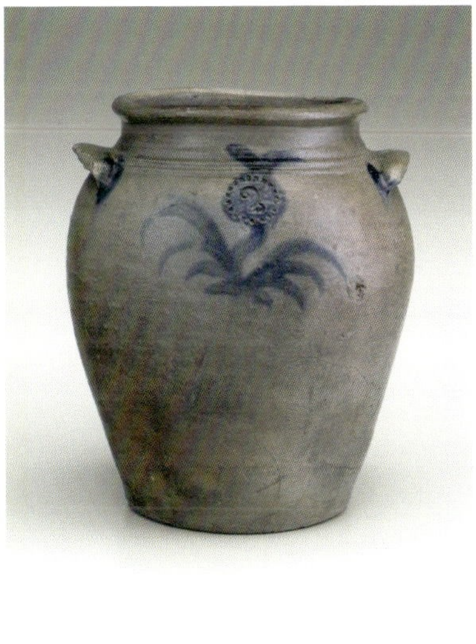

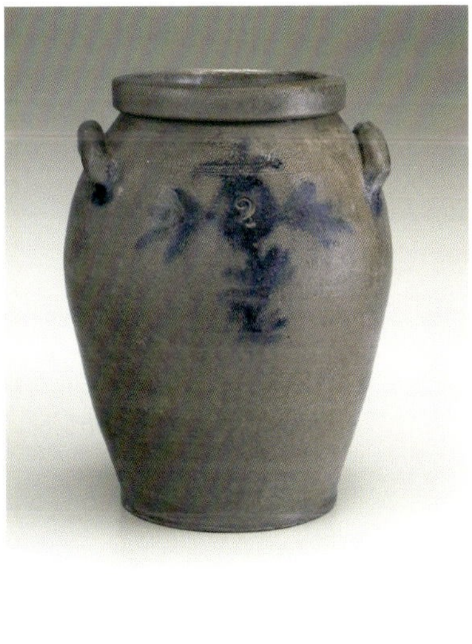

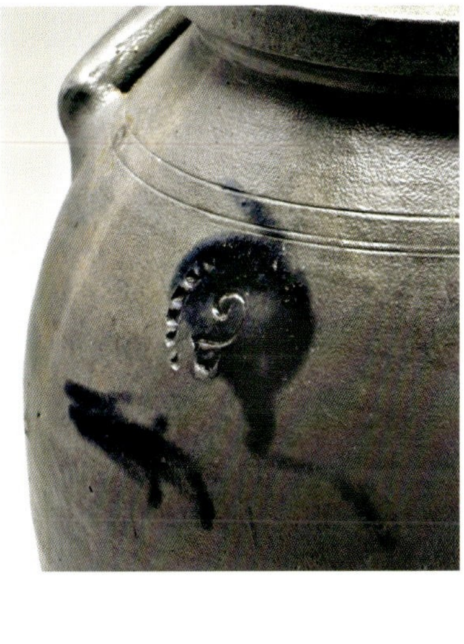

Figure 60 Storage jars, attributed to Samuel Wilson and/or associates, Richmond or Henrico County, Virginia, 1817–1850. Salt-glazed stoneware. H. (*left*) 11¼"; (*middle, right*) 12½". Capacity: 2 gallons. (Private collection; photo, Gavin Ashworth.)

This mark is typically enclosed within a ring of irregular long and short dashes; often it is incorporated into the cobalt motif. The jars have slightly curved and upwardly flaring applied tab handles with slightly tapered ends.

Connecting Groups I and II are three jars which share similarly styled capacity stamps, forms, handles, and decorative embellishments (fig. 60).

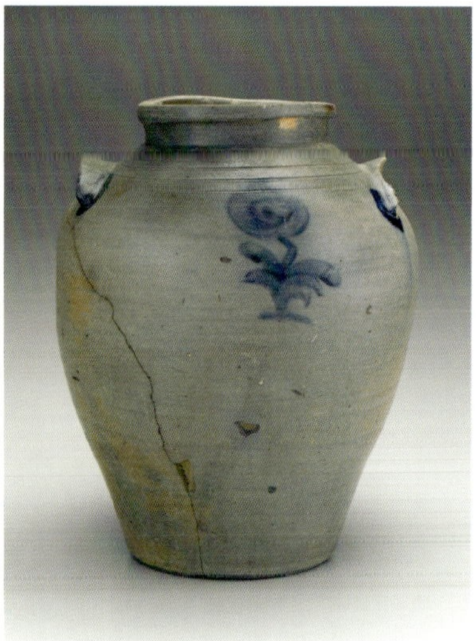

GROUP II "BENT TULIP."—This group consists of ovoid jars slightly more elongated than those from Group I, and slightly lighter in overall color, with a medium-gray clay body. They seem more finely potted, with a characteristic squared-off rim atop a short collar. Prominent, elongated, curved tab handles are highlighted at their blunt ends with cobalt (fig. 61). Decoration consists of a "bent tulip," sometimes paired with a capacity stamp. Alternatively, the "tulip" appears with a brushed swag-and-tassel on the reverse of one example, and the number of concentric incised lines that mark upper handle attachments also correspond to vessel capacity.

GROUP III "KELSO-CROSSHATCHED CAPACITY."—Representing this third grouping is a half-gallon pitcher stamped with a distinctive capacity mark and the name "KELSO" (fig. 62). The capacity marks are circular with a tightly spaced border of small, square teeth around a central, fine-lined

Figure 61 Storage jar, attributed to Samuel Wilson and/or associates, Richmond or Henrico County, Virginia, 1817–1850. Salt-glazed stoneware. H. 15½". (Private collection; photo, Gavin Ashworth.)

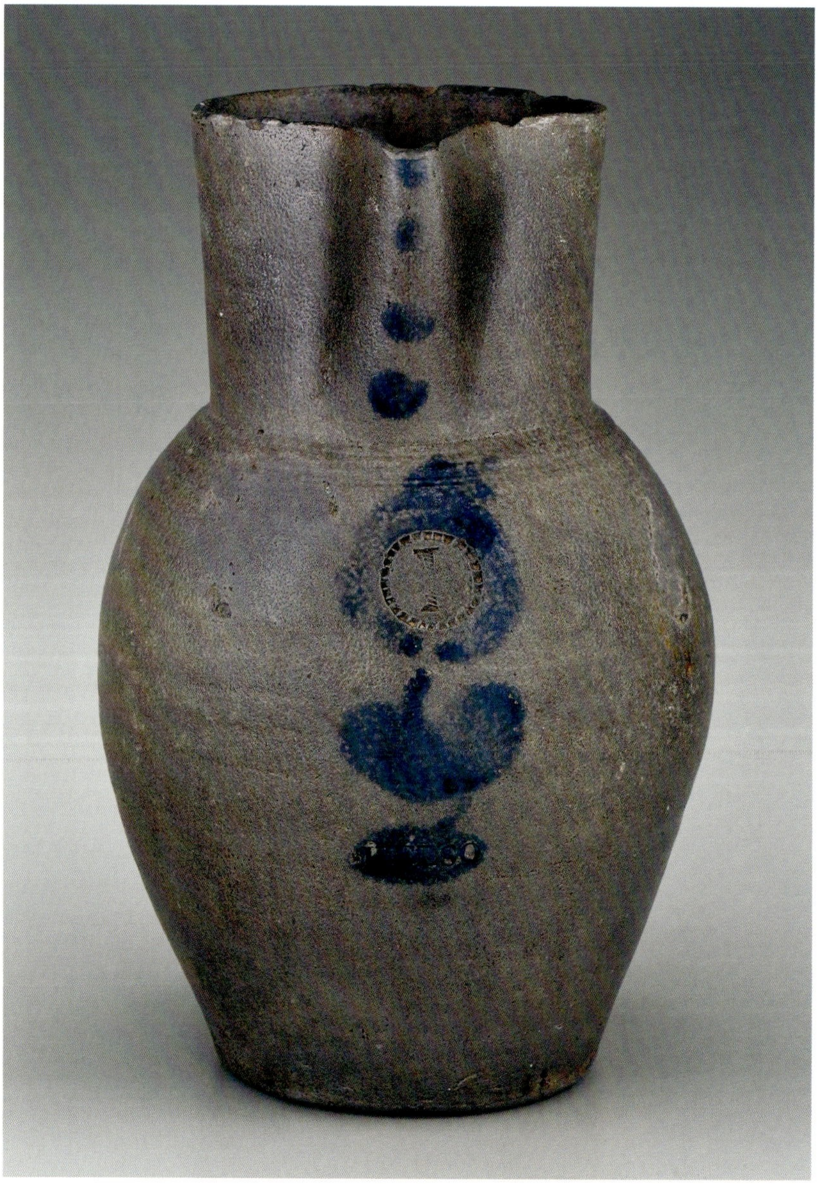

Figure 62 Pitcher, attributed to Samuel Wilson and/or associates, Richmond or Henrico County, Virginia, 1817–1850. Salt-glazed stoneware. H. 11⅛". Capacity: 1 gallon. Impressed: "KELSO" (Private collection; photo, Gavin Ashworth.) The floral decoration is virtually identical to that on the jar illustrated in fig. 63 (right).

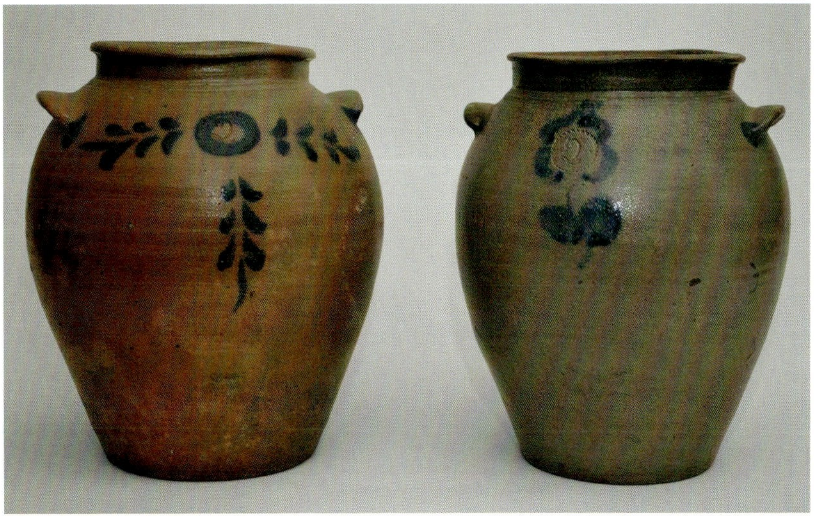

Figure 63 Storage jars, attributed to Samuel Wilson and/or associates, Richmond or Henrico County, Virginia, 1817–1850. Salt-glazed stoneware. H. of jar on right 11⅞". (Private collection; photo, Kurt Russ.)

numeral with a crossed-hatched interior. Decorative motifs either resemble that of the pitcher, with a broadly-brushed simple tulip or a simple brushed-blue open-circle motif surrounding the capacity mark on the vessel's upper shoulder. A fragment recovered from Four Mile Creek is decorated with this second element, suggesting the possibility that the responsible potter was in the immediate vicinity. While this group's curved tab handles with rounded edges are akin to those observed on both DuVal/Schermerhorn stoneware and that made by the Morgans in Rockbridge County (fig. 63), the flat tabs from Group V are more closely related. The vessels are generally well potted and relatively thin walled, with a somewhat squat, top-heavy, ovoid shape and highly variable clay coloration, which might be due to the clay source or the vagaries of firing.

GROUP IV "INVERTED TASSEL."—Another group of ovoid forms exhibits a simple inverted tripartite-"tassel" floral motif. Both applied "rolled" handles and curved blunt-ended, ear-like examples are noted (fig. 64). A four-gallon jar from this group features unique rouletting around its upper shoulder and exhibits handles identical to those from Group II. A half-

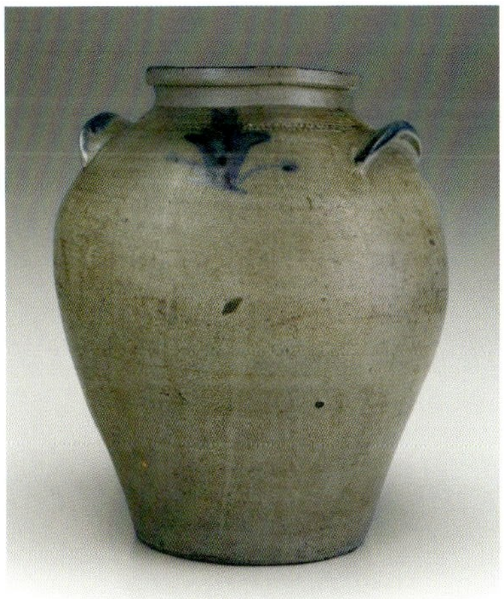

Figure 64 Storage jar, attributed to Samuel Wilson and/or associates, Richmond or Henrico County, Virginia, 1817–1850. Salt-glazed stoneware. H. 14½". (Private collection; photo, Gavin Ashworth.)

gallon jar has the rolled handles also seen from that group along with the inverted tassel decoration. That the decoration is unlike motifs from Group II but similar to those associated with Group I make it necessary to distinguish this as a separate group.

GROUP V "VERTICAL-COLLARED FLAT TAB."—Distinctive semi-ovoid vessel forms of a dark gray clay seen in various sizes (half- to four-gallon) all display the prominent, well-defined vertical collar with a protruding flat or lipped rim. Flat tab handles with rounded edges are applied sans cobalt, and single-stemmed flowers grow from between winglike leaves reminiscent of Amoss and terminate in a simple open bloom. Sometimes the bloom encloses a single dot. In the most common motif swags and tassels near the shouldertop intersect with the handles (fig. 65). Although sharing a combination of attributes with Groups I and III, the nature of the clay, potting skills, and similarities in both handles and decorative motif suggest a close association with Group I. This closeness may reflect the hand of an apprentice rather than a different maker, and hence it may be more appropriate to consider it a subgroup of Group I.

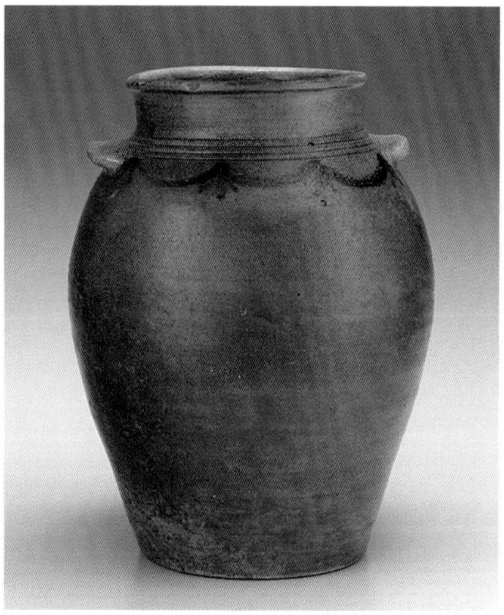

Figure 65 Storage jar, attributed to Samuel Wilson and/or associates, Richmond or Henrico County, Virginia, 1817–1850. Salt-glazed stoneware. H. 13¾". Capacity: 3 gallons. (Private collection; photo, Gavin Ashworth.)

The discovery of the Kelso pitcher may prove significant in unraveling the evolution of Samuel Wilson's participation in the area's stoneware production. John Kelso, a resident of Richmond, appears to have been involved in a manufacturing enterprise, according to the 1820 U.S. Census. In 1819 he is listed as the proprietor of a Richmond boardinghouse. The name on the pitcher is stamped in a large font, unusual for a merchant or grocer's mark, and neither profession is associated with a Richmond Kelso. Although John Kelso, "manufacturer," may have produced products sold in stoneware, there is a chance that he was a pottery owner. The situation is analogous to that of Richard Butt in Washington, D.C., who operated a "poor house" while employing a number of master potters in his nearby stoneware pottery.[84] The potential association of Schermerhorn and Kelso and the proximity of Schermerhorn's property to the Henrico Poorhouse are also intriguing. John Kelso's signature as special agent appears on Randolph's 1816 Mutual Assurance Company Declaration certifying the appraisal of the two buildings (pottery shop and dwelling) described therein.[85] This suggests both Kelso's familiarity with the valuation of properties of this nature and his larger business role in the area's growing economy.

Charles City County Stoneware

THE TREES POINT FACTORY AND A NEWLY DISCOVERED POTTERY AT WILSON'S (WHARF) LANDING

Another stoneware production area lies farther to the southeast, on the James River in Charles City County (fig. 66). Archeological test excavations at the Trees Point Factory have helped identify this factory, about which

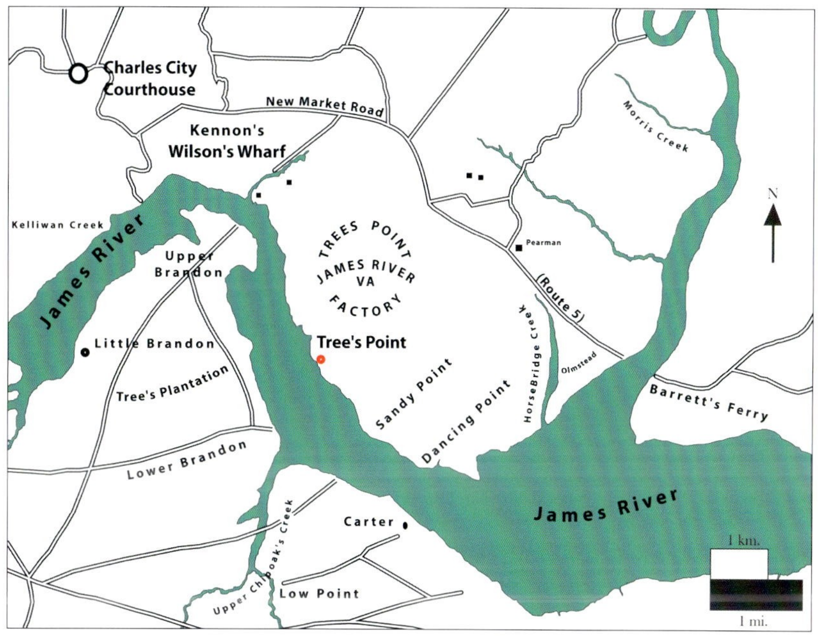

Figure 66 Map showing location of Trees Point Factory. (Courtesy, Oliver Mueller-Heubach.) Wilson's Landing is to the north in relation to the James River.

little is known despite its being the first James River pottery to be rediscovered (by Edwin Atlee Barber in the 1890s).[86] This short-lived manufactory was situated on lands containing "large deposits of suitable clays" for pottery production.[87] The remains of two circular kilns (figs. 67, 68) and a large, associated waster pile were identified on a bluff overlooking the James, just upriver from the promontory of Trees Point. Mid-nineteenth-century descriptions of the area also reference an associated wharf and store.

The operation produced utilitarian stoneware and specialized chemical distilling and storage wares. Of the more than 50,000 sherds recovered, a wide range of utilitarian vessel forms were noted, among them crocks, churns, milk pans/butter pots (and lids for same), jugs, pitchers, chamber pots, and bottles; the industrial wares included Woulfe's (Wolfe's) jars or acid condensers/receivers and related vessels of various forms and sizes as well as crucibles. These last were, significantly, also among the products being sold by the Parrs in Richmond and the Ducey's in Petersburg. Stamped or inscribed capacity marks are not uncommon and a few sherds

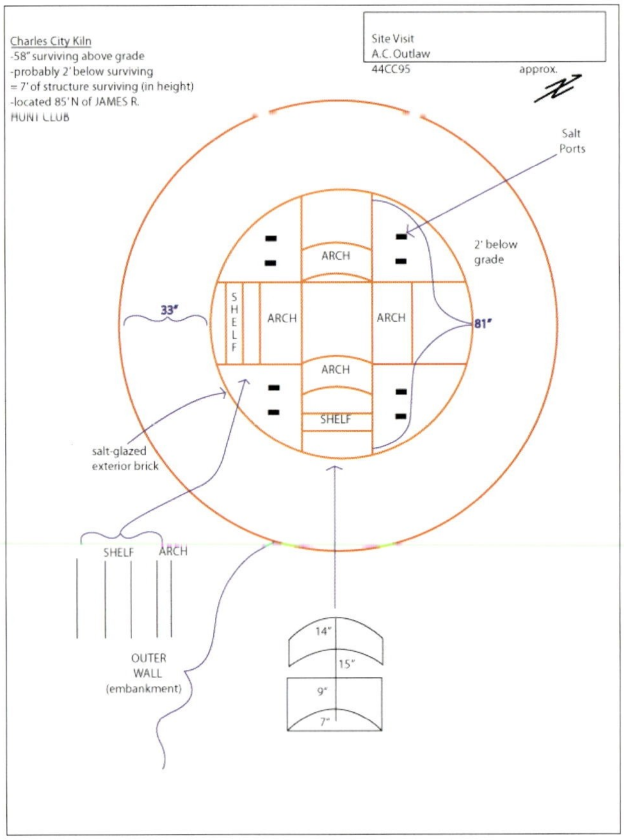

Figure 67 Field sketch or plan of Trees Point Factory kiln (1849–1853) by Alain Outlaw; traced by Oliver Mueller-Heubach

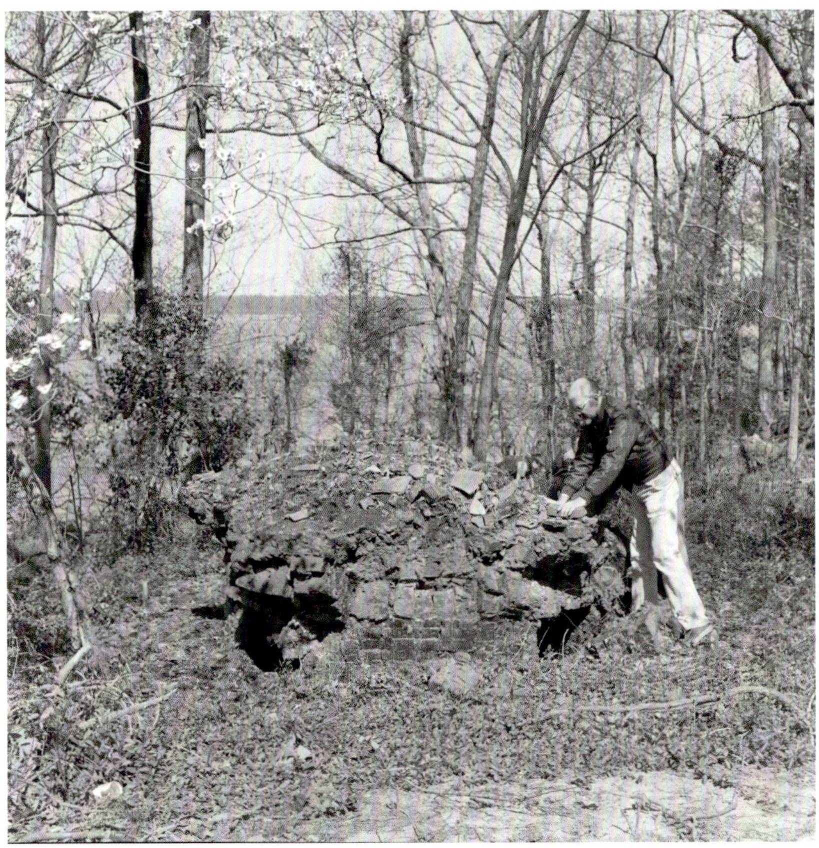

Figure 68 J. P. Hudson examining the remains of an above-ground kiln, 1978. (Courtesy, Department of Anthropology, College of William and Mary.)

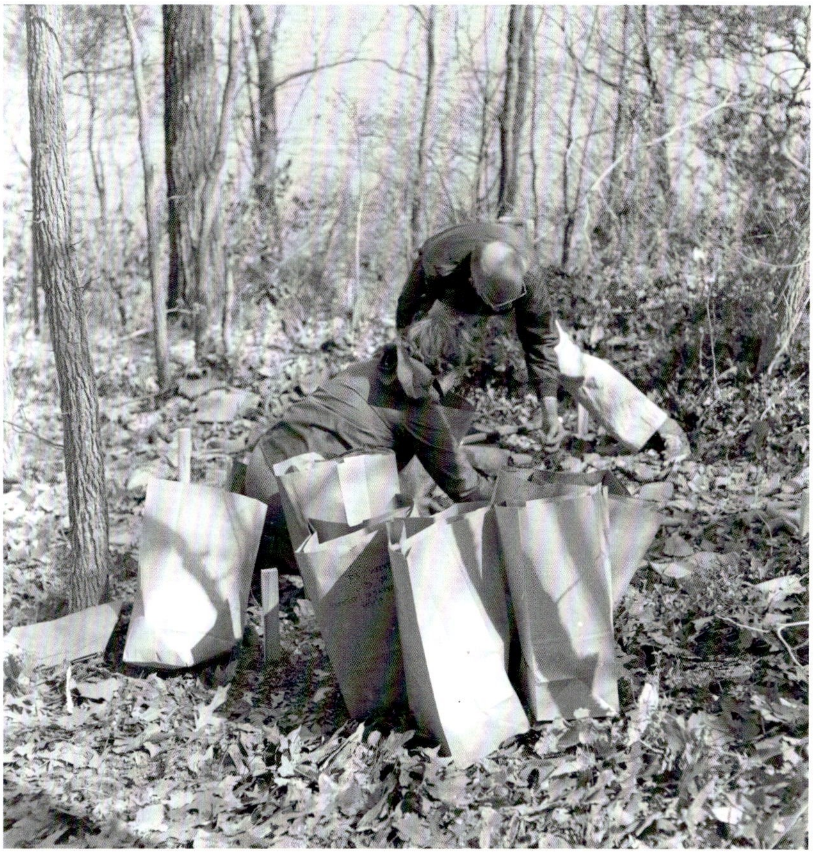

Figure 69 Stoneware fragments, Trees Point Factory, Charles City County, Virginia, 1849–1853. Salt-glazed stoneware. (Courtesy, William and Mary Archaeological Research Center.)

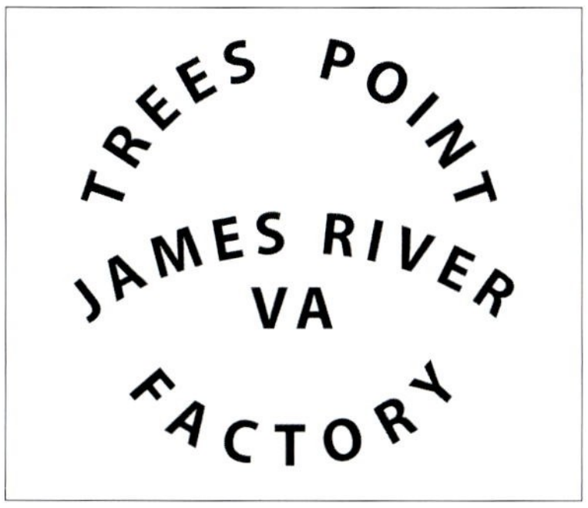

Figure 70 Rendering of maker's stamp of Trees Point Factory. (Artwork by Oliver Mueller-Heubach.)

exhibited utilize the stamp "TREES POINT / JAMES RIVER / VA / FACTORY" (figs. 69, 70). Other marks include simple cross marks and the partial inscription "WB . ." on a flask. Utilitarian vessels generally have straight to flaring sides and sometimes oversize and often bold rims. Decoration was limited to brushed-cobalt florals, observed on all vessel forms present (fig. 71). Although devoid of decoration or capacity marks, the chemical wares were decidedly darker in color, ranging from dark brown to gray-black, and appeared overfired: "extremely dense and highly vitrified." Interpreted by Dennis Pogue as possibly a conscious effort by the potter to ensure

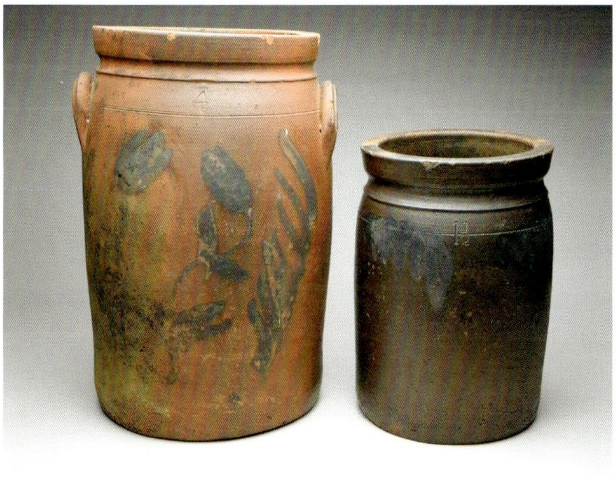

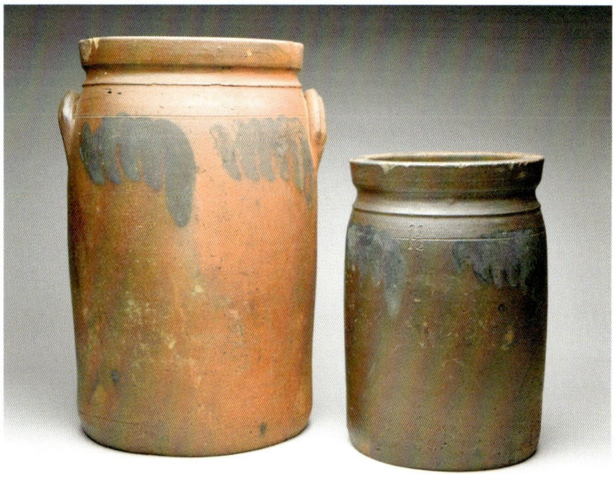

Figure 71 Storage jars, Trees Point Factory, Charles City County, Virginia, 1849–1853. Salt-glazed stoneware. H. 16" and 11½". (Private collection; photo, Kurt Russ.)

impermeability, the question whether the two kilns reflected a segregation of firing strategies is not addressed.[88]

Moro Phillips (1812–d. ca. 1885), the owner of the Trees Point Factory (ca. 1850–1853), hired Philadelphia master potter Sanford Perry. Phillips appears in Philadelphia directories in 1849 and 1850, but is then absent until 1855, at which time he is listed as a chemical stoneware manufacturer at the northwest corner of Chestnut and 31st streets. It is during this interim period that he tested Virginia as a production location. Phillips moved back to Philadelphia in 1853, however, and increases in the value of buildings on the Trees Point property between 1851 and 1856 suggest Sanford Perry was both master and manager. The assessments saw a sharp decline in 1858, coincident with Phillips's sale of 200 of the original 484 acres.[89]

Neighboring Wilson's Landing is also mentioned as a site of pottery production. It is unclear whether the Trees Point Factory has been confused with or was a successor to another pottery at Wilson's Landing, or whether they are in fact one and the same. Francis Wilson, who probably was related to both the potter Samuel Wilson and John Schermerhorn's wife, Sarah Christian Wilson, sold land to Schermerhorn, including Wilson's Landing by 1830.[90] Wilson likely influenced Schermerhorn's decision to visit and investigate the Charles City County area as a possible location for employment or business opportunity. Whether Wilson accompanied Schermerhorn during his initial visit in 1810–1811 is not confirmed, but Schermerhorn's subsequent marriage to Sarah Christian Wilson and his return to the area for full-time employment is reflective of this early association with Wilson. It is possible that one of their goals was to evaluate the suitability of the area for pottery manufacturing.

Wilson and Schermerhorn could have established a pottery here either prior to or post-1830, after moving from Richmond/Henrico, or perhaps maintained simultaneous operations here and at Montezuma. There is also speculation that they were partners in a firm here (either prior to or coincident with their Richmond and Henrico endeavors), with this concern later involving Perry and Phillips (now at Trees Point rather than Wilson's Landing). It is hoped that future research will clarify the role and relationship of these sites and individuals in the nineteenth-century pottery industry in Charles City County.

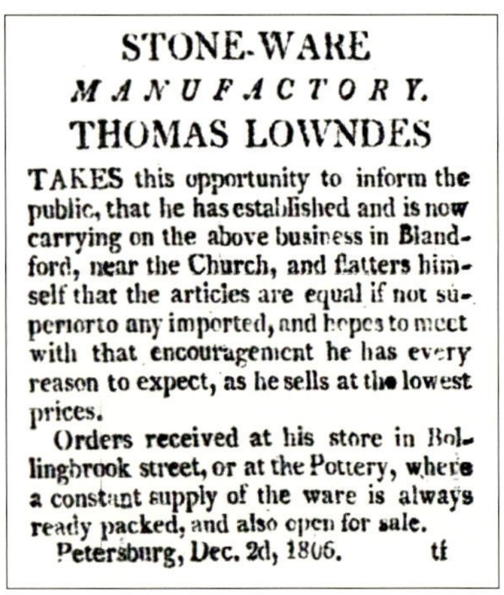
STONE-WARE
MANUFACTORY.
THOMAS LOWNDES
TAKES this opportunity to inform the public, that he has established and is now carrying on the above business in Blandford, near the Church, and flatters himself that the articles are equal if not superior to any imported, and hopes to meet with that encouragement he has every reason to expect, as he sells at the lowest prices.
Orders received at his store in Bollingbrook street, or at the Pottery, where a constant supply of the ware is always ready packed, and also open for sale.
Petersburg, Dec. 2d, 1806. tf

Figure 72 Advertisement, Thomas Lowndes Stone-Ware Manufactory, *Petersburg Intelligencer*, December 2, 1806, p. 3. This is an important early reference to a transplanted English stoneware concern about which little is known.

Petersburg and Dinwiddie County Potteries

With the exception of the Henry Lowndes Pottery, very little is known about stoneware production in the Petersburg/Dinwiddie County area of Virginia. As many as four different sites hosted at least six pottery businesses and some ten or more potters. Aided by journeymen, apprentices, and workers these shops operated through the first three-quarters of the nineteenth century.

THE LOWNDES/DUCEY POTTERY

By 1806 salt-glazed stoneware was being produced in the Blandford area of northeastern Petersburg, Virginia, by Thomas Lowndes and his family (fig. 72). An immigrant from the Staffordshire district in England, Lowndes

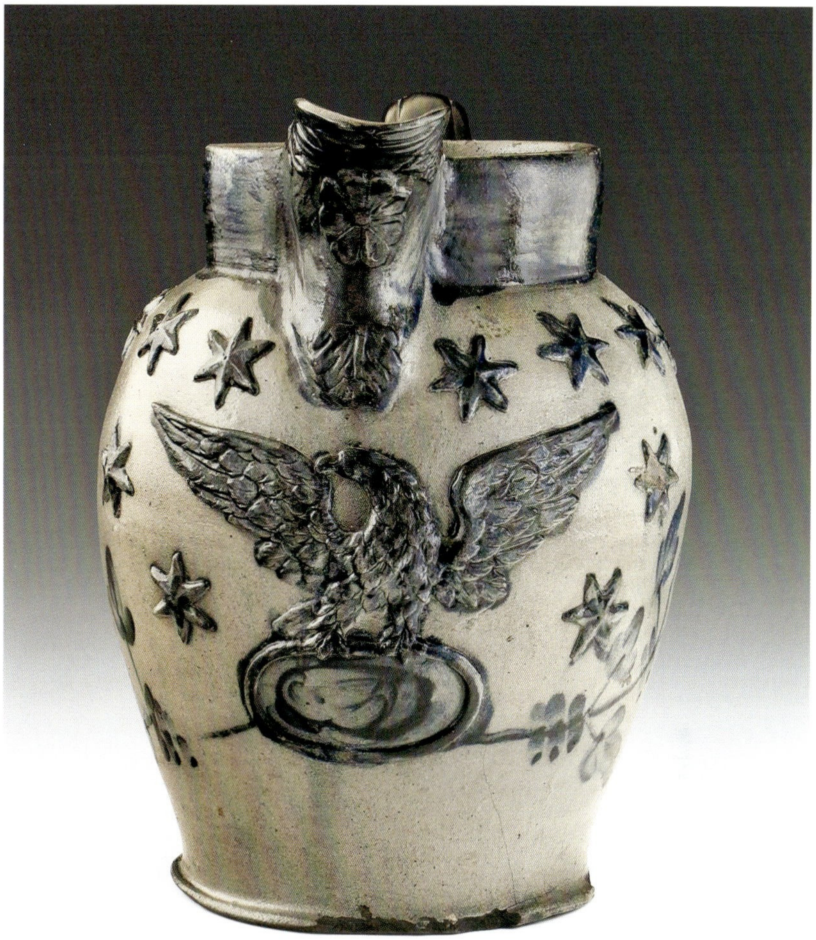

Figure 73 Pitcher, Henry Lowndes, Petersburg, Virginia, 1835–1840. Salt-glazed stoneware. H. 12". (Private collection; photo, Robert Hunter.) The classical shape of this elaborately decorated pitcher is typical of Staffordshire earthenware. The use of applied molding to the spout and the body is also influenced by British prototypes. The spread eagle, wreath, and stars all resemble decorative motifs incorporated in cast-iron decoration of the period. This pitcher is one of only two of this type known to exist.

must have brought a considerable repertoire of potter's skills, but very little evidence of his work has come to light. Better documented are products of his son, Henry Lowndes, who took over the operation of the pottery—with both interim and subsequent family involvement—after Thomas's death in 1811. Henry's two brothers, Thomas and John, were listed as potters in 1812, with Thomas being the proprietor immediately after his father's death.[91] By 1817 the children's full interest in the business had been legally transferred to their mother, Elizabeth. Newspaper advertisements suggest that the Lowndes made refined earthenware as well as stoneware, though no surviving examples of the former have been identified.[92]

The most ambitious of Henry Lowndes's pieces incorporated applied molded decoration. This type of decoration, perhaps reflecting Lowndes's Staffordshire heritage, was reserved for use on a series of presentation pitchers and coolers. In addition to these molded elements, some later pitchers show Henry's characteristic forms and decorative treatments (fig. 73). Lowndes's stoneware is easily recognizable by the frequently encountered one- to five-gallon storage jars and their distinctive brushed-cobalt floral designs and slip-trailed signatures, which adorn a large proportion of his work (fig. 74). A recently discovered diminutive churn is

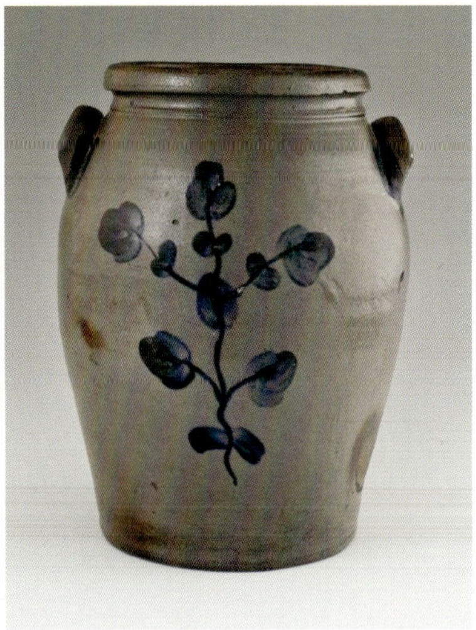

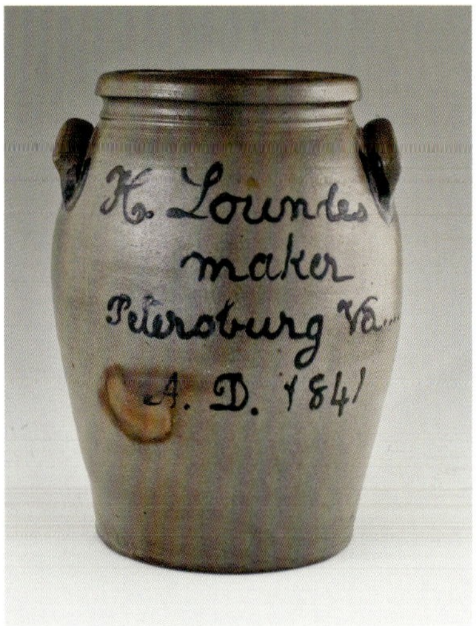

Figure 74 Storage jar, Henry Lowndes, Petersburg, Virginia, 1841. Salt-glazed stoneware. H. 12½". Capacity: 2 gallons. Slip-trailed inscription: "H. Lowndes / maker / Petersburg Va. / A.D. 1841" (Private collection; photo, Robert Hunter.)

embellished with the familiar applied eagle as well as characteristic brushed-cobalt florals. A number of other, less elaborate extant jars from Lowndes's (and related Petersburg) production are often misattributed to other area potteries. This is partly due to the brushed-cobalt capacities he employed, which resemble those that have come to characterize Schermerhorn's Richmond wares (fig. 75).

Another distinctive Lowndes motif is confirmed based on comparison with archaeological sherds from the pottery site showing opposing horizontal leafy vines. Though first appearing to be stenciled, the vines are in fact brushed cobalt. The series of leaves progressively decreases in size from large curved examples near the midline of the pitcher to small, almost dot-like specimens near the termination at either side of the vessel. In the past, a series of extant one- and two-gallon jars with this decoration were thought to have been made by Sweeney but can now be firmly associated with Petersburg wares.

The most common brushed-cobalt motif observed on Henry Lowndes products also appears on the reverse of the majority of his signature pots (i.e., those embellished with his name in slip-trailed script: "Henry (or H.) Lowndes/ Manufactor (or Maker)/ Petersburg/Va"). It resembles the "Baltimore clover" that has come to characterize Peter Hermann products. Lowndes's puffy, stylized blooms top multiple, symmetrically leafed branches emanating from a single stem. In this, suggests Robert Hunter, they more closely mimic a cotton plant. Given Petersburg's prominent textile industry and proximity to major cotton plantations, Lowndes's frequent application of the motif and our interpretation seem quite apropos. In overall appearance and impact, the motifs are strikingly similar. Vessel forms show more distinct variations, suggesting the physical movement of potters or journeymen from Lowndes to Sweeney rather than just imitation or borrowing.

Family members ran the pottery following Henry Lowndes's death, but ultimately it was sold to John and Thomas Ducey in 1855, brothers who had emigrated from Ireland. John was listed as a laborer living in a boarding-

Figure 75 Churn, attributed to Stephen B. Sweeney, Henrico County, Virginia, 1838–1863. Salt-glazed stoneware. H. 17½". (Private collection; photo, Gavin Ashworth.) Extant vessels have recently been identified that appear to have been produced by Sweeney but are decorated with a close variation on the Lowndes so-called Cotton Plant motif (see fig. 74). The branches are asymmetrical and the cotton blooms are less structured, more tightly defined, and sometimes executed with only two brushstrokes. These vessels may suggest an overlap between the two potteries either in the purposeful imitation of decoration or the exchange of workmen.

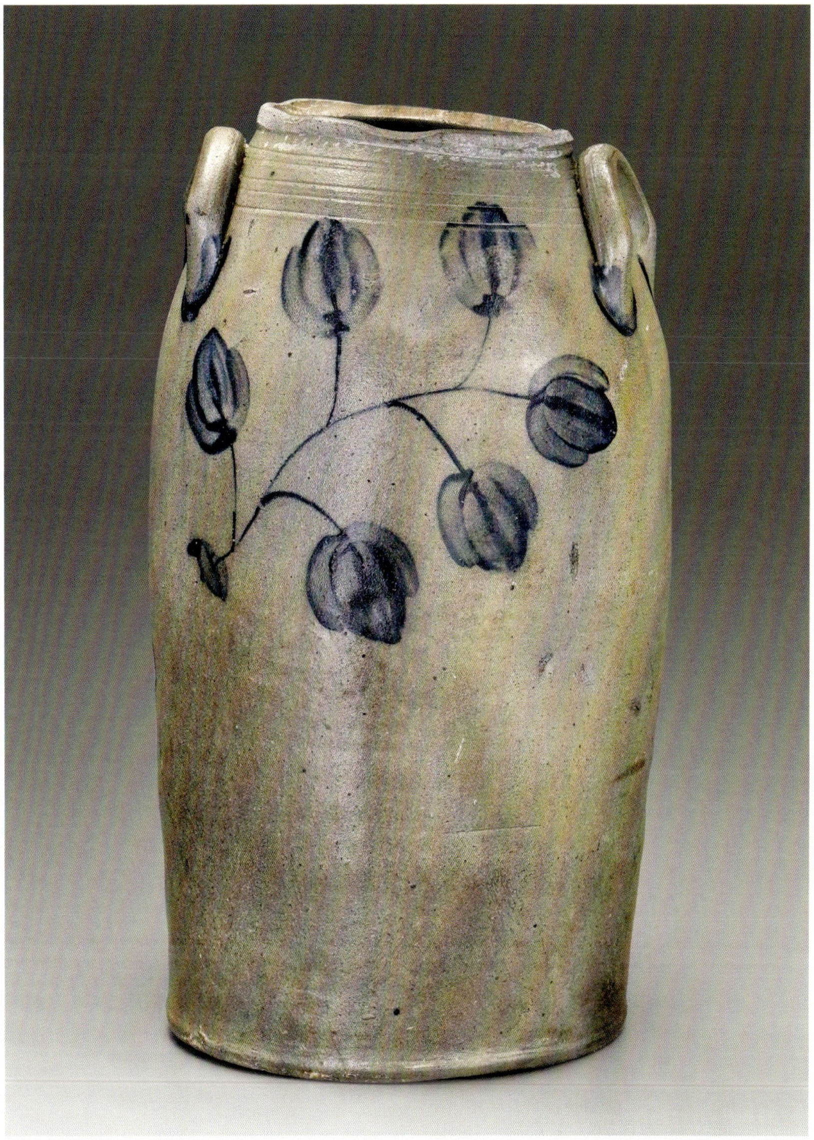

house in Cabell County, Virginia, in 1850, when he was twenty-one. In 1870 he is identified as a stoneware manufacturer living in Ward 3 in Petersburg City, unmarried but with an eighteen-year-old servant named George Smith. At the time his belongings were valued at $1,000 and his real estate, $7,000.[93]

The Duceys' operation was located at Watson and Wythe streets in Petersburg, and they continued producing stoneware there until about 1878. An article in the *Petersburg Index* discusses a visit to the "Petersburg Pottery and Stone Ware Manufactory of the Messrs. Duceys" in Blandford, describes the processes involved in potting, and offers valuable information about their clay source:

> The clay used by these gentlemen is all obtained from James River. There being none in this vicinity at all adapted to the purpose. It is of a far finer and smoother grain than any around Petersburg, and when mixed and ground and kneaded into balls presents a soft smooth surface ready for the plastic hand that is to fashion it into shape.[94]

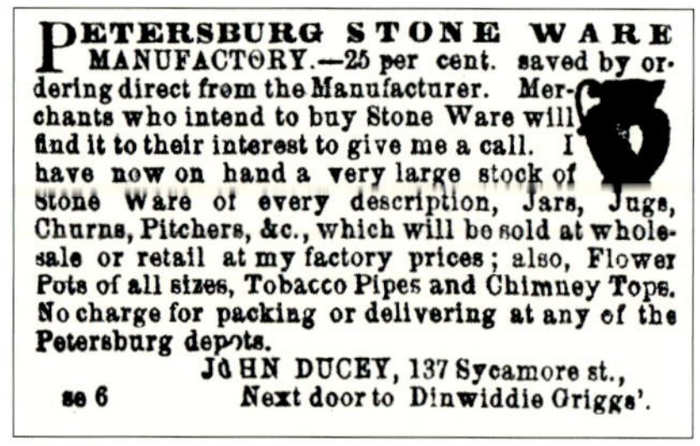

PETERSBURG STONE WARE MANUFACTORY.—25 per cent. saved by ordering direct from the Manufacturer. Merchants who intend to buy Stone Ware will find it to their interest to give me a call. I have now on hand a very large stock of Stone Ware of every description, Jars, Jugs, Churns, Pitchers, &c., which will be sold at wholesale or retail at my factory prices; also, Flower Pots of all sizes, Tobacco Pipes and Chimney Tops. No charge for packing or delivering at any of the Petersburg depots.

JOHN DUCEY, 137 Sycamore st.,
se 6 Next door to Dinwiddie Griggs'.

Figure 76 Advertisement for the Petersburg Stone Ware Manufactory, featuring the silhouette of a pitcher resembling earlier Lowndes Staffordshire-style ware. *Petersburg Index*, September 1872, p. 3, col. 3.

Little is known about the full range of the Duceys production, although a local advertisement in October 1865 indicates they were offering a full range of "the best quality of Stone Ware" products, including crucibles, the qualities and longevity of which were expounded on in affidavits from machinists and iron works to withstand 8–14 "heats."[95] Another ad provides a much more complete listing of the items offered and includes an illustration of a pitcher resembling earlier Staffordshire-styled Lowndes examples rather than the later forms of pitchers one would suspect they produced (fig. 76).[96] Extant vessels include brushed-cobalt storage jars with diminutive rims, applied crescent handles, and straight-sided or slightly curved forms often stamped with the Ducey mark (fig. 77). Among the decorative treatments are motifs that strongly resemble Lowndes florals as well as some tulips with anthers that seem to be styled after those of the potters in the Four Mile

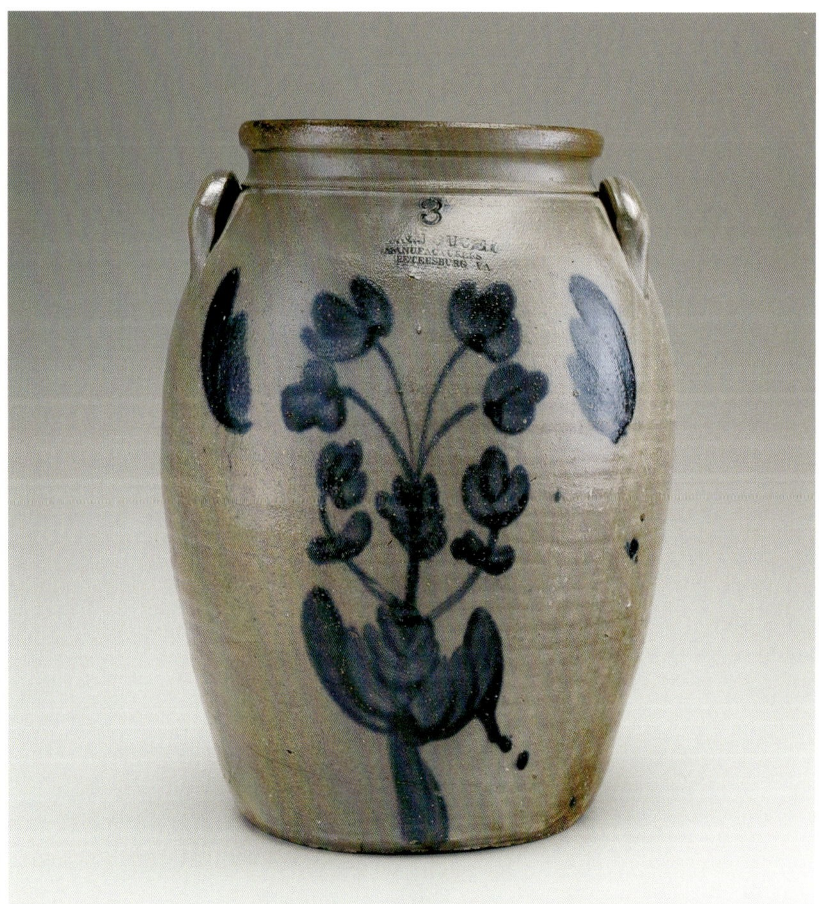

Figure 77 Storage jar, Thomas and John Ducey, Petersburg, Virginia, 1855–1872. Salt-glazed stoneware. H. 15". Capacity: 3 gallons. (Private collection; photo, Gavin Ashworth.) This well-made example of the Duceys' work displays a deep-cobalt–brushed floral decoration, following the Lowndes style. Although scant research has been conducted on this pottery, a number of examples exist, both marked and unmarked. The mid-nineteenth-century forms tend to be straight-sided or slightly ovoid, typical of the period, although the decoration recalls earlier Lowndes pieces.

Creek/Bailey's Creek area. Strikingly similar brushed-cobalt script capacity marks are observed beneath handles and in wreaths on both Lowndes- and Ducey-attributed wares, recalling similar numerals used by Schermerhorn at Rocketts and Montezuma. The form and appearance of capacities on Ducey wares are consistently slightly larger, as if executed with broader strokes of a more thickly applied cobalt oxide.

Archeological investigations recently conducted in the vicinity of the Lowndes nineteenth-century pottery operations by the William and Mary Center for Archaeological Research identified what is likely a portion of a kiln wall, evidenced by a few glazed bricks, a draw trial, and kiln furniture. Excavation expanded the known range of forms and decorative treatments produced at the pottery, including brushed, slipped, and stenciled cobalt decoration, several patterns of decorative rouletting, and limited incising. Only a small portion of the recovered stoneware assemblage exhibited decorative treatments that would allow for more precise attribution of both extant wares and those recovered from domestic archaeological contexts.[97]

THE POTTERIES OF HARWOOD AND HARWOOD & TURNER

In 1818 William Harwood established a "Stone Ware Manufactory" in Blandford, on the east side of Petersburg. An advertisement published in the *Halifax Compiler* on June 19, 1818, announced the opening of his operation and indicated the marketing of wares locally and to North Carolina:

> Stone Ware Manufactory,
> PETERSBURG, VIRGINIA.
> THE Subscriber wishing to make as great progress as possible, in Home Manufactures, has established in Petersburg, a Manufactory of Stone Ware, and pledges himself to the Merchants of Virginia and North Carolina, to furnish them with any quantities, well assorted, and of as good quality as as may be made in this Country.
> His prices are the same as those at the North: and will warrant all of his make to be equal to any from the Northern Towns.
> WM. HARWOOD
> Petersburg, May 8.[98]

By September 1818, within only a few months of his opening, he formed a partnership with one Mr. Turner.[99] After that brief but well-advertised association (it dissolved in mid-December), Harwood continued to operate for more than two and a half years.[100]

Given Harwood's experience in a wide range of ceramic technologies and production methods and his earlier association with DuVal and Schermerhorn, it seems likely that the wares he manufactured during his tenure in Petersburg resembled those produced in Richmond. Unfortunately, no extant marked examples from this period are known. Research into Harwood's life is underway in hopes of clarifying his role in the developing pottery industry in eastern Virginia.

A May 4, 1819, advertisement in the *Petersburg Republican* in which Harwood offered a $5 reward for the return of a runaway slave described as "ABRAHAM, 26 or 27 years of age, five feet six or seven inches high, stout made, has lost the greater part, if not all his toes, having been frost-bitten—had on when he went away homespun woolen clothes" suggests Harwood made use of slave labor at his pottery.[101]

In August 1821 Harwood, who was facing financial difficulty and "both acknowledging and desirous to pay his debts to others," sold to Enoch Burnett his Blandford concern: "my Bay horse, my pottery & all the ready made ware & said materials of every kind whatsoever, also the tools & fixtures attached thereto."[102]

ENOCH BURNETT

Enoch Burnett was an orphan who was taken on as an apprentice by Thomas Amoss in Baltimore on November 7, 1813. He became a skilled potter, plying his craft in Virginia and then Philadelphia before returning to Baltimore and ultimately settling in Washington, D.C.

Burnett (another Quaker) may have made his way to Virginia alongside Amoss, participating in his potting activities at Four Mile during the 1818–1821 period. Looking south to Petersburg, he seized the opportunity to assume the reins of his own operation. On August 27, 1821, he purchased Harwood's pottery, wares, fixtures, and tools. He simultaneously acquired several notes due Harwood and held an additional $100 note from him. Whether Burnett continued to operate the pottery in Petersburg, leased it to another potter, or simply used the fixtures elsewhere is not known, but it seems likely that he operated the concern until 1826 or 1827 before moving to Philadelphia.

On May 4, 1827, Enoch Burnett and his brother-in-law Henry Remmey Jr. purchased Branch Green's pottery in Philadelphia for $3,800. After selling his interest to Remmey in 1831, Burnett continued to be listed as a stoneware potter or merchant through 1836. By 1838 he was hired by Mauldine Perine in Baltimore, for whom he worked until at least 1842. Thereafter, he assumed ownership of Richard Butt's successful Washington, D.C., factory at 8th and I streets and there operated the Enoch Burnett Pottery (1846–ca. 1879) for the last three decades of his life.[103]

ROLAND WILSON'S STONE POTTERY

Roland Wilson, whose relationship to Samuel, Sarah Christian, and Francis is unknown, appears in the 1850 Industrial Census for Dinwiddie County as operating a "stone pottery." The capital invested is listed at $100, with labor being hand-powered by three male workers earning $60 per month in wages. The confusing entry in this record references the production of "600 per kind of products = rough granite" valued at $900.[104] Whether he was producing "stone pottery" made from clay or containers of "rough granite," or whether these are two different descriptive terms for the same product, remains a mystery. The location of his business is unknown.

JOHN DERING'S STONEWARE POTTERY

John Dering (During) is identified in the 1860 Industrial Census Records for Petersburg/Dinwiddie County as operating a stoneware pottery. It is described in the following manner: "$1,000 capital invested; 100 tons of clay worth $400; 150 cords of wood worth $450; hand-powered; 6 male workers for $240 wages; 16000 jugs and 3600 doz. Jars worth $5000."[105] Regrettably, the location of his operation remains a mystery and no signed wares are known.

Although no other pottery manufacturing sites or potters have been identified from the Petersburg/Dinwiddie County area, it is likely that many other individuals worked in the industry, either full-time or in the lag periods of their primary occupations.

Summary

Sixty years after the closure of America's pioneer stoneware factory at Yorktown, an influx of immigrant potters staged a remarkable revival of the craft. The founding of earthenware potteries in the young capital of Richmond by Lord and Park (with Robinson) and DuVal with Allinson during the last two decades of the eighteenth century foreshadowed this resurgence. Stoneware manufacturing in Virginia's Tidewater region from Richmond and along the James River to Petersburg began in 1806–1811. More than sixteen different pottery manufactories involving at least thirty-seven skilled potters operated in the region over the next eighty years.

At least four major production sites were active in Richmond from 1811 through 1865: DuVal's pottery, Schermerhorn's Port Mayo and Montezuma factories, and Parr's Rocketts pottery. More than thirteen documented potters and twenty employees/workers were involved. Southeast Henrico County contained another significant concentration of pottery manufacturing sites. Several Richmond potters were involved, either simultaneously or later, in stoneware production near New Market, located 10–13 miles southeast of the Henrico County courthouse and situated along Four Mile and Bailey's creeks. Here, no fewer than ten potters, along with their family members, employees, boys, and slaves worked at multiple nearby potteries, spin-offs from the initial Richard Randolph Manufactory of 1812/1813. It was these frequently anonymous individuals who formed the backbone of the growing industry and transformed significant local clay resources into usable products. They did so using skills they acquired as workers, indentures, and apprentices in pioneering late-eighteenth-century earthenware and early-nineteenth-century stoneware operations in Richmond just to the north.

The initial historical overview of the Four Mile Creek/Bailey's Creek area of southern Henrico County reveals it to be one of the most significant complexes of early-nineteenth-century stoneware manufacture on the eastern seaboard, beginning in 1813 with Randolph's Kaolin holdings and continuing through 1862 with Sweeney's Claymount concerns. Further documentary research in conjunction with archaeological field investigation is critically needed to more fully document, explore, and ultimately protect the well-preserved sites that remain, vestiges of this once-thriving industry.[106]

While the Charles City County area contained one well-documented pottery, Trees Point, and possibly one other, the Petersburg/Dinwiddie

County area involved as many as twelve potters between 1806 and 1878. The Lowndes Pottery, established as early as 1806, was in operation until at least 1842. Reopened by the Duceys sometime after 1855, the Blandford pottery continued as late as the elder brother's death, in 1878. William Harwood, his enslaved worker Abraham, and others were involved in early stoneware production circa 1818 to 1821 in Petersburg, an operation later purchased by Enoch Burnett. Two other poorly understood midcentury concerns have also been identified: the pottery of Roland Wilson, which might be connected to either a Charles City County or an earlier Richmond endeavor involving Wilson family members; and the pottery owned by German immigrant John Dering.

Exploration of these potters and their unique associations provides a better comprehension of individual production, interactions, and influences. It also raises new questions that can be answered only by further research into these Virginia stoneware potters and their fascinating wares. What is apparent, however, is the similarity of wares made by or attributed to different potters or potteries within this region. The cooperation between them and their close business, personal, and religious relationships linked potters and their vessels.

It is hoped that continuing research will provide a more comprehensive understanding of individual pottery sites, associations, and influences, leading to a more meaningful appreciation of the history of ceramics in Virginia.

1. Norman Barka, "Archaeology of a Colonial Pottery Factory: The Kilns and Ceramics of the Poor Potter of Yorktown," *Ceramics in America*, edited by Robert Hunter (Hanover, N.H.: University Press of New England, 2004), pp. 15–59.

2. Barbara H. Magid, "An Archaeological Perspective on Alexandria's Pottery Tradition," *Journal of Early Southern Decorative Arts* 21, no. 2 (Winter 1995): 41–82; Kurt C. Russ, "Exploring Western Virginia Potteries," *Journal of Early Southern Decorative Arts* 21, no. 2 (Winter 1995): 98–138.

3. "Stoneware of Eastern Virginia," an exhibit at the Virginia Historical Society in Richmond (September 1, 2004–March 1, 2005), was curated by Robert Hunter, Kurt C. Russ, and Marshall Goodman and highlighted significant wares from this region; Hunter, Russ, and Goodman, "Stoneware of Eastern Virginia," *Antiques* 167, no. 4 (2005): 126–33

4. "Kaolin to Claymount: Demystifying James River Valley Stoneware" was a recent (2012–2013) exhibit at the Museum of Early Southern Decorative Arts, curated by Kurt C. Russ, Robert Hunter, Oliver Mueller-Heubach, and Marshall Goodman, which focused on stoneware from the James River region. In addition, "No Stone (Ware) Left Unturned: Recent Discoveries in Virginia Stoneware," a day-long MESDA symposium in 2013, highlighted many of the new discoveries that are detailed in this article.

5. Oliver Mueller-Heubach, "From Kaolin to Claymount: Landscapes of the 19th-Century James River Stoneware Industry," PhD diss., Department of Anthropology, College of William and Mary, 2013.

6. It is likely that this James Brindley was related in some way to the well-known British canal engineer James Brindley (1716–1772), who spent much of his life in Leek, Staffordshire, and was a partner of his brother John in the Longport Pottery. He married Anne Henshall in 1765; her brother, Hugh Henshall, was also involved in canal construction.

7. Original handwritten letter by James Brindley in Wilmington, Delaware, to his unnamed uncle in England, dated August 10, 1784. It sold on eBay on February 9, 2011, for $306 (item no. 270699953066).

8. Bradford L. Rauschenberg, "B. DuVal & Co/Richmond: A Newly Discovered Pottery," *Journal of Early Southern Decorative Arts* 4, no. 1 (May 1978): 56; Robert Hunter and Marshall Goodman, "The Destruction of the Benjamin DuVal Stoneware Manufactory, Richmond,

Virginia," *Ceramics in America*, edited by Robert Hunter (Hanover, N.H.: University Press of New England for the Chipstone Foundation, 2005), pp. 37–60.

9. City of Richmond, Common Council Records, no. 1, 1782–1793, p. 291.

10. Ibid.; *Heads of Family at the First Census of the United States Taken in the Year 1790, Virginia* (Washington, D.C.: General Printing Office, 1908), p. 117.

11. *Virginia Gazette, and the Weekly Advertiser,* Richmond, November 29, 1783, p. 3, col. 3; *Maryland Journal and Baltimore Advertiser*, May 20, 1783, p. 3, col. 2.

12. Henrico County Deed Book, no. 1, 1781–1785, p. 98, April 1, 1783.

13. *Virginia Gazette, and the Weekly Advertiser*, Richmond, March 16, 1782, p. 4.

14. Ibid.

15. Bradford L. Rauschenberg, "B. DuVal & Co./ Richmond: A Newly Discovered Pottery," *Pottery Collectors Newsletter*, reprinted from *Journal of Early Southern Decorative Arts* (May 1978): 27 n. 1.

16. *Virginia Gazette and General Advertiser*, February 23, 1791, p. 3.

17. City of Richmond, Common Council Records, no. 1, 1782–1793, p. 235, June 20, 1791.

18. Edward Virginius Valentine, "The Dell" n.d. [189?]; *Allinson v. DuVal* 1795 Henrico County Court Suit Papers, July 16, 1791, April–May 1791 or 1792. The suit file housed at the Library of Virginia included a DuVal broadside dated July 1, 1791.

19. Kurt C. Russ and W. Sterling Schermerhorn, "Rocketts' Red Glare: John P. Schermerhorn and the Early Richmond-Area Stoneware Industry," in *Ceramics in America* (2005), pp. 61–92.

20. Ibid.

21. *Richmond Enquirer*, New Year's Day 1829, vol. 25, no. 73, p. 4.

22. Schermerhorn family papers, University of Missouri Library and Archives, courtesy of W. Sterling Schermerhorn. See also Russ and Schermerhorn, "Rocketts' Red Glare," p. 78, fig. 33, which illustrates a three-gallon Schermerhorn-attributed jar dated 1837. The date probably refers either to the last year of Andrew Jackson's presidency or to Schermerhorn's last year of producing pottery in Richmond at Port Mayo, Rocketts.

23. *Richmond City Business Directory*, 1856–1857, 1858–1859; Elizabeth J. Monroe, David W. Lewes, Joe B. Jones, and Michael D. Glascock, *Preliminary Archaeological Assessment of a Waster Pit Associated with the Parr Pottery Works (44HE0806), City of Richmond, Virginia* (Williamsburg, Va.: William and Mary Center for Archaeological Research, 2010), pp. 11–12. Report submitted to the Virginia Department of Historic Resources.

24. *Richmond City Business Directory*, 1860.

25. Henrico 1860 Industrial Census Records.

26. *Richmond Whig*, April 15, 1865.

27. Of particular note is a monumental twenty-gallon salt-glazed stoneware cooler decorated with an American eagle with outstretched wings and banners announcing "DAVID PARR" and "MANUFACTORY" extending from its beak. This important vessel was illustrated and discussed by Kurt Russ in his article "The Remarkable Stoneware of George N. Fulton," in *Ceramics in America*, edited by Robert Hunter (Hanover, N.H.: University Press of New England for the Chipstone Foundation, 2004), pp. 159–61.

28. Some support for an association between these individuals is that a J. B. Keesee was a witness to John Schermerhorn's will, which was written in 1848; whether this J. B. Keesee relates to Thomas W. Keesee is not known.

29. William and Mary Center for Archaeological Research, Williamsburg, Virginia.

30. Henrico County Land Tax Lists 1799–1803; Henrico Personal Property Tax Lists, 1804.

31. Norwich Mills was a eighteenth-century gristmill and bakery complex owned and operated by its namesake, Norwich, who was previously George Washington's Mount Vernon miller.

32. Mutual Assurance Society, vol. 16, no. 858, November 17, 1802. The insurance agents noted a deduction of $7,020 from the value of his mill complex due to decay or needed repairs, indicating an estimated cost of cost $21,900 to rebuild. See also Mutual Assurance Society, vol. 40, for a description of the property location; Mutual Assurance Company Declaration No. 858, ref. 1111, Insurance of David Ross's Norwich Mills on January 5, 1806.

33. Mutual Assurance Society, vol. 16, no. 858, November 17, 1802; Henrico County Land Tax Lists 1809–1820.

34. *Enquirer* (Richmond), October 22, 1813, p. 3, col. 4, and January 4, 1814, p. 4, col. 3; *Virginia Patriot, Richmond*, October 19, 1813, p. 3, col. 5.

35. Mutual Assurance Society, vol. 44, no. 1955, March 29, 1816.

36. Mutual Assurance Company, Declaration No. 858, ref. 1111, Insurance of David Ross's Norwich Mills on January 5, 1806.

37. Mutual Assurance Society, vol. 44, no. 1955, March 29, 1816.

38. Ibid.; Henrico County 1820 Census of Manufacturers.

39. Henrico County Land Tax Lists 1821–1850.

40. Henrico County Land Tax Lists 1809–1820; Henrico County Personal Property Tax Lists 1809–1820.

41. Henrico County Order Book No. 19, 1815–1816, p. 78.

42. Museum of Early Southern Decorative Arts research files identify both Joe Givin (52288) and Tom Givin (52289) as potter apprentices, Winston-Salem, North Carolina.

43. Henrico County Order Book No. 19, 1815–1816, p. 78, June 5, 1815: "Ordered that the Overseers of the poor bind out, according to law, Jack Givin aged eighteen years, and Frederick Givin aged fourteen years, free persons of colour, to Samuel P. Parsons"; Henrico County Order Book No. 19, 1815–1816, p. 78, June 5, 1815: "Ordered that the Overseers of the poor bind out Tom Givin aged fourteen years, and Joe Givin aged twelve years, free persons of colour, to Richard Randolph."

44. Henrico County Minute Book No. 20, 1816–1819, p. 226, July 7, 1817. This ad appeared only three days after Randolph's advertisement to sell his Kaolin property, suggesting a coincident shifting in the responsibilities of road maintenance from Randolph to Frayser.

45. Mutual Assurance Society of Virginia, Policy #1111; Mutual Assurance Society, vol. 44, no. 1955, March 29, 1816.

46. Thomas Jefferson to Richard Randolph, January 25, 1814, Jefferson Papers, Massachusetts Historical Society (hereafter Jefferson Papers).

47. Thomas Jefferson to Richard Randolph, September 27, 1814, Jefferson Papers. The story of Jefferson's brewing operation is fascinating and includes the establishment of a brew house at Monticello by the Fall of 1814 under the direction of Captain Joseph Miller, an English master brewer who was befriended by Jefferson and for a time resided at Monticello. Miller taught brewing skills to Peter Hemmings, who later took charge of these activities for Jefferson.

48. Randolph to Jefferson, February 18, 1814, Jefferson Papers, Massachusetts Historical Society; "The Dell," by Edward Virginius Valentine [189?]. Typed transcript in David Parr file, MESDA Craftsman Catalog; original lecture notes on file at Valentine Museum, Richmond.

49. William M. Kelso, *Archaeology at Monticello: Artifacts of Everyday Life in the Plantation Community*, Monticello Monograph Series (Charlottesville, Va.: Thomas Jefferson Memorial Foundation, 1997), pp. 101, 102, 119, fig. 68. Kelso referred interchangeably to these lidded, bottomless vessels as sea-kale pots, forcing pots, and blanching pots.

50. Thomas Jefferson to Bernard Peyton, February 21, 1821, Jefferson Papers, as referenced in Audrey Noël Hume, *Archaeology and the Colonial Gardener* (Williamsburg, Va.: Colonial Williamsburg Foundation, 1974), p. 54.

51. Thomas Jefferson, Account Book 1821–1826, Jefferson Papers, as referenced in A. Noël Hume, *Archaeology and the Colonial Gardener*, p. 54.

52. Kelso, *Archaeology at Monticello*, pp. 101, 102, 119; Edmund Peyton to Thomas Jefferson, dated March 29, 1821, Jefferson Papers.

53. Thomas Jefferson to Richard Randolph, May 13, 1822, Jefferson Papers.

54. Richard Randolph to Thomas Jefferson, May 30, 1822, Jefferson Papers.

55. It is interesting to note that a Mildred Hobson (possibly William's sister) was married to Jackson Frayser. Samuel was granted the property only one year before Mildred's death.

56. Henrico County Minute Book, No. 20, 1816–1819, p. 226, July 7, 1817.

57. Henrico County 1820 Census of Manufacturers, microcopy no. 279, roll no. 18, item no. 504.

58. *Richmond Commercial Compiler*, September 12, 1820, p. 3, col. 5.

59. Mutual Assurance Society, policy no. 4790, July 1822, and policy no. 7506, November 1829.

60. Henrico County Land Tax Lists 1818–1851; Henrico County Personal Property Tax Lists 1818–1850.

61. See Russ and Schermerhorn, "Rocketts' Red Glare," pp. 61–92; Henrico County 1820 Census of Manufacturers, Land Tax Lists 1818–1851, and Personal Property Tax Lists 1818–1850.

62. See Russ and Schermerhorn, "Rockett's Red Glare," pp. 61–92; Henrico County 1820 Census of Manufacturers, Land Tax Lists 1818–1851, and Personal Property Tax Lists 1818–1850.

63. Russ and Schermerhorn, "Rocketts' Red Glare," p. 76, fig. 29.

64. Charles E. Umstott illustrates this jar and attributes it to Thomas Lowndes in "The Lowndes Stoneware Pottery of Petersburg, Virginia," *Journal of Early Southern Decorative Arts* (1995): 88, figs. 2 and 2a.

65. Henrico County Land Tax Lists 1820–1821 and 1820 Census of Manufacturers.

66. Henrico County Deed, Dandridge to Amoss, and Indenture, dated September 8, 1820, recorded October 2, 1820, pp. 388–89.

67. Henrico County Land Tax Lists 1820–1821 and 1820 Census of Manufacturers.

68. Henrico County 1820 Census of Manufacturers.

69. *Richmond Enquirer*, October 1, 1819, p. 3, col. 5, *Richmond Commercial Compiler*, September 30, 1819, p. 3, col. 3. Since the Baltimore Thomas Amoss & Co. business was dissolved six months earlier, this could represent standing stock. Alternatively, it might be a sign that the name was briefly revived on Four Mile Creek; although Amoss did not have his own independent operation until 1820, he might have been using the same business name in the interim.

70. Henrico County Will Book No. 6, 1822–1827, August 7, 1822, p. 137; Henrico County Personal Property Tax Lists 1818–1825; Henrico County Land Tax Lists 1818–1832.

71. Survey of the land of "Richard Ellyson's Estate, Deceased," November 15, 1841, p. 268, showing "Factory Tract" and his wife's dower, Library of Virginia, Richmond, Va.

72. According to John N. Pearce, "The Early Baltimore Potters and Their Wares, 1763–1850," master's thesis, University of Delaware, 1959, p. 87, the firm of Thomas Amoss and Co. was operating from 1810 until 1817; more recent research by John Kille, however, indicates that the firm of Thomas Morgan and William produced stoneware at Pitt and Green streets from 1812 until 1815. Thomas Morgan subsequently partnered with his brother-in-law Thomas Amoss and established Thomas Amoss and Co., which produced stoneware at the same location until 1819. Thomas Morgan's son, William Morgan, then established together with Thomas Amoss another firm known as Morgan and Amoss, which operated until Amoss's death in late 1822, after which William Morgan continued as a sole proprietor until as late as ca. 1827. See Kille, "Distinguishing Marks and Flowering Designs: Baltimore's Utilitarian Stoneware Industry," in *Ceramics in America*, edited by Robert Hunter (Hanover, N.H.: University Press of New England for the Chipstone Foundation, 2005): 104–7.

73. Henrico County Land Tax Lists 1838–1844.

74. Henrico County Land Tax Lists 1830–1861.

75. "Sale of Claymount 8 Apr 1862 Richmond," *Daily Dispatch*, April 10, 1862, p. 1.

76. Henrico 1860 Industrial Census Records.

77. Henrico County Land Tax Lists 1844–1863.

78. Henrico County 1820 Census of Manufacturers, Land Tax Lists 1818–1851, and Personal Property Tax Lists 1818–1850.

79. *Richmond City Business Directory*, 1860, identifies an earthenware-related business in operation at this location by Sweeney.

80. A long inventory of other items round out the trappings of his successful life: a dining table and sideboard, silver tableware, fine table and tea china, "one curtain bedsted & furniture" valued at $50 dollars alone, two additional bedsteads at $30 each, one carriage and harness, and one buggy and harness, valued at $150 and $50, respectively.

81. Henrico County 1820 Census of Manufacturers, microcopy no. 279, roll no. 18, item 505.

82. Henrico County 1820 Census of Manufacturers and Personal Property Tax Lists 1819–1823.

83. Henrico County Land Tax Lists 1818–1851.

84. 1820 Census; *Richmond City Business Directory*, 1819; Crocker Farm Stoneware Auction Catalogue, p. 53, lot 279.

85. Mutual Assurance Society of Virginia, policy no. 1111; Mutual Assurance Society, vol. 44, no. 1955, March 29, 1816.

86. Dennis J. Pogue, "An Analysis of Wares from the Trees Point Stoneware Pottery (44Cc95), Charles City County, Virginia," *Quarterly Bulletin of the Archaeological Society of Virginia* 36, nos. 3–4 (1981): 122–36.

87. Edwin Atlee Barber, *The Pottery and Porcelain of the United States* (New York: G. P. Putnam's Sons, 1893), p. 178.

88. Pogue, "An Analysis of Wares," pp. 123, 129–30.

89. Ibid., pp. 123–24, 128.

90. Charles City Co., Va., Deed Book No. 7, September 16, 1830, p. 466 (101-acre Francis Wilson property purchased by John Schermerhorn).

91. Kenneth Scott, *British Aliens in the United States during the War of 1812* (Baltimore, Md.: Genealogical Pub. Co., 1979).

92. Umstott, "Lowndes Stoneware Pottery of Petersburg, Virginia," pp. 83–97.

93. 1850 U.S. Census, District 10, Cabell, Virginia; roll M432_938, p. 57, image 116; 1870 U.S. Census, Petersburg Ward 3, Petersburg (Independent City), Virginia; roll M593_1643, p. 274, image 343.

94. *Petersburg (Virginia) Index*, August 1, 1872, p. 3, col. 2.

95. Ibid., August 26, 1871, p. 2, col. 6/5.

96. Ibid., September 21, 1872, p. 3, col. 3.

97. Elizabeth J. Monroe, *Archaeological Assessment of the Lowndes Stoneware Pottery Site (44PG0475), City of Petersburg, Virginia* (Williamsburg, Va.: William and Mary Center for Archaeological Research, 2008).

98. Advertisement dated May 8, 1818, that appeared in the *Halifax Compiler* (North Carolina), June 19, 1818, p. 1, col. 1, announcing the establishment of Harwood's "Manufactory of Stone Ware."

99. *Petersburg (Virginia) Republican*, September 29, 1818, p. 3, col. 6.

100. Ibid., January 12, 1819, p. 1, col. 3.

101. Ibid., May 4, 1819, p. 3, col. 6.

102. Hustings Court Deed Book No. 6, 1819–21, p. 342, August 27, 1821.

103. A. Brandt Zipp, "Washington, D.C. Stoneware," *Antiques and Fine Art* (Autumn 2010): 202. See also Susan H. Myers, *Handcraft to Industry: Philadelphia Ceramics in the First Half of the Nineteenth Century*, Smithsonian Studies in History and Technology, 43 (Washington, D.C.: Smithsonian Institution Press, 1980), p. 68.

104. Dinwiddie County 1850 Industrial Census Records.

105. Dinwiddie County 1860 Industrial Census Records.

106. The vast majority of the archaeological sites associated with this significant area of stoneware production are located on private land. It is hoped that the individual landowners and organizations who hold this land and these precious cultural resources in their stewardship will, upon learning of the significance of these sites to our heritage and further understanding of eastern Virginia's historic stoneware industry and its products, be encouraged to support the proper full identification, definition, survey, study, and protection that these sites deserve.

Book Reviews

Christopher Benfey. *Red Brick, Black Mountain, White Clay: Reflections on Art, Family, and Survival*. New York: Penguin Press, 2012. 288 pp.; 33 b/w illus., index. $25.95.

> "It is a *delicious* Book; & like all *delicious* Things, you must take but a *little* of it at a time." Samuel Taylor Coleridge on William Bartram's *Travels* (quoted on pp. 249–50)

Christopher Benfey's book is difficult to pin down. In some ways it could be called a memoir, but the author includes lots of history from sources other than his own life. In some ways it could be considered a story of survival spanning three centuries, but despite its title reference to clays, you wouldn't read it as a text on ceramics history. Nor would you read this for the history of the now-famous Black Mountain College. Other sources give a more complete picture of these historical phenomena, and the author makes them easy to find through the chapter-specific resources he provides. Each of these aspects adds to the book's appeal for *Ceramics in America*'s audience, however; *Red Brick*'s also a really good "read."

According to Benfey, Mellon Professor of English at Mount Holyoke College, self-examination at middle age (he was fifty-four when he wrote the book) "is as much a task of recovery as of acquisition. The uncertain prospects ahead are weighed against the magnetic pull, stronger every day, of retrospect. To own one's past and the past of one's family takes on a peculiar urgency. The archeological impulse is a trust in the importance of origins, of beginnings" (pp. 11–12). Using key clay objects and scant knowledge of his family's past augmented with considerable additional research, Benfey set out on a journey that took him again and again to North Carolina as his mother lay dying in a nursing home in Greensboro. Turning the story around the state's clay history, going both forward and backward, he explores the family tales and myths learned in his childhood and the power of the Quaker life to craft self-awareness through contemplation and community participation.

I used to think it was pretty cool that I had managed to take possession of my grandmother's Quick Meal stove of the late 1920s when my father downsized his home in the 1990s. No one had used it in years. It was just waiting quietly in his basement for me to stumble across it, and now it has a place of honor in a remodeled room of my home, far removed from its original location. I don't plan to cook with it; rather, it's a piece of industrial sculpture dedicated to the memory of my hard-working grandmother, a single mom who supported my father and uncle by being a fancy top stitcher in a St. Louis shoe factory.

Now that I've read Benfey's book on the art and survival of his family, however, I see that my own family history is relatively ordinary. Benfey's ancestors descended from Apollo, survived encounters with Indians, and escaped Nazi tyranny. Still, his work challenges us readers to recognize the key patterns of American history in the stuff and stories of our own families.

Equally compelling is Benfey's poetic storytelling and his use of artifacts, especially those made of North Carolina clay, as central agents in his stories as well as bearers of history from one generation to the next. There is the simple red pitcher made at the Jugtown Pottery near Seagrove, from which he learned to recognize the persistence of tradition in art; the Sanford bricks made and laid by his grandfather, from which he learned to appreciate skill in even humble objects, and its transmission through generations; and the contours of mountain roads in western North Carolina, where he experienced the "hidden messages" on the "deepest lessons . . . in the contrast of textures" (p. 13) that émigré artists Josef and Anni Albers (his father's uncle and aunt) taught their Black Mountain College students. All of the people we meet in this book, from the Alberses to William Bartram, who tracked down and recovered white clay on Cherokee lands in western North Carolina for use in Wedgwood's jasper ware, are part of Benfey's kinship, both literally and metaphorically.

Bartram's work may well have offered a major impetus for Benfey's memoir. Benfey discusses at great length Bartram's *Travels and Other Writings* (1791) and its effect on the English Romantic poets, especially Samuel Taylor Coleridge. Bartram's book is ostensibly a tale of his botanical journeys through the American southeast wilderness just after the Revolution, but Coleridge thought Bartram presented "a series of poems, chiefly descriptive, *occasioned* by the Objects, which the Traveler observed" (quoted on pp. 249–50). Benfey notes that Coleridge viewed Bartram's descriptions of plants and soil as poetic metaphors for the people Coleridge knew, especially his friend William Wordsworth. Ultimately, Benfey relates Bartram's metaphors to the exotic plants and clays of the New World that were taken to England to thrive and inspire: "Just as there was a global trade in plants and materials, there was a parallel trade in verbal structures and ideas. And every once in a while, a restless genius came along—a Bartram, a Wedgwood, a Coleridge—who wandered from the familiar trail, risking falls and failure, and fused these new possibilities in unexpected ways, leaving lasting art for posterity" (p. 253).

In a promotional video produced by Mount Holyoke College when the book was released, Benfey summarized his feelings about the writing process and the eventual book:

> Part of the story of this book is the kind of story that all of us have as human beings. Why do we exist? What set of circumstances, coincidences, chance meetings ended up with us in the world? . . . I imagined that as I dug into this memoir more and more of it would be about me. In fact, the very opposite happened, that more and more of the book is about everything that is not about me [rather it is about my family and

> surroundings]. . . . Another deep truth in the book for me is about how much in our lives happens by chance . . . We do what we can, but chance has its way with us.[1]

Benfey wants his readers to find the parallels between objects and words in their own lives, mining the past to understand the present, and deriving a sense of satisfaction from a family's successful struggle for survival across time. Indeed, genes carry more than physical traits, and with enough knowledge and sensitivity to their potential we can understand the essence of our heritage as well as its facts. Perhaps Benfey will inspire more of his readers to mine their families' pasts and find poetry in the prosaic.

Ellen Paul Denker
ellen.denker@verizon.net

1. A video of Benfey reading a selection from his book in a public forum at Penland School of Crafts in the summer of 2012, when he was the Andrew Glasgow Writer-in-Residence at the school, can be viewed online at www.youtube.com/watch?v=H4zkOqfrayE. See also www.youtube.com/watch?feature=player_detailpage&v=_BFPFM5KZHI, in which he discusses what he learned while writing his memoir.

Christiaan J. A. Jörg, *Famille Verte: Chinese Porcelain in Green Enamels*. Schoten, Belgium: BAI/Groninger Museum, 2011. 192 pp.; 250 color and 9 b/w illus., bibliographical references. 29.50.

Porcelain decorated in the *famille verte* palette has long been viewed as one of China's finest and most luxurious export wares from its long history of ceramic production. But despite the ware's importance, there has never before been a publication specifically devoted to *famille verte*. Dr. Christiaan Jörg, retired curator of the Groninger Museum, professor of art history at the University of Leiden, and one of the most prolific scholars of export porcelain working today, makes a significant contribution toward filling this gap with his book on the subject.

This catalog accompanied an exhibit mounted in 2011 at the Groninger Museum in Groningen, the Netherlands. The pieces in the exhibit were drawn primarily from the four major public collections of Chinese porcelain in the Netherlands: the Rijksmuseum in Amsterdam, the Municipal Museum in The Hague, the Princessehof Ceramics Museum in Leeuwarden, and the Groninger Museum, with additional pieces coming from the Boijmans Van Beuningen Museum in Rotterdam and the J. M. van Diepen Foundation. The abundance of *famille verte* pieces in these collections reflects the importance of Chinese porcelain in the Dutch national patrimony and allows Jörg to illustrate the full depth and breadth of *famille verte*.

Famille verte (French for "green family") refers to porcelain decorated with a group of overglaze enamel colors dominated by a bright, translucent green along with iron red, yellow, black, and a purple-brown that was produced between 1685 and 1725 in Jingdezhen, the famed porcelain city of China. *Famille verte* porcelain was highly prized in Europe and was seen by many late-seventeenth- and early-eighteenth-century consumers as the most prized and desirable Chinese porcelain available. It eventually fell from favor about 1725, eclipsed by a new group of enamel colors, which became known as the *famille rose* colors. These had been developed in the early 1720s and included a rose red and an opaque white that allowed for shading of other colors and more naturalistic, complex depictions of flora, fauna, and figures.

Though seemingly straightforward, the term *famille verte* is somewhat problematic. It was coined in 1862 by the French ceramic historian Albert Jacquemart (who also coined the name *famille rose*). Ceramic scholars, Jacquemart among them, defined and grouped Chinese ceramics according to the material with which they were familiar—that is, predominantly porcelain that had been produced for export in the seventeenth and eighteenth centuries. They were not as aware of the overglaze enamel-decorated porcelain produced earlier, or the broad range of forms, designs, and decorative motifs produced for the Chinese domestic and non-European export markets. Traditional Chinese ceramic typologies describe ceramics decorated with overglaze enamels most closely corresponding to *famille verte* as *wucai* (five-colored) or *yingcai* (hard colors).

Jörg's book is arranged in a traditional format, with individual catalog entries for 178 objects. The introductory essay places *famille verte* within a broader context and provides a brief overview of the development and use of overglaze enamels on Chinese porcelain, from their development during the Tang dynasty (618–907), and their increasing popularity during the Ming (1368–1644) and early Qing dynasties (1644–1912), to their flowering during the reign of the Kangxi emperor (r. 1662–1722). The author also discusses how *famille verte* was made and how it fit into the export trade.

One of Jörg's insights is how the lack of underglaze blue or European designs on *famille verte* pieces may show how the porcelain industry in Jingdezhen operated. Enamels seem to have been painted in workshops unconnected to the potters who shaped, glazed, and fired pieces. As a result, the enamel workshops were not in direct contact with the merchants who commissioned pieces destined for the European export market and who could have provided Western designs to copy in enamels.

The individual catalog entries are grouped into twelve sections, each with an introductory essay. The sections have been defined according to design, form, usage, or market, such as "Flowers, Animals and Long Elizas," "European and Other Foreign Shapes," and "Verte for Asian Markets." Jörg describes these divisions as "informal" and "highly personal," but they are logical and reasonable. Each entry is illustrated, some with multiple images, and accompanied by text discussing the piece's significance. Attention is paid to the origins and meaning of many of the Chinese designs, such as the *Romance of the Western Chamber*, one of China's most famous dramatic works. Each of its twenty-four scenes is illustrated in the book, providing an easily accessible source for a popular design of export porcelain.

Though this is the first publication specifically devoted to *famille verte*, the subject has of course been covered in other publications. Many of the pieces illustrated here are among the finest in Dutch collections and have been discussed previously in print. But if the objects and ideas presented here are not necessarily brand-new discoveries, the synthesis of so much material and information into a single source makes this book an invaluable tool for scholars, curators, collectors, and dealers. It is sure to be a standard reference and a must-have for any Chinese porcelain library.

Ronald W. Fuchs II
Curator, Reeves Collection, Washington and Lee University

Maurice Hillis. *Liverpool Porcelain, 1756–1804*. Great Britain: Maurice Hillis, 2011. 564 pp.; 1,300 color illus., bibliography, index. £60.

Liverpool Porcelain, 1756–1804 is the culmination of more than thirty years of research by Maurice Hillis, an independent ceramics historian and collector who has lectured and published widely on eighteenth- and nineteenth-century English ceramics.[1] In this book, Hillis set for himself a formidable goal: to offer clarity and guidance on a complex subject that has long confused collectors and ceramics scholars alike. Porcelain was made at seven sites in Liverpool in the second half of the eighteenth century, yet none of the factories used a distinct mark, nor are there extant pattern and shape books for any of them. In addition, there are close resemblances among products made at many of the Liverpool factories, so it has often proven very difficult to attribute specific objects correctly. Earlier specialists on Liverpool porcelain did attempt to assign specific wares to each factory, but in the absence of factual evidence mistakenly included objects produced in Worcester, London, or Staffordshire.[2] Hillis builds on the important efforts of his predecessors by incorporating recent scholarship along with 1,300 color illustrations to provide a comprehensive, well-documented study of each factory and its output. As Geoffrey Godden vividly proclaims in the foreword to *Liverpool Porcelain*, "never again will the city name be used as a very convenient ceramic dustbin for all those pieces that did not fit the old, fashionable groupings" (p. ix).

Hillis divides his book into fifteen chapters, the first two of which offer a brief overview of eighteenth-century Liverpool, its ceramics industry in general, and its porcelain sites in particular. Nine chapters are devoted to individual factories. Each of these chapters follows a similar format, beginning with a biography of the factory's founder and a description of the factory's location, including a detail from a street map contemporaneous with the factory's operation. The author includes relevant discussion on business partners, workers, and apprentices. He details the factors utilized in making his attributions, including archeological evidence, dated objects, decoration, and type of porcelain body.[3] Hillis concludes these chapters with a thorough analysis of the wares made at each factory, presented according to form.[4] He carefully notes those elements—handles, spouts, molded patterns, and so on—that distinguish one factory's output from another. Each chapter is considerably enhanced by copious illustrations.

The most impressive aspect of the publication is the thoroughness of the author's research. He left no apparent stone unturned in his use of archival documents and archeological findings. In fact, as a member of the Liverpool

Porcelain Rescue Group, Hillis quite literally turned over a good number of those stones himself, specifically at the Brownlow Hill site of William Reid's factory and the Shaw's Brow site of Samuel Gilbody's factory.

Among the many archival sources he reviewed were corporation lease registers, rate books, fire insurance records, newspaper advertisements, polling lists, and freemen's books. Hillis verified the location of William Reid's factory on Brownlow Hill, for instance, by scrutinizing the lease register and a 1765 survey map, both preserved in the Liverpool Record Office (pp. 13–15). A Sun Fire Insurance policy dated October 22, 1746, provides the first mention of Richard Chaffers and Philip Christian as potters, plus documents their partnership in a pothouse on Shaw's Brow (p. 139). On the other hand, the fact that Hillis was unable to find any documentary evidence for William Ball's Ranelagh Street Pothouse, to which objects formerly were attributed, led him to conclude that the factory did not exist (pp. 494 95).

Hillis's equally fruitful examination of sherds and wasters was especially critical in helping him to identify wares from the William Reid and Samuel Gilbody factories.[5] Many photo captions in the chapters for those factories mention the existence of sherds matching a particular feature of the illustrated object. Regarding the output of Thomas Wolfe & Company, Hillis credits Alan Smith, whose 1968 recovery of sherds from the factory's site in Islington not only enabled later writers to attribute wares to this short-lived concern but also demonstrated that the factory produced a hybrid hard-paste porcelain and specialized in transfer-printed tea wares (pp. 477–78).

Hillis has visited many British and North American collections in order to inspect thousands of examples of Liverpool porcelain, as well as comparative examples from other factories. Every reader can benefit from the author's unparalleled knowledge based on this decades-long effort. What English ceramics enthusiast would not enjoy learning that one can tell the difference between a Worcester teapot and a teapot of similar form by Philip Christian's Liverpool factory because the underside of the latter's lid flange is glazed whereas the former's usually is not (p. 244)? Or that the high Chelsea ewers made by John Pennington can be distinguished from those made by Philip Christian, and later by Seth Pennington, because John's ewer usually has an oval foot rim rather than an eight-lobed one, as well as a handle that is thicker and set lower on the vessel (p. 357)? Such satisfying morsels of connoisseurship are available on virtually every page on which Hillis discusses the products of the Liverpool factories.

Moreover, when lack of archeological evidence requires Hillis to rely solely on extant objects, his extension of a few definitive traits of a given factory's wares into a fuller repertoire of characteristics identifying that maker is like watching a seasoned detective. Such was the case when Hillis endeavored to attribute wares to the factory of John Pennington, and later to his widow, Jane, who continued the factory after her husband's death. John and Jane Pennington's factory produced bone-ash porcelain. Hillis recognized that a number of bone-ash porcelain jugs had identical mask spouts and that some of the jugs bore dates ranging from 1772 to

1774. This period coincided with Philip Christian's years of production, but Hillis was able to eliminate Christian as the maker of these jugs because his factory made only soaprock porcelain. John Pennington's brother, James, was making bone-ash porcelain until 1773, but Hillis was able to eliminate him from consideration because one of the documentary jugs was dated 1790, long after James stopped making porcelain yet coinciding with the period when Jane Pennington was running the factory. Once Hillis was confident that the mask-spouted jugs were made at John and Jane Pennington's factory, he could then study them for other distinctive features, such as the type of handle employed, and look for those on other ceramic forms (pp. 319–24).

While Hillis's primary objective is to identify the full range of wares made at each factory, he loses no opportunity to showcase exceptional works, for another important goal of his is to challenge the ingrained perception among many collectors that Liverpool porcelain is of inferior quality to that made by other English factories. Hillis acknowledges that Liverpool factories, like many others, catered to a variety of markets, including those at the lower end of the economic scale. Nevertheless, he demonstrates that many Liverpool examples would hold their own against the best of its competitors. There is, for example, the covered vase, more than thirty inches tall and painted with a continuous Chinese landscape in underglaze blue, that was made at the Richard Chaffers factory (p. 215). Or the large Samuel Gilbody jug delicately painted in polychrome enamels and featuring a Chinese junk sailing from its jetty on one side and a tall British ship on the other (p. 303). Hillis also includes a number of very rare forms, such as a cream basket and spoon made by William Reid (p. 70), a holy-water stoup by Philip Christian (p. 267), and a James Pennington sauceboat in the shape of a duck (p. 124).

Hillis's comprehensive knowledge of Liverpool objects comes to bear significantly not only in the nine chapters focusing on individual factories but in three others as well. The first of these presents an illustrated list of all dated Liverpool porcelain known to the author, of which there are more than sixty examples. They are arranged in chronological order and briefly described, and for the reader's convenience Hillis indicates which ones are in public collections. The other two chapters address underglaze-blue and overglaze-enamel transfer-printed decoration on Liverpool porcelain.

In the chapter on underglaze decoration, Hillis describes some 120 patterns, the most comprehensive list ever published. He names and illustrates each pattern and, most important, indicates which Liverpool factory or factories used the pattern. He also indicates when different versions of the same print enable one to make an attribution to a specific factory. For instance, two versions of the Chinese Lady with a Bird in a Cartouche pattern were developed, but one was used only by John Pennington, while the other was used only by his brother Seth (pp. 428–29). This chapter alone will make Hillis's book an essential addition to the libraries of Liverpool porcelain collectors and other enthusiasts. The one regret is that the illustrations in this chapter are generally smaller than those in the individual factory chapters, so if the pattern under discussion happens to be fairly complex, its component parts may not be clear.

The author does not use the convenient list format for the enamel-printed patterns, but he reviews a wide variety of designs and always identifies the factory using the pattern. He focuses special attention on John Sadler, the renowned Liverpool printer who almost certainly bought Worcester and Longton Hall porcelain blanks to decorate and then resell but whose relationship with the Liverpool porcelain factories is more complicated. Sadler probably bought Liverpool blanks as well, but Hillis offers evidence indicating that Chaffers, Gilbody, and Reid commissioned Sadler to decorate their wares, which the respective factories would have then sold on their own. The Christian factory, meanwhile, might have been capable of doing its own enamel printing (p. 533; also p. 231). Hillis identifies Liverpool ceramic printers other than Sadler who possibly decorated porcelain produced in the city. He calls for more research to understand these Liverpool printer-factory relationships.

The last chapter addresses the marketing and distribution of Liverpool porcelain. Hillis specifies the newspapers, both local and as far away as in America, in which some of the Liverpool factories placed advertisements. He also lists place names inscribed on Liverpool porcelains or at which Liverpool sherds have been found, suggesting the geographic range of the factories' markets. *Ceramics in America* readers will be interested to know that these archeological sites include Colonial Williamsburg; Portsmouth, New Hampshire; and Fort Michilimackinac, Michigan.

Minor inconveniences arise in using *Liverpool Porcelain* as a reference. For instance, the index could be more complete. Josiah Wedgwood, Robert Hancock, and other individuals important to the history of English ceramics (if not necessarily to Liverpool porcelain) make guest appearances in the chapters but are not listed in the index. In addition, with the exception of the public-collection photographs in the chapter on dated Liverpool wares, names of the objects' owners appear only in a photograph-acknowledgments section. For readers wishing to see these pieces in person, credit lines in the captions would have been helpful, especially for objects owned by public institutions. Finally, it is not clear why the two chapters on printed decoration are not consecutive: underglaze patterns are the subject of chapter 9, enamel patterns are addressed in chapter 14.

These are trifling concerns that do not detract from the author's herculean accomplishment, made all the more impressive by virtue of the fact that Hillis published the work himself. He has produced a meticulously researched, profusely illustrated text that will help scholars, collectors, dealers, and others to understand, at last, the historically complex subject of Liverpool porcelain. His publication manifests the scholarship and careful analysis that lie behind it, ensuring that it will be the definitive source on Liverpool porcelain for many decades to come.

Brian Gallagher
Curator of Decorative Arts, The Mint Museum
brian.gallagher@mintmuseum.org

1. Previous publications by Hillis on topics related to Liverpool porcelain include Maurice Hillis, "The Analysis of Shards from the Brownlow Hill Site in Liverpool," *English Ceramic Circle Transactions* 18, pt. 2 (2003): 335–38; Maurice Hillis, "The Liverpool China Manufactory of Wm. Reid and Company: A Survey of Wares," *English Ceramic Circle Transactions* 18, pt. 1 (2002): 58–90; Maurice Hillis, "The Porcelain of John and Jane Pennington, c. 1771–1794," *Northern Ceramic Society Journal* 6 (1987): 1–22; Lyn Hillis and Maurice Hillis, "Late Christian or Early Pennington?," *Northern Ceramic Society Journal* 5 (1984): 29–32. Hillis also served as chairman of the Northern Ceramic Society from 2001 to 2012.

2. The first author to write an account of Liverpool porcelain was Knowles Boney, *Liverpool Porcelain of the Eighteenth Century and Its Makers* (London: B. T. Batsford, 1957). He included wares now known to have been made in London, Staffordshire, and Worcester. Bernard Watney, in his two-part article "Four Groups of Porcelain: Possibly Liverpool," *English Ceramic Circle Transactions* 4, pt. 5 (1959): 13–25, and *English Ceramic Circle Transactions* 5, pt. 1(1960): 42–52, made a groundbreaking effort to attribute works to specific factories, but he too included wares made in London and Staffordshire. He corrected this error in his *Liverpool Porcelain of the Eighteenth Century* (Somerset, Eng.: Richard Dennis, 1997).

3. With the exception of Thomas Wolfe's Islington China Works, which produced a hybrid hard-paste porcelain, eighteenth-century Liverpool factories exclusively made either soft-paste porcelain with bone ash or soft-paste porcelain with soaprock.

4. Forms include coffee pots, cups and mugs, teapots, jugs, sauceboats, plates, and dishes.

5. The Liverpool Porcelain Rescue Group (LPRG) conducted digs at Reid's factory site on Brownlow Hill in the late 1990s. Alan Smith excavated Gilbody's factory site at Shaw's Brow in 1966. LPRG conducted a second dig there in 1990.

Caroline Frank, *Objectifying China, Imagining America: Chinese Commodities in Early America*. Chicago: University of Chicago Press, 2011. 272 pp.; 49 halftones, 1 line drawing, 2 tables, notes, index. $25.00 (softcover).

Caroline Frank's highly readable new book on the presence and meaning of Chinese culture and commodities in colonial America is an important addition to ceramics scholarship and, more broadly, American studies. *Objectifying China, Imagining America* offers a synthetic approach that integrates visual culture and consumption theory with maritime and economic history. Frank brings a global perspective and provides a rigorous theoretical treatment of Chinese porcelain in colonial America. This in-depth analysis has been missing from the significant body of literature on China and the West despite the presence of Chinese export porcelain—arguably China's most durable commodity and signifier—in every colonial seaport.

Although it has long been known that tea and its inevitable accompaniment Chinese porcelain was present in colonial urban centers by the 1720s, in her opening chapter, "The First American China Trade," Frank makes a convincing case that Chinese commodities were entering the colonies with some frequency by the 1690s via a shadowy network of piracy and smuggling. Indeed, she recasts American colonists as enthusiastic China traders and full participants in global commercial networks long before the inauguration of direct U.S.–China trade in 1783, a fact that has been noted by historical archeologists as well. In the late seventeenth century, British imperial interests so successfully transformed these Anglo-American Indian Ocean traders from heroic privateering "Red Sea men" to marauding pirates that they were stripped not only of their Chinese goods but of their historical legitimacy (pp. 49–52). In part because of these (often ineffective) trade restrictions, Frank argues, Chinese goods achieved an outsize economic and symbolic place in the American consciousness over the next one hundred years.

In chapter 2, "Imagining China at Home," Frank presents a fascinating analysis of the japanned wall panels in the Vernon House, Newport, attributing them to sign painter William Gibbs, who owned the house by 1708 and died in 1729. More important is her insistence that they be included in the canon of early American painting. Here we have clear evidence that Americans were not only "actively and independently engaging with a Far Eastern aesthetic" at a much earlier date than previously thought, but that they were operating with the "same level of sophisticated prejudice" as any London practitioner of *chinoiserie* (p. 61). Colonists such as Gibbs were not naive provincials but British "citizens of the world." Using a variety of sources (and perhaps consulting Coromandel screens in his neighborhood), Gibbs created a series of paintings that blended a specifically New England technique with Chinese and European motifs to depict despotic and cruel infidels. In short, Gibb's imaginary China is a hybrid that visualized widely held negative stereotypes of the Chinese.

In chapters 3 and 4 Frank continues to make the case that Chinese commodities, despite restrictive British policies and the ascendancy of the East India Company, were much more widely available and appreciated in the eighteenth century than is apparent in official records. Yet colonial and federalist consumers rarely, if ever, offered specific interpretations of the blue and white China-scapes that graced their tables. The author inquires what porcelain meant to its original purchasers, before it acquired the weight of history and became treasured heirlooms of "Yankee mythology" that celebrated fortunes made through direct trade with China (p. 146). She delineates a long and compelling association among Chinese export porcelain, a mastery of overseas trade, and a cosmopolitan interest in the wider world.

Frank parses out the ambiguities, contradictions, and ironies that existed between a desire for Chinese goods and an ambivalence, even willful ignorance, about China and its people. She identifies "slippage" between the extensive factual information available to Americans and their attitudes toward Chinese commodities, which served as ever-shifting receptacles for Western fantasies, desires, fears, and prejudices, overwhelming any desire for real knowledge about the Chinese. Frank argues that we must attend to the politics of the American context. Early owners of porcelain in the colonies were often thwarting British imperial authority. Frank cites both Dutch influence and an "embrace of commercial capitalism" as key factors in the taste for Chinese goods, a taste in tension with Puritanical fears of licentious freedoms, seductive ornament, and perhaps ill-gotten riches (pp. 146–56).

Chinese commodities entered and circulated in the colonies in a variety of ways, some legal and many not, but always invested with a panoply of contradictory meanings: booty or badges of gentility, feminine extravagance or manly trading prowess? According to Frank, ownership of Chinese porcelain in America initially "invoked a masculine liberty to make one's own rules and mastery at sea and on land" (p. 161). Nonetheless, the same anxieties about porcelain, and Chinese goods in general, evident in

Britain discourse—as *femme fatale*, as vanity, or as a symbol of ruinous extravagance that stoked mercantilist fears of dependency—appeared in the colonies as well and grew with increasing consumption. However, the earlier associations persisted, and British restrictions on colonial trade ultimately invested Asian commodities with "a powerful ideology of imperial control and domination" (p. 108). In the post-Revolutionary period, freedom from Britain meant freedom to trade with China and the means to secure the country's future prosperity. This is one reason the inauguration of official U.S.–China trade in 1783 has loomed so large in our collective imagination. In this view Chinese porcelain was completely disengaged from its production but had everything to do with its procurement, forever "entangled with the disposition of American maritime capital and the men who controlled it" (p. 172).

In Frank's concluding chapter, "Manly Tea Parties," she reconsiders the meaning of Boston's "destruction of the tea," one of the best-known events of the Revolution (p. 181). Building on the work of Robert St. George, T. H. Breen, and others, Frank argues that the American response to Chinese commodities, particularly tea, ultimately offers a key index to the moment we became un-British. The century-long history of Chinese commodities in America inevitably politicized tea as an agent of debilitating enslavement wielded by a tyrant mother country who had disowned her children, casting them as "lesser" colonized subjects who had somehow gone native in the enervating climate of America. Tea was such a potent symbol precisely because of latent but powerful fears about the possibly emasculating, even fatal, effects of Chinese contamination. Charged with alcohol consumed from a Chinese punch bowl, the tea-partiers resisted becoming "servile Indians" by donning their garb and destroying "the colonizer's tool" (p. 201).

This thought-provoking book covers a lot of ideological ground. The author probes the long and ambiguous relationship between America and China and finds racism and fear of contagion at its core. The West found Chinese commodities both irresistible and potentially dangerous; controlling trade with China was thus of major importance in the run-up to the Revolution, in its aftermath, and to the present day. Quibbles are minor. Some may question whether "china," "chaney," and its variants *always* meant Chinese export porcelain in the more than one thousand probate inventories that stand at the center of Frank's research (pp. 135–36). Certainly in ship manifests and commercial records, the term *china* may be taken at face value, but how discriminating were inventory takers when it came to classifying ceramics? Frank rightly calls her probate evidence "impressionistic" rather than definitive (p. 139). An appendix charting more of this prodigious research would have been welcome, as would color plates, but this is not a picture book, and given the economic realities of academic publishing, the book's reasonable price aptly compensates for these admittedly minor deficiencies.

Jessica Lanier
Salem State University

Jerzy Gawronski, ed., with contributions by Jerzy Gawronski, Ranjith Jayasena, Ab Lagerweij, Sebastiaan Ostkamp, Ron Tousain, and Jørgen Veerkamp. *Amsterdam Ceramics: A City's History and an Archaeological Ceramics Catalogue, 1175–2011*. Amsterdam: Uitgeverij Bas Lubberhuizen/Bureau Monumenten & Archeologie, 2012. 335 pp.; 1,500+ color illus. and numerous b/w line drawings, 2 appendixes, references. □39.50 (softcover).

Recovery of seventeenth-century Dutch artifacts in the former New Netherland, an area centered in modern New York City, became a reality in the 1970s with excavations in Lower Manhattan and Albany. Archeologists working in New Netherland recognized that they needed to develop competence in identifying/interpreting Dutch artifacts beyond measuring pipestems and perusing art-history publications on "delftware." At the time, however, reference works on Dutch utilitarian ceramics were practically nonexistent, and the Dutch language was a substantial barrier. Since the late 1980s, however, lavishly illustrated archeological reports/catalogs published by Dutch municipal archeologists and museums have enhanced American access to these data.

Contact between American historical archeologists and the public archeologists in Amsterdam has been decades-long and cordial, and the Amsterdam office quickly became a pilgrimage destination for archeologists. From 1988 to 1989, I was funded to do dissertation research based in that office, hosted by then-head Jan Baart and his excellent team, Ab Lagerweij and Wiard Krook. The collections were astounding, and since then have been augmented by an additional twenty-three years of labor. Oddly, despite this embarrassment of riches, relatively few publications were generated by the office, although the 1977 *Opgravingen in Amsterdam* (Excavations in Amsterdam) achieved a certain notoriety.[1] This is about to change, however. When interviewed by the newspaper *Het Parool*, Amsterdam city archeologist Jerzy Gawronski, editor of the present work, indicated that *Amsterdam Ceramics* will be followed by publications on shoes, glass, and tools.

This English-language catalog presents 1,247 ceramic vessels recovered from 213 sites. The 580 vessels (46 percent) from the period 1575–1700 are directly relevant to many archeologists in former Dutch colonies and in those areas with historical trade links to the Netherlands, where this volume will be welcome.

First, however, come nine chapters on the history and development of Amsterdam, from "Amsterdam 1175–1300" through "Amsterdam 1850–2011." These offer fascinating discussions of Amsterdam's evolving environment, population, and government, as well as technological, social, and political trends. Both documentary and artifactual evidence tell Amsterdam's story. The text is sprinkled with photos of artifacts ranging from a Neolithic stone axe to a cell phone. The high-water table has resulted in excellent preservation of both leather and wooden artifacts, and in situ assemblages, features, and trench profiles are illustrated. Almost every chapter includes a photo of a spoon, a shoe, a drinking vessel, and a lighting device, allowing the reader to follow a very simple typology across periods.

The catalog is divided into the same time periods used for the history chapters. The bridge between history and catalog is somewhat disappointing, with short "ceramics" paragraphs at the end of each history chapter. These describe the general ceramic assemblage of the particular period with a few insights into changes and new ware types.

A review of the introductory pages explaining the catalog's organization, especially the description of the reasoning behind the ordering of the artifact entries, is advisable before browsing. American archeologists should study the Dutch definitions of the terms *majolica* and *faience* and note that the word *delftware* occurs nowhere.

Entries within each time period are divided by ceramic type and then subdivided by vessel form. Each has a crisp color photograph, an English "object name," the item's height or diameter measurement, and a place of origin; sometimes there is also a cross-sectional drawing. Despite the large, elegant white spaces between the entries, additional data are relegated to Appendix 1. Since the entries within each period are not in chronological order, one cannot follow the evolution of a particular form without flipping pages. Appendix 1 provides the site name, a tighter date range, and a code in the Deventer System (DS).

Named after its city of origin, the DS—officially the "Dutch Classification System for Late- and Post-Medieval Ceramics and Glass"—has emerged as the dominant typological system in Dutch archeology. The DS offers a desirable standardization of terms and date ranges, not idiosyncratic to one city. The code consists of a fabric/material designation, a Dutch vessel-type name generally based on historical usage, and a code number for the specific form. Unfortunately, the catalog provides only the three-letter DS vessel codes and does not supply the corresponding Dutch historical terms, which have been derived from documentary research and vetted by the DS board (a free pdf of the DS is available, however, if one emails info@archeologie.nl). Some of the English-only catalog terms are unhelpful overtranslations, such as no. 601, "cauldron," for the large cooking pots the DS designates r-gra-11 (redware-*grape* [cooking pot]-11). The catalog's "cauldron" conjures scenes from *Macbeth*, and "marmite," a French historical term, is out of place. Another awkward translation is calling the pedestal salts (nos. 906–908) "salt tubs." The format also has no flexibility to provide additional, needed information for some of the entries. A few objects might have been better illustrated to make all their attributes visible, for example the "birds' drinking bowl" (nos. 409, 627), which has one flattened side so it can be placed flush against a birdcage.

The occasional difficulties with the translated text are more than compensated for by the book's ability to reach a large English-speaking audience, however. For those interested, the Dutch version of the history section is available on the publisher's website, www.lubberhuizen.nl. For readers unfamiliar with the Dutch language or Amsterdam, the many tongue-twisting toponyms used are sometimes hard to follow, and the map provided is not really adequate to the task of pinpointing the locations.

These drawbacks aside, the catalog is a valuable resource. It provides images and general date ranges for a wide variety of ceramic artifacts, reflective of Amsterdam's position as premier trading center and immigration destination. The volume makes accessible a large number of examples of the range of typical Dutch utilitarian earthenware forms (vessels, handles, feet, bodies, rims, bases) as well as contrasting non-Dutch examples. With this book in hand, there will be no excuse for misidentifying a late-seventeenth-century, lead-glazed, buff-bodied Dutch *grape* as North Devon gravel-tempered (something I encountered recently). Beyond the usefulness of illustrating many everyday earthenware forms, the variety of decoration—mainly of Dutch, Italian, Portuguese, and French origin—on the majolica and faience entries should prove priceless to archeologists and connoisseurs. Given the quality of this production, one will look forward to the promised future volumes.

Richard G. Schaefer
Historical Perspectives, Inc.

1. Jan Baart, Wiard Krook, Ab Lagerweij, Nina Ockers, Hans Van Regteren Altena, Tuuk Stam, Henk Stoepker, Gerard Stouthart, and Monika van der Zwan, *Opgravingen in Amsterdam* (Excavations in Amsterdam) (Haarlem: Fibula-Van Dishoek, 1977).

Index

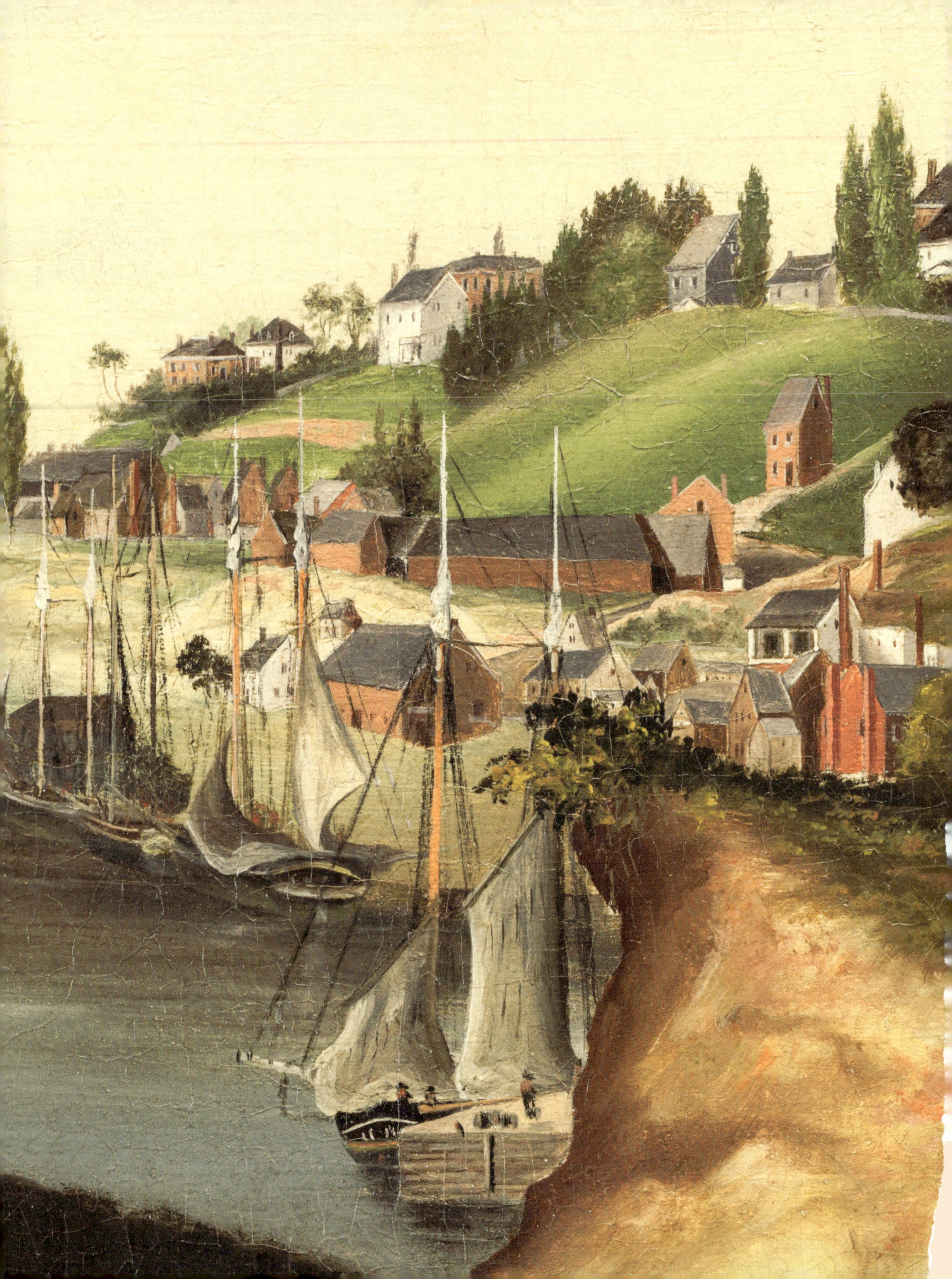